Introductory Linear Electrical Circuits and Electronics

Introductory Linear Electrical Circuits and Electronics

Michael C. Kelley
Cornell University

Benjamin Nichols
Cornell University

WILEY
John Wiley & Sons
New York   Chichester   Brisbane   Toronto   Singapore

Library of Congress Cataloging in Publication Data:

Kelley, Michael C.
    An introduction to linear circuits and electronics / Michael C.
Kelley, Benjamin Nichols.

    p.   cm.
    Bibliography: p.
    Includes index.

    ISBN 0-471-61251-0
    1. Electric circuits, Linear.   I. Nichols, Benjamin.   II. Title.
TK454.K45 1988
621.3815'3--dc19   88-18818
                                                                CIP

Preface

The study of electrical engineering is often described under two major headings: *electrophysics* and *electrosystems*. An electrical system in its most general definition consists of physical devices connected together to perform a particular function. A detailed understanding of such a system requires knowledge of the physical principles that underlie the properties of the devices. These include the elements of electrodynamic theory, material science, solid state and quantum physics, to name just a few. The understanding also requires knowledge of the principles that govern the interconnections of these devices into a system and the response of the system to certain inputs ("signals"). These include the mathematical tools required to solve the equations of the system and the choice and description of the particular time functions to represent the signals.

In this introductory text the emphasis is on the system approach, and no previous study of electrical circuits is assumed. It is expected, however, that students have taken (or are taking concurrently) introductory physics courses including electrodynamics. Knowledge of introductory calculus is also assumed. Since the text is primarily geared toward the normal sophomore level, some mathematical methods are developed as needed. An example is the algebra of complex numbers which the typical student may not yet fully understand.

At the end of each chapter there is a section entitled "Practice Problems and Illustrative Examples." These sections are an integral part of the text. Engineering is the process of investigating how to solve problems. Consequently, understanding of the concepts is developed by the practice of applying the concepts to the solution of problems. The "Illustrative Examples" demonstrate that application and bring out some of the features of problem solution that are not discussed specifically in the body of the chapter. There are also a number of examples worked out within the text itself which are meant to reinforce the material as students encounter it. Some sections are marked with asterisks, which means that the material is not necessary for the logical flow of the text. They will provide students with some insights that may be valuable later.

The term "linear" in the title of the text is important. As with most fields of technical and scientific study, the systems with which we deal often display the property of linearity. One important consequence of this property is that if two solutions satisfy the equations which govern the system, then their sum will also satisfy the same equations. This leads to a number of simplifications in the mathematics and analysis. In fact, the mathematical benefits are so great that we create linear models of those devices we wish to describe in the text, but which are themselves inherently nonlinear. Elementary examples of these include diodes and transistors. We also make extensive use of an important integrated circuit element termed the *operational amplifier*. When a system includes such devices as diodes, transistors, and operational amplifiers, the system is usually called "electronic." It makes sense therefore, to begin the study of electrical engineering with an introduction to linear electrical circuits and electronics.

In Part A we present the fundamental elements of circuit theory. Although most students have had some contact with electrical circuits, we spend some time in Chapter 1 on fundamental concepts to define the basis for the text material. The standard circuit laws and theorems are developed and applied in Chapters 2 and 3, and Chapter 4 introduces dependent sources, which are used throughout the text. Two-port systems are also described in Chapter 4 and the ideal operational amplifier is introduced as a circuit element.

In Part B we discuss the concept of energy storage elements and show that their use in electrical circuits leads to the necessity of mathematical analysis using differential equations. The solutions to such equations are well known and lead in a straightforward manner to the introduction of the parameter s and the impedance function $Z(s)$. The latter is used extensively in solutions to the natural and forced response of first-order electrical systems in Chapter 6. Emphasis is placed upon the concept that the natural response characterizes the response of a system when the energy sources are set equal to zero. Phantom sources of value zero are used to illustrate this and the pole-zero method then evolves very naturally. The circuit response to pulses and pulse trains is also presented.

In the last part of Chapter 6 a second order system is analyzed using the impedance method and it is found that an imaginary value for s arises. This leads to the use of complex algebra, which is reviewed in Chapter 7. This chapter is the first in a series of three chapters dealing with the response of electrical systems to signals characterized by a single frequency ω. Since electrical utilities are the most important example of such systems, they are studied in some detail in Chapters 8 and 9. Power factors, complex power, ideal transformers, generators, and three-phase power are all discussed and numerous examples presented in the text and the problem sets.

In Part D students can extend their capabilities to include analysis of the response of linear systems to signals which have a more complicated time dependence than that characterized by a single frequency. Fourier's theorem is at the heart of this analysis and is introduced as a fundamental mathematical concept. The transfer function plays an essential role in the development and is used to implement the analysis methods. The detailed study of Fourier–Laplace methods are beyond the scope and needs of the present text, but the fundamental role they play in the analysis is made clear and should provide a solid conceptual basis for upper-level courses. We show that the transfer function may be used to determine both the natural and the forced response. Again the pole-zero method arises in a natural fashion. We continue with a discussion of the frequency response of electrical systems. The db unit is introduced as are Bode plots. A general solution is found in terms of the poles and zeroes of the transfer function which is then specialized to the case of real poles and zeroes. The latter case is analytically tractable and many examples are discussed including the design of simple Op-Amp filters. In Chapter 12 we consider second-order systems and such concepts as critical damping and the quality factor are described. We also describe analytical and graphical solutions for the case of complex poles or zeroes and the resonant response of second-order systems.

The linear elements used up to this point are now supplemented with non-linear elements that are based upon the use of semiconductor materials. The response of a *pn* junction is crucial to the usefulness as well as the nonlinear behavior of such electronic devices. In Chapter 13 a simple description of the *pn* junction physics is provided, but emphasis is placed upon the I-V characteristics of the diode. Diode models and rectifier circuits are discussed. A linear model of the diode is described as background material for the more important linear models needed to understand transistor amplifiers. In Chapter 14 two generic transistors are introduced: the *n* channel enhancement MOSFET and the *npn* bipolar junction transistor. There is no attempt to describe the entire range of transistor types available. Rather, two important and qualitatively different types are chosen and analyzed in detail. This basis should be sufficient to allow students to understand other devices as they are introduced in later courses or in the laboratory. Linear small signal models are introduced in Chapter 14 along with the various dc biasing methods used. The full frequency response of *RC* transistor amplifiers is presented in Chapter 15 including coupling and stray capacitors, as well as parallel capacitors used to deliberately limit the frequency response of the amplifier. A discussion of voltage followers completes the small signal amplifier discussion.

The material can be covered in one semester, but the pace is quite fast and it may be necessary to leave out some material. The three-phase power section of Chapter 9 is fairly self-contained and could be eliminated. Also the section in Chapter 12 on second-order systems could be presented as lecture material only, something of a breadth component, but material for which the student is not fully responsible. Similarly in the last chapter on practical *RC* transistor amplifiers, it simplifies matters if students are not required to analyze circuits with self-bias resistors. Once again the lecture material could include a discussion of the practical benefits of including this feature. Accompanying Introductory Linear Electrical Circuits and Electronics is a special SPICE supplement written by Joseph Tront from Virginia Polytechnic University. Please contact your local sales representatives for more details.

The authors gratefully acknowledge the efforts by the numerous students who struggled through the early version of this text. We are particularly indebted to Andrew Noel, Jon Schoenberg, and Dianne Umpiere who worked on the early text and Stephanie Berg who helped with the solutions manual. Special thanks are due Laurie Shelton whose cheerful competence is a marvel to behold and Ali Avcisoy whose skill and attention to detail made the artwork task feasible.

Michael C. Kelley

Benjamin Nichols

Contents

*Sections marked with an asterisk are not required for the logical flow of the text.

ix

Chapter 9
Electric Power Systems / 295

SECTION D
FREQUENCY RESPONSE OF ELECTRICAL SYSTEMS / 321

Chapter 10
Transfer Functions, Spectral Synthesis, and Complete Response Revisited / 323

Chapter 11
Frequency Response—Analysis and Design / 365

Chapter 12
Second-Order Circuits / 413

SECTION E
ELEMENTARY ELECTRONIC DEVICES / 451

Chapter 13
Semiconductor Diodes and Diode Circuits / 453

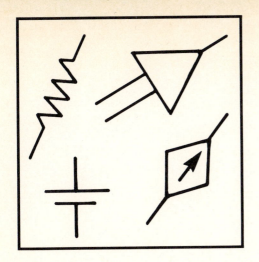

SECTION
A
CIRCUIT
ANALYSIS

The field of electrical engineering uses concepts from physics, mathematics, material science, and a number of other disciplines. In this section we begin by relating the two fundamental parameters used by electrical engineers, namely, voltage and current, to principles from electromagnetic theory. The physical and mathematical properties of a number of circuit elements are then defined based upon their current and voltage characteristics. These are enough to begin our study and, as the text progresses, we shall add new circuit elements as they naturally arise. Conservation principles from physics, along with the mathematical properties of the equations that describe electrical phenomena, are then used to derive several laws and theorems which may be used to analyze circuits. These include Ohm's and Kirchhoff's laws, the principles of linearity and superposition, Norton's and Thevenin's theorems, and so on. In Chapter 4 dependent sources and ideal operational amplifiers are added to the circuit elements with which we deal, and further practice is given in circuit analysis methods. We also introduce the important electrical system concept of a two-port, which accepts some voltage or current at the input, modifies it in some way, and presents a modified parameter at the output.

Chapter 1

Fundamental Elements of Circuit Theory

In this chapter we review elements of electrostatics and the response of charged particles to forces. The concepts of voltage, current, electrical energy, and electrical resistance are introduced and analogies are drawn to other physical systems. Various symbols are introduced, and so are the systems of units used in the text. Ideal voltage and current sources are defined, as are idealized instruments used to measure voltage and current.

Although the principles in this chapter are elementary, their mastery is very important for the subsequent material. Considerable care has been taken in the definition of the various parameters, particularly with respect to the sign conventions used in analysis. Rationalized MKS units are used throughout the text.

1.1 VOLTAGE AND CURRENT

Electric charges have units of coulombs and can be positive or negative. Charged particles exert forces on each other which are conveniently described by introduction of the electric field, $\mathbf{E}$. The boldface denotes that $\mathbf{E}$ is a vector quantity and has a magnitude and direction. The force on a charged "test" particle of charge q is given by $\mathbf{F} = q\mathbf{E}$ (Fig. 1.1). Inasmuch as there are particles with both positive and negative charge values, the particle motion can either be parallel or antiparallel to $\mathbf{E}$. The units of $\mathbf{E}$ are newtons per coulomb. A close analogy exists with the gravitational force, which is given by $\mathbf{F} = M\mathbf{G}$, where M is the particle mass and $\mathbf{G}$

FIGURE 1.1.
Separated charges produce an electric field pointing from positive toward negative charge. The units for _E_ are newtons/coulomb.

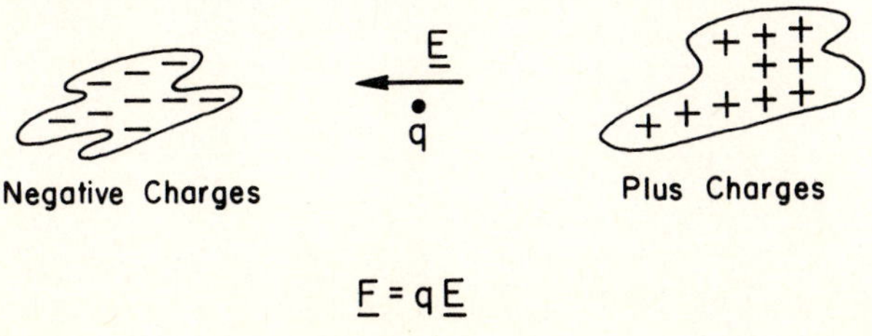

Negative Charges   Plus Charges

$$\underline{F} = q\,\underline{E}$$

is the gravitational field. As far as we know there are no negative mass particles, hence the motion due to a gravitational force is always parallel to **G**. The units of **G** are newtons per kilogram.

In dealing with electrical circuits it is more convenient to work with an energy parameter, the voltage, rather than the electric field itself. This parameter is measured in joules per coulomb, which corresponds to energy per unit charge. The unit is the volt. In the case of gravity it is easy to see how energy may be stored or released. When a cart is pushed up a hill, gravitational energy is stored (Fig. 1.2). This stored energy may then be released during the descent. Note that energy is stored by moving the cart against the gravitational field and is released when the cart is allowed to fall down with a component of velocity parallel to **G**. For any reasonable-size hill of height h, the earth's gravitational field may be considered to be a constant value g, and the gravitational potential energy is given by Mgh. If we divide this quantity by M, we obtain the gravitational potential gh, which is equal to the potential energy per unit mass, in joules per kilogram.

Likewise, in an electrical system, energy is stored when charges are caused to move against the electrical force exerted upon them. The agent that accomplishes this might be chemical, as in a battery, or it might be supplied by photons as in a solar cell. Quantitatively, the amount of electrical energy given to a charge of q when it is moved against a constant electric field from point b to point a through a distance l is given by force times distance, or

$$W = qEl \quad \text{(joules)} \tag{1.1}$$

This expression is analogous to the gravitational energy Mgh discussed above. If we divide by the charge q, we obtain a quantity analogous to the gravitational potential, which we call the electrical voltage:

$$V = \frac{W}{q} = El \quad \text{(joules/coulomb)} \tag{1.2}$$

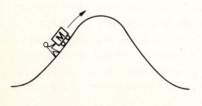

Energy Stored

FIGURE 1.2.
Gravitational energy is stored when the cart is pushed up the hill and released when it rolls down.

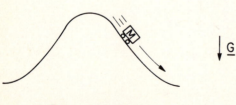

Energy Released

Unlike mass, electric charge comes in two forms, which we term positive and negative. This means we must pay particular attention to sign conventions in defining the voltage. Consider the simple representation of an electrical battery in Fig. 1.3. The figure carrying a positive charge q represents chemical forces inside the battery which transport charges in the direction opposite to the electrical forces to which they are subjected. This is analogous to the figure in Fig. 1.2 pushing the mass against gravity. The electrical potential at point a with respect to point b we therefore define as positive, just as we consider the gravitational potential to be positive at the top of a hill. The voltage difference across the battery then is $\Delta V = V_a - V_b$.

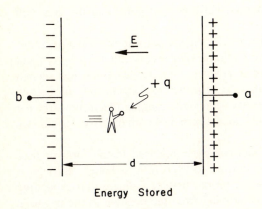

Energy Stored

FIGURE 1.3.
Electrical energy is stored either when positive charges are moved in the direction opposite to an electric field or when negative charges are moved parallel to an electric field.

It is important to realize that only voltage differences have physical meaning. This can be understood again with reference to a gravitational analogy such as the one illustrated in Fig. 1.4. The inclined plane and the ball of mass m have the same properties in both halves of the figure, but the entire unit on the right-hand side has been elevated by the distance h_0. The energy available after the ball rolls down the plane, $mg\Delta h$, is the same in both cases and only depends upon the height, Δh, of the inclined plane. The extra energy (mgh_0) of the unit does not contribute at all to the velocity of the ball. It is the energy difference due to the height of the inclined

FIGURE 1.4.
An elevated massless inclined plane yields an identical energy output, independent of the "reference voltage" gh_0.

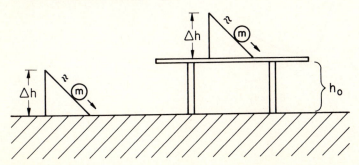

plane which is analogous to the voltage difference ΔV across a battery. An arbitrary additive constant V_0 can therefore be added to every voltage in an electrical system without changing the properties of the circuit.

This feature may be used to advantage in circuit analysis since it allows us to arbitrarily define the voltage at one point equal to zero volts. This reference point is usually termed ground potential because it is often physically connected by a wire to the earth's surface. Once such a reference point, b, is defined in a circuit we often refer to the voltage at some other point, a, in the circuit such as V_a. By this we mean the voltage $\Delta V = V_{ab} = V_a - V_b$, which just equals V_a since the voltage at b is defined to be zero. This is such a common practice in circuit design that henceforth we will often discuss the "voltage at a point" with the tacit understanding that we mean the difference in potential between that point and the zero-volt reference.

The flow of charge, the electrical current I, is the second important parameter of electrical engineering. The units of I are coulombs per second, that is, the amount of charges per unit time that flow past a given point. One coulomb per second is called an ampere, which is abbreviated A. In principle, current may be carried by charges of either sign. In most materials negatively charged electrons carry the current because they are very light and move much faster than heavy positively charged particles.

In Fig. 1.5 a circuit is shown in which energy is released in the form of heat and light when an external path allows current to flow from a battery through a light bulb. The positive and negative signs on the battery indicate that a positive charge would give up electrical energy in moving from the positive to the negative terminal along an external path. Likewise, a negative charge would give up its stored electrical energy in moving from the negative terminal to the positive terminal along an external path. In this example the electrical energy drawn from the battery appears as light and heat from the light bulb. Note that if the battery voltage is to remain constant, current must also flow through the battery to maintain the charge separation at the terminals of the battery. This current must be driven *against* the electric field inside the battery by the chemical reactions in the battery. Commonly used voltage and current values are listed in Table 1.1.

FIGURE 1.5.
Charges falling off the "electrical hill" can release stored electrical energy and, for example, light a bulb.

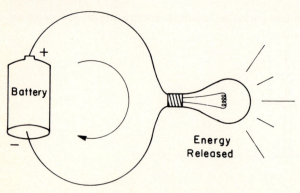

TABLE 1.1
Common Voltage and Current Units

picovolt	pV	10^{-12} Volt	picoamp	pA	10^{-12} Ampere
nanovolt	nV	10^{-9} Volt	nanoamp	nA	10^{-9} Ampere
microvolt	μV	10^{-6} Volt	microamp	μA	10^{-6} Ampere
millivolt	mV	10^{-3} Volt	milliamp	mA	10^{-3} Ampere
volt	V	1 Volt	amps	A	1 Ampere
kilovolt	kV	10^{3} Volt	kiloamp	kA	10^{3} Amperes
megavolt	MV	10^{6} Volt	megamp	MA	10^{6} Amperes

1.2 ELECTRICAL RESISTANCE

To use stored electrical energy in a sensible way, the engineer must be able to control the rate at which current flows from an energy source such as a battery. In the gravitational analogy the mass current down the hill is determined not only by the height of the hill, but by friction in the wheels and by the resistance of the air. Likewise, the current flow from a battery can be limited by the friction or resistance to current flow of the elements attached to the battery. Elements used to control the flow of electric charge are termed resistors.

Materials can be divided into categories depending upon how easily they allow the passage of electric charges. If a material is not a good conduit for charged particles it is called an insulator, while if it easily passes current it is labeled a conductor. For example, metals such as copper and silver are quite good conductors, while rubber and many plastics are good insulators. The relative conductivities of common materials differ by 22 orders of magnitude. This enormous range is indicative of the great control an engineer can have over the flow of current.

Experiments show that most materials behave in such a way that the current through them is directly proportional to the voltage across them. This property of materials is most often expressed as Ohm's law,

$$V = IR \tag{1.3}$$

where R is termed the resistance of the material. In Section 1.7 at the end of this chapter a very simple mathematical model is constructed to represent a crystalline material. For this model we are able to derive Eq. 1.3 and to relate it to the geometry and internal makeup of the material. Section 1.7 is marked with an asterisk to indicate that it need not be considered as required reading but rather as background material for the rest of the book. Such sections appear throughout the text and act as a linkage with physics, materials science, mathematics, or other course work.

Equation 1.3 is a linear relationship between V and I. The expression may also be written in the form

$$I = \frac{V}{R} \tag{1.4}$$

in which the current I may be considered to be a function of the voltage V. If we plot the current as a function of the voltage for a resistor, the result is a straight line such as the one illustrated in Fig. 1.6. Such a plot is called the I–V curve for the

FIGURE 1.6.
The *I-V* **curve for a 10-Ω resistor.**

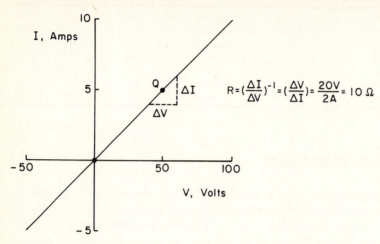

element. The derivative of Eq. 1.4, dI/dV, equals the reciprocal of the resistance, R^{-1}. Since the derivative of a linear function is a constant, it can be evaluated from the slope of the line at any point on the line. For example, the triangle construction shown in the figure yields the slope

$$\frac{dI}{dV} = \frac{\Delta I}{\Delta V} = \frac{1}{R}$$

We have specified a particular point Q in the figure; but for a linear function, the slope is independent of position on the line.

1.3 IDEAL CIRCUIT ELEMENTS

In this text most of the problems and examples deal with ideal circuit elements, some of which are defined in this section. In many cases real circuit elements are very well represented by these idealized models.

Resistor: The circuit element that denotes resistance is called a resistor and has the symbol shown in Fig. 1.7. Common units are Ω (ohms), kΩ (kilohms, or thousands of ohms) and MΩ (megohms, or millions of ohms). An ideal resistor has a constant value independent of the current through it or the voltage across it. In the real world if the current is too high, the element can heat up, which usually changes the resistance value. Care must thus be taken in designing circuits to ensure that the element is not subject to heating beyond its specified limits.

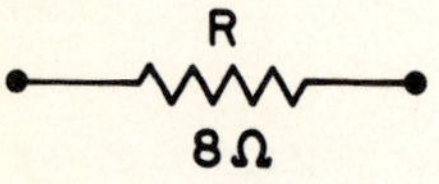

R
●——WW——●
8Ω

FIGURE 1.7.
Symbol for a resistor of value $R = 8\Omega$.

Voltage Source: An ideal voltage source is shown in Fig. 1.8, where V is the voltage between points a and b. The plus sign defines V_{ab} as being positive when

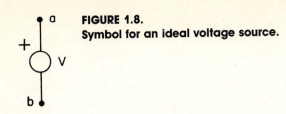

FIGURE 1.8.
Symbol for an ideal voltage source.

the number V is positive. If V is constant in time, it is called a dc voltage source. The term "dc" originated from the term "direct current" but now refers to either a voltage or a current which is constant in time as opposed to a time-varying quantity. If the voltage varies in time in an oscillatory fashion the source is called "ac" for "alternating current." Voltage sources of this type are discussed in detail in later chapters. An ideal voltage source supplies the prescribed voltage V, or $v(t)$ if it varies in time, independent of the current that flows through it. Throughout the text we shall use lowercase letters to indicate time-varying quantities. To reiterate, the plus symbol next to a voltage source is necessary to define the polarity of V. For the source shown in Fig. 1.8,

$$V = V_a - V_b$$

where V_a is the voltage at point a and V_b is the voltage at point b.

We often use the symbol shown in Fig. 1.9 to denote an ideal dc voltage source. The symbol also represents a battery. The longer line always denotes the positive side of the battery. A real battery, as opposed to an ideal one, cannot supply a constant voltage independent of current. For example, if an automobile starter is engaged with the headlights on, the lights grow dimmer due to the decrease in voltage at the battery terminals during high current drain on the battery.

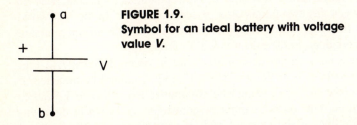

FIGURE 1.9.
Symbol for an ideal battery with voltage value *V*.

Current Source: An ideal current source is denoted by the symbol shown in Fig. 1.10, where I is the current from a to b. This is sometimes written I_{ab} to show specifically that the current flows from a to b. An ideal current source supplies the current I independent of the voltage across the source. That is, the current from a to b is I no matter what the voltage $(V_a - V_b)$. (Note the similarity to the definition

FIGURE 1.10.
**Symbol for an ideal current source car-
rying current *I* from point *a* to point *b*.**

of an ideal voltage source.) An analogy to a current source is an automobile with a "cruise control" which maintains a constant speed that is independent of the terrain. When going up a hill the engine must do more work but the speed remains constant.

Other symbols we shall use are the ideal short circuit (Fig. 1.11*a*), which has zero voltage across it ($V_a - V_b = 0$), no matter what the current I_{ab} is; the open circuit (Fig. 1.11*b*), which has zero current through it no matter what the voltage across it is; and the switch, which acts as either an open circuit or a short circuit (Fig. 1.11*c*). The short circuit and the open circuit can be considered as limits of zero and infinite resistance, respectively, since the latter stops all current and the former will pass charges with no friction. It is useful to note that the short circuit can also be considered as the limit of an ideal voltage source with a value of zero volts, since it has that value independent of the current from *a* to *b*. Likewise, an open circuit can be considered to be a current source of value zero amperes, since it has that value independent of the voltage difference $V_a - V_b$. We use these concepts in later chapters.

FIGURE 1.11.
Symbols for a short circuit (a), open circuit (b), and switch (c).

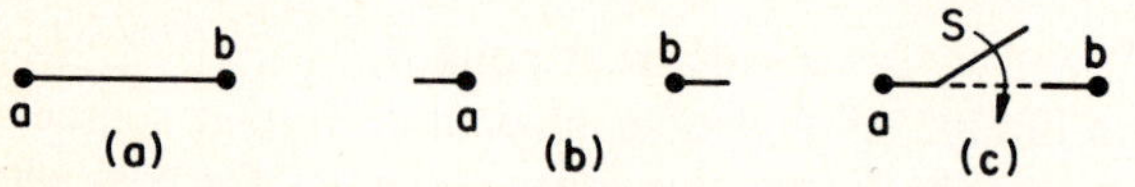

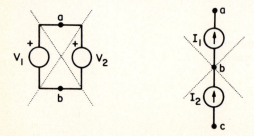

Electrical wires, for our purpose, are considered to yield a short circuit between any points to which they are connected. When we draw circuit diagrams and connect various points with lines, these represent perfectly conducting wires which are thus short circuits. The voltage defined at any point on such a line is always the same with respect to a common reference potential (ground potential), since there is no voltage difference between any two points on a perfect conductor. In the real world long wires and/or very thin wires (see Section 1.7 below) will have a finite nonzero resistance and a corresponding voltage drop when a current flows.

Some other comments concerning ideal sources are in order. Referring to Fig. 1.12, two ideal voltage sources of different values V_1 and V_2 cannot be attached to the same points (a, b), since $V_a - V_b$ cannot be simultaneously equal to V_1 and to V_2. Likewise, two ideal current sources cannot be connected as shown in the right-hand side of Fig. 1.12. Since a short circuit and an open circuit are the limits of ideal sources, a short circuit cannot be attached where voltage source V_2 is located in the figure, nor can I_2 be replaced by an open circuit.

FIGURE 1.12.
Ideal sources cannot be connected in the two fashions illustrated.

1.4   THE ELEMENTAL CIRCUIT

A simple electrical circuit is diagrammed in Fig. 1.13. Because the lines represent wires that are short circuits, $V_a - V_c = 0$ and $V_b - V_d = 0$. The voltage difference V due to the battery thus also exists across the resistor. The charges separated from each other by the chemical forces in the battery are free to flow toward each other through the resistor. We now know that the electrons actually move through the wires and through the resistor from b to d to c to a as shown in the left-hand side of Fig. 1.13. This constitutes a flow of negative charge counterclockwise around the loop. This fact was not known in the early days of electrical measurements, and the convention arose to define the current as flowing in the clockwise direction as if the charge carriers were positively charged. This conventional usage persists, and we define the current I to flow in a clockwise direction as shown in the right-hand side of Fig. 1.13.

FIGURE 1.13.
Electron flow and "conventional current" flow in the elemental electrical circuit. *V* and *I* are positive numbers.

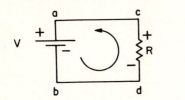

Electron Flow Direction   Conventional Current Flow

Ohm's law is now written more precisely in the form

$$I_{ab} = \frac{V_a - V_b}{R} \tag{1.5}$$

where the algebraic sign relationship refers to Fig. 1.14 and I_{ab} denotes the current from a to b. This way of writing Ohm's law for a resistor always yields the correct direction for the current arrow provided we are given the voltages V_a and V_b referenced to a common ground potential. If V_a is less than V_b, then the quantity I_{ab} is a negative number. If a current has a negative magnitude, it means that the actual current flows opposite to the direction of the arrow.

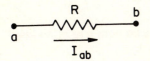

FIGURE 1.14.
Schematic diagram of the sign convention for Ohm's law.

Since an electric field points from positive charges toward negative charges, the electric field directions inside the battery and the resistor are as illustrated in Fig. 1.15. As we shall discuss in more detail in Chapter 2, the current loop must be continuous around a closed circuit in a steady state. Thus, the current flows parallel to the electric field in the resistor (i.e., from high to low potential) and antiparallel to the electric field in the battery (from low to high potential). It is a general result

FIGURE 1.15.
The direction of the electric field inside a battery and inside a resistor for the elemental circuit.

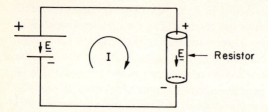

that in a receiver of electrical energy such as a resistor, current flows parallel to the electric field (from high to low potential), just as the cart in Fig. 1.1 receives gravitational energy when it moves down the hill. When devices such as a battery, a solar cell, or a motor generator act as sources of electrical energy, current flows from low to high potential through them. The relationship between voltage and current for receivers and sources of energy are illustrated in Fig. 1.16.

FIGURE 1.16.
The general relationships for receivers and sources of electrical energy.

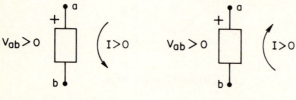

1.5 MEASURING DEVICES

Devices that measure electrical current are called ammeters. Devices that measure electrical voltage are called voltmeters. In practice it is necessary to know the limitations of any measuring instrument. To avoid that problem here, we shall define "ideal" meters and use the symbols shown in Fig. 1.17. The meters shown are ideal in the sense that their insertion does not change the system in any way. Thus, there is no work done by charges moving from a to b through the ammeter and the voltage

FIGURE 1.17.
An ideal ammeter (A_m) measures the current I through the wire from a to b. An ideal voltmeter (V_m) measures V, the voltage drop from b to c.

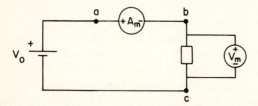

between a and b is zero. An ideal ammeter therefore acts like a short circuit. Also, there can be no current through the ideal voltmeter and thus no energy involved in the operation of the meter. An ideal voltmeter therefore acts like an open circuit.

Since the current through an ammeter may be in either direction it is necessary to know not only the magnitude of the current but also its direction. The polarity signs of the ideal ammeter in Fig. 1.17 mean that if the meter reading is positive, the current is flowing through it from + to −, that is, from the positive to the negative terminal. If the meter reading is negative, the current is flowng through it from − to +. Similarly, if the voltmeter reading is positive, there is a voltage drop from its + to its − terminal. If the voltmeter reading is negative, there is a voltage drop from its − to its + terminal.

TEXT EXAMPLE 1.1

Determine the readings of the ammeter and voltmeter in Fig. 1.17 if $V_0 = -12$ V and the element between b and c is a 6-kΩ resistor.

Solution: Since the ammeter acts like a short circuit, the voltage at a and b will be the same, and

$$V_{bc} = -12 \text{ V}$$

From Eq. 1.5,

$$I_{bc} = \frac{V_{bc}}{(6000)}$$

and thus

$$I_{bc} = \frac{-12}{6000} = -2 \text{ mA}$$

The readings are −12 V by the voltmeter and −2 mA by the ammeter.

A familiar example of an ammeter is the automobile display that shows whether the car battery is discharging (giving up energy) or is being charged (receiving energy) by the generator. When the current I in Fig. 1.18 is positive the meter reading

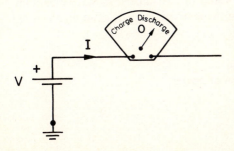

FIGURE 1.18.
Part of an automobile electrical system. Note the symbol used to show the "ground" reference, or zero potential. Here, the battery is being discharged.

is positive. The battery is discharging and acts as a source of electrical energy. When the current I is negative, the meter reading is negative. The battery is being charged and acts as a receiver of electrical energy. It is a common misconception that current must always flow from a battery as shown in Fig. 1.15. This is not necessarily true if other sources exist in the circuit. We reiterate, however, that a resistor is a passive element and can only be a receiver of energy. The current in a resistor always flows in the direction of the voltage drop. The three bar symbol in Fig. 1.18 denotes the ground reference point.

1.6 ELECTRICAL POWER

The MKS unit of energy is the joule (J). The rate at which energy is being delivered or received is called power. The basic unit of power is the watt (W), which equals 1 joule per second. How power is determined in an electrical system may be understood with reference to Fig. 1.19.

In the figure the voltage drop from a to b is V_{ab} volts. Every coulomb of charge transported from a to b through the element gives up V_{ab} joules of energy. The current I_{ab} is the number of coulombs per second going from a to b through the element. The total rate at which energy is being received by the element is thus the product of V_{ab} and I_{ab}. To be precise, if V_{ab} and I_{ab} as shown are both positive numbers or if they are both negative numbers, then the element is a receiver of energy. Algebraically, the power received, P_R, is given by

$$P_R = V_{ab}I_{ab} \tag{1.6}$$

which has the units (joules/coulomb)·(coulomb/second) = joules/second = watts. If V_{ab} and I_{ab} have opposite signs, the power received will be negative, which means that the element is acting as a source of energy.

When the element in Fig. 1.19 is a resistor of value R ohms, then $V = IR$ and Eq. 1.6 can be written (by substituting either for V or I) in two useful alternate forms:

$$P_R = V_{ab}I_{ab} = I_{ab}^2 R = \frac{V_{ab}^2}{R} \tag{1.7}$$

Since the current through a resistor must always be in the direction of the voltage drop, each expression in Eq. 1.7 always yields a positive number. Thus, electrical power will always be delivered to a resistor. The electrical energy delivered to the resistor is converted into heat, which will result in some increase of the resistor temperature and a possible change in the value of resistance. If the temperature increase is excessive, the resistor may be permanently damaged, as in an electrical

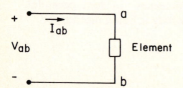

FIGURE 1.19.
Computation of power.

fuse. In practice, the power rating for a given resistor (e.g., $\frac{1}{8}$ watt, $\frac{1}{4}$ watt, 2 watts, etc.) is included in the specifications along with the actual resistance value. Unless otherwise specified, we shall assume in the text that any temperature change is sufficiently small so that the resistance value is constant.

TEXT EXAMPLE 1.2

Determine the power supplied by the battery in Text Example 1.1, as well as the power dissipated by the resistor.

Solution: We found the reading of the ammeter to be negative so current flows from b to a through it and therefore (see Chapter 2) from a to c through the battery. From Eq. 1.6 the power received by the battery is

$$P_R = V_{ac}I_{ac} = (-12)(+2 \times 10^{-3})$$

$$P_R = -24 \ \text{mW}$$

where mW stands for milliwatts. Since P_R is negative, the battery is supplying power. This power is received by the resistor, that is,

$$\frac{(V_{ab})^2}{R} = \frac{144}{6000} = 24 \ \text{mW}$$

or

$$(I_{ab})^2 R = (2 \times 10^{-3})^2 (6000) = 24 \ \text{mW}$$

As found in text Example 1.2, for elements other than resistors, the product in Eq. 1.6 may not be positive. For the case of the automobile battery of Fig. 1.18, if I is positive, the current through the battery is in the direction of the voltage rise. The battery is then converting chemical energy into electrical energy and is acting as a source (the battery is discharging). If I is negative, electrical energy is being delivered to the battery (the battery is either being charged or getting hot!).

In any self-contained system the law of the conservation of energy holds. The total energy supplied by those elements acting as sources must equal the total energy being delivered to all other elements. We shall use this conservation law in the next chapter to develop a fundamental relation among the voltages in a system.

1.7   A SIMPLE MODEL FOR ELECTRICAL RESISTANCE*

To understand how the flow of current is controlled we consider a very simple model for a solid material. It is now known that the negative charge is carried by the least massive stable particle, the electron, which carries $-(1.6 \times 10^{-19})$ coulombs of

---

* Sections marked with an asterisk are not required for the logical flow of the text.

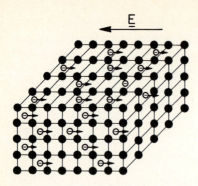

FIGURE 1.20.
In a metal, electric current is carried by negatively charged electrons moving through a crystal lattice.

charge. The relative conductivity of most materials is determined by how many free electrons are available to carry current and by how fast they can move when a voltage is applied. Both of these parameters are determined by the structure of the material in question. This accounts for the wide range of possible conductivity values since all materials are different in this regard. Figure 1.20 illustrates the drift of free electrons through a crystal lattice. To be specific, suppose we wish to determine the *I–V* characteristic of a piece of uniform crystalline material such as the one illustrated in Fig. 1.21, which is subject to an applied voltage V_{ab}. Since we are interested in the current, we must calculate

$$I_{ab} = -ne <w> A \tag{1.8}$$

where *n* is the number density of free electrons, (m^{-3}), *e* is the electron charge (coulombs), and $<w>$ is the average drift velocity (m/sec) of the electrons. Checking the units, we have

$$I_{ab} = (m^{-3})(\text{coulombs})(m/\text{sec})(m^2) = \text{coulombs/second} = \text{amperes}$$

The drift velocity $<w>$ can be found from one of Newton's laws, $F = ma$. Two forces are important, the electrical force due to the applied voltage, and the frictional force felt by the electrons when they collide with the atoms in the crystal lattice structure of the material. We are not interested in the details of how one electron moves, but only in the average velocity of an electron which in turn yields the average current. Newton's law for the average electron of mass *m* is

$$m\frac{d <w>}{dt} = -e <E> - \frac{(m <w>)}{\tau} \tag{1.9}$$

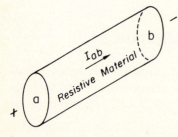

FIGURE 1.21.
A cylindrical material of cross-sectional area A and length l.

where $-e <E>$ is the average electrical force on the electron and τ is the time it takes for the electron to collide with the atoms in the lattice. By an average "collision" time we mean the time required for the electron to come to rest from a velocity $<w>$. In such a collision, the electron loses an amount of momentum equal to $m <w>$ in the time τ. This constitutes a frictional force in the direction opposite to the velocity, and the average frictional force term is then $-m <w>/\tau$, as indicated in Eq. 1.9.

To determine $<w>$ we can set the left-hand side of 1.9 equal to zero since for a constant applied electric field the average velocity does not change with time. Then,

$$<w> = \frac{-e <E> \tau}{m}$$

Substituting this result into Eq. 1.8,

$$I_{ab} = \frac{ne^2 <E> \tau A}{m}$$

This equation is not quite in final form, since we require that the current be expressed in terms of the voltage rather than in terms of the electric field. However,

$$<E> = \frac{V_{ab}}{l}$$

where l is the length of the piece of material; and finally

$$I_{ab} = \left(\frac{ne^2 \tau A}{lm} \right) V_{ab} \tag{1.10a}$$

or

$$V_{ab} = RI_{ab} \tag{1.10b}$$

where

$$R = \frac{lm}{ne^2 \tau A} \tag{1.11}$$

is the resistance. Equations 1.10a and 1.10b are just Ohm's law.

This very simple model has yielded a linear relationship between voltage and current which is valid for many applications. The constants n and τ are determined by the material, while l and A are determined by the geometry of the piece. It is clear that considerably different values for R can be constructed by varying these four quantities. The resistance is directly proportional to the length, since for the same voltage, $<E>$ and $<w>$ are inversely proportional to l. R is inversely proportional to the area, since more electrons can pass through the material if it has a larger cross section, analogous to the flow of water through a pipe.

The electrical energy given to each electron due to the force $e <E>$ is lost into heat each time the electron collides with the atoms or molecules in the material. This is the energy dissipation we refer to above and have shown to be given by I^2R or V^2/R. In an electrical heating system or an oven, we directly use this heat as the "end product." In most other systems the heat dissipated in resistors is lost energy, and effort is usually required to minimize the power used in this manner.

To characterize the electrical nature of a given type of material a parameter is needed that is independent of the geometric factors A and l in Eq. 1.11. We thus define the resistivity, ρ, of a given material as

$$\rho = \frac{m}{ne^2\tau} \quad [\text{ohm} \cdot \text{m}] \tag{1.12}$$

Then, $R = \rho l/A$. Equivalently, we may define the conductivity, σ, as

$$\sigma = \frac{ne^2\tau}{m} \quad [\text{ohm} \cdot \text{m}]^{-1} \tag{1.13}$$

in which case $R = l/\sigma A$. The mho (ohm spelled backwards!) has units of inverse ohms. Except for the universal constants m and e, these two parameters depend solely on the number of free electrons and the lattice collision time constant. Materials handbooks usually list either Eq. 1.12 or Eq. 1.13.

TEXT EXAMPLE 1.3

A resistor is constructed from pure carbon which has conductivity $\sigma = 2.85 \times 10^4$ [ohm $\cdot$ m]$^{-1}$. If the piece of carbon has a cylindrical shape 0.2 mm in diameter and 10 mm long, find its resistance.

Solution: From Eq. 1.11,

$$R = \frac{l}{A\sigma} = \frac{10 \times 10^{-3}}{\pi 10^{-8} (2.85 \times 10^4)} = 11 \ \Omega$$

Carbon composite materials are commonly used in the construction of resistors.

PRACTICE PROBLEMS
AND
ILLUSTRATIVE EXAMPLES

EXAMPLE 1.1

Two metal plates in the battery illustrated below are separated by 1 cm and have an average electric field of 150 V/m between them. What is the voltage measured by the voltmeter? If the switch S is closed and the chemical action in the battery maintains the voltage, what is the reading of the ammeter?

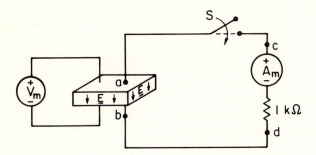

Solution: From Eq. 1.2 the voltage magnitude is the average electric field (150 V/m) times the separation distance (0.01 m), which yields 1.5 V. Since the electric field points downward, a positive charge must move against it "up the hill" to get from point b to point a. The voltage at a is thus higher than at b, and the voltmeter reads +1.5 V. When the switch is closed, current flows from high to low potential through the resistor, so the ammeter will read a positive value for the current the way it is hooked up in the figure. Since all the wires and the ideal ammeter behave like short circuits, the full voltage appears across the resistor. Using Ohm's law, the ammeter reading will be $+[1.5 \text{ V}/1000 \text{ } \Omega] = 0.0015 \text{ A} = 1.5 \text{ mA}$.

Problem 1.1

Find the power supplied by the battery in the circuit of Example 1.1 and show that it is equal to the power dissipated in the 1 kΩ resistor.

Problem 1.2

Suppose that the ideal ammeter in Example 1.1 is replaced by a resistor which has a resistance of 100 Ω. Is the following statement true or false? Since the new resistor must dissipate some power, the overall power supplied by the battery is increased over the result in Problem 1.1. Explain your reasoning.

Problem 1.3

Find the voltage across the 1-kΩ resistor in Problem 1.2. Why is this value less than 1.5 V?

Problem 1.4

Suppose the voltage between two hemispheres of a spherical battery with radii $r_1 = 1$ m and $r_2 = 1.2$ m is given by the expression $V(r) = 1000(r_1^2/r)$ volts. This voltage is due to a radial electric field between the hemisphere which is pointed outward as shown in the figure. Evaluate the voltage at the two plates ($r = r_1$ and $r = r_2$) and determine the magnitude and sign of the voltage measured by the voltmeter. What is the average value of the electric field between the plates?

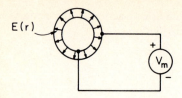

$$\text{Ans.: } -167 \text{ V}, \ 835 \text{ V/m}$$

Problem 1.5

A 5-kΩ resistor is inserted across the terminals of the voltmeter in Problem 1.4. Find the power supplied by the spherical battery.

Problem 1.6

A 1-meter-long bar of copper with a circular cross-sectional area of 10 cm² has a resistance of about 17×10^{-6} ohm. If the same bar is heated and pulled to a length of 100 meters, find the new resistance of the "wire." (*Hint:* Use the expression 1.11 in the text where l is the length and A is the area of the wire.)

$$\text{Ans.: } R = 0.17 \ \Omega$$

Problem 1.7

A thin film of rectangular cross section which is 1 millimeter wide and 1 millimeter long is to be used to construct a resistor with a value of 1000 Ω by depositing a material with $\sigma = 1000 \ [\text{ohm} \cdot \text{m}]^{-1}$. How thick must the material be? (*Hint:* $R = l/\sigma A$.)

$$\text{Ans.: } 1 \text{ micron } (10^{-6} \text{ m})$$

Problem 1.8

The same material is used as in Problem 1.7 but the length deposited is increased to 1 cm (keeping the width the same). How thick must the new film be to yield the same resistance?

Problem 1.9

A digital wristwatch draws a steady current of 0.2 microamperes from a 1.5-volt cell. If 10 joules of energy are available in the cell, how long can the watch operate? What is the average power supplied by the battery?

$$\text{Ans.: } 3.3 \times 10^7 \text{ sec } (\approx 1 \text{ year}), \ 0.3$$
$$\text{microwatts}$$

Problem 1.10

In the circuit shown below, $R_1 = R_2 = 10 \ \text{k}\Omega$. It is also known that the current through each of two resistors is 1 mA ($I_1 = I_2 = 1$ mA). What is the voltage across each resistor? Find V_1.

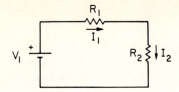

Problem 1.11

In the circuit shown below, the current from the source must flow through both resistors (see Chapter 2).

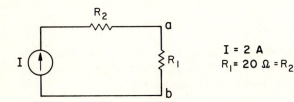

$$I = 2 \text{ A}$$
$$R_1 = 20 \ \Omega = R_2$$

(a) What is the voltage $V_a - V_b$?   Ans.: 40 V

(b) What is the power delivered to R_1?   Ans.: 80 W

(c) What is the voltage across the current
source and how much total power
does it deliver?   Ans.: 80 V; 160 W

Problem 1.12

Suppose the resistor R_1 in Problem 1.11 is reduced to 10 Ω. Find the new voltages and powers requested in parts a, b, and c.

Problem 1.13

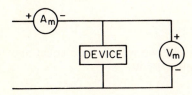

In the circuit shown here, what is the power received by the device

(a) If the ammeter reads -2 A and the
voltmeter reads $+100$ V?   Ans.: -200 W (The device is supplying
power.)

(b) If the ammeter reads -2 A and the
voltmeter reads -100 V?   Ans.: $+200$ W

Note: Even though you may know Kirchoff's current law from other courses do not use it in the problems below.

EXAMPLE 1.2

In the network shown below, we say the resistors are hooked up in parallel. Use Ohm's law to find the current through the two resistors. Calculate the power dissipated in the two resistors. Use this power result to find the total current I drawn from the battery. Show that $I = I_1 + I_2$.

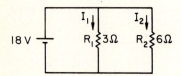

Solution: From Ohm's law, $I_1 = 18/3 = 6$ A and $I_2 = 18/6 = 3$ A. The power dissipated in the jth resistor is given by $P_j = I_j^2 R_j$ so $P_1 = 108$ W and $P_2 = 54$ W. Since the total power supplied by the battery is VI, we have

$$P_T = 108 + 54 = 162 \text{ W} = VI$$

and since $V = 18$ V,

$$I = 9 \text{ A}$$

Indeed, the total current equals the sum of the currents supplied to the two elements in parallel.

Problem 1.14

Suppose in Example 1.2 that the resistor R_1 is 6 Ω and the resistor R_2 is 12 Ω. Find the current through and the power dissipated in the two resistors. Use this to find the total current supplied by the battery. Check to see that the total current equals the sum of the two currents.

Problem 1.15

In the figure for Problem 1.10 above, assume that the two resistors are equal and that $V_1 = 12$ V. Find R_1 and R_2 if the battery supplies 18 W.

Problem 1.16

In part (a) of Problem 1.13, suppose a 10-Ω resistor is inserted in parallel with the device and that the device maintains the voltage as measured by the voltmeter ($+100$ V) and the current as measured by the ammeter (-2 A). How much power is delivered to the resistor? Make an argument which shows that this power must come from the device.

Problem 1.17

Ohm's law, $V = IR$, may be considered as a linear function relating the voltage V to the current I. Likewise, we may write I as a function of V, $I = (1/R) V$. Both of these expressions are of the form

$$y = f(x) = ax$$

Make a plot of the current I versus the voltage V if R is a 5-Ω resistor. A plot of current versus voltage for a device is often called an I–V curve.

Problem 1.18

Sketch the I–V curve for a resistor twice as large (10 Ω) and for a resistor one-half as large (2.5 Ω) as used in Problem 1.17. How does the magnitude of the slope of the line relate to the resistor value?

EXAMPLE 1.3

In the next chapter we shall show that the current must be the same through any elements hooked up in "series." For example, the currents I_1, I_2, and I_3 are all equal in the circuit shown here. Use this fact along with power arguments to find the current $I = I_1 = I_2 = I_3$. (You may use this fact in subsequent problems as well.)

Solution: The power dissipated in each resistor is $I^2R = 5I^2$. Since there are two resistors, the total power dissipated is $P_T = 10I^2$. There is only one source, so it must supply this power. The power P_s supplied by the 10-V source with current I flowing from low to high potential is given by $P_s = VI = 10I$. Setting these two expressions equal

$$P_s = 10I = 10I^2 = P_T$$

we have $I = 1$ A.

Problem 1.19

The circuit below contains a fixed 10-V battery and a 5-Ω resistor. Make a plot of all possible currents I_{ab} through the load resistor R_L as a function of all possible voltages $V_{ab} = (V_a - V_b)$ across the resistor R_L. (*Hint:* This is a linear system and you may assume that the function is a straight line. Since two points determine a line, use the result from Example 1.3 to find one point. The second point may be found by setting $R_L = 0$, which is a short circuit. Then, use Ohm's law and the fact that $V_{ab} = 0$ for a short circuit.)

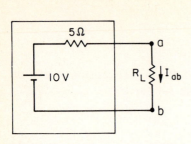

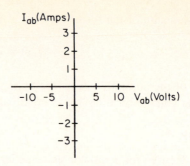

Problem 1.20

Use the axes below and make a plot of the power delivered to the load resistor R_L as a function of R_L keeping the 10-V battery and the 5-Ω resistor fixed in the circuit for Problem 1.19. Estimate from the plot the value of R_L that yields the maximum power transfer to the load resistor. (You might try using techniques from calculus to prove that $R_L = 5\ \Omega$ is the correct answer. We prove this in Chapter 3.)

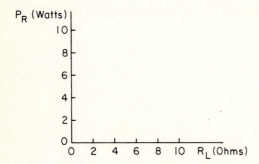

Problem 1.21

The I–V "curve" (actually a straight line in this case) generated in Problem 1.19 represents all possible voltage and current pairs that can characterize the connection between a and b in the circuit. The curve generated in Problem 1.17 represents all possible (I, V) pairs for an ideal 5-Ω resistor. Suppose a 5-Ω resistor is attached between a and b as shown below. Once this is accomplished, the curves in Problems 1.17 and 1.19 must be simultaneously satisfied since only one value of the pair (I_{ab}, V_{ab}) can occur. Find the intersection of the two curves by plotting them on the same graph and hence the value of V_{ab} and I_{ab} for the circuit shown below. Check the result using analytical circuit theory and the fact that the current must be the same through both resistors, which is proven in Chapter 2.

Ans.: $V_{ab} = 5$ V

$I_{ab} = 1$ A

EXAMPLE 1.4

In the figure shown, let $I_s = 8$ A, $R = 3\,\Omega$, and $V_s = 16$ V.

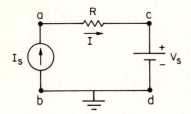

The sources are ideal and hence maintain their respective values. Find the total power supplied by the current source, the power delivered to R, and the power delivered to the battery. Find the voltage $V_a - V_b = V_a$. You may use the fact that $I = I_s$ for a series circuit, which is proved in the next chapter.

Solution: The current I_s must flow through the resistor and the battery ($I = I_s = I_{cd}$). The voltage across the resistor ($V_a - V_c$) must equal $I_s R$, or 24 V. The power delivered to R is $(V_a - V_c)I_s = I_s^2 R = 192$ W. Since current flows from high to low potential, power is delivered to the battery in the amount $I_s V_s = 128$ W. The total power supplied by the current source is the sum of the power dissipated in the resistance and the power supplied to the battery, and thus equals $192 + 128 = 320$ W.

The power P supplied by the current source equals $(V_a - V_b)I_s$. Since the point b is grounded, $V_b = 0$, and we have $P = V_a I_s = 320$ W. Therefore, $V_a = 320/8 = 40$ V. Note that the voltage drop from a to b (40 V) equals the sum of the voltage drops from a to c (24 V), the voltage drop from c to d (16 V), and the voltage drop from d to b (0 V). This is an example of Kirchhoff's voltage law, which is discussed in Chapter 2.

Problem 1.22

In the figure of Example 1.4, let $I_s = 5$ A, $R = 2\,\Omega$, and $V_s = 10$ V. Let the system operate for 1 hour, and assume that the battery is ideal and hence maintains its voltage for the entire time interval.

(a) What is the power supplied
    by the current source?　　　　　　　Ans.: 100 W

(b) What is the voltage across
    the current source ($V_a - V_b$)?　　　Ans.: 20 V

(c) How much energy is dissipated in
    the resistor?　　　　　　　　　　　Ans.: 180,000 J

(d) How much extra energy is dissi-
    pated by or stored in the battery in
    that hour?　　　　　　　　　　　　Ans.: 180,000 J

Problem 1.23

In the figure of Example 1.4, let $I_s = 2$ mA, $R = 3$ kΩ, and $V_s = 10$ V.

(a) What is the voltage across
the resistor ($V_a - V_c$)?   Ans.: 6 V

(b) What is the voltage across the
current source?   Ans.: 16 V

(c) What is the power supplied by
the current source?   Ans.: 32 mW

(d) Suppose R is reduced by factor of 2.
What is the power supplied by the
current source?   Ans.: 26 mW

Problem 1.24

Suppose the resistors in Example 1.4 and Problem 1.22 are both increased by a factor of 2. Find the voltage across the current source in each case.

EXAMPLE 1.5

In the figure shown, let $I_s = 2$ A, $R = 4 \, \Omega$, and $V_s = 12$ V. The sources are ideal. Find the power supplied to the resistor and the power supplied by each source. Find the current I_2.

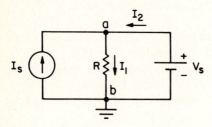

Solution: Since the voltage $V_a - V_b = V_a = V_s$, the current $I_1 = V_s/R = 3$ A. The power dissipated in R is $I_1^2 R = 36$ W. The power supplied by the current source is $I_s(V_a - V_b) = 24$ W. Thus, the power supplied by the battery must be $36 - 24 = 12$ W. The power supplied by the battery is $V_s I_2 = 12$ W. Therefore, $I_2 = 12/12 = 1$ A. Note that $I_s + I_2 = 2 + 1 = 3$ A $= I_1$.

Problem 1.25

In the figure of Example 1.5, let $I_s = 4$ A, $R = 4 \, \Omega$ and $V_s = 12$ V.

(a) What is the current I_1 and the power
dissipated in the resistor?   Ans.: 3 A and 36 W

(b) What is the power supplied by the
current source?   Ans.: 48 W

(c) What is the power supplied by
the battery?   Ans.: −12 W (i.e., power is delivered
to the battery)

(d) What is the current I_2?   Ans.: -1 A

(e) Check that $I_s + I_2 = I_1$.   Ans.: $4 - 1 = 3$

Problem 1.26

In the figure of Example 1.5, let $I_s = 3$ mA and the resistor be 1 kΩ. If the power dissipated in the resistor is 100 mW, what is the voltage difference $(V_a - V_b)$? Assume the sources are ideal. (Note that this voltage must be equal to the voltage V_s.) Also find the current supplied by the battery, and show that the currents add up.

Problem 1.27

Suppose the resistor in Problem 1.25 is reduced to 2 Ω. Find the power supplied by each source.

Problem 1.28

In the figure of Example 1.5, suppose $V_s = 20$ V and the battery supplies 200 mW (0.2 W). If $R = 1$ kΩ, what is I_s? Show $I_s + I_2 = I_1$.

Chapter 2
The Circuit Laws

In this chapter conservation principles are used to generate two circuit analysis laws. These are then applied to derive some very useful relationships which allow quick solution of a wide variety of circuit problems. More complex circuits require a methodical approach, leading to systems of simultaneous equations that can be constructed by direct application of the conservation laws. More elegant solution methods are discussed in Chapter 3.

2.1 KIRCHHOFF'S VOLTAGE LAW

Conservation of energy is a fundamental law of physics and plays an important role in circuit analysis. Returning again to the gravitational analogy, the potential gained going up a hill is converted to kinetic energy and frictional heat on descent. If the cart makes a round trip, that is, if it rolls down the hill and is pulled back up again as shown in Fig. 2.1, the amount of the gravitational energy gained going up the hill equals the energy released going down.

   The electrical analogy to this conservation principle is known as Kirchhoff's voltage law, or KVL in shorthand notation. One way to state the law is as follows: If an electrical circuit is completely traversed in a closed path, the sum of all the energy gains ("voltage rises") in the path must equal the sum of all the energy losses ("voltage drops"). As an example, consider again the simple circuit shown in Fig. 2.2. Suppose the loop is traversed clockwise starting at c. The voltage at that point

FIGURE 2.1.
In each traversal of the gravitational loop, the energy gained equals the energy lost.

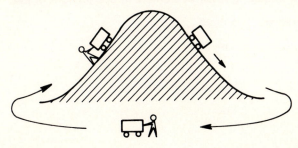

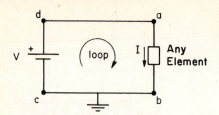

FIGURE 2.2.
Application of KVL to the elemental circuit.

is defined as zero. The voltage rises going from c to d by an amount equal to the battery voltage V. There is no voltage change along the perfectly conducting wires from d to a nor from b to c. Thus, by Kirchhoff's voltage law, the voltage must decrease by the amount V across the element. If the element is a resistor of R ohms, we can state an important result: The voltage falls by an amount $+IR$ from point a to point b when a current I flows from a to b through a resistance R. This is just Eq. 1.3 in words.

Successful application of Kirchhoff's voltage law demands close attention to details concerning the algebraic signs of the parameters. Furthermore, it is essential to develop a standardized procedure for attacking the problems. In practice one can equally well generate correct algebraic equations using KVL by any of three methods. One can add up all the voltage rises encountered in traversing a loop and set them equal to all the voltage drops:

$$\Sigma \text{ Voltage rises } = \Sigma \text{ Voltage drops}$$

where Σ denotes a sum. Alternatively, one can add up all the voltage rises around the loop and set them equal to zero:

$$\Sigma \text{ Voltage rises } = 0$$

noting that a voltage drop is a negative rise. Finally, all the voltage drops around a circuit can be summed and set equal to zero:

$$\Sigma \text{ Voltage drops } = 0$$

The latter turns out to be the most convenient form, and we shall use it in most cases. The student should choose the method with which she or he is most comfortable. For resistors, the proper algebraic signs for summing voltage rises are given on the left side of Fig. 2.3, and for summing voltage drops, on the right. Similarly, the

FIGURE 2.3.
Sign conventions for current flow through a resistive element in applications of KVL.

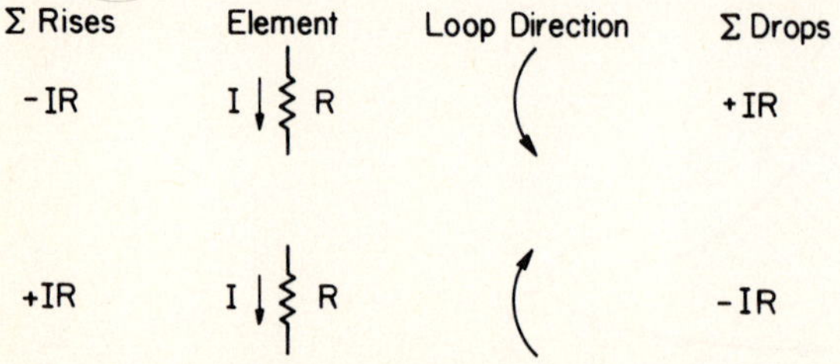

Σ Rises	Element	Loop Direction	Σ Drops
$-IR$	$I \downarrow\ R$	(	$+IR$
$+IR$	$I \downarrow\ R$	(	$-IR$

FIGURE 2.4.
Sign conventions for traversal of a voltage source in applications of KVL.

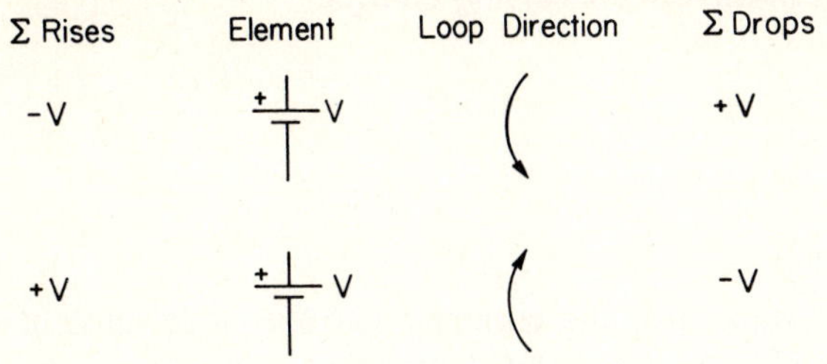

Σ Rises	Element	Loop Direction	Σ Drops
$-V$	$\pm\!\!\!\!\top\, V$	$\curvearrowleft$	$+V$
$+V$	$\pm\!\!\!\!\top\, V$	$\curvearrowleft$	$-V$

contributions of voltage sources to the summations for the two loop directions are given in Fig. 2.4.

A simple example shows how this law can be applied to analyze a circuit problem (Fig 2.5). The loop direction has been chosen arbitrarily, as have the directions of the currents through resistors R_1 and R_2. (We prove in the next section that the current in a series circuit such as this must be the same through both resistors.) Referring to Figs. 2.3 and 2.4 and adding up the voltage drops around the loop yields

$$\Sigma \text{ Voltage drops } = -V_0 + I_1 R_1 + I_2 R_2 = 0$$

which may be written

$$V_0 = I_1 R_1 + I_2 R_2 \tag{2.1}$$

This analysis yields one equation with two unknowns, I_1 and I_2. We need a second physical principle to solve the problem, namely, Kirchhoff's current law.

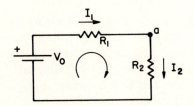

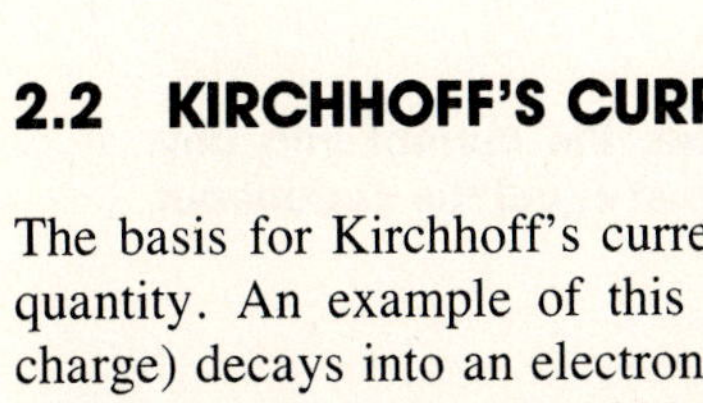

FIGURE 2.5.
Circuit to illustrate the KVL analysis method.

2.2 KIRCHHOFF'S CURRENT LAW

The basis for Kirchhoff's current law (KCL) is that electric charge is a conserved quantity. An example of this occurs when a gamma ray photon (which has no charge) decays into an electron and a positron (Fig. 2.6). These two particles carry equal but oppositely signed charges so that the net charge is still zero.

KCL in its elementary form states that in a circuit with no time variations (and hence a constant current flow) the current which flows into any single point or "node" of a circuit along a conducting wire must be equal to the current which leaves that point along any other wire or wires attached to the node. Consider

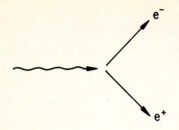

FIGURE 2.6.
Gamma ray annihilation converting energy into mass, but conserving charge.

the case shown in Fig. 2.7. KCL states that the current I into the node must be equal to the sum of all the currents $(I_1 + I_2 + I_3 + I_4)$ leaving that point. As a proof consider what would happen if there were a small difference ΔI. Since the units of ΔI are coulombs per second, there would be a continuous increase of charge at the node. This is clearly not a steady state, since the charge density at the node and resultant electric forces would increase without bound, which is impossible. Thus, ΔI must be zero.

As with KVL, we can write Kirchhoff's current law in several ways. For example at any given node we can write

$$\Sigma \text{ Current entering node} = \Sigma \text{ Current away from node}$$

Alternatively, this can be expressed

$$\Sigma \text{ All currents entering a node} = 0$$

or

$$\Sigma \text{ All currents leaving a node} = 0$$

Note that a current leaving a node is equivalent to a negative current entering the node and vice versa.

Because the term node is used quite often, we must be more precise in defining it. A node is any portion of an electrical circuit that is linked by perfectly conducting wires. In other words, if we can enclose some portion of a circuit by a surface such that only conducting wires lie inside, that portion of the circuit can be "collapsed" to a single node. For example, the circuit in Fig. 2.8*a* can be redrawn as in Fig. 2.8*b* by collapsing the apparent multiple nodes to only two nodes.

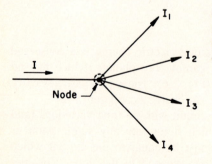

FIGURE 2.7.
In a steady state, the current into any point or node must equal the current out of that point.

FIGURE 2.8.
Reconstruction of apparent multiple nodes (a) to a system with only two nodes (b).

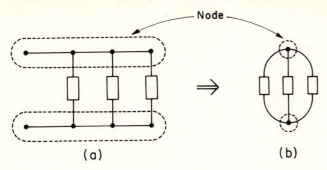

Using KCL, we can now complete the solution of the example in Fig. 2.5. At node a, we can add up all the currents to the node to give

$$I_1 + (-I_2) = 0$$

so

$$I_1 = I_2 = I$$

Equation 2.1 can now be solved since

$$V_0 = I(R_1 + R_2)$$

and finally,

$$I = \frac{V_0}{R_1 + R_2}$$

One of the convenient features of this formulation is that neither the direction of the loop nor the choice of the current directions matters; the algebra takes care of any incorrect assumption. For example, suppose we started this problem by assigning currents in the opposite direction, as shown in Fig. 2.9. By KCL, $I_1' = I_2' = I'$, and soving yields

$$I' = \frac{-V_0}{R_1 + R_2}$$

The numbers I_1' and I_2' are thus negative. The actual currents would be in the direction opposite to the arrows in Fig. 2.9, which is the same result obtained previously.

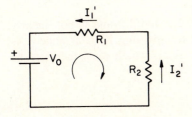

FIGURE 2.9.
Figure 2.5 revisited.

Suppose we try a more complicated example: determining the current through the resistor R_2 in the circuit shown in Fig. 2.10a.

FIGURE 2.10a.
Find the current I_2 through resistor R_2.

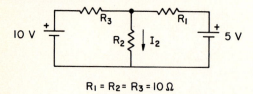

$$R_1 = R_2 = R_3 = 10 \ \Omega$$

First, we label the current in the other two resistors with a name and a direction. If the solution for any of the currents turns out to be negative, it merely means that the current is flowing in the direction opposite to the original arrow (Fig. 2.10b).

FIGURE 2.10b.
Assignment of loop and current directions.

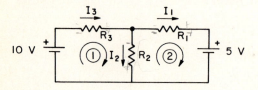

Adding up the voltage drops around loop 1 yields,

$$-10 + 10I_3 + 10I_2 = 0 \qquad 10\left(I_1 + I_2\right) \qquad (2.2)$$

$$10I_1 + 10I_2$$

since each resistor is 10 Ω. Traversing loop 2 gives

$$-10I_2 + 10I_1 + 5 = 0 \qquad (2.3)$$

(To verify these signs, refer again to the right side of Figs. 2.3 and 2.4.) One slight problem remains: there are three unknowns but only two equations. We need one more independent equation. If we construct a loop around the entire perimeter of the circuit, the equation that results is

$$-10 + 10I_3 + 10I_1 + 5 = 0$$

However, this equation is simply the sum of Eqs. 2.2 and 2.3 and is not linearly independent. The required third equation is found by applying KCL at node a as shown in Fig. 2.11. The current entering node a is I_3, which must therefore be equal to the sum of the two currents leaving the node:

$$I_3 = I_1 + I_2 \qquad (2.4)$$

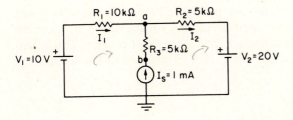

$$I_3 = I_1 + I_2$$

FIGURE 2.11.
The currents at node *a*.

This completes a set of three equations and three unknowns which can be solved for I_2. Substituting Eq. 2.4 into Eq. 2.2 yields

$$-10 + 10I_1 + 10I_2 + 10I_2 = 0$$

$$10I_1 + 20I_2 = 10$$

From Eq. 2.3,

$$10I_1 - 10I_2 = -5$$

Subtracting these two equations,

$$30I_2 = 15$$

and hence

$$I_2 = +0.5 \text{ A}$$

TEXT EXAMPLE 2.1 ──────────────────────────────────

Find the currents I_1 and I_2 in the circuit below and the power supplied by each of the three power sources. Show specifically that the power supplied by all sources equals the total power dissipated in the resistors.

Solution: By KCL, the current to point *b* from the current source must pass through resistor R_3 as well and thence to node *a*. KCL at node *a* yields

$$I_2 = I_1 + 1 \tag{1}$$

where the currents are all measured in milliamperes. There are three possible loops about which to employ KVL. However, the two small loops both include a current source, and we do not yet know the voltage across that source. Choosing a clockwise loop around the perimeter, which has no current source, we add up all the voltage drops and set them to zero, yielding the equation

$$-10 + (10 \text{ k}\Omega)(I_1) + (5 \text{ k}\Omega)(I_2) + 20 = 0 \tag{2}$$

Now, if I_1 and I_2 are measured in milliamperes (10^{-3}A), the factor 10^{+3} implied by the kΩ symbol will multiply to unity with this factor, and this equation becomes, after some manipulation,

$$10I_1 + 5I_2 = -10 \tag{3}$$

where again the currents I_1 and I_2 are measured in milliamperes. Solving the two simultaneous equations (1) and (3),

$$I_1 = -1 \qquad I_2 = 0$$

Note that the values of I_1 and I_2 are independent of resistor R_3. To determine the power supplied by each source, we need to know the voltage across the current source and the current through each voltage source. For the voltage sources we already have enough information. The power received by the 10-volt battery is given by P_1, where

$$P_1 = (10 \text{ V})(1 \text{ mA}) = 10 \text{ mW}$$

P_1 is positive since the current flows from high to low potential through the battery. It is a receiver of energy. The power received by the 20-volt battery is given by P_2, where

$$P_2 = (20)(0) = 0$$

The 20-volt battery is neither receiving nor providing power. To find the power associated with the current source we need to find the voltage across it. Since the current $I_2 = 0$, there is no voltage difference across the 5-kΩ resistor. This means that the voltage at point a, V_a, must also equal 20 V. Using Ohm's law we can find V_b,

$$\frac{V_b - V_a}{5 \text{ k}\Omega} = 1 \text{ mA}$$

$$V_b = 5 + 20 = 25 \text{ V}$$

Since current flows from low to high potential through the current source, it is a source of energy and the power supplied is P_3, where

$$P_3 = (25 \text{ V})(1 \text{ mA}) = 25 \text{ mW}$$

The total power supplied by the sources is

$$P_T = 25 - 10 = 15 \text{ mW}$$

This must equal the power dissipated in all the resistors,

$$\Sigma I_j^2 R_j = I_1^2 R_1 + I_2^2 R_2 + I_s^2 R_3 = 10 + 0 + 5 = 15 \text{ mW} = P_T$$

and the calculations are consistent.

2.3 SOME USEFUL CONSEQUENCES OF KIRCHHOFF'S LAWS

When circuit elements are connected as shown in Fig. 2.12a, we say they are "in series"; and when connected as in Fig. 2.12b, we say they are "in parallel." Two elements are in series when there is no other element connected to their common node. Then, by KCL, the current through the elements in series must be exactly the same. Elements are in parallel when they are connected between the same two nodes. Then, by KVL, the voltage across elements in parallel must be exactly the same.

A number of important results now follow from Kirchhoff's laws which are very useful in their own right and should be remembered. To study resistors in series we return to Fig. 2.5. We have already found that $I = V_0/(R_1 + R_2)$, and we note that R_1 and R_2 are in series. We can find the voltage across each of the resistors using Ohm's relation, $V = IR$. Across resistor R_1,

$$V_1 = IR_1 = \frac{R_1}{R_1 + R_2} V_0 \qquad (2.5a)$$

Likewise,

$$V_2 = \frac{R_2}{R_1 + R_2} V_0 \qquad (2.5b)$$

These two equations illustrate the *voltage divider rule*, which shows how a given voltage V divides across two resistors in series. Each resistor has across it a "percentage" of the total voltage, proportional to the magnitude of the resistor. This

FIGURE 2.12.
Series (a) and parallel (b) connection of circuit elements.

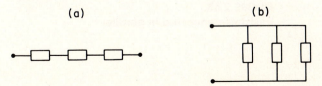

(a)   (b)

example also shows that we may replace the two series resistors with an equivalent resistance, R_E, where

$$R_E = R_1 + R_2 \tag{2.6}$$

By equivalence we mean the following: Two networks are deemed equivalent if, for any voltage V across them, an identical current I is drawn.

Suppose the two resistors are connected in parallel, as shown in Fig. 2.13. The voltage across both of the resistors is the same, namely, V, since both are connected across the battery. What is the current through each of them? Using Ohm's law:

$$I_1 = \frac{V}{R_1}$$

$$I_2 = \frac{V}{R_2}$$

applying KCL at node a,

$$I = I_1 + I_2$$

combining

$$I = \frac{V}{R_1} + \frac{V}{R_2} = V \left(\frac{1}{R_1} + \frac{1}{R_2} \right)$$

or

$$V = I \left(\frac{1}{R_1} + \frac{1}{R_2} \right)^{-1}$$

Our first conclusion is that the equivalent resistance in this case is

$$\frac{V}{I} = R_E = \left(\frac{1}{R_1} + \frac{1}{R_2} \right)^{-1} \tag{2.7a}$$

which can also be written

$$R_E = \frac{R_1 R_2}{R_1 + R_2} \tag{2.7b}$$

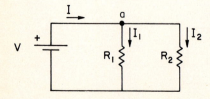

FIGURE 2.13.
Resistors connected in parallel.

For reference, note that Eq. 2.7*a* can also be written,

$$G_E = \frac{1}{R_E} = \frac{1}{R_1} + \frac{1}{R_2} = G_1 + G_2$$

where the quantities labeled by G are called the conductances. The units of G are "inverse ohms," often called mhos. A more recent designation for this measurement unit is the Siemen. It is often more convenient to work out parallel networks using conductances, since they simply add up for elements in parallel. The equivalent resistance can always be found at the end of the computation by taking the reciprocal of the equivalent conductance.

As an example of how to determine R_E for a network with several resistors, we can find the equivalent or total resistance between points *a* and *b* as shown in Fig. 2.14*a*. Several steps are necessary with alternate application of the series and parallel resistance equations (Fig. 2.14*b*); the final R_E is equal to 6 Ω. It is worth remembering that two equal resistors in parallel yield an equivalent resistor half as large.

A shorthand notation is often used for parallel networks; that is, if R_1 is parallel to R_2, we can denote this by

$$R_E = R_1//R_2 = \frac{R_1 R_2}{R_1 + R_2}$$

If several resistors are in parallel, we can write

$$R_E = R_1//R_2//R_3, \text{ etc.}$$

The network can be solved in small steps. For example, if

$$R_E = R_1//R_2//R_3$$

we may determine $R' = (R_2//R_3)$, then find $R_1//R'$, that is,

$$R_E = R_1//(R_2//R_3)$$

Alternatively, the total conductance can be found,

$$G_E = G_1 + G_2 + G_3$$

and then the reciprocal, since $R_E = G_E^{-1}$.

FIGURE 2.14*a*.
A complex resistor network and its equivalent circuit representation.

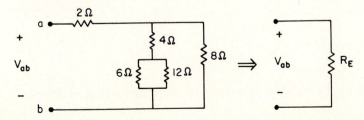

FIGURE 2.14b.
Steps in simplifying the network in Fig. 2.14a.

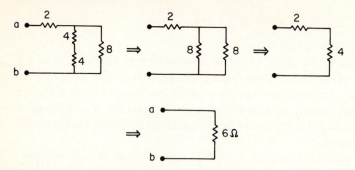

The second conclusion from the analysis associated with Fig. 2.13 is that in a parallel configuration the current divides between the two resistors. In this circuit the total current is

$$I_T = \frac{V}{R_E} = V\frac{R_1 + R_2}{R_1 R_2} \tag{2.8}$$

The individual resistors carry currents V/R_1 and V/R_2, respectively. Solving Eq. 2.8 for V, and noting that the current through R_1 is given by $I_1 = V/R_1$, yields

$$I_1 = \frac{R_2}{R_1 + R_2}I_T \tag{2.9a}$$

Likewise, the current through R_2 is

$$I_2 = \frac{R_1}{R_1 + R_2}I_T \tag{2.9b}$$

The fractions in front of I_T in Eqs. 2.9a and 2.9b are analogous to the fractions in Eqs. 2.5a and 2.5b and define the *current divider rule*. Note that, unlike the voltage divider rule, the current divider rule has the opposite index in the numerator; that is, the formula for I_1 has R_2 in the numerator while I_2 has R_1. Inspection shows that the smallest resistor draws the largest current. If one of the resistors is a short circuit ($R = 0$), the entire current flows through it.

Before leaving this section two general ways of expressing these results can be noted. If there are more than two resistors in series, as in Fig. 2.15, then the voltage across resistor R_i is

$$V_i = \left(\frac{R_i}{R_E}\right)V \tag{2.10}$$

where R_E is the sum of the resistors. Similarly, if there are several resistors in

FIGURE 2.15.
Application of the voltage divider rule for multiple series resistors.

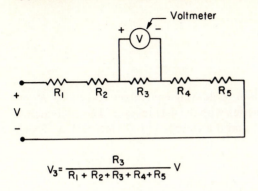

$$V_3 = \frac{R_3}{R_1 + R_2 + R_3 + R_4 + R_5} V$$

parallel, as in Fig. 2.16, we can write

$$I_i = \left(\frac{R_E}{R_i} \right) I_T \tag{2.11}$$

Using the conductance concept, this result can also be written as

$$I_i = \left(\frac{G_i}{G_E} \right) I_T \tag{2.12}$$

where R_E is the equivalent resistance and $G_E = R_E^{-1}$.

FIGURE 2.16.
Application of the current divider rule for multiple parallel resistors.

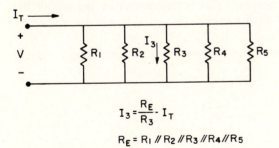

$$I_3 = \frac{R_E}{R_3} \cdot I_T$$

$$R_E = R_1 \mathbin{/\!/} R_2 \mathbin{/\!/} R_3 \mathbin{/\!/} R_4 \mathbin{/\!/} R_5$$

TEXT EXAMPLE 2.2 ──────────────────────────────

Find the equivalent resistance to the right of the two terminals a and b and use this result to determine the current I. Use the current divider principle to find I_1 and the voltage divider principle to find V_{ac}. Show that the power supplied by the battery equals the power dissipated in the resistors.

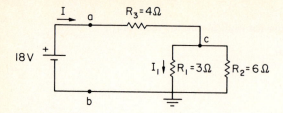

Solution: First, the parallel combination yields an equivalent resistance $R_p = (3 \times 6)/(3 + 6) = 2\ \Omega$. This resistor is in series with the 4-Ω resistor. The equivalence between points a and b is $4 + 2 = 6\ \Omega$, and the current is $I = 18/6 = 3$ A. Using the current divider rule,

$$I_1 = \frac{R_2}{R_1 + R_2} 3 = 2\ \text{A}$$

Using the voltage divider rule,

$$V_{ac} = \frac{4}{4 + 2} 18 = 12\ \text{V}$$

The power supplied by the battery is $P = (18)(3) = 54$ watts. The powers dissipated are $I^2 R_3 + V_c^2/R_1 + V_c^2/R_2 = 36 + 12 + 6 = 54$ watts, which checks.

Thus far, the examples in this section have dealt entirely with voltage sources. The simplest circuit employing a current source is as shown in Fig. 2.17. By KCL, the current I must also flow through the resistor, hence the voltage across the resistor is given by $V = IR$. This number can take on any value merely by changing the resistor. In the elemental voltage circuit of Fig. 1.13, any current can be made to flow through an ideal voltage source merely by changing the resistor. An ideal current source will have a voltage across it that is determined by the rest of the circuit. An ideal voltage source will have a current through it that is determined by the rest of the circuit.

Applying the current divider rule to the following circuit containing a current source (Fig. 2.18), we have

$$I_1 = \frac{R_2}{R_1 + R_2} I$$

FIGURE 2.17.
An elemental circuit employing a current source.

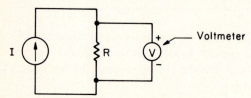

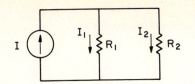

FIGURE 2.18.
The current divider.

$$I_2 = \frac{R_1}{R_1 + R_2} I$$

These equations constitute the current divider law as applied to a current source. Equivalently, using conductances,

$$I_1 = \left(\frac{R_T}{R_1}\right) I = \left(\frac{G_1}{G_T}\right) I$$

$$I_2 = \left(\frac{R_T}{R_2}\right) I = \left(\frac{G_2}{G_T}\right)$$

where once again

$$R_T = \frac{R_1 R_2}{R_1 + R_2}$$

2.4 APPLICATIONS OF KVL AND KCL TO MORE COMPLICATED CIRCUITS: A STANDARDIZED APPROACH

Consider the circuit shown in Fig. 2.19a. What are the currents through the two resistors? As was mentioned earlier, we need to generate a set of equations equal to the numbers of unknowns. First, some definitions are necessary for the current directions, and a choice of loops must be made for the application of KVL. If we take a counterclockwise loop around the right-hand side of the circuit and add up the voltage drops (see Fig. 2.19b),

$$-20 + 15I_{cd} + 5I_{ab} = 0$$

This yields one equation and two unknowns. Note that the current I_{cd} flows through the battery and the 15-Ω resistor since they are in series. The second equation comes from KCL applied at point a. The two currents which enter a are I_{cd} and the 8 A from the current source. These must add up to yield the outflowing current, so

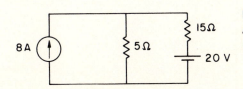

FIGURE 2.19a.
A circuit used to illustrate Kirchhoff's laws.

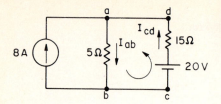

FIGURE 2.19b.
Current and loop definitions for
Fig. 2.19a.

$$8 + I_{cd} = I_{ab}$$

which completes the set. Substituting,

$$-20 + 15I_{cd} + 40 + 5I_{cd} = 0$$

so $I_{cd} = -1$ A. The negative sign means that the actual choice of current direction through the 15-Ω resistor is from d to c. In this circuit the battery is being charged by the current source. Continuing, I_{ab} must be $+7$ A.

Since it is quite easy to mix up the signs of the various quantities involved, some students prefer a standard procedure for carrying out such analysis. For example, the following sequence of steps has proved to be quite useful:

1. Define current arrows everywhere.
2. For currents through resistors, write in small plus signs at the base of each arrow and minus signs at the tip. That is, the arrow points from plus toward minus.
3. Write in plus and minus signs to show the given or assumed polarity of any voltage sources.
4. Draw in the loop direction, being sure that it is not confused with the current arrows.
5. Develop the KVL equations by assigning an algebraic sign to each element which corresponds to the first sign you reach when that element is approached in going around the loop.
6. Apply KCL at enough node points to yield a complete set of independent equations.

The circuit given in Fig. 2.20a can be used to illustrate this method, where the currents through R_1 and R_2 are to be found. Following the method, arrows are drawn for each current and algebraic signs are generated as shown in Fig. 2.20b. Making a clockwise loop around the perimeter starting at the lower left-hand corner,

FIGURE 2.20a.
Circuit for illustrating the standardized approach.

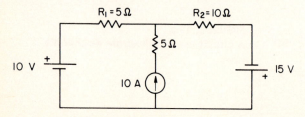

FIGURE 2.20*b*.
Assignment of algebraic signs.

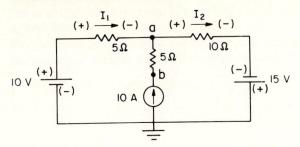

the first sign is negative. This is followed by two positive signs and a negative sign, and we have, from KVL,

$$-10 + 5I_1 + 10I_2 - 15 = 0$$

Applying KCL at node *a*,

$$I_1 + 10 = I_2$$

These two equations have the solution $(I_1, I_2) = (-5 \text{ A}, +5 \text{ A})$. Note that the 10-A current from the current source must (by KCL) also flow through the 5-Ω resistor in series with it.

Observe that the loop we employ for KVL analysis is the outer one, which does not include the current source. This is necessary because the voltage across the current source cannot be found until the problem is solved. To complete the solution, knowing that I_1 is -5 A, we can calculate that the voltage at *a* (with respect to ground) is

$$V_a = (5)(5) + 10 = 35 \text{ V}$$

The voltage at point *b* can be found from the equation

$$\frac{V_b - V_a}{5 \, \Omega} = 10 \text{ A}$$

which, after substituting $V_a = 35$ V, yields

$$V_b = 85 \text{ V}$$

This is the voltage across the current source.

2.5 TIME-VARYING VOLTAGES AND CURRENTS

Although our examples in these early chapters involves dc circuits, it is interesting to note that Ohm's law for an ideal resistor holds even if *v* and *i* are functions of time, that is, $v(t) = Ri(t)$.

As long as the elements used are only ideal sources and resistors, we could solve circuits involving time-varying sources as easily as those with constant sources. As we shall see in later chapters, however, the inclusion of other (nonresistive) elements such as capacitors and inductors in circuits with time-varying sources will lead to a set of differential equations rather than the simple algebraic equations we are considering here. Then, the analysis becomes more involved and new mathematical tools must be developed.

PRACTICE PROBLEMS AND ILLUSTRATIVE EXAMPLES

EXAMPLE 2.1

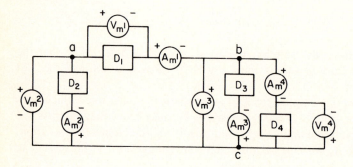

In the circuit shown, D_1, D_2, D_3, and D_4 are any two-terminal devices. The ammeters (A_m) and voltmeters (V_m) are ideal.

(a) If A_m1 reads $+3$ A and A_m4 reads $+1$ A, what are the readings of A_m2 and A_m3?

(b) If V_m1 reads $+20$ V and V_m3 reads -40 V, what are the readings of V_m2 and V_m4?

Solution: (a) Since A_m1 reads $+3$ A, there is 3 A entering node b through D_1. Since A_m4 reads $+1$ A, there is 1 A leaving node b through D_4. By KCL, the current leaving node b through D_3 must be $+2$ A. (Note that there is no current through the ideal voltmeter V_m3.) Since A_m3 reads the current into node b through D_3, A_m3 reads -2 A.

Applying the same reasoning at node c, A_m2 reads $+3$ A. Alternatively, we could apply KCL at node a to get the same result. (In general, if we apply KCL at every node, one of the equations will be superfluous.)

(b) Going clockwise around the loop $abca$, by KVL the sum of the voltage drops must equal zero. From a to b the voltage drop is $+20$ V. From b to c the voltage drop is -40 V. The voltage drop from c to a must be $+20$ V. Since V_m2 reads the voltage rise from c to a, V_m2 reads -20 V.

Since V_m3 reads the voltage drop from b to c and V_m4 reads the voltage rise from b to c they must be of opposite signs. Therefore, V_m4 reads $+40$ V.

Problem 2.1

In the circuit of Example 2.1, which devices are supplying power and which devices are receiving power? If the power receivers are all resistors, what are the resistor values?

$$\text{Ans.: } D_1 \text{ and } D_2 \text{ are receivers}$$
$$D_3 \text{ and } D_4 \text{ are sources}$$
$$R_1 = R_2 = 20/3 \ \Omega$$

Problem 2.2

How much power is supplied by each of the sources in Problem 2.1? Assume D_3 is an ideal voltage source and R_1 and R_2 are the resistors found above. What kind of source must D_4 be? What values do the sources D_3 and D_4 have?

Problem 2.3

The circuit shown below is exactly the same form as that of Example 2.1, where the I and V values are the meter readings.

(a) If $I_2 = 2$ A and $I_3 = -4$ A, find I_1 and I_4.

$$\text{Ans.: } I_1 = 2 \text{ A}$$
$$I_4 = -2 \text{ A}$$

(b) If $V_2 = 10$ V and $V_4 = 6$ V, find V_3 and V_1.

$$\text{Ans.: } V_3 = -6 \text{ V}$$
$$V_1 = 16 \text{ V}$$

Problem 2.4

Determine which elements in the circuit of Problem 2.3 are receivers and which are sources. Show that the power supplied equals the power received.

Problem 2.5

If the receivers of energy in Problem 2.3 and Problem 2.4 are resistors, draw the circuit diagram using standard symbols for the resistor and sources. Label each element fully.

Problem 2.6

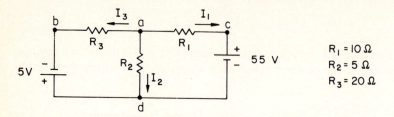

In the circuit shown (note the similarity to Fig. 2.10a in the text), find I_1, I_2, and I_3.

$$\text{Ans.: } I_1 = -4 \text{ A}$$
$$I_2 = 3 \text{ A}$$
$$I_3 = 1 \text{ A}$$

Problem 2.7

What are the voltages across the three resistors in Problem 2.6? That is, find V_{ad}, V_{ab}, and V_{ac}. Find the power supplied by the two batteries and show that this value equals the power dissipated in all the resistors.

EXAMPLE 2.2

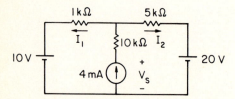

In the circuit above, find the currents I_1 and I_2.

Solution: As noted in the text when dealing with resistors in kilohms, it is often very convenient to treat all the currents in milliamperes since then the product IR is given in volts. We use this technique in this solution. KCL at node a yields the equation

$$4 = I_1 + I_2$$

There are only two unknowns so we need only one more equation. Since the voltage across a current source is not known until the problem is completely solved, we use KVL around the outer loop. Adding up the voltage drops yields

$$-10 - I_1 + 5I_2 + 20 = 0$$

Substituting,

$$-10 - 4 + I_2 + 5I_2 + 20 = 0$$

Therefore

$$I_2 = -1 \text{ mA}$$

and

$$I_1 = 5 \text{ mA}$$

Problem 2.8

Find the voltage V_s across the current source in Example 2.2.

EXAMPLE 2.3

Find the total power dissipated in all the resistors in the following circuit.

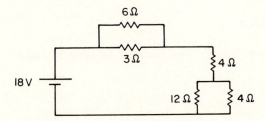

Solution: If we can find the total current I out of the battery, then the total power will just be $(18)I$. To find the equivalent total resistance, we first resolve the two parallel networks as follows: $(6//3) = 2 \, \Omega$ and $(12//4) = 3 \, \Omega$. The total series resistance is then $9 \, \Omega$ and thus $I = 2$ A. The total power is 36 W.

Problem 2.9

Find the power supplied by each source in Example 2.2. Show that the power dissipated in all the resistors is equal to the power supplied by the sources.

Problem 2.10

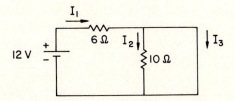

Find the currents I_1, I_2, and I_3 and the voltage across the 10-Ω resistor.

$$\text{Ans.: } I_1 = 2 \text{ A}$$
$$I_2 = 0$$
$$I_3 = 2 \text{ A}$$
$$V_{10} = 0$$

(Note that the voltage across the 10-ohm resistor equals 0 due to the short circuit.)

Problem 2.11

If the wire short-circuiting the 10-Ω resistor in Problem 2.10 is cut, find the new set of currents and the voltage across the 10-Ω resistor.

Problem 2.12

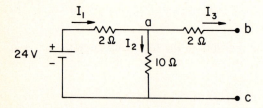

Find all the currents and the voltage $V_{bc} = V_b - V_c$. This is called the open-circuited voltage at the terminal bc.

$$\text{Ans.: } I_1 = I_2 = 2 \text{ A}$$
$$I_3 = 0$$
$$V_b - V_c = V_a - V_c = 20 \text{ V}$$

Note that since no current flows through the 2-Ω resistor, there is no voltage drop across it, and $V_{bc} = V_{ac}$.

Problem 2.13

If a short circuit is placed across the terminal b–c, find the new set of currents and the voltage V_{bc}. Since the new I_3 flows through the short circuit, we call it the short circuit current.

EXAMPLE 2.4

Find the current I using the current divider rule.

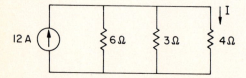

Solution: First note that the 3-Ω and 6-Ω resistors can be replaced by $R_E = (3 \times 6)/(3 + 6) = 2\ \Omega$. Now the current through the 4-Ω resistor (call it R_4) is given by the current divider rule:

$$I = \frac{2}{2 + 4}(12) = 4\ \text{A}$$

An alternative solution, using the "conductances," is (from Eq. 2.12)

$$I = \frac{(1/4)}{(1/6 + 1/3 + 1/4)}(12) = 4\ \text{A}$$

Problem 2.14

Find the voltage across the 12-A current supply in Example 2.4 and show that it is consistent with each *IR* drop across the three resistors. Find the power supplied by the source and show that it equals the sum of all the powers dissipated in the resistors.

Problem 2.15

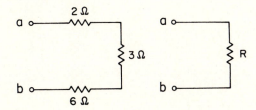

The two circuits shown are equivalent at *ab* when $R =$

Ans.: $R = 11\ \Omega$

Problem 2.16

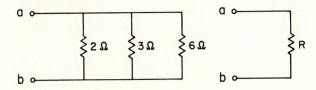

The two circuits are equivalent at *ab* when $R =$

Ans.: $R = 1\ \Omega$

Problem 2.17

In the circuit below, find the voltage V_2 and the equivalent resistance across the 12-V supply.

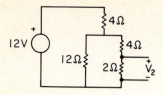

Problem 2.18

Find the current I in the circuit shown below and the equivalent resistance across the 48-A supply.

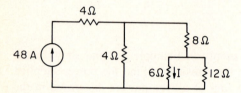

Problem 2.19

Find the power supplied by the sources in Problems 2.17 and 2.18.

Problem 2.20

Find R_E for the circuit below.

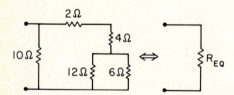

Problem 2.21

If a 20-V supply is inserted across the 10-Ω resistor in Problem 2.20, find the current through the 12-Ω resistor.

Problem 2.22

Find R_E for the circuit below.

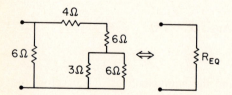

Problem 2.23

If an unknown voltage source is connected across the open terminals in the circuit of Problem 2.22 and the voltage across the 3-Ω resistor is measured to be 3 V, what is the value of the voltage source?

Problem 2.24

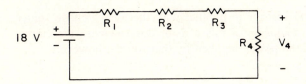

If $R_1 = 20\ \Omega$, $R_2 = 6\ \Omega$, $R_3 = 8\ \Omega$, and $R_4 = 20\ \Omega$, find V_4.

Ans.: $V_4 = 6.67$ V

Problem 2.25

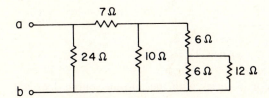

Find the equivalent resistance at the terminals ab.

Ans.: 8 Ω

Problem 2.26

A 9-A current supply is inserted between a and b in the circuit of Problem 2.25 such that current flows from b to a. What is the voltage $V_a - V_b$? Find the current through the 24-Ω resistor, the 7-Ω resistor, the 10-Ω resistor, and the 12-Ω resistor.

EXAMPLE 2.5

Find the voltage V in the following circuit.

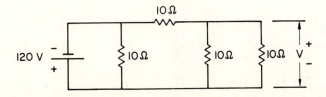

Solution: We can use the voltage divider rule after some manipulation of the resistors. The voltage V is across both of the 10-Ω resistors on the right-hand side. Their parallel combination equals 5 Ω. That 5-Ω resistor is in series with the "top" 10-Ω resistor. Since -120 V are across this series combination, we can use the voltage divider rule, which says

$$V = \frac{5}{10 + 5}(-120) = -40 \text{ V}$$

Note that the value of the 10-Ω resistor in parallel with the ideal voltage source does not affect the result. The inclusion of this resistor does increase the current through the voltage source and the power supplied by that source, however.

Problem 2.27

Find V_1 in the circuit below.

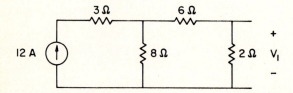

Ans.: $V_1 = 12$ V

(Note that the value of the 3-Ω resistor in series with the current source does not affect the result.)

Problem 2.28

Find the voltage across the 8-Ω resistor in Problem 2.27. Check that this equals the value $8I_8$, where I_8 is the current through the 8-Ω resistor.

Problem 2.29

What voltage source would yield the same voltage V_1 across the 2-Ω resistor in Problem 2.27 if it replaced the 12-A supply? Which supply provides the most total power?

Problem 2.30

Find V_2 in the circuit below. What current source would be required in place of the 18-V battery to yield the same voltage? Draw the circuit so that the polarity is clear.

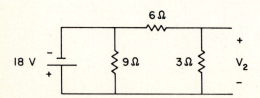

Problem 2.31

In the circuit below, derive an algebraic expression for V_{ab} in terms of I_s, R_1, R_2, R_3, and R_4.

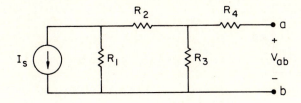

Ans.: $V_{ab} = -I_s \dfrac{R_1 R_3}{R_1 + R_2 + R_3}$

Problem 2.32

Find I in the circuit below.

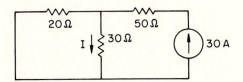

Ans.: $I = 12$ A

Problem 2.33

What voltage source is needed to yield the same current I through the 30-Ω resistor if it replaced the 30-A current source in Problem 2.32?

Problem 2.34

Find V_{ab} for the circuit below.

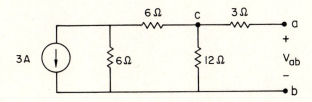

Ans.: $V_{ab} = V_{cb} = -9$ V

Problem 2.35

Suppose a short circuit is applied across the terminal a–b in the circuit of Problem 2.34. Find the short-circuited current from a to b.

Problem 2.36

Find *I* in the circuit below.

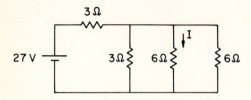

Ans.: 1.5 A

Problem 2.37

If the 27-V battery in Problem 2.36 is replaced by a current source of value I_s, what must the current I_s be to yield the same current *I*? Find the power supplied by the two sources in their respective circuits.

Problem 2.38

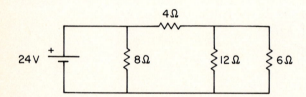

Find the power dissipated in the 6-Ω resistor and the total power supplied by the battery.

Ans.: 24 W

144 W

Problem 2.39

If all the resistors in Problem 2.38 are changed to kΩ (e.g., 8 kΩ, 4 kΩ, etc.), find the same two power levels.

Problem 2.40

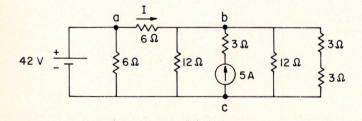

Find *I* and the power delivered by each source.

Problem 2.41

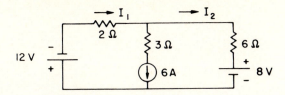

(a) Find I_1 and I_2.

Ans.: $I_1 = 2$ A, $I_2 = -4$ A

(b) Find the power supplied by
each source.

Ans.: $P_{12} = -24$ W
$P_6 = +204$ W
$P_8 = 32$ W

(c) Find the power dissipated in
each resistor.

Ans.: $P_2 = 8$ W
$P_3 = 108$ W
$P_6 = 96$ W

Problem 2.42

In the circuit below, find the power supplied to the 8-Ω resistor and the total power
supplied by the battery.

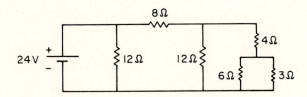

Problem 2.43

In the circuit below, find the power supplied by each source.

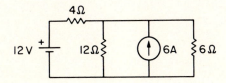

Ans.: $P_V = -18$ W
$P_I = 108$ W

Problem 2.44

Suppose the positions of the two sources are reversed in Problem 2.43. Find the
powers supplied by the two sources.

Chapter 3
Circuit Theorems and Analysis Methods

Any circuit that includes only known independent sources and resistors can be solved using the method developed in Chapter 2. After defining the current through every element and the voltage across every element, KVL and KCL can be applied to write a set of independent simultaneous equations equal to the number of unknowns. Given sufficient time and energy, it is possible to solve for all the unknowns. Of course, given sufficient time and energy, one can dig any size hole using only hand tools. In this chapter more powerful tools for circuit analysis are presented and applied.

3.1 LINEARITY AND SUPERPOSITION

All the equations resulting from the application of Kirchhoff's laws to circuits of sources and resistors are linear; that is, the voltages and currents enter only to the first power and there are no cross products of the variables. The linear nature of the equations makes it possible to use many simplifying methods for the solution of complex circuits. As an example, consider the circuit shown in Fig. 3.1a. Using KVL,

$$V_a = 2I_2 + 12I_1 \text{ and } V_a = 3I_3$$

Using KCL,

$$I_b + I_2 = I_1$$

FIGURE 3.1a.
Circuit used to illustrate superposition.

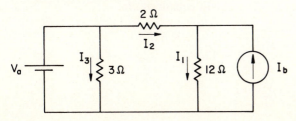

Solving the above equations for I_1, I_2 and I_3,

$$I_1 = \frac{V_a}{14} + \frac{I_b}{7}$$

$$I_2 = \frac{V_a}{14} - \frac{6I_b}{7}$$

$$I_3 = \frac{V_a}{3}$$

It is clear from this way of writing the solutions that I_1, I_2, and I_3 may each be considered as the sum of contributions from the two sources V_a and I_b. For example, if we let I_1' be the contribution from V_a with I_b set equal to zero, then

$$I_1' = \frac{V_a}{14}$$

Likewise, if I_1'' is the contribution from I_b with V_a set equal to zero, then

$$I_1'' = \frac{I_b}{7}$$

Finally $I_1 = I_1' + I_1''$ is the sum of the two contributions. This is called the *method of superposition*.

To emphasize the linear nature of this result, we can write the I_j as

$$I_1 = a_1 V_a + b_1 I_b$$

$$I_2 = a_2 V_a + b_2 I_b$$

$$I_3 = a_3 V_a + b_3 I_b$$

where the a_j and b_j terms are constants ($b_3 = 0$ in this case). Thus, any current in the problem (or any voltage, for that matter!) can be written as a linear superposition of the sources V_a and I_b. Note that these constants may or may not have units. The parameters b_1 and b_2 are dimensionless, being the relationship between two currents, but the a_j terms have units of mhos (inverse resistance).

In practice, the method could be applied to the example using circuit methods as follows. The circuit used to determine the contributions I_1', I_2', and I_3' due only to the voltage source V_a is shown in Fig. 3.1b. The current source is set equal to zero by replacing it with an open circuit. The circuit equations are

$$I_1' = I_2' = \frac{V_a}{2 + 12} = \frac{V_a}{14}$$

$$I_3' = \frac{V_a}{3}$$

FIGURE 3.1b.
Finding I_1', I_2', and I_3' due to V_a.

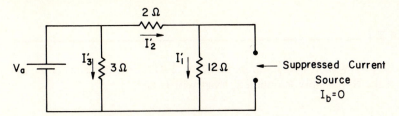

Likewise, the circuit for finding I_1'', I_2'' and I_3'' due to the current source I_b is as shown in Fig. 3.1c, where the voltage source has been set equal to zero by replacing it with a short circuit. The circuit equations are

$$0 = 3I_3''$$

$$0 = 2I_2'' + 12I_1''$$

$$I_b = I_1'' - I_2''$$

which may be solved to give

$$I_1'' = \frac{I_b}{7}, \quad I_2'' = \frac{-6I_b}{7}, \quad I_3'' = 0$$

Finally, the sums

$$I_1 = I_1' + I_1'' \qquad I_2 = I_2' + I_2'' \qquad I_3 = I_3' + I_3''$$

yield the same results as found before.

The general statement of the principle of superposition is that any current or voltage in a linear system can be found as the sum of the contributions from each source acting alone, with all the other sources set equal to zero, or, as we sometimes say, with the sources "suppressed." Setting a current source equal to zero means that the source is replaced by an open circuit; setting a voltage source equal to zero means that the source is replaced by a short circuit. The reader is reminded of the earlier comment in Chapter 1 that a short circuit may be considered to be a voltage source of value zero volts and an open circuit may be considered to be

FIGURE 3.1c.
Finding I_1'', I_2'', and I_3'' due to I_b.

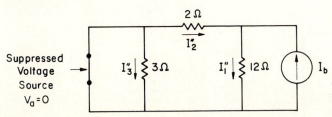

a current source of value zero amperes. Note that the contribution of each source is proportional to the magnitude of that source, a property that is typical of linear systems.

TEXT EXAMPLE 3.1 _____

In the circuit below, which was used in Problem 2.41, find the current I_1 by superposition. If the 8-V supply is doubled, what will I_1 be?

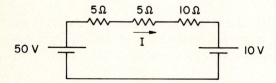

Solution: Three sources are involved, and we can consider $I_1 = I_1' + I_1'' + I_1'''$, where each primed current is due to a single source acting alone. To solve for I_1', we suppress all but the 12-V supply.

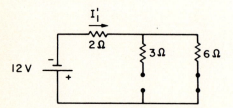

No current flows through the open circuits, so there is a single series resistance of 8 Ω and $I_1' = -12/8 = -1.5$ A. The contribution due to the current source can be found from the circuit

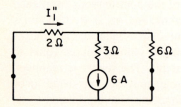

The 6-A current divides between the two resistors such that $I_1'' = (6/8)(6) = 4.5$ A. Finally, for the 8-V source we have the circuit

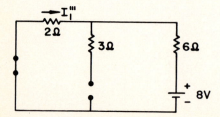

and $I_1''' = -8/8 = -1A$. The total current

$$I_1 = I_1' + I_1'' + I_1''' = 2 \text{ A}$$

as found in the answer to Problem 2.41. Without the superposition principle we would have to start over to find the current I_1 if the 8-V supply were doubled. Using the result above for I_1''', however, we see the contribution due to a 16-V battery would be -2 A. The other values do not change, and the answer is $I_1 = 1$ A.

Another use of the linearity principle, where only a single source is involved, is the analysis of so-called ladder circuits, an example of which is shown in Fig. 3.2a. First, the problem might be attacked by "brute force" methods, by redrawing an equivalent set of series and parallel resistors (Fig. 3.2b), and thence solving the system step by step as in Fig. 3.2c. Now, to solve this by linearity arguments we can use the so-called *unit current method* (or, if more appropriate, the *unit voltage method*). Using this technique in this case we assume a convenient value for the current in the last resistor, say 1 A, and find out what battery at the front end would yield that current. Then, by superposition any other battery will yield a proportional result. For example, in Fig. 3.2d, if I_1 is 1 A, then V_1 must be 10 V, and the current I_2 must also be 1 A. Proceeding from right to left in this manner, the voltage V_2 equals 10 V plus the voltage I_2R across the 10-Ω resistor,

$$V_2 = 10I_2 + V_1 = 10 + 10 = 20 \text{ V}$$

FIGURE 3.2a.
Ladder circuit.

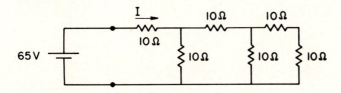

FIGURE 3.2b.
Another way to draw the circuit in Fig. 3.2a.

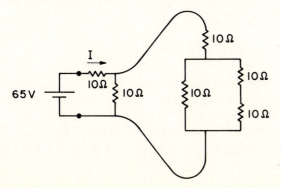

FIGURE 3.2c.
Consecutive equivalent circuits showing that $I=4$ A.

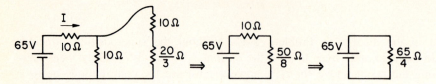

FIGURE 3.2d.
Step-by-step analysis of the circuit assuming $I_1=1$ A.

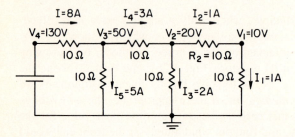

Note that we have labeled the node with this voltage value. Now a current will flow to ground from this point through the 10-Ω resistor. This current is labeled I_3 and has a magnitude

$$I_3 = \frac{V_2}{10} = 2 \text{ A}$$

The current I_4 is the sum

$$I_4 = I_3 + I_2 = 2 + 1 = 3 \text{ A}$$

Repeating these steps,

$$V_3 = 10I_4 + V_2 = 30 + 20 = 50 \text{ V}$$

$$I_5 = \frac{V_3}{10} = 5 \text{ A}$$

$$I = I_5 + I_4 = 5 + 3 = 8 \text{ A}$$

$$V_4 = 10I + V_3 = 80 + 50 = 130 \text{ V}$$

Now we have worked our way back to the front end and found that a 130-V battery yields a total current $I = 8$ A. By linearity we may conclude that the current I driven by a 65-V battery will be 4 A. In the process we have also solved for all the currents and voltages in the ladder. For a 65-V battery, each current and voltage will be half of that produced by a 130-V battery, yielding $I_1 = 0.5$ A, $V_3 = 25$ V, $I_4 = 1.5$ A, and so on.

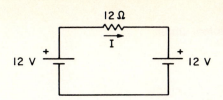

FIGURE 3.3.
Simple circuit illustrating that power calculations are not linear operations.

Finally, some comments on power calculations are necessary. Since power calculations involve either the product of voltage and current, the square of the current, or the square of the voltage, they are not linear operations and the superposition principle cannot be applied blindly to determine the power. The example illustrated in Fig. 3.3 shows this quite clearly. Since there is no voltage difference across the 12-Ω resistor, no current flows and no power is dissipated in the resistor. If we determine the current I from superposition, we find the current I' due to the left-hand battery to be 1 A, while the current I'' due to the right-hand battery is -1 A. The total current is

$$I = I' + I'' = 0$$

which agrees with the result above. We cannot merely add the powers associated with I' and I'', however, since they yield a total

$$(I')^2 R + (I'')^2 R = 24 \text{ W}$$

which is incorrect, since no power is dissipated in the resistors or supplied by the batteries at all.

3.2 THEVENIN AND NORTON EQUIVALENT CIRCUITS

Any electrical system that has only one pair of wires emerging is called a one-port system. If the "black box" illustrated in Fig. 3.4 contains only linear elements such as resistors and ideal voltage and/or current sources, then the current I must be a linear function of the voltage V across the terminals. In equation form, we must have a linear functional relationship between I and V, say, $I = AV + B$, where A and B are constants. Considering I as a function of V in this equation, it follows that the graph of this function is a straight line with slope A and intercept $I = B$ when $V = 0$. We use this result in the proof below.

Thevenin's theorem states that such a circuit may be replaced by a single voltage source in series with a single resistor, as illustrated in Fig. 3.5a. The proof of this important result proceeds as follows. Consider the set of all possible voltage and

FIGURE 3.4.
A one-port system.

Black Box

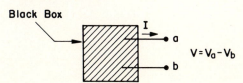

$$V = V_a - V_b$$

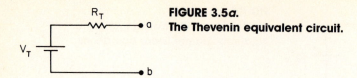

FIGURE 3.5a.
The Thevenin equivalent circuit.

current pairs (V, I) that could characterize the output of the system, as illustrated in Fig. 3.5b. Since the black box contains only linear elements and ideal sources, the (V, I) pairs must all fall on a straight line. Two points determine a line and we concentrate on finding the two values where the line intercepts the I axis and the V axis. The former is called the short-circuit current I_{sc} since the voltage across $a–b$ will be zero if the output is short circuited. The corresponding (V, I) pair is labeled $(0, I_{sc})$ and is plotted in Fig. 3.5c. When the element is an open circuit, no current flows and $I = 0$. The voltage in this state is called the open circuit voltage V_{oc}. The corresponding (V, I) pair is labeled $(V_{oc}, 0)$, and is also plotted in Fig. 3.5c. The straight line plotted through these two points yields all possible (V, I) pairs. The equation for this line is also given in the figure.

Our task now is to find the values of V_T and R_T in the circuit of Fig. 3.5a that would yield the same set of (V, I) pairs as the original black box. Referring to that figure, if there is an open circuit at $a–b$, no current flows through R_T and the voltage $V_a - V_b = V_T$. Therefore, $V_T = V_{oc}$. To determine R_T, suppose a short circuit is placed across the output $a–b$ of Fig. 3.5a. Since $V_T = V_{oc}$, it follows that

$$I_{sc} = \frac{V_{oc}}{R_T}$$

or

$$R_T = \frac{V_{oc}}{I_{sc}}$$

and we have found the required value for R_T.

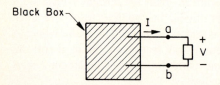

Black Box

FIGURE 3.5b.
Circuit for determining all possible (V,I) pairs.

FIGURE 3.5c.
The $I–V$ curve for the output of Fig. 3.5b.

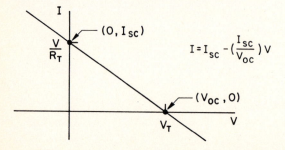

$$I = I_{sc} - \left(\frac{I_{sc}}{V_{oc}}\right) V$$

FIGURE 3.6.
Measurement setup to determine open-circuited voltage (a) and short circuited current (b), for a one-port system.

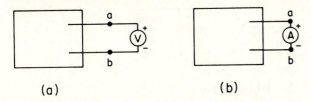

(a)   (b)

To summarize, the Thevenin equivalent circuit for a black box containing linear elements and sources can be determined by making two measurements at the terminals, the open-circuited voltage and the short-circuited current (Fig. 3.6). Then, we have

$$V_T = V_{oc}$$

$$R_T = \frac{V_{oc}}{I_{sc}}$$

If the circuit inside the box is known, we can also make an analytical determination of these parameters. As an example, suppose we wish to find the Thevenin equivalent to the circuit in Fig. 3.7a at the terminals a and b. In the open-circuited case, which is already illustrated in Fig. 3.7a, the voltage can be found by noting first that the parallel combination 10 Ω//(5 Ω + 5 Ω) = 5 Ω is in series with the remaining 5-Ω resistor. Six volts appear across the parallel combination. One-half of this appears across the points a and b, so $V_{oc} = V_T = 3$ V. The short-circuited current can be found from Fig. 3.7b by noting that inclusion of the short circuit yields a total of only 5 Ω //10 Ω. The total current is thus $I_T = 12/(5 + (10//5)) = 36/25$ A. The current divider law yields

$$I_{sc} = \left(\frac{10}{15}\right) I_T = \frac{24}{25} \text{ A}$$

So,

$$V_T = 3 \text{ V}$$

$$R_T = \frac{V_{oc}}{I_{sc}} = \frac{25}{8} \text{ Ω}$$

and the Thevenin equivalent circuit is as shown in Fig. 3.7c.

The equation for the line in Fig. 3.5c can be used to generate an alternate equivalent circuit which yields the same function at the output a and b. Since the

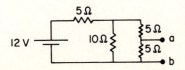

FIGURE 3.7a.
Circuit to illustrate Thevenin's Theorem.

FIGURE 3.7b.
Circuit that illustrates how to find I_{sc}.

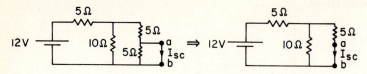

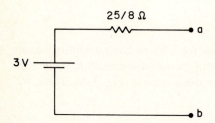

FIGURE 3.7c.
The Thevenin Equivalent Circuit at the port a–b.

term V_T/R_T is a constant with the dimensions of a current, we may write the equation for the straight line,

$$I = \frac{-V}{R_T} + \frac{V_T}{R_T}$$

in the alternative form

$$I = I_N - \frac{V}{R_T} \tag{3.1}$$

where $I_N = V_T/R_T$. If we apply KCL to the circuit illustrated in Fig. 3.8, we see that

$$I_N = \frac{V}{R_N} + I$$

or

$$I = I_N - \frac{V}{R_N}$$

FIGURE 3.8.
The Norton Equivalent Circuit.

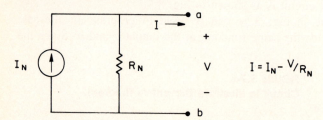

This equation is identical to Eq. 3.1 provided that $R_N = R_T$. In other words, the circuit in Fig. 3.8 is also an equivalent circuit that employs a current source and a parallel resistor rather than a voltage source with a series resistor. Such an equivalent circuit is termed a *Norton equivalent circuit*. The Norton equivalent current source has the value

$$I_N = I_{sc}$$

and the Norton equivalent resistance is given by

$$R_N = \frac{V_{oc}}{I_{sc}}$$

Note that the Norton equivalent resistance is equal to the Thevenin value,

$$R_N = R_T = R_o$$

The symbol R_o is sometimes used to emphasize that the resistors R_N and R_T are equal. As far as the outside world is concerned, the black box in Fig. 3.4 can be replaced by either of the two circuits shown in Fig. 3.9. Returning to the example above and using these results, the Norton equivalent to the circuit in Fig. 3.7a is as shown in Fig. 3.10.

Sometimes it is easy to find either I_{sc} or V_{oc}, but not both. Then it may be convenient to find R_o directly without taking the ratio. The so-called R_o method yields the value of the resistor directly as follows:

1. Suppress all the independent sources inside the box by replacing all voltage sources with short circuits and all current sources with open circuits.
2. Find the equivalent resistance R_o between points a and b. The resistance found between a and b then equals R_N and R_T.

As an example we evaluate the resistance R_o for the circuit given in Fig. 3.7a. After suppressing the source the circuit is as shown in Fig. 3.11a, which can be redrawn and analyzed to find R_o as shown in Fig. 3.11b. The resistance R_o is $(25/8)\ \Omega$, as found before for R_T and R_N.

FIGURE 3.9.
The Thevenin and Norton Equivalent Circuits.

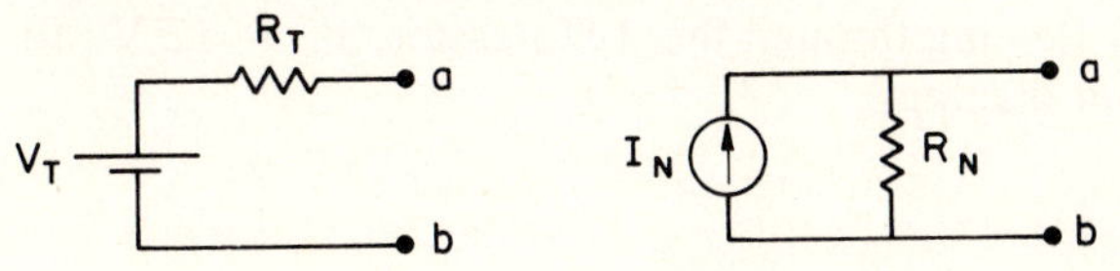

$R_T = R_N = R_o = (V_{oc}/I_{sc})$

$V_T = V_{oc}$

$I_N = I_{sc}$

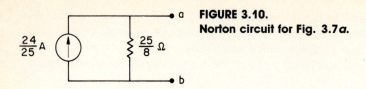

FIGURE 3.10.
Norton circuit for Fig. 3.7a.

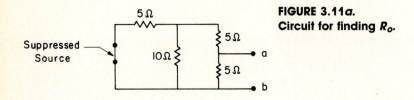

FIGURE 3.11a.
Circuit for finding R_o.

FIGURE 3.11b.
Steps for finding R_o.

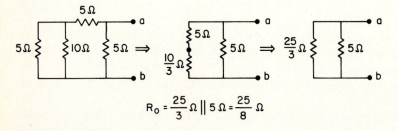

$$R_o = \frac{25}{3}\,\Omega \,\|\, 5\,\Omega = \frac{25}{8}\,\Omega$$

TEXT EXAMPLE 3.2 _____

Find the Thevenin and Norton Equivalent Circuits at the terminals a–b for the circuit below.

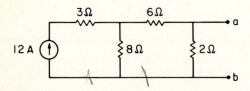

Solution: To find the open circuited voltage ($V_{oc} = V_a - V_b$) we can use the circuit diagram above. Since the 6-Ω and 2-Ω resistors are in series, the 12-A current splits equally between the 8-Ω resistor and the series 8-Ω resistance formed by the 6-Ω and 2-Ω resistors. With 6 A flowing through the 2-Ω resistor, $V_{oc} = 12$ V. To find I_{sc} we need a different circuit diagram.

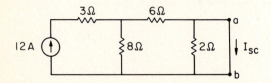

The short circuit effectively removes the 2-Ω resistor from the circuit. The current divider law now yields

$$I_{sc} = \left(\frac{8}{14}\right)(12) = \frac{48}{7} \text{ A}$$

With V_{oc} and I_{sc} determined we have

$$R_T = R_N = \frac{V_{oc}}{I_{sc}} = \frac{7}{4} \Omega$$

and the two equivalent circuits are

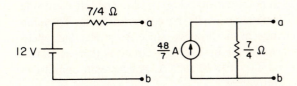

Finally, using the R_o method to check these results, we have, after suppressing the current source,

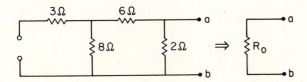

The open circuit effectively removes the 3-Ω resistor from the circuit. The 6-Ω and 8-Ω resistors are resistors in series, equaling 14 Ω. This is in parallel with the 2 Ω resistor, so

$$R_o = \frac{28}{16} = \frac{7}{4} \Omega$$

which does indeed agree.

One final note on the R_o method is in order. The student should be aware that it is not always possible to reduce a given resistor network to an equivalent resistor using only parallel and series resistor analysis methods. A quick perusal of Example 3.6 and subsequent problems at the end of this chapter will verify this statement. In such a situation one proceeds as follows. Any sources inside the box are suppressed. An external voltage is applied to the terminals and the resulting current I is determined.

Then the ratio V/I equals R_o for the system. This method will always work, but it is usually easier to find R_o using series and parallel analysis methods if they apply. Furthermore, when dependent sources are encountered as is shown in the next chapter, R_o *must* be found by applying a voltage V and determining I in this manner.

3.3 APPLICATIONS OF EQUIVALENT CIRCUITS—THE SOURCE TRANSFORMATION METHOD AND THE MAXIMUM POWER TRANSFER THEOREM

The source transformation method takes advantage of the equivalence of the Norton and Thevenin circuits, which can be interchanged at will. Sometimes circuits are more easily analyzed in one form than in another, and the so-called source transformation method is used. As a case in point consider Fig. 3.12a. The left-hand side is in the Norton configuration and hence can easily be converted to its Thevenin equivalent. The resistor does not change in the process, since $R_N = R_T$. The open-circuited voltage across the 5-Ω resistor is 50 V. The new circuit is then as shown in Fig. 3.12b, and the current is $I = (50 - 10)/20 = 2$ A. A Thevenin circuit can also be transformed into its Norton equivalent.

A useful theorem can be easily proved with the Thevenin equivalent circuit. The theorem states that maximum power is transferred to the resistor R_L in Fig. 3.13 when that resistor has a value equal to the Thevenin resistance of the circuit to which it is connected. We often refer to R_L as the load resistor in such a configuration.

Using Eq. 1.7, the load power is

$$P_L = I^2 R_L = \left[\frac{V_T}{R_T + R_L} \right]^2 R_L$$

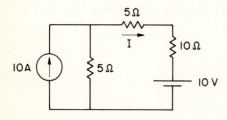

FIGURE 3.12a.
Find the current I by source transformation.

FIGURE 3.12b.
The "Thevenized" circuit.

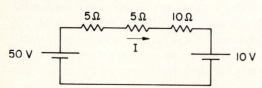

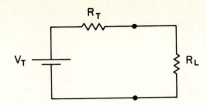

FIGURE 3.13.
What value should R_L take to yield the
maximum power dissipated in R_L?

To maximize this function with respect to the parameter R_L, we can take its first derivative and set the result equal to zero:

$$\frac{dP_L}{dR_L} = \frac{(R_T + R_L)^2 V_T^2 - 2V_T^2 R_L (R_T + R_L)}{(R_T + R_L)^4} = 0$$

which simplifies to

$$(R_T + R_L) - 2R_L = 0$$

Thus,

$$R_T = R_L$$

This result is known as "matching the load." Maximum power transfer occurs when the load resistor matches the Thevenin resistance of a given system. This is the reason that a stereo system should have an 8-Ω output resistance feeding power to an 8-Ω speaker and that a radio transmission system should employ a transmitter which is matched to the antenna. Then the maximum possible power is transferred to the speaker and to the radiation pattern, respectively.

3.4 INTRODUCTION TO THE NODE VOLTAGE AND LOOP CURRENT METHODS

The object of the node voltage and loop current approaches to circuit analysis is to reduce the number of equations that must be solved simultaneously. This is accomplished by applying Kirchhoff's laws in the process of the solution rather than in a step-by-step "brute force" approach.

The *node voltage method* assigns (unknown) voltages with respect to a reference (usually ground potential) at each node and then applies Kirchhoff's current law at each node. These node voltages become the independent variables of the problem. The key is that, since we have symbols for the voltage at each node, we can formally determine all the currents. Consider the particular example shown in Fig. 3.14a, which has four nodes. We are always free to assign one node to zero or ground potential, which is done in Fig. 3.14b. The node between the battery and the 4-Ω resistor is a known value which we label as $+10$ V with respect to ground potential. The two unknown node voltages are labeled V_a and V_b. Applying KCL at nodes a and b,

FIGURE 3.14a.
A circuit used to find the current _I_ via the node voltage method.

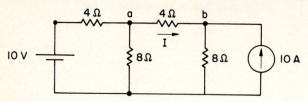

FIGURE 3.14b.
Currents needed to apply KCL.

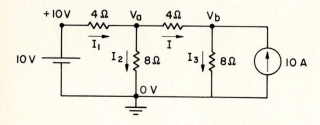

$$I_1 = I_2 + I$$

$$I + 10 = I_3$$

From Ohm's law we have

$$I_1 = \frac{10 - V_a}{4}$$

$$I_2 = \frac{V_a - 0}{8}$$

$$I = \frac{V_a - V_b}{4}$$

$$I_3 = \frac{V_b - 0}{8}$$

Substituting these in the KCL equations, we reduce the problem to only two equations in the two unknowns V_a and V_b:

$$\frac{10 - V_a}{4} = \frac{V_a}{8} + \frac{V_a - V_b}{4}$$

$$\frac{V_a - V_b}{4} + 10 = \frac{V_b}{8}$$

These can be solved for V_a and V_b, the node voltages, which are found to be 20 V and 40 V, respectively. To finish the problem, we find

$$I = \frac{V_a - V_b}{4} = -5 \text{ A}$$

Any of the other three currents labeled could also be found using V_a and V_b.

The *loop current method* is very similar. Here, we assign a loop current, often called a mesh current, to each closed contour in the circuit and apply Kirchhoff's voltage law to each such contour, generating simultaneous equations for the loop currents. As an example, consider the circuit shown in Fig. 3.15a. There are four loops in this circuit. However, whenever a loop goes through a current source, that current is known. To make the actual number of unknown loop currents obvious, two source transformations can be used first to yield the equivalent circuit shown in Fig. 3.15b. (Note that at the left we have added the two 4-Ω resistors to yield 8 Ω.) We now apply KVL around the two loops using i_1 and i_2 as independent variables. The "rote" method of Chapter 2 is used to keep the algebraic signs straight. The signs are written in Fig. 3.15b for each battery and for all resistors including the 4-Ω resistor, which is shared by the two loop currents. Care must be taken in this leg of the KVL loop, since it forms an interface between the two current loops. The currents through the various elements are often termed the branch currents. At an interface between two loops, the loop currents superpose to yield the branch current in that interface element or elements.

First, traversing the left-hand loop clockwise (in the direction of the loop current), we have from KVL

$$+24 + 8i_1 + 4I + 8 = 0$$

Since $I = i_1 - i_2$ we have

$$+24 + 8i_1 + 4(i_1 - i_2) + 8 = 0$$

FIGURE 3.15a.
Circuit to find the current I using loop currents.

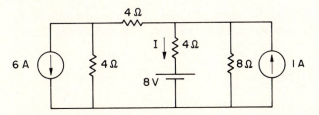

FIGURE 3.15b.
Loop currents and sign conventions for evaluation of KVL around the loops.

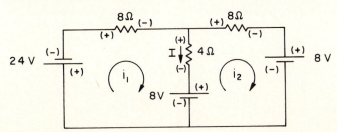

Likewise, traversing the right-hand side in a clockwise loop and applying KVL yields

$$-8 + -4I + 8i_2 + 8 = 0$$

which becomes

$$-8 - 4(i_2 - i_1) + 8i_2 + 8 = 0$$

These two equations have the simultaneous solution

$$i_1 = -3 \text{ A}$$

$$i_2 = -1 \text{ A}$$

and hence the branch current is $I = i_1 - i_2 = -2$ A.

TEXT EXAMPLE 3.3

Use both the node voltage and loop current methods independently to solve the circuit in Text Example 3.1, which is reproduced here.

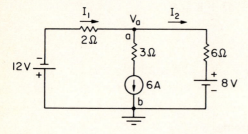

Solution: First, using node voltage, we define the node a to have the voltage V_a and the voltage at node b to be zero volts. Then, using KCL at node a and Ohm's law as expressed in Eq. 1.5,

$$\frac{-12 - V_a}{2} = 6 + \frac{V_a - 8}{6}$$

$$-36 - 3V_a = 36 + V_a - 8$$

$$V_a = -16 \text{ V}$$

Knowing V_a, we have

$$I_1 = \frac{-12 + 16}{2} = 2 \text{ A}$$

and

$$I_2 = \frac{-16 - 8}{6} = -4 \text{ A}$$

which check with answers to Problem 2.41. Defining the currents i_1 and i_2 as shown below, we can use the loop current method as well.

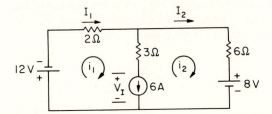

This circuit is not particularly well suited for the loop current method since KVL must be used and we do not know the voltage across the 6-A current source, V_I. With some additional effort we can grind out the solution, however. First, we derive the loop equations as usual:

$$+12 + 2i_1 + 3(i_1 - i_2) + V_I = 0$$

$$-V_I + -3(i_1 - i_2) + 6i_2 + 8 = 0$$

Because of the problem with V_I these are two equations with three unknowns. A suitable third equation is

$$i_1 - i_2 = 6$$

since the branch current in the center is determined by the current source. We can then proceed to find the solutions for i_1 and i_2, which are 2 A and -4 A, respectively. Note that the branch currents I_1 and I_2 are equal to the corresponding loop currents i_1 and i_2.

3.5 FURTHER COMMENTS ON THE NODE VOLTAGE AND LOOP CURRENT METHODS

In the node voltage method we identify all the nodes on the circuit. Choosing one of them as the reference (i.e., zero), we specify as the unknowns the node voltages at each of the other nodes. Each node voltage is the voltage at that node with respect to the reference. If the circuit has n nodes there are $n - 1$ unknowns. At each of these $n - 1$ nodes we can write a KCL equation, which automatically results in a set of $n - 1$ independent simultaneous equations in $n - 1$ unknowns. Each KCL equation can be conveniently written since the sum of the currents leaving the node through resistors is equal to the current entering the node from sources.

If the element connecting node i to node j is a resistor, R_{ij}, the current leaving node i through R_{ij} can be written in terms of the node voltages and equals $(V_i - V_j)/R_{ij}$. If the element connecting node i to node j is a current source, that current is of course known. If the element connecting node i to node j is a voltage source $V_{ij} = V_s$, the current cannot be determined directly since the voltage V_s is independent of the current. However, the number of unknowns is decreased by one, since in this case $V_j = V_i + V_s$. For example, in the circuit of Fig. 3.16, the voltage of the central node is known since it is equal to $V_a + 10$. The KCL equations can now be written directly in terms of the two unknown node voltages V_a and V_b.

$$\text{At node } a: \quad 6 - \frac{V_a}{10} = I_s = \frac{V_a + 10}{20} + \frac{V_a + 10 - V_b}{60}$$

$$\text{At node } b: \quad \frac{V_a + 10 - V_b}{60} = \frac{V_b}{10} + 3$$

which can be solved to find $V_a = 30$ V and $V_b = -20$ V. Knowing V_a and V_b we can return to the original circuit of Fig. 3.16 to find directly the voltage across every element and the current through every element.

In the loop current method we identify the minimum number of loop currents necessary to specify the current through each element. We then write a KVL equation around each loop to produce a set of simultaneous equations equal in number to the number of unknown loop currents. For ease of writing the voltage drops in the KVL equation, it is common practice to define a clockwise direction for each loop current and to traverse each loop in the direction of the loop current. In writing the equation for loop i, the voltage drop across a resistor R_{ij}, which is common to loops i and j, will be $(i_i - i_j)R_{ij}$.

If there are current sources in the circuit, the current is known in that branch, but the voltage drop across the current source must be found from the rest of the circuit. If there is a resistor in parallel with the current source as in Fig. 3.15a, the loop containing that source may be eliminated by converting the current source and parallel resistance to their Thevenin equivalent as in Fig. 3.15b. Alternatively, the branch current due to a current source may be used to define a known loop current which is then used in the same manner as the unknown loop currents. As an example we again consider the circuit of Fig. 3.15a using the loop currents designated in Fig. 3.17. Writing the sum of the voltage drops around loops 1 and 2,

FIGURE 3.16.
Example of a circuit with a voltage source between two nodes.

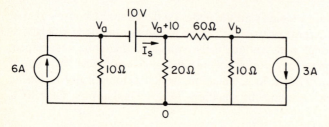

FIGURE 3.17.
Circuit with four loops, two of which are known.

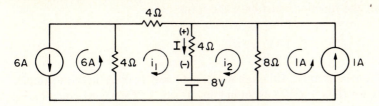

$$\text{Loop 1:} \quad + 4(i_1 + 6) + 4l_1 + 4(i_1 - i_2) + 8 = 0$$

$$\text{Loop 2:} \quad - 8 - 4(i_1 - i_2) + 8(i_2 + 1) = 0$$

This set reduces to

$$12i_1 - 4i_2 = -32$$

$$-4i_1 + 12i_2 = 0$$

which can be solved to find $i_1 = -3$ A and $i_2 = -1$ A, as before. Having found i_1 and i_2, we can find directly the current through every element and the voltage across every element.

In any particular circuit the preferred method of solution is one that leads to the least number of simultaneous equations. When there are many elements in parallel, there are many loops but few nodes and the node voltage method is preferable. Conversely, when there are many elements in series, there are few loops but many nodes and the loop current method is preferable.

PRACTICE PROBLEMS
AND
ILLUSTRATIVE EXAMPLES

EXAMPLE 3.1

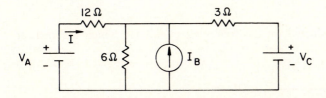

In the circuit shown it is found that when the voltage V_A is equal to 7 V, the current I is $-7/6$ A. If V_A is changed to 28 V, what will be the new value of I?

Solution: The current *I*, or any other current or voltage in the circuit, can be considered to be a function of the three sources, that is, $I = f(V_A, I_B, V_C)$. Since the elements are linear, the function must have the form

$$I = AV_A + BI_B + CV_C$$

where A, B, and C are constants. Although we do not know I_B or V_C in this example, we can treat the factor $(BI_B + CV_C)$ as an unknown, say X. Then, one equation we can derive from the information given is

$$\frac{-7}{6} = 7A + X \tag{1}$$

and another is

$$I = 28A + X \tag{2}$$

yielding two equations and three unknowns. We can generate another equation by suppressing the sources I_B and V_C since then $X = 0$. The circuit for this situation is

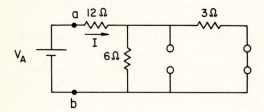

and the equation is

$$I = AV_A + 0$$

Now, $6//3 = 2\ \Omega$, so $I = V_A/14$, which shows that $A = 1/14$. (Note that A^{-1} is just the Thevenin or Norton equivalent resistance R_o for the one-port system to the right of terminals a and b; that is, the resistance $R_o = R_T = R_N$ found by suppressing the sources.) Substituting $A = 1/14$ into Eq. 1 yields $X = -10/6$, and putting both A and X into Eq. 2 yields $I = 1/3$ A, which is the required solution.

Problem 3.1

Two experiments were conducted on the following circuit for which the resistor values are unknown:

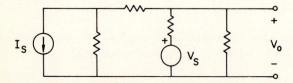

For $I_s = 0$ and $V_s = 3$ V, $V_o = 1$ V. For $V_s = 0$ and $I_s = 2$ A, $V_o = -1$ V. Find the output voltage V_o if $I_s = 12$ A and $V_s = 6$ V.

Ans.: $V_o = -4$ V

Problem 3.2

In the circuit of Example 3.1, use superposition to find the current I if $V_A = 7$ V, $I_B = 14$ A, and $V_C = 28$ V.

Ans.: $I = -17/6$ A

Problem 3.3

If I_B is increased to 28 A in Problem 3.2, what is the new value for I? What value of I_B is required to yield $I = 0$ if V_A and V_C are kept fixed at 7 V and 28 V, respectively?

Problem 3.4

In the circuit of Problem 3.1, let I_s remain at 12 A and the (unknown) resistors be unchanged. Use the superposition principle to find the value of V_s required such that $V_o = 0$ V.

Problem 3.5

Use superposition to find the three currents in Problem 2.6. Show explicitly the contributions of each source in your answer.

Problem 3.6

Use the superposition principle to find the voltage V_a and the current I in the circuit below. Show explicitly the contribution of each source in your answer.

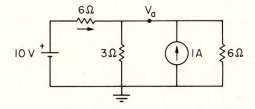

Problem 3.7

Find the voltages V_a and V_b in the circuit below using superposition. Show explicitly the contributions from each source.

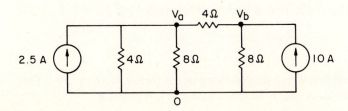

Problem 3.8

Treat Example 2.2 as a superposition problem and find the three contributions to the current I_2 due to the three sources. Check that they add up to -1 mA.

Problem 3.9

Use superposition to find the two contributions to the current I in Problem 2.40 from the two sources. Check that they add up to 3 A.

Problem 3.10

Use superposition to find the current I_2 in Problem 2.41.

Problem 3.11

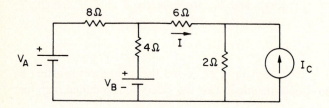

In the circuit shown, the values of V_A and I_C are fixed. When $V_B = 0$, the current $I = 2$ A. Find the value of I when $V_B = 16$ V.

Ans.: $I = 3$ A

Problem 3.12

In the circuit below, it follows from linearity that we can write

$$V_x = AI_A + BV_A + CV_B$$

where A, B, and C are constants. Find the value of A.

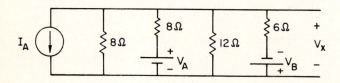

Ans.: $A = -2\,\Omega$

Problem 3.13

For the circuit of Problem 3.12, find B and C. Then find V_x if $I_A = 1$ A, $V_A = 10$ V, and $V_B = 8$ V.

Problem 3.14

The current I in the circuit below can be expressed in the form $I = AI_A + BV_B$. Find the algebraic expressions for A and B in terms of R_1, R_2, R_3, and R_4.

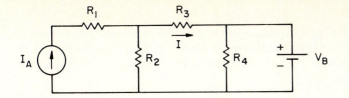

$$\text{Ans.: } A = \frac{R_2}{R_2 + R_3}$$

$$B = \frac{-1}{R_2 + R_3}$$

Problem 3.15

If $V_B = 10$ V and all the resistors are 1 kΩ in the circuit of Problem 3.14, what value of I_A will yield a current $I = 4$ mA? What value of I_A will yield zero current?

Problem 3.16

Suppose a 1-ampere current is known to flow through the 8-Ω resistor in the circuit below. Find the source V_s. Some useful currents are shown and should be checked as the problem is worked by the ladder method.

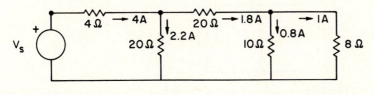

$$\text{Ans.: } V_s = 60 \text{ V}$$

Problem 3.17

A different voltage source is attached to the same set of resistors as in Problem 3.16, and the voltage across the 8-Ω resistor is found to be 1 V. What value does the new voltage source have?

Problem 3.18

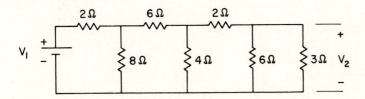

Use the unit voltage method to find the ratio V_2/V_1 in the circuit below.

$$\text{Ans.: } \frac{V_2}{V_1} = \frac{1}{12}$$

Problem 3.19

The voltage source is removed from the circuit in Problem 3.16, and a 12-A current source is attached to the same set of resistors as shown in the circuit diagram. Use the linearity principle and the current values listed in that problem to deduce the voltage across the 8-Ω resistor.

Problem 3.20

The voltage source V_1 in Problem 3.18 is replaced by a current source of value 2 A. Use the unit current method to find the current through the 3-Ω resistor. (*Hint:* 1 A through the 3-Ω resistor yields 3 V. This voltage also appears across the 6-Ω resistor, so a total of 1.5 A flows through the 2 Ω resistor, and so forth.)

EXAMPLE 3.2

In the circuit for Example 3.1, let $V_A = 7$ V, $I_B = 7$ A and $V_C = 14$ V. Construct the Thevenin equivalent circuit for the portion of the circuit to the right of the 12-Ω resistor. Then find the current I and check it against the previous result.

Solution: First, we can use source transformation to convert V_C and the series 3-Ω resistance to a current source with a parallel resistance.

Since the 3-Ω and 6-Ω resistances are in parallel, they have an equivalent resistance of 2 Ω. The two current sources are in parallel and can be added to give an equivalent current source of $I_B + (14/3) = (35/3)$ A. Thus, the Norton and Thevenin equivalent circuits are, respectively,

Now, to find I, we may use the equivalent circuit

$$I = \frac{7 - 70/3}{14} = -7/6 \text{ A}$$

which agrees with the answer given in the example.

Problem 3.21

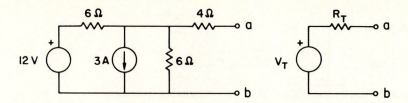

The two circuits shown are equivalent at the terminals a and b. Find V_T and R_T.

$$\text{Ans.: } V_T = -3 \text{ V}$$
$$R_T = 7 \text{ }\Omega$$

Problem 3.22

The two sources in Problem 3.21 are exchanged keeping the same polarity (for example, the current arrow remains downward). Find the Thevenin equivalent circuit.

Problem 3.23

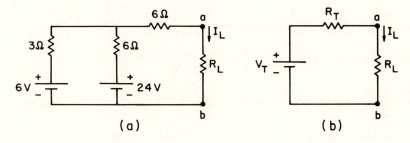

(a)   (b)

Find the values of V_T and R_T in circuit (b) above that will result in the same current I_L as in circuit (a).

$$\text{Ans.: } V_T = 12 \text{ V}$$
$$R_T = 8 \text{ }\Omega$$

Problem 3.24

Find the Thevenin and Norton equivalent circuits at the port a–b in Problem 2.12. Check that $R_o = V_T/I_N = V_{oc}/I_{sc}$.

Problem 3.25

In Problem 2.31 a mathematical expression was found for V_{ab} which equals the Thevenin equivalent voltage at the port a–b. Find analogous expressions for I_N and $R_o = R_N = R_T$.

Problem 3.26

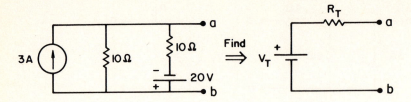

Find the Thevenin equivalent circuit for the network above.

$$\text{Ans.: } V_T = 5 \text{ V}$$
$$R_T = 5 \ \Omega$$

Problem 3.27

Find the Norton equivalent circuit at the port $a–b$ for Problem 3.26.

Problem 3.28

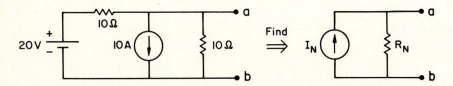

Find the Norton equivalent circuit at the port $a–b$ for the network above.

Problem 3.29

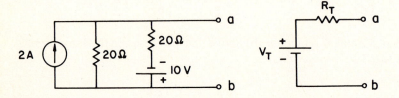

The two circuits shown are equivalent at the terminals a and b. Find V_T and R_T.

$$\text{Ans.: } V_T = 15 \text{ V}$$
$$R_T = 10 \ \Omega$$

EXAMPLE 3.3

In the circuit of Problem 3.29, a resistance of R_x is connected between a and b. What is the maximum power that can be delivered to R_x?

Solution: The maximum power transfer theorem states that a matched load yields the maximum power transfer. Thus, $R_x = R_T = 10 \ \Omega$. For this value of R_x the current through R_x is $15/20 = (3/4)$ A. The power transferred to R_x is $I^2R_x = (9/16)(10) = 5.63$ W.

Problem 3.30

Find the value of the load resistor R_L in the circuit below that yields the maximum transfer of power to that load resistor. Find the value of the maximum power that can be transferred.

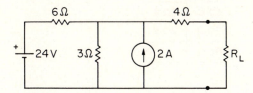

Problem 3.31

Find the maximum power that can be supplied to the port b–c for the circuit below.

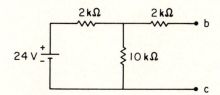

Problem 3.32

Determine the maximum power which can be supplied to the load resistor R_L in the circuit below.

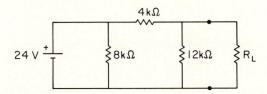

Problem 3.33

For the circuit below, find the value of R_L that will result in maximum possible power delivered to R_L and find that power if $V_A = 12$ V and $I_B = 1$ A.

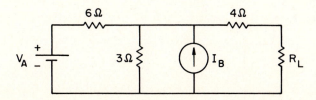

Problem 3.34

The box shown consists of *dc* sources and resistors. Measurements made at terminals *a* and *b* yield the results shown in the table. Predict the reading of the voltmeter when $R = 80 \, \Omega$.

Ans.: $V_m = 48$ V

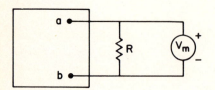

R	V_m
10 Ω	20 V
20 Ω	30 V
80 Ω	?

EXAMPLE 3.4

In the circuit shown, write two equations from which the loop currents i_1 and i_2 may be found in terms of the resistances R_1, R_2, R_3, and R_4.

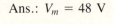
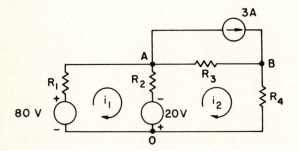

Applying KVL clockwise around the left-hand side, we have

$$-80 + R_1 i_1 + R_2(i_1 - i_2) - 20 = 0$$

Note that $(i_1 - i_2)$ is the total current downward through resistor R_2. This equation can also be written

$$(R_1 + R_2)i_1 - R_2 i_2 = 100 \qquad (1)$$

Before applying KVL clockwise around the right-hand side, we can use a source transformation to convert the current source to a voltage source. The loop then becomes

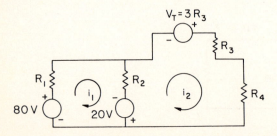

and we have

$$+20 - R_2(i_1 - i_2) - 3R_3 + R_3 i_2 + R_4 i_2 = 0$$

Gathering terms,

$$-R_2 i_1 + (R_2 + R_3 + R_4)i_2 = 3R_3 - 20 \qquad (2)$$

This completes the set of two equations.

Problem 3.35

Use the loop current method to solve Problem 2.6. Show explicitly the values for i_1 and i_2 and that $I_2 = i_1 - i_2$, where i_1 and i_2 are clockwise mesh currents.

Problem 3.36

Use the loop current method to solve Example 2.2. In this circuit the voltage V_s should be treated as an unknown, which means that a third equation must be generated to solve the problem. Give values for the loop currents explicitly.

Problem 3.37

Use the loop current method to find the currents I_1 and I_2 in Problem 2.41. Treat the voltage across the 6-A current source as an unknown.

Problem 3.38

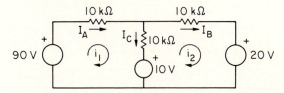

Write two loop current equations and solve for i_1, i_2, I_A, I_B, and I_C.

$$\text{Ans.: } 20i_1 - 10i_2 = 80$$
$$20i_2 - 10i_1 = -10$$
$$i_1 = 5 \text{ mA} = I_A$$
$$i_2 = 2 \text{ mA} = I_B$$
$$I_c = 3 \text{ mA}$$

EXAMPLE 3.5

For the circuit shown, use the node voltage method to find V_b and the current I.

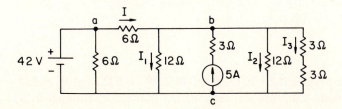

Solution: First let $V_c = 0$; that is, set it equal to ground potential. Then, $V_a = 42$ V, and the only unknown is V_b. Applying KCL at node b,

$$I + 5 = I_1 + I_2 + I_3$$

which can be written

$$\frac{42 - V_b}{6} + 5 = \frac{V_b}{12} + \frac{V_b}{12} + \frac{V_b}{6}$$

for which the solution is $V_b = 24$ V. Then, $I = (42 - 24)/6 = 3$ A.

Problem 3.39

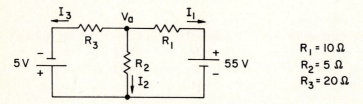

R$_1$ = 10 Ω
R$_2$ = 5 Ω
R$_3$ = 20 Ω

Use the node voltage method to find V_a and then I_1, I_2, and I_3.

$$\text{Ans.: } V_a = 15 \text{ V}$$
$$(I_1, I_2, I_3) = (-4 \text{ A}, 3 \text{ A}, 1 \text{ A})$$

Problem 3.40

Use the node voltage method to find the two unknowns V_a and V_b.

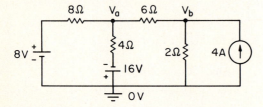

Problem 3.41

For the circuit of Example 3.4, write two equations from which the node voltages V_A and V_B may be found in terms of the resistances R_1, R_2, R_3, and R_4.

Ans.:

$$V_A(1/R_1 + 1/R_2 + 1/R_3) - (V_B/R_3) = (80/R_1) - (20/R_2) - 3$$
$$-(V_A/R_3) + V_B(1/R_3 + 1/R_4) = 3$$

Problem 3.42

In the circuit of Example 3.4, let $R_1 = 10 \ \Omega$, $R_2 = 40 \ \Omega$, $R_3 = 5 \ \Omega$, and $R_4 = 2 \ \Omega$. Find the loop currents, node voltages, and the currents through R_3 (from A to B)

and through R_2 (from A to 0). (Note that the current through R_3 is not equal to i_2.)

$$\text{Ans.: } i_1 = 6 \text{ A}, \ i_2 = 5 \text{ A}$$
$$V_A = 20 \text{ V}, \ V_B = 10 \text{ V}$$
$$I_{AB} = 2 \text{ A}, \ I_A = 1 \text{ A}$$

Problem 3.43

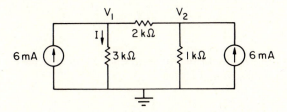

Write two independent equations from which V_1 and V_2 may be found. Find the node voltages V_1 and V_2 and the current I.

$$\text{Ans.: } 5V_1 - 3V_2 = 36$$
$$-V_1 + 3V_2 = 12$$
$$V_1 = 12 \text{ V}; \ V_2 = 8 \text{ V}; \ I = 4 \text{ mA}$$

Problem 3.44

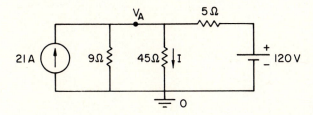

Find V_A using the node voltage method. Find I using the superposition principle and show that it equals $V_A/45$.

$$\text{Ans.: } V_A = 135 \text{ V}$$
$$I = 3 \text{ A}$$

Problem 3.45

For the circuit of Problem 3.12, use the node voltage method to solve directly for V_x in the form

$$V_x = AI_A + BV_A + CV_B$$

and find the values of A, B, and C.

$$\text{Ans.: } A = -2 \ \Omega, \ B = 1/4, \ C = -1/3$$

Problem 3.46

Find the node voltages V_a and V_b in the circuit below.

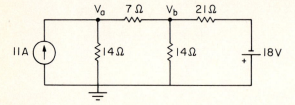

Problem 3.47

Find the loop currents i_1, i_2, and i_3 in the circuit of Problem 3.46. Show that they are consistent with the node voltage values found above.

Problem 3.48

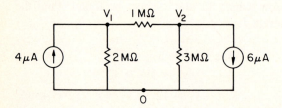

For the circuit above, find V_1 and V_2. Note that the resistors are megohm values $(10^6 \ \Omega)$.

Problem 3.49

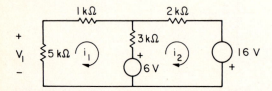

In the circuit above, find i_1, i_2, and the voltage V_1.

EXAMPLE 3.6

What is the equivalent resistance R_{EQ} to the right of a and b for the circuit shown below?

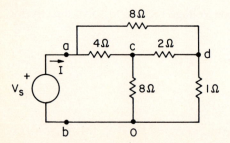

Solution: Since there are no sets of resistors in pure series or parallel combinations to the right of a and b, we cannot find the equivalent resistance by series-parallel combinations. Instead, assume a source V_s as shown and compute I. Then, $R_{EQ} = V_s/I$. Using the node voltage method, let node b be the reference node and write the node voltage equations at the unknown nodes c and d. The node voltage V_a equals V_s. At node c,

$$\frac{V_c - V_s}{4} + \frac{V_c}{8} + \frac{V_c - V_d}{2} = 0$$

At node d,

$$\frac{V_d - V_s}{8} + \frac{V_d - V_c}{2} + \frac{V_d}{1} = 0$$

which can be reduced to

$$7V_c - 4V_d = 2V_s$$

$$-4V_c + 13V_d = V_s$$

for which the solutions are $V_c = 0.4V_s$ and $V_d = 0.2V_s$. Then, at node a,

$$I = \frac{V_s - V_c}{4} + \frac{V_s - V_d}{8} = 0.15V_s + 0.1V_s = 0.25V_s$$

and $R_{EQ} = V_s/I = 4\,\Omega$.

Another way to solve this problem is to choose a convenient numerical value for V_s at the start and then to find the numerical value of I. The ratio of course will be independent of the choice of the value of V_s but will yield R_{EQ}. In this example, if V_s had been chosen to be 20 V, then $V_c = 8$ V and $V_d = 4$ V. For this case, $I = (20 - 8)/4 + (20 - 4)/8 = 5$ A and $R_{EQ} = 20/5 = 4\,\Omega$ as before.

Problem 3.50

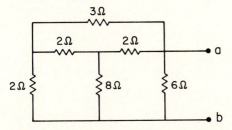

Find the equivalent resistance R_{EQ} at the terminals a and b.

Ans.: $R_{EQ} = 2\,\Omega$

Problem 3.51

Find the equivalent resistance at the terminals a and b for the circuit below.

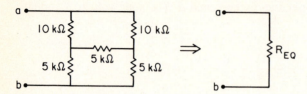

Problem 3.52

Find the equivalent resistance at the terminals a and b for the circuit below.

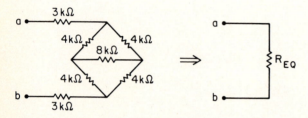

Problem 3.53

Find the equivalent resistance across the diagonal in the circuit below (i.e., between the two nodes indicated by the dots).

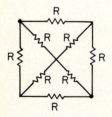

Problem 3.54

Find the current I in the circuit below.

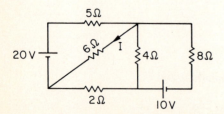

Problem 3.55

Find the current flowing from right to left through the 20-Ω resistor in the figure below.

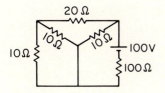

Problem 3.56

Predict the ammeter reading in the circuit below.

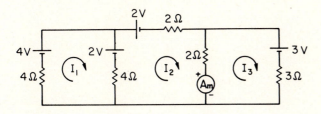

Chapter
4
Controlled Sources
and Two-Port Systems

In the previous chapters we have discussed the fundamental principles of circuit analysis and illustrated their use with circuits containing resistors and sources. These principles have application far beyond the limited possibilities of purely resistive networks. In this chapter we extend the circuit elements available to our use by introducing controlled sources. These can be either current or voltage sources and have the property that their value depends upon some other current or voltage in the circuit. A bipolar junction transistor is a good example of a device that behaves in this manner since the collector current is controlled by the base current. Since we do not introduce transistors until much later in the text, this chapter is somewhat abstract. However, from a system standpoint the concept of a voltage or a current amplifier is straightforward and does not depend upon the details of how one might be built. From this viewpoint we are able to discuss amplifiers and how they may be modeled using controlled sources. A very important practical device, the operational amplifier, is also introduced; its successful use in circuit design is often possible without a deep understanding of its internal properties.

4.1 CONTROLLED SOURCES

Thus far, we have introduced two types of independent sources: the voltage and the current source. Very often in circuit design and analysis the concept of a controlled source is quite useful. Such a source is also referred to as a dependent source. To distinguish between controlled and independent sources a diamond shape is used rather than the circular one we have adopted for independent sources (Fig. 4.1).

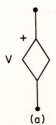

FIGURE 4.1.
Symbols for controlled voltage (a) and current (b) sources.

(a)

(b)

A controlled source must ultimately be linked to some independent source. For example, the voltage across the controlled voltage source in Fig. 4.1 could be determined either by some independent voltage source V_s,

$$V = \mu V_s$$

or by some independent current source I_s,

$$V = kI_s$$

where μ and k are constants of proportionality. The parameter μ is dimensionless since the controlled and the independent sources are both measured in volts. The units of k are volts/ampere corresponding to control of a voltage by a current. Likewise, a controlled current source can be controlled by another current source,

$$I = \beta I_s$$

or by a voltage source,

$$I = gV_s$$

where β is dimensionless and g has units of amperes/volt.

At the moment such controlled sources are abstractions. However, we will soon encounter examples of devices that can be accurately modeled in this way. Controlled sources are handled nearly identically to independent sources. Consider the circuit shown in Fig. 4.2, in which we are interested in the value of the current I. Since there is only one loop here, the current I must flow through resistor R_1 as well as resistor R_2. Applying KVL,

$$-V_s + IR_1 + \mu V_s + IR_2 = 0$$

Hence,

$$I = \frac{V_s(1 - \mu)}{R_1 + R_2}$$

This result is somewhat bizarre since it indicates that if $\mu > 1$, the circuit to the right of the port a–b has an effective resistance, $R = V_s/I$, which is negative. Clearly,

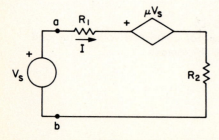

FIGURE 4.2.

An example of a simple circuit including a controlled source.

controlled sources bring some interesting possibilities to circuit design which do not exist in Chapters 1 through 3.

The dependent sources we use need not be controlled by independent sources. In the circuit shown in Fig. 4.3, for example, the dependent voltage source is controlled by the voltage V_1 which exists across the 1-kΩ resistor.

FIGURE 4.3.
A circuit illustrating the use of a voltage controlled voltage source.

TEXT EXAMPLE 4.1

Find the voltage V_{ab} in the circuit given in Fig. 4.3 above.

Solution: We use the node voltage method at node c. KCL at that node yields

$$\frac{10 - V_c}{2\ k\Omega} = \frac{V_c}{1\ k\Omega} + \frac{V_c - 10V_c}{2\ k\Omega}$$

where we have used the fact that $V_1 = V_c$. Solving, $V_c = -1.66$ V. Since $V_{ab} = 10V_1 = 10V_c$, we have $V_{ab} = -16.6$ V.

An example with two controlled sources is shown in Fig. 4.4a. Here, we have a current-controlled voltage source in the center leg of the circuit ($V_2 = 10I_2$) and a current-controlled current source in the right-hand leg ($I_3 = 10I_1$). To apply a loop analysis, it is convenient to source-transform the current source to a voltage source by changing from a Norton-type to a Thevenin-type circuit. This reduces the number of loops from three to two. The open-circuited voltage across the parallel 5-Ω resistor is $I_3R = 50I_1 = V_3$. The Thevenin resistance equals the Norton resistance (5-Ω), and, combining the two 5-Ω resistors, we have the equivalent circuit of Fig. 4.4b. Using the loop currents i_1 and i_2 and applying KVL around the first loop,

$$-30 + 10i_1 + 5(i_1 - i_2) + 10i_2 = 0$$

FIGURE 4.4a.
Find the current I_A using loop analysis.

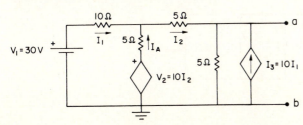

FIGURE 4.4b.
Transformed circuit with loop currents shown.

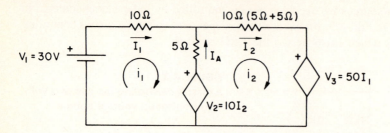

Applying KVL around the second loop (noting that $I_1 = i_1$ and $I_2 = i_2$),

$$-10i_2 + 5(i_2 - i_1) + 10i_2 + 50i_1 = 0$$

Combining terms,

$$15i_1 + 5i_2 = 30$$

$$45i_1 + 5i_2 = 0$$

and thus $i_1 = -1$ A and $i_2 = 9$ A. The current I_A equals $(i_2 - i_1) = 10$ A.

As a further example using circuits with controlled sources, suppose we wish to determine the Thevenin equivalent circuit at the port $a - b$ in Fig. 4.4a. The open-circuited voltage, $V_{oc} = V_a - V_b$, is just the voltage across the 5-Ω resistor in Fig. 4.4a which is in parallel with the dependent source I_3. The current to ground through this resistor, I_5, is

$$I_5 = I_2 + I_3 = 9 + 10(-1) = -1 \text{ A}$$

The open-circuited voltage is $V_{oc} = (-1)(5) = -5$ V. The short-circuited current must be found from a different circuit diagram, namely, as shown in Fig. 4.4c. The short circuit between a and b forces the voltage at point a to be zero volts. In this case, $I_5 = 0$ since the voltage across that 5-Ω resistor is $0 - 0 = 0$ volts. Thus, we have

$$I_{sc} = I_2 + I_3 = I_2 + 10I_1$$

FIGURE 4.4c.
Circuit needed to determine the short-circuited current.

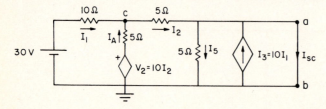

To find I_1 and I_2, we can use the node voltage method to find the only unknown node voltage V_c. Writing KCL at node c,

$$\frac{30 - V_c}{10} = \frac{V_c - V_2}{5} + \frac{V_c}{5}$$

But $V_2 = 10I_2 = 10(V_c/5)$, so

$$\frac{30 - V_c}{10} = \frac{V_c}{5} - \frac{10V_c}{25} + \frac{V_c}{5}$$

Hence, $V_c = +30$ V, $I_1 = 0$, $I_2 = 6$ A, and finally $I_{sc} = 6$ A. The Thevenin and Norton equivalent resistances are $R_T = R_N = V_{oc}/I_{sc} = -(5/6)$ Ω, and the corresponding Thevenin and Norton equivalent circuits are those shown in Fig. 4.4d. In this example the Thevenin resistance is negative! This result stems from the fact that there are controlled sources inside the "black box." A detailed analysis of the meaning and application of negative resistance circuits is beyond the scope of this text. We do note that the power I^2R associated with the Thevenin or Norton resistance in such a circuit is also a negative number. This implies that there must be some source of energy associated with the controlled source or sources in the circuit. Later, when we study transistors, which are often modeled using such controlled sources, we shall see that such devices usually draw power from batteries or other power sources which provide the energy required to amplify signals or create "negative" resistors.

FIGURE 4.4d.
Thevenin and Norton Equivalent Circuits.

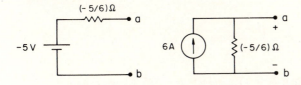

Finally, a word of caution is in order regarding the R_o method in cases involving controlled sources. If we merely suppress all the sources inside the box and determine the equivalent resistance, a negative number will never result. Instead, to use the R_o method we must suppress only the independent sources. Then, as discussed in Chapter 3, we apply a voltage V to the terminals, determine I, and then calculate $R_o = V/I$. This technique was discussed briefly at the end of Section 3.2 and is illustrated in Text Example 4.2 using the circuit in Fig. 4.4a.

TEXT EXAMPLE 4.2 ————————————————————

Use the R_o method to find the Norton and Thevenin equivalent resistances for the circuit given in Fig. 4.4a which has two dependent sources.

Solution: The only independent source is the 30-V battery. Suppressing this with a short circuit yields the circuit to the left of the port $a - b$ illustrated below.

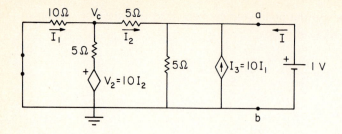

If we apply a 1-V source across the terminals a and b as illustrated and calculate the current I, then the ratio $R_o = 1/I$ will yield the desired resistance value in ohms. Applying KCL at node c, we have

$$\frac{0 - V_c}{10} + \frac{10I_2 - V_c}{5} = \frac{V_c - 1}{5}$$

But since a 1-V source has been placed across $a - b$, $I_2 = (V_c - 1)/5$. Substituting this into the equation above and solving for V_c yields $V_c = -2$ V. Applying KCL at node a gives

$$I + I_3 + I_2 = \frac{1}{5}$$

where the term on the right-hand side is the current through the 5-Ω resistor to ground due to the 1-V source. Now the current source I_3 depends upon the current I_1, which is given by

$$I_1 = \frac{0 - V_c}{10} = \frac{2}{10}$$

and hence $I_3 = 10I_1 = 2$ A. Since $I_2 = (V_c - 1)/5 = -3/5$ A, we obtain

$$I = -\frac{6}{5} \text{ A}$$

Finally,

$$R_o = \frac{1}{I} = -\frac{5}{6} \, \Omega$$

which is the same result found earlier.

To summarize this section, the analysis methods learned in the previous chapters may be used on circuits with controlled sources provided that some care is exercised.

Further examples are given in the problem set at the end of this chapter. We use such sources extensively in Chapters 14 and 15.

4.2 TWO-PORT SYSTEMS

The two-port is an essential tool of electrical engineering design. In its most basic form a two-port is a system which accepts some signal at its input, modifies it in some way, and then presents the modified version of the signal at its output (Fig. 4.5). The black box could be a single resistor, or it could be as complex as a stereo radio system, where the input is a weak incoming electromagnetic wave and the output is a much stronger signal in the audio frequency range. In this text we deal with two-port systems that have no independent sources inside. Commonly used types of models are presented in this section and related to voltage and current amplifiers.

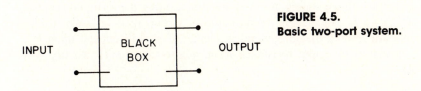

FIGURE 4.5.
Basic two-port system.

We will frequently encounter the specific type of two-port system that is shown symbolically in Fig. 4.6. If the actual elements in the two-port are nonlinear devices such as transistors or diodes, the "linear elements" in the diagram refer only to the small-signal linear models of those devices. Such models are discussed in some detail in Chapters 14 and 15. To the left of $a - b$ we may use Thevenin's theorem to replace that one-port with an equivalent circuit consisting of a voltage source (V_s) in series with a resistance (R_s) that will result in the same V_1 and I_1 as in the actual system. To the right of $c - d$ we may replace the one-port by its Thevenin equivalent, which in this case consists only of a resistance (R_L), that will result in the same V_2 and I_2 as in the actual system. There is no voltage source in the Thevenin equivalent circuit because there are no independent sources to the right of $c - d$ and the open-circuit voltage of that block will therefore be zero. Using the equivalent circuits to the left of $a - b$ and to the right of $c - d$, we may thus reduce any system of the form shown in Fig. 4.6 to the following form without loss of

FIGURE 4.6.
Two-port showing the source and load specifically.

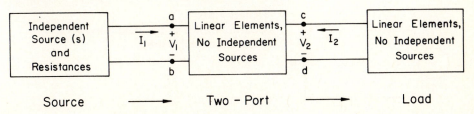

FIGURE 4.7.
Circuit equivalent to Fig. 4.6 showing the two-port as a generalization.

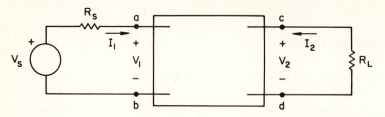

generality and produce the same values of I_1, V_1, I_2, and V_2 as in the actual system (Fig. 4.7).

Our task now is to create a model for the two-port itself. Consider first the (input) terminals a–b. Since there are no independent sources in the box, we can replace the *entire* system to the right of a–b by a single equivalent resistance $R_i = V_1/I_1$. This resistance is called the "input resistance" of the two-port. It may be measured by applying a known voltage source at a and b and measuring the resultant current. The values of V_1 and I_1 can then be found from the circuit model shown in Fig. 4.8a. It is important to note that in general the value of R_i will depend upon R_L. If the load is changed, the input resistance R_i may also change. (It usually will change, in fact.)

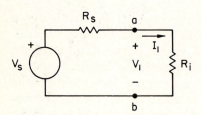

FIGURE 4.8a.
Definition of the input resistance R_i.

Considering now the parameters V_2 and I_2 at c–d, we may use Thevenin's theorem to replace the *entire* system to the left of c–d by a voltage source V_T in series with a resistance R_T. As always, V_T is the open-circuited voltage at c–d, that is, the voltage V_2 when R_L is replaced by an open circuit. Following our prescription for the "R_o method," we may determine R_T by applying a known voltage source at c–d and measuring the resultant current when V_S is replaced by a short-circuit. The values of V_2 and I_2 can then be found using the circuit model shown in Fig. 4.8b. The input and output models described above can now be drawn in the same

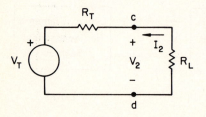

FIGURE 4.8b.
Characteristics of the output model if a Thevenin circuit is used.

FIGURE 4.8c.
Final intermediate step in constructing a model.

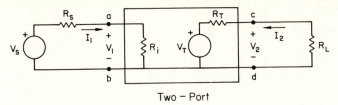

Two – Port

diagram, shown in Fig. 4.8c. In the remainder of this section and the next, we modify this intermediate form to create a model suitable for the study of voltage amplifiers. Current amplifiers are discussed in Section 4.4.

The intermediate model in Fig. 4.8c is not very useful because V_T and R_T depend upon V_s and R_s. We prefer a model that could be used for any source voltage and any source resistance. This can be accomplished by relating the voltage V_T to V_1, the voltage at the *input* to the two-port system. The argument goes as follows. By the voltage divider law, V_1 is proportional to V_s. Likewise, since there are no independent sources in the box, by the linearity principle V_T must be proportional to V_s. It follows, therefore, that V_T is proportional to V_1. We can now produce a useful model by replacing the output portion of the model in Fig. 4.8c by a dependent source proportional to V_1 in series with a resistance. This gives us the model shown in Fig. 4.9. As we shall see in the next section, this model can be used very nicely to study voltage amplifiers.

FIGURE 4.9.
Model using an input resistance (R_i), a voltage controlled voltage source (μV_1), and an output resistance (R_o).

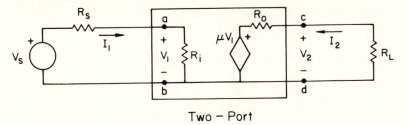

Two – Port

4.3 VOLTAGE AMPLIFIERS

We now apply the model developed above to a discussion of voltage amplifiers. An ideal voltage amplifier accepts an input signal from some source V_s and multiplies it by some constant factor μ to create an output $V_o = \mu V_s$, no matter what values the source resistance R_s or the load resistance R_L take on. Such an ideal amplifier can be modeled using the circuit shown in Fig 4.10. In this version of Fig. 4.9, the input resistance is $R_i = \infty$ (an open circuit) and the output resistance is $R_o = 0$ (a short circuit). Since there is an open circuit at the input, no current is drawn, and $V_i = V_s$. Since the output resistance is zero ohms, the full voltage μV_s appears across

FIGURE 4.10.
Model for an ideal voltage amplifier.

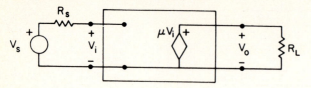

the load resistor and

$$V_o = \mu V_s$$

no matter what value R_L takes on. This is the equation for an ideal amplifier of gain μ.

Any real-world voltage amplifier model is characterized by a finite value for R_i and a nonzero value for R_o; that is, by a circuit of the form shown in Fig. 4.11. We call μ the intrinsic gain of the amplifier. Now the voltage V_i will depend upon the value of R_i via the voltage divider which exists on the input. V_o will also be decreased since the controlled source voltage is μV_i, not μV_s. Similarly, since the output is the voltage across the load resistor, it is reduced further by another voltage divider. To be specific, the output voltage divider in Fig. 4.11 yields

$$V_o = \left(\frac{R_L}{R_L + R_o} \right) \mu V_i$$

But at the input

$$V_i = \left(\frac{R_i}{R_i + R_s} \right) V_s$$

Substituting and dividing by V_s the amplifier gain is given by

$$A_v = \frac{V_o}{V_s} = \mu \left(\frac{R_i}{R_i + R_s} \right) \left(\frac{R_L}{R_L + R_o} \right)$$

Written in this manner it is clear how the input and output resistances of the amplifier affect the gain. On the input side it is desirable that R_i be much greater than the source resistance, since

FIGURE 4.11.
Model for a real voltage amplifier.

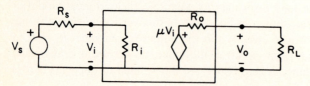

$$\lim_{R_i \to \infty} \left(\frac{R_i}{R_i + R_s} \right) = 1$$

A good voltage amplifier thus has a high input resistance. On the output side a low output resistance is desired, since

$$\lim_{R_o \to 0} \frac{R_L}{R_L + R_o} = 1$$

To summarize, if $R_i \gg R_s$, and $R_o \ll R_L$, the amplifier gain will be nearly equal to the intrinsic gain μ. The limiting values $R_i = \infty$ and $R_o = 0$ correspond to the ideal amplifier in Fig. 4.10.

TEXT EXAMPLE 4.3

Suppose a 1-V battery with a source resistance of 50 Ω is connected to an amplifier whose output is attached to a 1000-Ω (1-kΩ) load. If the input resistance of the amplifier is 1000 Ω, the amplifier has an intrinsic gain of 10, and the output resistance of the amplifier is 100 Ω, what is the voltage gain $A_v = V_o/V_s$?

Solution: The model is

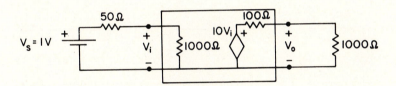

Applying the voltage divider principle at the input and output,

$$V_i = \left(\frac{R_i}{R_i + R_s} \right) V_s \text{ and } V_o = \mu \left(\frac{R_L}{R_L + R_o} \right) V_i$$

Substituting the latter into the former,

$$A_v = \frac{V_o}{V_s} = \mu \left(\frac{R_i}{R_i + R_s} \right) \left(\frac{R_L}{R_L + R_o} \right) = 10 \left(\frac{1000}{1050} \right) \left(\frac{1000}{1100} \right)$$

and

$$A_v = 8.66$$

This shows that the amplifier gain is not equal to 10 since both the input voltage divider, due to a finite input resistance, and the output voltage divider, due to a finite output resistance, affect the overall gain.

Often it is sufficient to estimate the effects of finite input and output resistances. Here, we could say that the source resistance is 1/20, or 5% of the input resistance, while the output resistance is 10% of the load. Roughly speaking, we "lose" 5% at the input and 10% at the output, so the total gain will be about 8.5, not too far from the exact value.

4.4 FURTHER EXAMPLES OF THREE PARAMETER MODELS

There are only four different types of models possible using an input resistance R_i. These are illustrated in Fig. 4.12. Each of these models is characterized by a certain input resistance R_i, which is determined in an identical fashion as shown in "Two-Port Systems" (Section 4.2). The polarities of the voltage sources and the directions of the current sources are somewhat arbitrary but conform to conventions that are appropriate for the systems we study in this text (e.g., Chapters 14 and 15). We shall use models of these types in the remainder of the chapter. Models (c) and (d) will have considerable application later in the text for modeling transistors. The next example illustrates the use of model (d) in Fig. 4.12.

FIGURE 4.12.
Four possible three-parameter models.

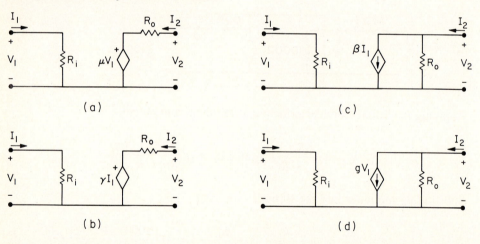

(a)   (c)

(b)   (d)

TEXT EXAMPLE 4.4

The circuit model illustrated below may be used to describe a MOSFET transistor response to small signals. Find an expression for the voltage gain $A_v = V_o/V_s$ of the system in terms of the model parameters. Evaluate the gain numerically if $R_s = 10$ kΩ, $R_i = 10$ MΩ, $r_d = 20$ kΩ, $R_L = 2$ kΩ, and $g_m = 10^{-2}$ mho. Note that the entire circuit is treated as a voltage amplifier even though the dependent source is a voltage-controlled current source. Notations vary in the applications you will meet. Here, we have used V_i at the input and V_o at the output, which for reference correspond to V_1 and V_2 in Fig 4.12.

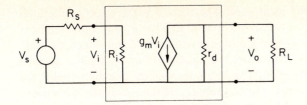

Solution: As we shall see in Chapter 14, a MOSFET transistor can be modeled with a voltage-controlled current source as shown here. The current source depends upon the voltage V_i. Our goal is to find the voltage gain

$$A_v = \frac{V_o}{V_s}$$

but we must first find V_i in terms of V_s. Using the voltage divider principle

$$V_i = \left(\frac{R_i}{R_i + R_s}\right) V_s$$

The dependent current $I' = g_m V_i$ is then

$$I' = g_m \left(\frac{R_i}{R_i + R_s}\right) V_s$$

This current divides between r_d and R_L. The current through the load resistor is then

$$I_L = \left(\frac{r_d}{r_d + R_L}\right) I'$$

Since the current is "upward" through the load resistor, the load voltage is, after substituting for I',

$$V_o = -R_L I_L = -R_L \left(\frac{r_d}{r_d + R_L}\right) g_m \left(\frac{R_i}{R_i + R_s}\right) V_s$$

and

$$A_v = \frac{V_o}{V_s} = -g_m R_L \left(\frac{r_d}{r_d + R_L}\right) \left(\frac{R_i}{R_i + R_s}\right)$$

Using the parameters given above, the resulting voltage gain is found to be $A_v = -18.2$. It is instructive to investigate the three terms separately:

$$A_v = -(20)(0.909)(0.999) = -18.2$$

In some sense the first term is the ideal gain. The second is the effect of the load resistor upon the amplifier. The ratio shows that if $r_d \gg R_L$ or equivalently if $R_L \ll r_d$, the ratio would be unity and the load resistor would not affect the gain. Likewise, if $R_i \gg R_s$, the second ratio would be equal to 1 and the input resistance of the two-port would not affect the gain. This example, and the one discussed in the text above, illustrate one of the most common problems that arises in the laboratory, namely, the effect of finite input and output resistance values on the operation of a two-port. Note that the sign of the voltage gain is negative. Such a circuit is termed an inverting amplifier.

Model (c) in Fig. 4.12 above may be used to study current amplifiers in much the same way voltage amplifiers were treated in Section 4.3. For example, amplification of a current source I_s that has a finite source resistance R_s may be studied using the circuit diagram given in Fig. 4.13. The amplifier can be analyzed in a totally analogous manner to the voltage amplifiers discussed above. The Norton equivalent models for the input and output one-ports are used along with the current divider rule. The controlled current source is related to the input current by β. Using the current divider principle,

$$I_i = \left(\frac{R_s}{R_s + R_i} \right) I_s$$

FIGURE 4.13.
Three-parameter model suitable for a current amplifier.

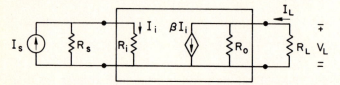

The load current is related to the controlled current source by the expression

$$I_L = \left(\frac{R_o}{R_o + R_L} \right) \beta I_i$$

and finally the current gain $A_I = I_L / I_s$ is given by

$$A_I = \beta \left(\frac{R_s}{R_s + R_i} \right) \left(\frac{R_o}{R_o + R_L} \right)$$

An ideal current amplifier is represented in Fig. 4.14, where $R_i = 0$ and $R_o = \infty$. That is, the ideal current amplifier has zero input resistance so that the entire source current flows into the amplifier. Likewise, the output Norton resistance is infinite so that the entire output current flows through the load. Then, $I_L = \beta I_s$.

FIGURE 4.14.
Ideal current amplifier.

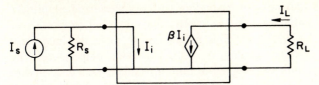

TEXT EXAMPLE 4.5

In the current amplifier circuit illustrated in Fig. 4.13, let $R_s = R_L = 5$ kΩ, $R_i = 100$ Ω; $R_o = 100$ kΩ, and $\beta = 50$. Find the current gain $A_I = I_L/I_s$. If $I_s = 50$ μA, what is the voltage across the load resistor R_L?

Solution: The two current divider ratios are, at the input,

$$\frac{R_s}{R_s + R_i} = \frac{5000}{5100} = 0.98$$

and at the output,

$$\frac{R_o}{R_o + R_L} = \frac{10^5}{1.05 \times 10^5} = 0.95$$

The load current is then

$$I_L = \beta(0.98)(I_s)(0.95)$$

and the current gain is

$$A_I = 46.6$$

For $I_s = 50$ μA, $I_L = 2.3$ mA, and the load voltage is

$$V_L = -(2.3 \times 10^{-3})(5 \times 10^3) = -11.6 \text{ V}$$

One final note is in order concerning two-port models. We have emphasized models in this text using a single resistor at the input since they are very useful for transistor analysis. Each model in Fig. 4.12 utilizes three parameters, for example, (R_i, μ, R_o) in the case of model a. One conceptual drawback in such a methodology is that, in general, such a model is only valid in so far as the load resistor is not changed. In practice this is not a severe constraint. However, it remains true that if a model is to be constructed which is valid no matter what the source and load characteristics are, it requires *four* independent parameters. For further information on four-parameter models the reader is referred to Problems 4.17[*]–4.20[*]. These entries are starred and hence not crucial to the rest of the material.

4.5 THE OPERATIONAL AMPLIFIER

One problem with electronic devices corresponding to the generalized amplifiers described above is that the gains, A_v or A_I, depend upon internal properties of the two-port system (μ, β, R_i, R_o, etc.). This makes design difficult since these parameters usually vary from device to device, as well as with temperature. The operational amplifier, or Op-Amp, is designed to minimize this dependence and to maximize the ease of design. An Op-Amp is an integrated circuit that has many component parts such as resistors and transistors built into the device. At this point we will make no attempt to describe these inner workings.

A totally general analysis of the Op-Amp is beyond the scope of this text. We will instead study one example in detail, then present the two Op-Amp laws and show how they can be used for analysis in many practical circuit applications. These two principles allow one to design many circuits without a detailed understanding of the device physics. Hence, Op-Amps are quite useful for researchers in a variety of technical fields who need to build simple amplifiers but do not want to design at the transistor level. Later in the text we shall also show how to build simple filter circuits using Op-Amps. The transistor amplifiers, which are the building blocks from which Op-Amp integrated circuits are constructed, will be discussed in Chapters 14 and 15.

The symbol used for an ideal Op-Amp is shown in Fig. 4.15. Only three connections are shown: the positive and negative inputs, and the output. Not shown are other connections necessary to run the Op-Amp such as its attachments to power supplies and to ground potential. The latter connections are necessary to use the Op-Amp in a practical circuit but are not necessary when considering the ideal Op-Amp applications we study in this chapter. The voltages at the two inputs and the output will be represented by the symbols V^+, V^-, and V_o. Each is measured with respect to ground potential. Operational amplifiers are differential devices. By this we mean that the output voltage with respect to ground is given by the expression

$$V_o = A(V^+ - V^-) \tag{4.1}$$

where A is the gain of the Op-Amp and V^+ and V^- the voltages at inputs. In other words, the output voltage is A times the difference in potential between the two inputs.

Integrated circuit technology allows construction of many amplifier circuits on a single composite "chip" of semiconductor material. One key to the success of an operational amplifier is the "cascading" of a number of transistor amplifiers to create a very large total gain. That is, the number A in Eq. 4.1 can be on the order of 100,000 or more. (For example, cascading of five transistor amplifiers, each with a gain of 10, would yield this value for A.) A second important factor is that these circuits can be built in such a way that the current flow into each of the inputs

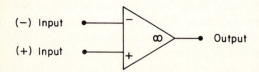

(−) Input

(+) Input

Output

FIGURE 4.15.
Schematic diagram of an ideal operational amplifier. Not shown are connections to power supplies and the ground potential reference.

is very small. A third important design feature is that the output resistance of the operational amplifier (R_o) is very small. This in turn means that the output of the device acts like an ideal voltage source.

We now can analyze the particular amplifier circuit given in Fig. 4.16 using these characteristics. First, we note that the voltage at the positive input, V^+, is equal to the source voltage,

$$V^+ = V_s$$

Various currents are defined in part (b) of the figure. Applying KVL around the outer loop in (b) and remembering that the output voltage, V_o, is measured with respect to ground, we have

$$-I_1R_1 - I_2R_2 + V_o = 0 \qquad (4.2)$$

Since the Op-Amp is constructed in such a way that no current flows into either the positive or negative input, $I^- = 0$. KCL at the negative input terminal then yields

$$I_1 = I_2$$

Using Eq. 4.2 and setting $I_1 = I_2 = I$,

$$V_o = (R_1 + R_2)I \qquad (4.3)$$

We may use Ohm's law to find the voltage at the negative input, V^-, noting the assumed current direction and the fact that ground potential is zero volts:

$$\frac{V^- - 0}{R_1} = I$$

So,

$$V^- = IR_1$$

FIGURE 4.16.
An operational amplifier circuit (a) and the currents we define to analyze the system (b).

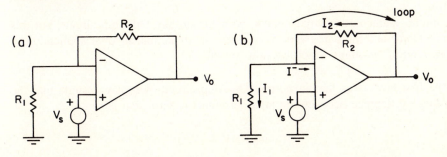

and from Eq. 4.3,

$$V^- = \left(\frac{R_1}{R_1 + R_2}\right) V_o$$

Since we now have expressions for V^+ and V^-, Eq. 4.1 may be used to calculate the output voltage,

$$V_o = A(V^+ - V^-) = A\left(V_s - \frac{R_1 V_o}{R_1 + R_2}\right)$$

Gathering terms,

$$V_o\left[1 + \frac{AR_1}{R_1 + R_2}\right] = AV_s \tag{4.4}$$

and finally,

$$A_v = \frac{V_o}{V_s} = \frac{A(R_1 + R_2)}{R_1 + R_2 + AR_1} \tag{4.5a}$$

This is the gain factor for the circuit. If A is a very large number, large enough that $AR_1 >> (R_1 + R_2)$, the denominator of this fraction is dominated by the AR_1 term. The factor A, which is in both the numerator and denominator, then cancels out and the gain is given by the expression

$$A_v = \frac{R_1 + R_2}{R_1} \tag{4.5b}$$

This shows that if A is very large, then the gain of the circuit is independent of the exact value of A and can be controlled by the choice of R_1 and R_2. This is one of the key features of Op-Amp design—the action of the circuit on signals depends only upon the external elements which can be easily varied by the designer and which do not depend upon the detailed character of the Op-Amp itself. Note that if $A = 100,000$ and $(R_1 + R_2)/R_1 = 10$, the price we have paid for this advantage is that we have used a device with a voltage gain of 100,000 to produce an amplifier with a gain of 10. In some sense, by using an Op-Amp we trade off "power" for "control."

A similar mathematical analysis can be made on any Op-Amp circuit, but this is cumbersome and there are some very useful shortcuts that involve application of the *two laws of Op-Amps* which we now present.

I. The first law states that in normal Op-Amp circuits we may assume that the voltage difference between the input terminals is zero, that is,

$$V^+ = V^-$$

II. The second law states that in normal Op-Amp circuits both of the input currents may be assumed to be zero:

$$I^+ = I^- = 0$$

The first law is due to the large value of the intrinsic gain A. For example, if the output of an Op-Amp is 1 V and $A = 100,000$, then $(V^+ - V^-) = 10^{-5}$ V. This is such a small number that it can often be ignored, and we set $V^+ = V^-$. The second law comes from the construction of the circuitry inside the Op-Amp which is such that almost no current flows into either of the two inputs. In Chapter 14 we discuss certain transistors that behave in this fashion. For now this property must be taken somewhat on faith.

We now revisit the previous example to show how these two laws can be used. Using law I, $V^- = V^+ = V_s$. Since $V^- = V_s$, the current I_1 through R_1 is equal to V_s/R_1. By law II of Op-Amps, this entire current must flow through resistor R_2. The voltage at the output is related to V_s by the Ohm's law expression $(V_o - V_s)/R_2 = I_2$, but $I_2 = I_1$, so

$$V_o = V_s + I_1 R_2 = V_s + \left(\frac{V_s}{R_1}\right) R_2$$

Factoring out V_s and dividing,

$$A_v = \frac{V_o}{V_s} = 1 + \left(\frac{R_2}{R_1}\right) = \frac{R_1 + R_2}{R_1}$$

as found before. Application of the two laws of Op-Amps has yielded the correct result without the complications of including the gain parameter A and subsequently taking its limit as $A \to \infty$. As shown in Fig. 4.15, we indicate an ideal Op-Amp by an ∞ sign inside the symbol.

The circuit analyzed above is called a noninverting amplifier since the output has the same sign as the input signal (A_v is positive). Consider now the Op-Amp circuit shown in Fig. 4.17. To analyze this example, we use the two Op-Amp laws. Since the positive input is grounded, $V^+ = 0$ and by the first law $V^- = 0$. The current I is thus V_s/R_1, and by the second law this current must also flow through resistor R_2 as shown. This current flows from ground potential ($V^- = 0$) toward V_o, so we have, using Ohm's law,

$$\frac{0 - V_o}{R_2} = I = \frac{V_s}{R_1}$$

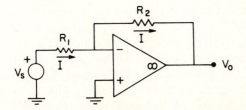

FIGURE 4.17.
The Op-Amp circuit for an inverting amplifier.

and

$$A_v = \frac{V_o}{V_s} = \frac{-R_2}{R_1}$$

This circuit is called an inverting amplifier since A_v is a negative number. In all of these circuits V_o is assumed to be referenced to ground.

The ideal Op-Amp seems at first to contradict Kirchhoff's current law at the output. For example, suppose, as shown in Fig. 4.18, we build an inverting amplifier and attach a load resistor to the output. We have noted earlier that the output resistance R_o of an Op-Amp is very small, so the full voltage V_o appears across the load resistor. In this circuit $A_v = -R_2/R_1 = -1$ so $V_o = -1$ V. Then, $I_L = 0.2$ mA while $I = 0.1$ mA. Both currents flow to the output node and must therefore flow into the Op-Amp. However, we have said that no current can flow into or out of the two input terminals. Where does the 0.3 mA go? The answer lies in the fact that we have neglected to show the connections which bring electrical power to the Op-Amp as well as the ground connection. The current that is supplied to the load comes from these voltage sources. Later in the text we discuss how to build the transistor amplifiers that are buried inside the Op-Amp and we shall see how the power supplies are used and why they are needed. For the applications in this chapter we use the ideal Op-Amp with three terminals and to assume that any current into or out of the output is supplied by "magic."

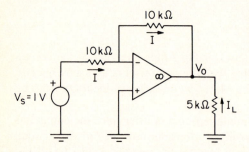

FIGURE 4.18.
Inverting amplifier with a load resistor attached.

A practical limitation on the use of the Op-Amp is set by the dc power supplies mentioned above which are needed to power the transistors inside. The output voltages cannot exceed the voltage of these power supplies. For example, if the power supplies used are 15 volts and the magnitude of the gain is 10, the magnitude of the source voltage V_s may not exceed 1.5 volt if we wish the output to be ten times the source. In practice, it would be better to limit the maximum magnitude of the source voltage to, say, 1.0 volt or to make the gain less than 10, since often the amplifier response becomes nonlinear near its limits.

Finally, we construct a three-parameter model patterned after model (a) of Fig. 4.12 for the noninverting and inverting amplifiers illustrated in Figs. 4.16 and 4.17. For the former, the source is attached directly to the positive input which draws no current; thus $R_i = \infty$. We have noted above that the Op-Amp has a very small output impedance, so $R_o = 0$. The model is shown in Fig. 4.19, where $R_i = \infty$, $R_o = 0$, and $\mu = (R_1 + R_2)/R_1$. In the diagram we have replaced V_2 with V_o to conform to the standard usage of Op-Amp circuits. The current I_1 is zero by the se-

FIGURE 4.19.
Three-parameter model for the non-inverting amplifier of Fig. (4.16).

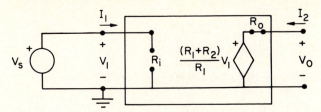

cond law of Op-Amps, but I_2 need not be zero. Once again we note that this current must be supplied from the voltage sources that power the device.

The inverting amplifier may be modeled as shown in Fig. 4.20. Unlike the noninverting amplifier, there is a finite input resistance R_i in this voltage amplifier. Since $V^- = 0$ V, the current I_s drawn from the source V_s is equal to V_s/R_1. The input resistor then equals $V_s/I_s = R_1$.

FIGURE 4.20.
Three-parameter model for the inverting amplifier of Fig. 4.17.

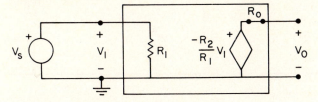

TEXT EXAMPLE 4.6 _____

Use the two laws of Op-Amps to find the voltage gain $A_v = V_o/V_s$ of the circuit diagrammed below.

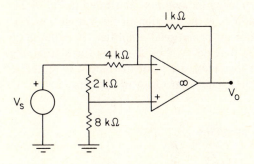

Solution: Since no current can flow into the positive input, it acts like an open circuit. The voltage V^+ is related to V_s by the voltage divider rule:

$$V^+ = \left(\frac{8 \text{ k}\Omega}{8 \text{ k}\Omega + 2 \text{ k}\Omega} \right) V_s = 0.8 V_s$$

By the first law of Op-Amps, $V^- = V^+ = 0.8V_s$ as well. The current through the 4-kΩ resistor is then

$$\frac{V_s - 0.8V_s}{4 \text{ k}\Omega} = \frac{0.2V_s}{4 \text{ k}\Omega}$$

This same current must flow from the negative input to the output, so we have

$$\frac{V^- - V_o}{1 \text{ k}\Omega} = I = \frac{0.2V_s}{4 \text{ k}\Omega}$$

But substituting $V^- = V^+ = 0.8V_s$ and multiplying both sides by $4 \text{ k}\Omega$ yields

$$4(0.8V_s - V_o) = 0.2V_s$$

$$V_o = \left(\frac{3}{4}\right) V_s$$

and

$$A_v = \frac{V_o}{V_s} = \frac{3}{4}$$

4.6 NEGATIVE FEEDBACK AND OP-AMP CIRCUITS*

In each operational amplifier circuit studied, there has been a connection from the output back to the negative input of the device. Circuits constructed in this manner are said to employ negative feedback. We can make an algebraic model that describes a negative feedback system using the notation illustrated in Fig. 4.21. In this model, A is the amplifier gain and F is the fraction of the output which is "fed back" to the input of the device. For an operational amplifier, A is the large positive number defined by Eq. 4.1. The circled symbol Σ stands for a summing point, and the minus sign indicates that the fraction FV_o of the output is routed to the negative input as has been the case in each of our Op-Amp circuits.

FIGURE 4.21.
Block diagram for a negative feedback amplifier.

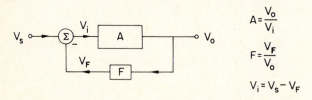

$$A = \frac{V_o}{V_i}$$

$$F = \frac{V_F}{V_o}$$

$$V_i = V_s - V_F$$

* Sections marked with an asterisk are not required for the logical flow of the text.

Because the summing operation subtracts the feedback voltage V_F from the signal voltage V_s, the output voltage is given by

$$V_o = AV_i = A(V_s - V_F)$$

But $V_F = FV_o$, so

$$V_o = A(V_s - FV_o) \tag{4.6}$$

Gathering terms,

$$V_o(1 + AF) = AV_s$$

and

$$V_o = \left(\frac{A}{1 + AF}\right) V_s$$

We define the amplifier gain with feedback as A_F, so

$$A_F = \frac{V_o}{V_s} = \frac{A}{1 + AF} \tag{4.7}$$

When $AF \gg 1$, the gain with feedback equals $1/F$, which is independent of A. Recollect that one of the reasons for designing the Op-Amp amplifier circuits was to yield properties that were independent of the detailed nature of the active device. In this case the exact value of A is immaterial, it just has to be large.

We apply this feedback model to the noninverting amplifier Op-Amp circuit discussed in the previous section. Referring to Fig. 4.22, since no current flows into the negative input, the fraction of the voltage V_o present at the negative input can be found from the voltage divider principle:

$$V^- = \frac{R_1}{R_1 + R_2} V_o$$

and the feedback parameter $F = V^-/V_o$ is given by

$$F = \frac{R_1}{R_1 + R_2}$$

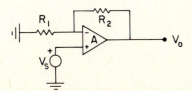

FIGURE 4.22.
Noninverting amplifier circuit with finite gain A.

From (4.7)

$$A_F = \cfrac{A}{1 + \cfrac{AR_1}{R_1 + R_2}}$$

which is equivalent to Eq. 4.5*a*. Again, in the limit that AF is very large,

$$A_F = \frac{R_1 + R_2}{R_1}$$

as in Eq. 4.5*b*. The noninverting amplifier Op-Amp circuit in Fig. 4.22 and Fig. 4.16*a* therefore employs the negative feedback principle and can be described by the block diagram in Fig. 4.23.

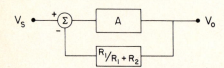

FIGURE 4.23.
Model in Fig. 4.21 applied to the circuit in Fig. 4.22.

As noted earlier, we may say that negative feedback sacrifices "power" for "control." The gain of the full circuit when AF is large, $(R_1 + R_2)/R_1$, is much less than the intrinsic gain A of the Op-Amp. However, the designer has full control of the gain since she or he is free to choose values of the resistors R_1 and R_2 from a wide range of possible values.

PRACTICE PROBLEMS
AND
ILLUSTRATIVE EXAMPLES

EXAMPLE 4.1

The circuit shown below includes a voltage-controlled current source. Find the voltage V_2 across the 2-kΩ resistor and the current I_2 through the 2-kΩ resistor if $g_m = 2 \times 10^{-3}$ mho and $V_s = 10$ V.

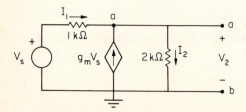

Solution: The voltage at node a equals V_2. If we apply KCL at this node we have

$$\frac{V_s - V_2}{1 \text{ k}\Omega} + g_m V_s = \frac{V_2}{2 \text{ k}\Omega}$$

Using given values for V_s and g_m and multiplying through by $2 \text{ k}\Omega$ yields

$$20 - 2V_2 + 40 = V_2$$

and finally,

$$V_2 = 20 \text{ V}$$

From Ohm's law,

$$I_2 = 10 \text{ mA}$$

Problem 4.1

Find V_2 and I_2 in the circuit above if the dependent source is a current-controlled current source having the value $I_s = \beta I_1$ where $\beta = 3$. I_1 is the current through the 1-kΩ resistor with the polarity shown above. The voltage V_s remains at 10 V and the resistors are unchanged.

Problem 4.2

Find the Thevenin and Norton equivalent circuits at the port $a - b$ for the circuit given in Example 4.1.

Problem 4.3

Use loop currents to find V_{ab} for the circuit given in Fig. 4.3 of the text. Show equations for i_1 and i_2 explicitly.

Problem 4.4

Write two loop current equations and solve for i_1, i_2, I_A, I_B, and I_C.

Ans.:

$$-100 + 10i_1 + 20i_2 + 10(i_1 - i_2) + 10i_1 = 0 \qquad I_A = i_1 = 3A$$
$$-10i_1 + 10(i_2 - i_1) + 10i_2 + 40 = 0 \qquad I_B = i_2 = 1 \text{ A}, I_C = 2 \text{ A}$$

Problem 4.5

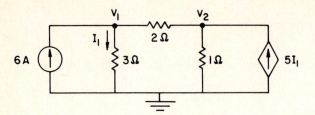

Write two independent equations from which V_1 and V_2 may be found. Solve the equations for V_1 and V_2 and determine I_1.

$$\text{Ans.: } 5V_1 - 3V_2 = 36$$
$$-13V_1 + 9V_2 = 0$$
$$V_1 = 54V; V_2 = 78V; I_1 = 18 \text{ A}$$

Problem 4.6

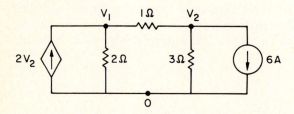

For the circuit above, find V_1 and V_2.

$$\text{Ans.: } V_1 = 18V; V_2 = 9 \text{ V}$$

Problem 4.7

Find the input resistance R_i "seen" by the 6-A current source in Problem 4.5.

$$\text{Ans.: } R_i = 9 \ \Omega$$

Problem 4.8

Set up and solve the loop current equations for the circuit below. Find i_1, i_2, and V_a.

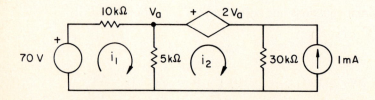

EXAMPLE 4.2

For the circuit below, use the node voltage method to find V_x and find the Thevenin and Norton equivalent circuits at the port a–b.

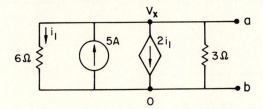

Solution: The node voltage method uses KCL at the point labeled V_x. There is a dependent current source equal to $2i_1$ in one of the legs, so we must first find i_1 in terms of V_x; that is, $i_1 = V_x/6$. The KCL equation is

$$5 = \frac{V_x}{6} + 2\frac{V_x}{6} + \frac{V_x}{3}$$

which has the solution $V_x = 6$ V. To determine the Thevenin and Norton equivalent circuits, we will find V_{oc} and I_{sc}. Now, $V_x = V_{oc}$, so we need only find I_{sc}. When the port a–b is short-circuited, the point labeled by V_x equals 0 volts. Thus, $i_1 = 0$. Writing KCL at node a,

$$5 = 0 + 0 + I_{sc}$$

and the resistance $R_T = R_N = 6/5 = 1.2\ \Omega$. The Thevenin and Norton equivalent pairs are then $(V_T, R_T) = (6\ \text{V}, 1.2\ \Omega)$ and $(I_N, R_N) = (5\text{A}, 1.2\ \Omega)$.

Problem 4.9

Find the Thevenin and Norton resistance in Example 4.2 using the R_o method.

$$\text{Ans.: } R_o = 1.2\ \Omega$$

Problem 4.10

Find the Thevenin and Norton equivalent circuits at the port a–b below.

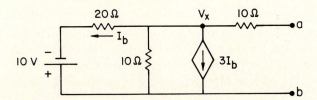

Problem 4.11

Find the Thevenin and Norton resistance in Problem 4.10 using the R_o method.

Problem 4.12

For the circuit shown in the figure below, find the Thevenin and Norton equivalent circuits.

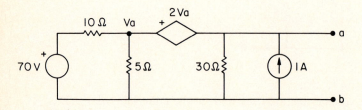

Ans.: $V_T, R_T = -30$ V, -3.75 Ω

$I_N, R_T = 8$ A, -3.75 Ω

EXAMPLE 4.3

Show that the circuit inside the box shown below is such that $V_2/V_s = (2/15)$ using circuit analysis methods learned in Chapter 2.

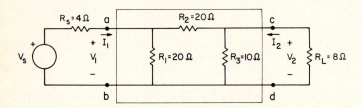

Then, show that the three-parameter model illustrated next yields the same relationship if $R_i = 11\Omega$, $\mu = 1/3$ and $R_o = 20/3\Omega$.

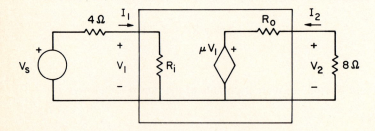

Solution: We can use the unit voltage method to find V_2/V_s in the original circuit. If $V_2 = 1$ V, then the current through the resistor R_3 equals 0.1 A. The current from left to right through R_2 is $0.1 + 0.125 = 0.225$ A, and hence the voltage at point a equals $1 + 20(0.225) = 5.5$ V. The current through resistor R_1 then equals 0.275 A, and hence $I_1 = 0.275 + 0.225 = 0.5$ A. Finally, this implies that $V_s = 5.5 + 4(0.5) = 7.5$ V and

$$\frac{V_2}{V_s} = \frac{1}{7.5} = \frac{2}{15}$$

We now use the model to make the same calculation. The circuit is

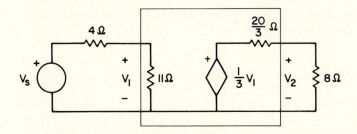

Using the voltage divider principle at the output,

$$V_2 = \left(\frac{R_L}{R_o + R_L}\right) \mu V_1$$

$$V_2 = \left(\frac{24}{20 + 24}\right) \frac{1}{3} V_1 = \frac{8}{44} V_1$$

The voltage divider rule at the input yields

$$V_1 = \left(\frac{R_i}{R_i + R_s}\right) V_s = \frac{11}{15} V_s$$

Substituting V_2 for V_1 using the expression above yields

$$V_2 = \frac{2}{15} V_s$$

and hence,

$$\frac{V_2}{V_s} = \frac{2}{15}$$

as found earlier.

Problem 4.13

Suppose that the 4 Ω resistor in the circuit of Example 4.3 is changed to 8 Ω. Find the voltage ratio V_o/V_s in the actual circuit and show that the same value results from the model. (This is a general result. The three-parameter model will give the correct result for any value of V_s and R_s.)

Problem 4.14

Again referring to Example 4.3, leave the 4-Ω resistor the same but change the load resistor R_L to 16 Ω. Find the voltage ratio V_o/V_s for the actual circuit. Now, show that the model does *not* yield the same result. (This illustrates the comment in the text that for a three-parameter model to be valid, the load resistor cannot be changed.)

Problem 4.15

For the two-port in the circuit of (a) below, suppose that the values in the equivalent three-parameter model shown in (b) are $(R_i, g_m, R_o) = (3\ \Omega, -0.25\ \text{mho}, 2\ \Omega)$.

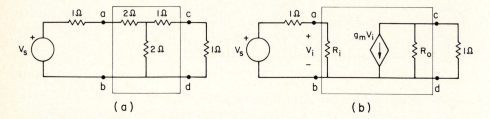

Find the ratio V_{cd}/V_s and show that it is the same for both circuits.

Problem 4.16

For a certain linear two-port, the three-parameter model is identical to part (d) of Fig. 4.12. The parameters are given by the set $(R_i, g_m, R_o) = (20\ \text{k}\Omega, 10^{-3}\ \text{mho}, 10\ \text{k}\Omega)$ when a 10-kΩ load resistor is placed across the output. If a source V_s in series with a resistance of 5 kΩ is connected to the input terminals and a resistance of 10 kΩ is connected across the output terminals, find V_2/V_s.

$$\text{Ans.:}\quad V_2/V_s = -4$$

Problem 4.17[*]

In the text we have discussed only three-parameter models thus far. In Chapter 14 we shall develop a four-parameter model for a bipolar junction transistor which looks like (b) below. This is called the h-parameter model.

As an exercise in using dependent sources, show that the model (b) yields the same output voltage V_2 as the actual circuit shown in (a) when a voltage source $V_s = 12$ V

---

[*] The problems with asterisks are not necessary for continuity.

with source resistance $R_s = (45/18)\Omega$ is connected to the input and a load resistance $R_L = 18\ \Omega$ is attached to the output. The particular case to be analyzed is shown below with either (a) or (b) inserted in the box.

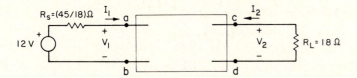

The model parameter set is given by $(h_i, h_r, h_f, h_o) = (2\ \Omega, (1/3), -(1/3), (1/6)$ mho).

Problem 4.18[*]

Show that the model (b) and the actual circuit (a) in Problem 4.17 yield the same output voltage V_2 if R_L is changed to 9 Ω while V_s and R_s are fixed. Note that this is quite different from the three-parameter model for which the load resistor cannot in general be changed without changing the model. In other words, a four-parameter model of this type has sufficient information to represent a two-port system (with no independent sources) without reference to the "outside world."

Problem 4.19[*]

Another type of four-parameter model is shown in part (b) below. Show that the model (b) yields the same output voltage as the actual circuit (a) when a voltage source $V_s = 10$ V with a source resistance $R_s = 2\ \Omega$ is connected to the input and a load resistance $R_L = 18\ \Omega$ is connected to the output as shown below.

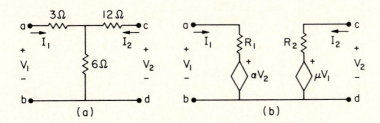

The four parameters are $(R_1, \alpha, R_2, \mu) = (7\ \Omega, 1/3, 14\ \Omega, 2/3)$.

Problem 4.20[*]

Show that the model in Problem 4.19 remains valid if the load resistor is changed to 12 Ω by evaluating V_2/V_s using the actual circuit and the model. The source V_s and source resistor R_s are unchanged.

Problem 4.21

In the voltage amplifier circuit below, $\mu = 100$ and R_i is 20 kΩ. What is the upper limit on R_o so that the system gain $AV = V_2/V_s$ will exceed 85?

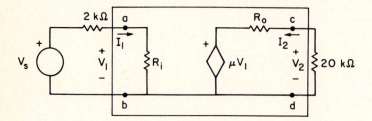

Problem 4.22

For the same voltage amplifier circuit as used in Problem 4.21, suppose $\mu = 100$ and $R_o = 1$ kΩ. What is the minimum value for R_i required if the voltage gain $A_V = V_2/V_s$ is to exceed 90?

EXAMPLE 4.4

The circuit shown below is typical of amplifier circuits using bipolar junction transistors. It employs a current-controlled current source which models the collector current I_C. The latter is proportional to the base current I_B, that is, $I_C = \beta I_B$. Some source V_s supplies the base current through the resistance R_s.

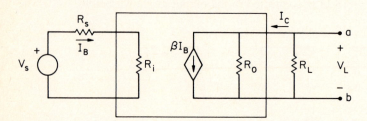

Find the voltage V_L in terms of V_s and the voltage gain $A_v = V_L/V_s$ if $R_s = 9.6$ kΩ, $R_i = 0.4$ kΩ, $\beta = 100$, $R_o = 12$ kΩ, and $R_L = 6$ kΩ.

Solution: The output voltage equals the current βI_B times the equivalent resistance 12 kΩ//6 k$\Omega = 4$ kΩ. The polarity is negative since the current flows from node b to node a. That is,

$$V_L = V_a - V_b = -(V_b - V_a) = -\beta(I_B)(4 \text{ k}\Omega)$$

We determine I_B from the input circuit

$$I_B = \frac{V_s}{R_s + R_i} = \frac{V_s}{10 \text{ k}\Omega}$$

Finally,

$$V_L = -(100)\left(\frac{V_s}{10 \text{ k}\Omega}\right)(4 \text{ k}\Omega)$$

and

$$A_v = \frac{V_L}{V_s} = -40$$

Problem 4.23

For the circuit in Example 4.4 above, find A_v if the transistor remains the same (that is, R_i, β and R_o remain the same), but $R_L = 12 \text{ k}\Omega$ and $R_s = 19.6 \text{ k}\Omega$.

Problem 4.24

By what factor is the voltage gain in Example 4.4 changed if the transistor β is reduced to 50 and R_s is reduced to 4.6 kΩ.

Problem 4.25

Suppose the voltage source V_s and the resistance R_s in the circuit of Example 4.4 are both replaced by a current source $I_s = 100 \ \mu\text{A}$. The rest of the circuit is the same. What is the output voltage?

Problem 4.26

If we consider the circuit in Example 4.4 as a current amplifier with gain

$$A_I = \frac{I_L}{I_B}$$

where I_L is the current through the load resistor R_L from node a to node b, find A_I.

Problem 4.27

Using the same definition of I_L as in the previous problem, find A_I in Problem 4.25, where

$$A_I = \frac{I_L}{I_S}$$

EXAMPLE 4.5

The circuit below includes the linear model of a transistor that we will treat later. The elements inside the "box" correspond to the transistor. The 10-kΩ and 4-kΩ resistors arise when the transistor is attached to power supplies, while the 5.1-kΩ and 1-kΩ resistors correspond to the source and load resistors, respectively. We wish to find the "output voltage" v_o in terms of the input voltage v_i for the circuit and to determine the amplifier gain A_v of the circuit.

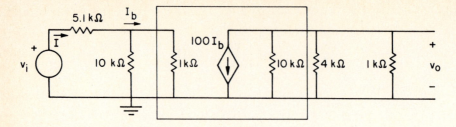

Solution: Using the current divider rule,

$$I_b = \frac{10}{11} I$$

and using the equivalent resistance $10 \text{ k}\Omega // 1 \text{ k}\Omega \approx 0.9 \text{ k}\Omega$

$$I = \frac{v_i}{5.1 \text{ k} + 0.9 \text{ k}} = \frac{v_i}{6 \text{ k}}$$

Thus,

$$I_b = \left(\frac{10}{11} \right) \left(\frac{v_i}{6 \text{ k}} \right)$$

(In the transistor this is known as the "base" current and controls the dependent source in the output circuit.) For the output circuit,

$$v_o = -100 I_b R_E$$

where

$$\frac{1}{R_E} = \frac{1}{10 \text{ k}\Omega} + \frac{1}{4 \text{ k}\Omega} + \frac{1}{1 \text{ k}\Omega}$$

and $R_E = 741 \ \Omega$. Substituting for I_b,

$$v_o = -100 \left(\frac{10}{11} \right) \left(\frac{v_i}{6 \text{ k}} \right) (741)$$

which reduces to

$$A_v = \frac{v_o}{v_i} = -11.2$$

Problem 4.28

The circuit below includes a linear model for a transistor. The circuit behaves as a current amplifier. Find an algebraic expression for the "current gain"

$$A_I = \frac{I_L}{I_s}$$

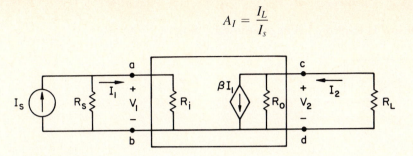

Problem 4.29

In the circuit below, which includes a linear model for a certain transistor, let $\beta = 80$, $R_s = R_o = R_L = 2 \text{ k}\Omega$ and $R_i = 1 \text{ k}\Omega$. Using both circuits, find V_o/I_s and I_L/I_s.

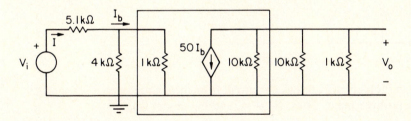

Problem 4.30

Find the voltage gain $A_v = V_o/V_i$ for the circuit below.

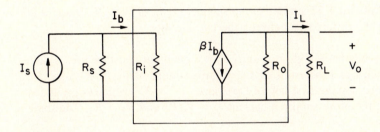

Problem 4.31

Find an algebraic expression for the gain, $A_V = V_o/V_s$, of the two-Op-Amp circuit shown below.

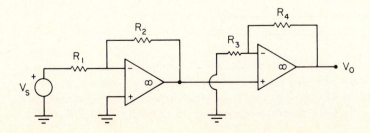

Problem 4.32

The circuit below is termed a "voltage follower." Find $A_v = V_o/V_s$.

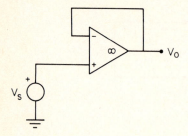

EXAMPLE 4.6

Construct three-parameter models for the circuit in Problem 4.31 if $R_3 R_4 = R_1 R_2$ and for circuit in Problem 4.32 similar to the models given in Figs. 4.19 and 4.20 in the text. Discuss the differences between the two circuits.

Solution: Both circuits have a gain of unity, but their input resistance R_i is quite different. In the circuit of Problem 4.31 the negative input of the first stage is at ground potential since, by the first law of Op-Amps, $V^- = V^+$ which is grounded. The source V_s must therefore supply a current $I_s = V_s/R_1$ and $R_i = R_1$. The three-parameter model is

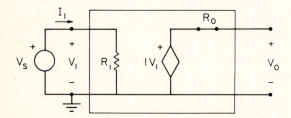

The circuit in Problem 4.32 draws no current from the source since it is connected directly to the positive input. Thus, $R_i = \infty$, and the model is

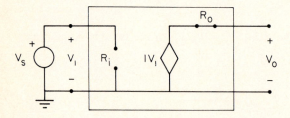

If V_s had some finite internal resistance R_s, the first amplifier circuit would not yield a total gain of unity since the amplifier would "load down" the source.

The circuit in Problem 4.32 is very important since it provides a voltage output equal to V_s but has an "infinite" input resistance and an output resistance R_o which is near zero. Such a circuit is sometimes called a "buffer amplifier," since it isolates the source and its possible finite resistance R_s from further electronics which must be driven by the output voltage V_o.

Problem 4.33

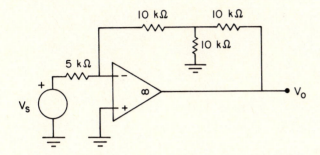

In the ideal Op-Amp circuit shown, find V_o/V_s.

$$\text{Ans.: } V_o/V_s = -6$$

Problem 4.34

Find the gain that would result in the circuit above if the 10-kΩ resistors are all replaced by 20-kΩ resistors.

Problem 4.35

In the ideal Op-Amp circuit shown, find V_o/V_s if $R_1 = R_3 = 1$ kΩ, $R_2 = 4$ kΩ, and $R_4 = 9$ kΩ.

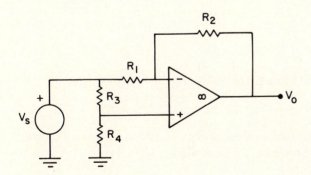

$$\text{Ans.: } V_o/V_s = 1/2$$

Problem 4.36

Repeat Problem 4.35 if the resistors are all 10 kΩ except $R_2 = 5$ kΩ.

Problem 4.37

In the ideal Op-Amp circuit shown, find V_o/V_s.

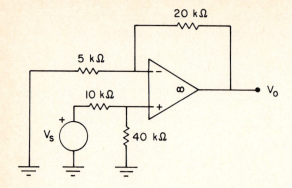

Ans.: $V_o/V_s = 4$

Problem 4.38

Repeat Problem 4.37 if the 20-kΩ resistor is changed to 45 kΩ.

Problem 4.39

For the ideal Op-Amp circuit shown, find V_o/V_s.

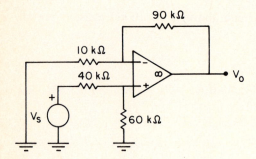

Problem 4.40

Repeat Problem 4.39 if the 60-kΩ resistor is replaced by a 40-kΩ resistor.

Problem 4.41

Construct three-parameter models such as those in Figs. 4.19 and 4.20 and Example 4.6 for the Op-Amp circuits in Problems 4.33 and 4.35.

Problem 4.42

Construct a three-parameter model for the Op-Amp circuit in Problem 4.37.

EXAMPLE 4.7

The Op-Amp circuit shown below is known as a "differential amplifier" because the output voltage is proportional to the difference between the signals V_2 and V_1. Find V_o in terms of V_1 and V_2 and show that this name makes sense.

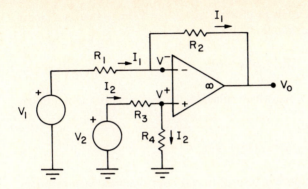

Solution: The circuit may be solved by superposition. We consider V_o to be the sum of a contribution V_o' due to the source V_1 and a contribution V_o'' due to the source V_2. To find V_o' we suppress V_2. Inspection then shows that $(V^+)' = 0$ and the resulting circuit is identical to an inverting amplifier

$$V_o' = -\frac{R_2}{R_1}V_1$$

If we now suppress V_1, the resulting amplifier is similar to a non-inverting amplifier. Then

$$V_o'' = \left(\frac{R_2 + R_1}{R_1}\right)(V^+)''$$

But in this case

$$(V^+)'' = \left(\frac{R_4}{R_3 + R_4}\right)V_2 \text{ and}$$

$$V_o'' = \left(\frac{R_2 + R_1}{R_1}\right)\left(\frac{R_4}{R_3 + R_4}\right)V_2$$

Finally,

$$V_o = \left(\frac{R_2 + R_1}{R_1}\right)\left(\frac{R_4}{R_3 + R_4}\right)V_2 - \frac{R_2}{R_1}V_1$$

This equation shows we may build a circuit with a response of the form

$$V_o = aV_2 - bV_1$$

For the special case $R_4/(R_3 + R_4) = R_2/(R_1 + R_2)$ (i.e., $R_4/R_3 = R_2/R_1$), we have

$$V_o = \left(\frac{R_2}{R_1}\right)(V_2 - V_1)$$

Thus, we may either use this circuit to produce an output voltage that is a linear combination of V_1 and V_2, for example,

$$V_o = aV_2 - bV_1$$

(where a and b are constants) or one which subtracts the two signals and multiplies the result by a constant,

$$V = a(V_2 - V_1)$$

(when b is equal to a). Such circuits can be used to carry out algebraic manipulations as parts of an analog computer.

Problem 4.43

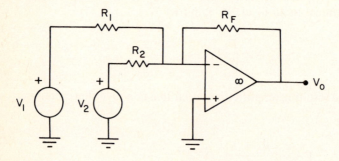

Use superposition to show that for the circuit above,

$$V_o = -\left[\left(\frac{R_F}{R_1}\right) V_1 + \left(\frac{R_F}{R_2}\right) V_2\right]$$

Note: In the design problems below, all resistors should be greater than or equal to 1 kΩ to keep the power low.

Problem 4.44

Design a circuit using only one Op-Amp which has a gain of $+1.5$.

Problem 4.45

Design a circuit using one or more Op-Amps that have a gain of $+0.5$.

Problem 4.46

Design a circuit using one Op-Amp that yields minus the average of two voltages V_1 and V_2, that is,

$$A_v = -\left(\frac{V_1 + V_2}{2}\right)$$

Problem 4.47

Design a circuit using one Op-Amp that yields the computation

$$V_o = 4(V_1 - V_2)$$

Problem 4.48

In the ideal Op-Amp circuit shown, find V_o in terms of V_a and V_b using superposition.

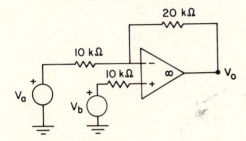

Ans.: $V_o = 3V_b - 2V_a$

Problem 4.49

Use superposition to find the output voltage V_o in terms of (V_1, V_2, V_3) and (R_1, R_2, R_F).

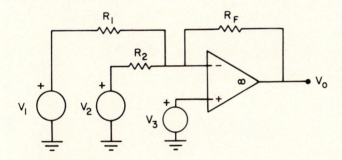

Problem 4.50

Design a circuit using only one Op-Amp that yields an output voltage $V_o = 4V_3 - V_1 - 2V_2$.

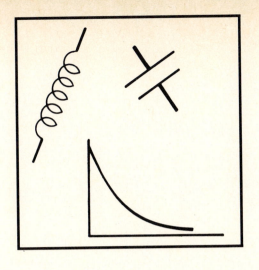

SECTION
B

CIRCUITS INCLUDING ENERGY STORAGE ELEMENTS

In this section we introduce two new circuit elements—the inductor and the capacitor—which are capable of storing electrical energy. This storage capability leads to current–voltage relationships that involve time dependence in a fundamental manner. Application of the circuit laws to networks including energy storage elements leads to a mathematical description involving differential equations. These equations are linear, ordinary differential equations with constant coefficients and have well-known solutions involving exponential functions. This mathematical result leads naturally to the introduction of the impedance concept, which is then used as a tool to solve for a circuit response without the need to derive the appropriate differential equation in every case. The natural response of first-order circuits is studied in some detail, as is the forced and complete response to dc sources. A cursory study of a second-order circuit shows that the formalism leads to solutions of the form $e^{j\omega t}$ where $j = \sqrt{-1}$. The meaning and implications of this result are explored in Section C.

Chapter 5
Inductors and Capacitors

Thus far we have dealt with systems in which the currents and voltages were either independent of time or in which the current and voltage displayed the same time dependence except for a constant multiplicative factor [e.g., $v(t) = R \times i(t)$]. We now introduce two other basic circuit elements: the inductor and the capacitor. The voltage/current relationships for these elements involve time derivatives or integrals, rather than the algebraic relation $V = IR$ for resistors. KVL and KCL still apply, but they result in differential equations rather than the algebraic equations we have solved in previous chapters.

5.1 INDUCTANCE AND INDUCTORS

The fundamental equations of electromagnetic field theory are called Maxwell's equations. In their full glory they demonstrate how electric and magnetic fields are related to each other and how these fields vary in space and time. These equations show that if there are no time variations at all, the electric field equations separate entirely from those of the magnetic field (B). That is, the electric field is determined entirely by electric charges and the magnetic field is determined entirely by electric currents. We now describe a situation in which E and B are interrelated when the associated voltages and currents vary in time. This discussion is not intended to be complete and the student is referred to more advanced courses on electromagnetic theory. Here our purpose is to motivate examination of the current-voltage relationship for an inductor.

If we have a steady current flowing around a loop of wire, then a magnetic field will exist as illustrated in Fig. 5.1. If the current is decreased to zero, the magnetic field will also decrease. However, the system is now time varying and Maxwell's equations show that the magnetic and electric fields are no longer unrelated. Experiments show that as the magnetic field dies out, an electric field and its associated voltage are generated which are proportional to the rate at which the magnetic field changes with time. Since the magnetic field is determined by the current and the voltage is related to the electric field, the fact that the electric field is proportional to the time derivative of the magnetic field implies that the voltage is also proportional to di/dt:

$$v \propto \frac{di}{dt} \qquad (5.1a)$$

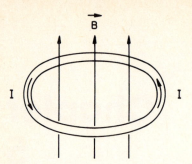

FIGURE 5.1.
A current around a loop of wire creates a magnetic field through it.

This relationship forms the basis for our analysis which, as usual, is entirely presented in terms of the circuit parameters i and v. Note that we use lowercase letters because the voltage and current are linked fundamentally through a time-dependent expression. The proportionality constant that converts the expression 5.1a into an equality is called the self-inductance L. Including this constant, the fundamental $I - V$ relationship for an inductor is,

$$v = L\left(\frac{di}{dt}\right) \tag{5.1b}$$

The self-inductance parameter L can vary by many orders of magnitude because it depends upon the geometry of the circuit element, the material from which it is constructed, the number of loops through which the magnetic field penetrates, and so on. This is, of course, analogous to resistor construction, and allows an enormous range in the values of such circuit elements. The symbol and the sign convention associated with inductors are illustrated in Fig. 5.2a, that is,

$$v_{ab} = L\left(\frac{di_{ab}}{dt}\right) \tag{5.1c}$$

FIGURE 5.2a.
Conventions for the current through and the voltage across an inductor.

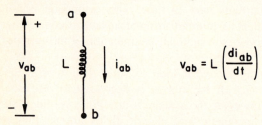

This equation for inductors is analogous to Ohm's law for resistors. The units of inductance are henries, and common circuits often employ millihenry (mH) or microhenry (μH) inductors. Inductors are sometimes referred to as "chokes." In order to use inductors in circuit analysis we need to investigate how they behave when more than one is present in a circuit. Consider the series configuration illustrated in Fig. 5.2b. In this series case the current is common to the two

FIGURE 5.2*b*.
Inductors configured in series.

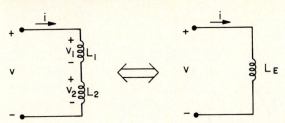

inductors, so the total voltage drop is equal to the sum $v = L_1 di/dt + L_2 di/dt = (L_1 + L_2)di/dt$. The equivalent inductance, L_1, is equal to the sum and we discover that inductors add like resistors in series:

$$L_E = L_1 + L_2 \qquad (5.2a)$$

A parallel configuration is shown in Fig. 5.2*c*. For the circuit on the right-hand side,

$$v = L_E \left(\frac{di}{dt} \right)$$

or

$$\frac{v}{L_E} = \frac{di}{dt}$$

By KCL, $i = i_1 + i_2$ and

$$\frac{v}{L_E} = \frac{di_1}{dt} + \frac{di_2}{dt}$$

Using $v_1 = L_1 di_1/dt$ and $v_2 = L_2 di_2/dt$

$$\frac{v}{L_E} = \frac{v_1}{L_1} + \frac{v_2}{L_2}$$

FIGURE 5.2*c*.
Inductors configured in parallel.

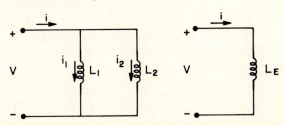

But, for a parallel circuit, $v = v_1 = v_2$. Factoring out yields,

$$\frac{1}{L_E} = \frac{1}{L_1} + \frac{1}{L_2}$$

which may also be written

$$L_E = \frac{L_1 L_2}{L_1 + L_2} \tag{5.2b}$$

These formulas are identical to those used for evaluation of parallel resistors. To summarize, inductors in a series or parallel configuration behave analogously to resistors.

Before proceeding, we need to be more precise in defining the terms "steady state" and "transient" in the electrical engineering context. Of course, dc circuits such as those studied in earlier chapters are in a "steady state" since the currents and voltages do not change in time. We also use the term steady state to describe currents and voltages that display repetitive temporal waveforms. Thus, systems in which the currents and voltages can be described by constant-amplitude, constant frequency sinusoidal functions are also considered to be in a steady state. The key here is the fact that neither the amplitude nor the frequency of the sinusoid changes in such a steady state circuit. The term "transient" refers to the behavior of the voltage or current when it is in transition between one steady state and another.

In a dc circuit, which has no time variation, an inductor behaves like a short circuit. This is clear from the defining Eq. 5.1c, since $v_{ab} = 0$ when the time derivative vanishes. In other words, for a dc current, I, $v_{ab} = 0$ no matter what the value of I. This is the definition of a short circuit. In the real world the loops of wire that make up an inductor will have a small resistance which sometimes has important consequences. For an ideal inductor, however, the resistance at dc is zero. Unless specifically stated otherwise, the student should assume that ideal inductors are intended in the remainder of the text.

Equation 5.1c also shows that the current through an inductor cannot change instantaneously since, if it did, the voltage across the element would be infinite. In mathematical terms, the current through an inductor must be a continuous function of time. This property of inductors is important since it will be used to determine "boundary conditions" when solving differential equations. As an example of this continuity principle, suppose a switch is opened at time $t = 0$, as shown in Fig. 5.3. The continuity principle for inductors requires that the current through the inductor at a time $t = 0^-$ (just before the switch is opened), equals the current at $t = 0^+$ (just after the switch is opened),

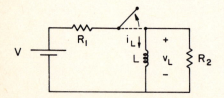

FIGURE 5.3.
Circuit illustrating the effect of opening a switch while an inductor is carrying current.

$$i_L(0^-) = i_L(0^+) \qquad\qquad (5.3)$$

To be specific, consider the above circuit with $V = 10$ V, $L = 1$ mH, $R_1 = 5$ Ω, and $R_2 = 5$ Ω. Before the switch is opened we assume it has been closed for a long time and that the system is in a steady state. All voltages and currents are dc, $di/dt = v_L = 0$ and the inductor acts as a short circuit. The current through the inductor is the same as that drawn by a short circuit and $i_L = 2$ A. The current through the inductor must be a continuous function of time, so the current through the inductor at $t = 0^+$ is also 2 A,

$$i_L(0^-) = 2 \text{ A} = i_L(0^+)$$

We shall discuss "what happens next" in Chapter 6 when the differential equation describing this circuit is derived and solved for all $t \geq 0$.

This example reveals what at first glance seems to be an unphysical result. The voltage across the inductor *does* change at $t = 0$. This may be seen as follows. For $t \leq 0$, the inductor behaves as a short circuit and $v_L(0^-) = 0$. At $t = 0^+$, by KCL, the inductor current must flow through the resistor and

$$v_L(0^+) = -i_L(0^+)R_2 = -10 \text{ V}$$

where the second term is the voltage drop across the resistor due to the series 2-A current. This paradox is explained as follows. The voltage does not change instantaneously but occurs in a time roughly equal to the length of the 5-Ω resistor divided by the speed of light. This is the time required for a voltage pulse (i.e., a voltage change) to travel across the resistor. Such short intervals are usually ignored in circuit theory, but they can be important in modern computers where "delay line" effects can seriously affect fast computations. Henceforth, in this text we shall consider changes of this type to be instantaneous.

Equation 5.2 can be integrated to yield the current through the inductor as a function of time:

$$i_{ab}(t) = \left(\frac{1}{L}\right)\int_0^t v_{ab}(t')dt' + i_{ab}(t = 0) \qquad\qquad (5.4)$$

In this equation t' is a dummy variable. The integral is carried out from 0 to t, and is therefore a function of t. Equations 5.1c and 5.4 are needed to apply KCL and KVL in problems involving inductors.

Finally, we come to the most important property of an inductor, its ability to store electrical energy and therefore act as a source of energy at a future time. The proof comes from the general result given earlier concerning the relationship between energy sources and energy receivers. When $di_L/dt > 0$ for an inductor, the associated voltage polarity is as illustrated in Fig. 5.4. The inductor in this case is a receiver of energy since the current flows from high to low potential (a resistor *always* satisfies this relationship and is always a receiver of energy). This received energy is stored in the magnetic field which fills the space between the coils of the inductor and may be released at a later time. When $di_L/dt < 0$ but i_L is still downward, current flows from low to high potential and the inductor acts as an

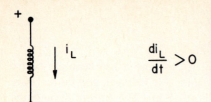

FIGURE 5.4.
Voltage polarity and current direction when $di_L/dt > 0$. The inductor is an energy receiver since the current flows from high to low potential.

$$\frac{di_L}{dt} > 0$$

energy source. This situation is illustrated in Fig. 5.5. Energy storage by inductors and capacitors is discussed in more detail in Section 5.3 Note that the circuit in Fig. 5.3 is such that for $t \geq 0$, current flows from low to high potential through the inductor (e.g., at $t = 0^+$, $v_L(0^+) = -10$ V). The inductor acts like a current source after the switch is thrown. The energy required to power this current source ultimately came from the battery which drove the current i_L for $t \leq 0$.

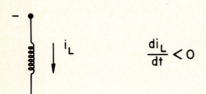

FIGURE 5.5.
Voltage polarity and current direction when $di_L/dt < 0$. The inductor is an energy source since current flows from low to high potential.

$$\frac{di_L}{dt} < 0$$

TEXT EXAMPLE 5.1

In the circuit below find the current $i(0^+)$, the instant after the switch closes. Determine whether the inductor is an energy source or energy receiver at $t = 0^+$. How much power is available from the inductor at that instant?

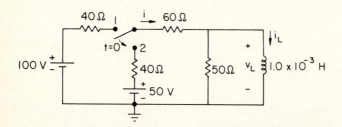

Solution: The fundamental principle we use here is that the current through the inductor must be a continuous function of time, $i_L(0^-) = i_L(0^+)$. To use this fact, we must first find the current at $t = 0^-$ from the appropriate circuit, which is

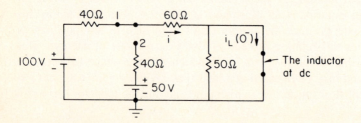

At dc, the inductor acts like a short circuit, so we have replaced it in the figure with a wire. The entire current i flows through the short circuit, so $i_L(0^-) = 100/100 = 1$ A. Now at the instant of time $t = 0^+$ the circuit looks as follows:

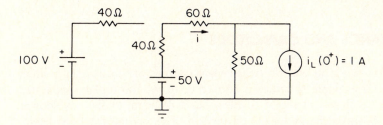

where we have replaced the inductor with a current source of value 1 A since that value current must flow at $t = 0^+$. We can find $i(0^+)$ by superposition. The current due to the 50-V battery is

$$i' = \left(\frac{50}{150}\right) = \frac{1}{3} A$$

The current i'' due to the current source is

$$i'' = \left(\frac{50}{150}\right) i_L(0^+) = \frac{1}{3}A$$

The total current $i(0^+) = 2/3$ A. Note that the current i is not continuous at $t = 0$! It need not be continuous since it does not flow through an inductor. To determine the power relationship we need to find the voltage v_L across the inductor at $t = 0^+$. It is given by the solution of Ohm's law,

$$\frac{50 - V_L}{100} = \frac{2}{3}$$

which yields

$$V_L = -16.7 \text{ V}$$

The current $i_L(0^+)$ is therefore flowing from negative to positive potential, and the inductor is acting as a source of energy. Just before the switch is thrown, the magnetic field inside the inductor due to the 1-A current constitutes a reservoir of stored energy. At $t = 0^+$ that energy is available to drive currents until such a time that the energy is dissipated by friction in the resistors. In the next chapter we shall see just how to determine the entire time history of the current by solving the differential equation which describes the system. We can determine the instantaneous power supplied by the inductor at $t = 0^+$ which is

$$v_L(0^+)i_L(0^+) = (16.7)(1) = 16.7 \text{ W}$$

As noted above, the power supplied is positive since current flows from low to high potential, in this case, from -16.7 V to 0 V through the inductor.

5.2 CAPACITANCE AND CAPACITORS

Energy can also be stored in electric fields when charges are located in a small volume. Suppose, for example, a battery is hooked up to two metal plates separated by some distance, as shown in Fig. 5.6. Electrons will begin to flow as shown, building up a net positive charge $(+Q_t)$ on the top plate (as the electrons leave) and a negative charge $(-Q_b)$ on the bottom plate as electrons enter. The electrons will continue to flow until the voltage across the plates equals the battery voltage V, when the current ceases. An electric field will build up inside the capacitor due to these charges. According to KCL, the current corresponding to the electron flow through the top and bottom wires must be the same. This in turn requires that the two sets of charges must be equal, that is, $Q_t = Q_b$. Since the charges are equal, we shall set $Q_t = Q_b = Q$. This result does not depend upon the geometry of the two pieces of metal or the material between them. The electric field between the plates is proportional to the charge Q. Since the voltage across the capacitor is also proportional to the electric field, it follows that the voltage is proportional to Q. The constant of proportionality is called the capacitance C, and

$$Q = Cv \tag{5.5}$$

The parameter C thus measures the ability or "capacity" of the system to store charge. The value of C is the amount of positive or negative charge measured in coulombs which may be stored per volt of potential difference across the capacitor.

The unit used for capacity is the farad (F) for which one coulomb of charge is held for every volt of potential. This is a very sizable amount of charge, requiring roughly 6×10^{18} electrons. More useful values of capacitance are measured in microfarads (μF) or picofarads (pF). A 1-picofarad capacitor (10^{-12} F) holds about 6×10^6 electrons per volt of potential.

If we take the time derivative of both sides of Eq. 5.5, we have

$$\frac{dQ}{dt} = C\frac{dv}{dt}$$

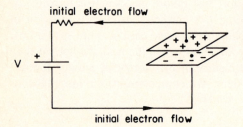

initial electron flow

initial electron flow

V

FIGURE 5.6.
Schematic diagram of the initial charge flow when a capacitor is hooked up directly to a battery.

However, dQ/dt has units of coulombs per second, and is equal to the current i associated with the charge flow in the wires leading to and from the capacitor. The equation analogous to the inductor Eq. 5.1c is then

$$i = C\frac{dv}{dt}$$

To be more precise, the sign convention is given by the relationship between current and voltage shown in Fig. 5.7, which yields

$$i_{ab} = C\frac{d(v_a - v_b)}{dt} = C\frac{dv_{ab}}{dt} \tag{5.6}$$

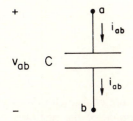

FIGURE 5.7.
Sign convention for the current and voltage relationship for a capacitor.

This result makes physical sense since as current flows onto the top of the capacitor (and off the bottom), the plus and minus charges build up with time, and hence v_{ab} increases with time. Note that since $Q_t = Q_b$, the current to the top plate must always equal the current away from the bottom plate, and we are justified in using the single symbol i_{ab} in Fig. 5.6. This also implies that the current is *effectively continuous* through the capacitor even though no charge moves across the space. As far as the external circuit is concerned, KCL is satisfied and we can consider the current from a to b to be equal to i_{ab}. We return to the question of current flow "through" a capacitor later. The integral of Eq. 5.6 yields the voltage across a capacitor at any time after $t = 0$ as

$$v_{ab}(t) = \frac{1}{C}\int_0^t i_{ab}(t')dt' + v_{ab}(0) \tag{5.7}$$

which is analogous to Eq. 5.4. For steady voltages the current through a capacity is zero, $dv_{ab}/dt = 0$, and the capacitor acts like an open circuit.

Capacitors do not add like resistors and inductors when hooked up in series and parallel. When in parallel the voltage across the capacitors is the same (Fig. 5.8), but the two capacitors store different amounts of charge, Q_1 and Q_2. The total charge stored is

$$Q_T = Q_1 + Q_2 = C_1V + C_2V = (C_1 + C_2)V$$

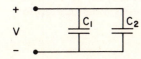

FIGURE 5.8.
Hooked up in parallel, the voltage across the two capacitors is identical.

Defining the equivalent capacity as $C_E = Q_T/V$, it follows that

$$C_E = C_1 + C_2 \qquad (5.8)$$

Capacitors in parallel add together like resistors or inductors in series.

Capacitors in the series configuration shown in Fig. 5.9 are such that the total voltage is given by

$$V_T = V_1 + V_2 = \frac{Q_1}{C_1} + \frac{Q_2}{C_2}$$

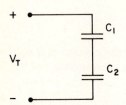

FIGURE 5.9.
Capacitors hooked up in series.

If we examine the central region of the figure (Fig. 5.10), no net charge can cross the open space between the two sets of plates, so the magnitude of the negative charge on the bottom of C_1 must equal the magnitude of the positive charge on the top of C_2,

$$Q_1 = Q_2 = Q$$

Thus,

$$V_T = \frac{Q}{C_1} + \frac{Q}{C_2}$$

or

$$V_T = \left(\frac{1}{C_1} + \frac{1}{C_2} \right) Q$$

and the equivalent capacity is defined as

$$C_E = \frac{Q}{V_T} = \left(\frac{1}{C_1} + \frac{1}{C_2} \right)^{-1}$$

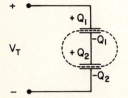

FIGURE 5.10.
Inside the dashed volume no charge has been created or destroyed, so the two charges $-Q_1$ and $+Q_2$ must add up to zero.

$-Q_1 + Q_2 = 0$
$\Rightarrow Q_1 = Q_2$

which can also be written

$$C_E = \frac{C_1 C_2}{C_1 + C_2} \tag{5.9}$$

This relation for series capacitors is identical in form to the formula for resistors and inductors in parallel. A voltage divider principle can now be derived for capacitors. Since $V_T = V_1 + V_2$,

$$V_1 = V_T - V_2 = \frac{Q}{C_E} - \frac{Q}{C_2}$$

$$V_1 = \frac{Q}{C_E}\left(1 - \frac{C_E}{C_2}\right)$$

But $Q/C_E = V_T$, and using Eq. 5.9 for C_E yields

$$V_1 = \left(\frac{C_2}{C_1 + C_2}\right) V_T$$

Note that this voltage divider principle for series capacitors has the same form as the current divider equation for resistors and inductors.

Since Eq. 5.6 is identical in form to Eq. 5.1c, arguments similar to those in Section 5.1 can now be used to show that the voltage across a capacitor cannot change instantaneously and that, at any time t_1,

$$v_C(t_1^-) = v_C(t_1^+)$$

This is the continuity principle for capacitors. Often we start or change some system at $t_1 = 0$, so

$$v_C(0^-) = v_C(0^+) \tag{5.10}$$

This result is important for problems involving differential equations. The result makes physical sense since the voltage cannot change until some current has time to flow onto or off the capacitor. Such a change cannot occur instantaneously and the voltage must be a continuous function of time.

As an example, consider the circuit shown in Fig. 5.11a, in which the switch opens at $t = 0$. The circuit has been established for a long time prior to $t = 0$, and, since all the voltages are steady, the capacitor acts like an open circuit. The voltage

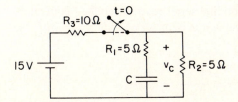

FIGURE 5.11a.
Circuit displaying time-dependent voltages and currents using a capacitor.

across the capacitor is therefore equal to the voltage across R_2. The voltage $v_C(0^-)$ is given by the voltage divider rule

$$v_C(0^-) = \left(\frac{5}{15}\right)(15) = 5 \text{ V} = v_C(0^+)$$

In the last step we have applied the requirement that the voltage across a capacitor must be a continuous function of time. For $t \geq 0$, the circuit looks like that shown in Fig. 5.11b, and the capacitor will "discharge" through the two 5-Ω resistors. Note that the current i_N through the resistor R_2 at $t = 0^+$ is

$$i_N(0^+) = \frac{v_C(0^+)}{10} = 0.5 \text{ A}$$

while the current at $t = 0^-$ was 1 A. Thus, the current is not a continuous function of time. We labeled the current i_N, since in Chapter 6 we refer to such a current as a natural current.

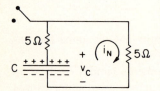

FIGURE 5.11b.
Circuit diagram corresponding to Fig. 5.11a after the switch is thrown.

TEXT EXAMPLE 5.2 _____

For the circuit shown below, find the equivalent capacitance C_E to the right of the port $a-b$; the values of the current i and voltage v' at $t = 0^+$, just after the switch closes; and the power supplied or received by the capacitor C_1 at $t = 0^+$. Point b is at ground potential.

Solution: The two capacitors in parallel add like series resistors so their equivalent is just $C_2 + C_3 = 6\mu F = C'$. This $6\mu F$ capacitor is in series with $C_1 = 3\mu F$, so the total is

$$C_T = C_1 C'/(C_1 + C') = \frac{3 \times 6}{3 + 6} = 2\mu F$$

When the switch is in position 1, the voltage at terminal 1 is equal to 12 V. This holds since at dc the equivalent capacitor C_E acts like an open circuit, therefore no current flows through the 5-kΩ resistor and there is no voltage drop across it. When the switch closes, the voltages across all the capacitors remain unchanged. The voltage at point 2 is then 12 V, and hence

$$i(0^+) = \frac{22 - 12}{10^4} = 1 \text{ mA}$$

The voltage v' also remains unchanged, but we need to find its value. From the voltage divider principle for capacitors derived above,

$$v'(0^+) = \left(\frac{C_1}{C_1 + C'}\right) 12 = \left(\frac{3}{9}\right) 12 = 4 \text{ V}$$

To determine the power associated with C_1 at $t = 0^+$, we also need to know the voltage across it, which is 8 V. The instantaneous power is $i(0^+)v_1(0^+) = 8$ mW. Since the current i flows from high to low potential "through" the capacitor C_1, the element is receiving power. In fact, a moment's thought shows that the current $i(t)$ will continue to flow until the voltage across the set of capacitors builds up to 22 V. Exactly how this occurs and how long a time is required is the subject of Chapter 6.

There need not be, and in fact there usually is not, air between the plates of a capacitor. Most capacitors have an insulating material inside which increases the ability of the device to store charge. The size and shape of the unit can also be varied. These two variable design features allow for a large range in capacity values. Most commonly used devices range from 10^{-12} F (1 picofarad) to 10^{-4} F (100 microfarad).

5.3 ENERGY STORAGE

The rate at which energy is received by a circuit element at any instant of time is the product of the voltage drop across the element and the current through the element in the direction of the voltage drop,

$$p_R = vi \tag{5.11a}$$

Applying this formula to our three circuit elements, we have

$$p_R = (iR)i = i^2 R \qquad \text{for a resistance} \tag{5.11b}$$

$$p_R = \left(L\frac{di}{dt}\right) i \qquad \text{for an inductance} \tag{5.11c}$$

$$p_R = v \left(C \frac{dv}{dt} \right) \qquad \text{for a capacitance} \qquad (5.11d)$$

Because the voltages and currents may be time dependent, we use lowercase letters here.

From equation 5.11b, it is clear that the power received by a resistor must always be positive. However, the power to an inductance or capacitance may be positive or negative, depending on whether the current (through an inductor) or the voltage (across a capacitor) is increasing or decreasing with time. When the product $p_R = vi$ is positive, such circuit elements receive and store electrical energy in the same way that a mass and spring system store mechanical energy when the spring is being compressed.

To find the total energy supplied to a circuit element between two times t_1 and t_2, we perform the following integral,

$$W_T = \int_{t_1}^{t_2} v(t)i(t)dt \qquad (5.12)$$

Consider first the resistor. Replacing $v(t)$ by $Ri(t)$ yields

$$W_T = R \int_{t_1}^{t_2} i^2(t)dt \qquad (5.13)$$

Since i^2 is always positive, the integrand is always positive and the electrical energy supplied to a resistor can only increase with time. Energy cannot be stored in a resistor, only dissipated. For an inductance,

$$W_T = L \int_{t_1}^{t_2} i\left(\frac{di}{dt}\right) dt = L \int_{i(t_1)}^{i(t_2)} i\, di = \left(\frac{1}{2}\right) L[i^2(t_2) - i^2(t_1)] \qquad (5.14)$$

where a *change of variable* has been made in the second step. Two conclusions can be drawn. The total energy supplied to an inductor in some time interval can be positive *or* negative depending upon the magnitude of the currents at t_1 and t_2. If $i^2(t_2) > i^2(t_1)$, then a net energy has been supplied *to* the inductor. For the opposite inequality, a net energy has been supplied *by* the inductor. The quantity $(\frac{1}{2}) L i^2(t)$ thus equals the total energy stored in the inductor at any time t, no matter what its past history. For a capacitor,

$$W_T = C \int_{t_1}^{t_2} v\left(\frac{dv}{dt}\right) dt = C \int_{v(t_1)}^{v(t_2)} v\, dv = \left(\frac{1}{2}\right) C[v^2(t_2) - v^2(t_1)] \qquad (5.15)$$

and the energy stored in a capacitor is $(\frac{1}{2}) Cv^2(t)$ at any time t. Ideal inductors and capacitors are passive elements and neither generate nor dissipate energy themselves. To store energy they must be "charged up" by some true voltage or current source. The energy is stored in the magnetic field which exists inside an inductor (threading the coils) and in the electric field which exists inside a capacitor (between the plates).

TEXT EXAMPLE 5.3 _____

Find the total energy stored in the inductor at $t = 0$ in the circuit illustrated in Fig. 5.3 as well as the power it supplies at the instant when the switch opens.

Solution: The total stored energy at $t = 0$ is equal to $(\frac{1}{2}) i^2(0^+)L = 2 \times 10^{-3}J = 2$ mJ. The power supplied by the inductor at that instant is $2 \times 10 = 20$ W.

_____

5.4 APPLICATIONS OF KIRCHHOFF'S LAWS

Kirchhoff's laws still hold for time-varying circuits although they now generate expressions that include derivatives and/or integrals. As an example, suppose we have the following circuit to which we wish to apply KVL and for which the source $v_s(t)$ is a known function of time (Fig. 5.12).

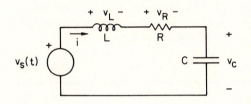

Figure 5.12.
A circuit described by a differential equation which is derivable using KVL.

By KVL,

$$-v_s(t) + v_L(t) + v_R(t) + v_C(t) = 0$$

Using Eq. 5.1c and Eq. 5.7,

$$-v_s(t) + L\left(\frac{di(t)}{dt}\right) + i(t)R + \frac{1}{C}\int_0^t i(t')dt' + v_C(0) = 0$$

This can be considered as a mixed integral/differential equation for $i(t)$, since $v_s(t)$ is a known function of time. We can remove the integral by taking the derivative of every term. Taking the derivative of an integral yields the function in the integral, while the derivative of a constant vanishes. These operations then yield the equation

$$L\frac{d^2i(t)}{dt^2} + R\frac{di(t)}{dt} + \left(\frac{1}{C}\right)i(t) = \frac{dv_s(t)}{dt}$$

This is a linear, second-order, ordinary, inhomogeneous differential equation with constant coefficients. In the next chapter we discuss what these terms mean as well as various methods of solving such equations.

As another example, we may use KCL at node a in Fig. 5.13 to derive an equation for $v_a(t)$. By KCL, currents i_L, i_C, and i_R must add up to the source $i_s(t)$. Using Eqs. 5.4 and 5.6 we have

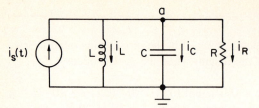

FIGURE 5.13.
A circuit described by a differential equation which is derivable using KCL.

$$i_s(t) = \frac{1}{L} \int_0^t v_a(t')dt' + i_L(0) + \frac{Cdv_a(t)}{dt} + \frac{v_a(t)}{R}$$

After taking the derivative of each term, we have the differential equation

$$\left(\frac{1}{L}\right) v_a(t) + C\frac{d^2v_a(t)}{dt^2} + \left(\frac{1}{R}\right)\frac{dv_a(t)}{dt} = \frac{di_s(t)}{dt}$$

Solutions to differential equations always involve application of boundary conditions. In this regard the continuity principles for inductors (Eq. 5.3) and capacitors (Eq. 5.10) are very important. The problems at the end of this chapter include a number of examples involving determination of initial conditions.

TEXT EXAMPLE 5.4 _____

For the circuit shown, write two coupled differential equations from which the two node voltages v_A and v_B may be found for $t > 0$. $I_L(0)$, $V_1(0)$, and $V_2(0)$ are initial values associated with the inductor and two capacitors. Do not attempt to solve the two equations.

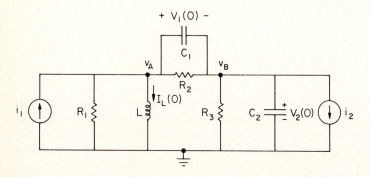

Solution: To use the node voltage method we apply KCL at the two nodes A and B. The result will be two equations in the two unknown voltages v_A and v_B. These will be functions of time since the energy storage elements all initially contain stored energy at the time $t = 0$. At node A, KCL yields

$$i_1 = \frac{v_A}{R_1} + \left[\left(\frac{1}{L}\right)\int_0^t v_A(t')dt' + I_L(0)\right] + \frac{v_A - v_B}{R_2} + C_1\left(\frac{d(v_A - v_B)}{dt}\right)$$

The term in the brackets is the current through the inductor including the initial current at $t = 0$. Applying KCL at node B yields

$$C_1 \frac{d(v_A - v_B)}{dt} + \frac{v_A - v_B}{R_2} = \frac{v_B}{R_3} + C_2 \frac{v_B}{dt} + i_2$$

The second equation is already a differential equation. If we wish to have two coupled differential equations to deal with, we must take the derivative of each term in the first equation which yields

$$\frac{di_1}{dt} = \left(\frac{1}{R_1}\right) \frac{dv_A}{dt} + \left(\frac{1}{L}\right) v_A(t) + \left(\frac{1}{R_2}\right) \frac{d(v_A - v_B)}{dt} + C_1 \frac{d^2(v_A - v_B)}{dt^2}$$

The result is two coupled linear ordinary inhomogeneous differential equations with constant coefficients for the voltages $v_A(t)$ and $v_B(t)$. One of these equations is second order since it includes a second derivative term. The other is of first order.

5.5 CAPACITORS AND INDUCTORS IN OP-AMP CIRCUITS

Inductors and capacitors can be used in operational amplifier circuits to perform the operations of integration and differentiation on signals which are functions of time. As an example, consider the circuit shown in Fig. 5.14. From the first law of Op-Amps $v^+ = v^- = 0$; hence, using KVL,

$$v_o = -v_C \tag{5.16a}$$

Also,

$$i(t) = \frac{v(t)}{R} \tag{5.16b}$$

By the second law this current must also flow through the capacitor (onto the left-

FIGURE 5.14.
Op-Amp circuit which can be used to integrate a signal *v(t)* over time.

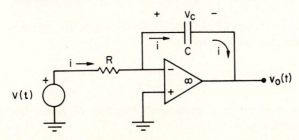

hand capacitor plate and from the right-hand capacitor plate to the output). Using Eq. 5.7,

$$v_C = \frac{1}{C} \int_0^t i(t')dt' + v_C(0)$$

Substituting Eqs. 5.16a and 5.16b into this expression yields

$$v_o(t) = -\frac{1}{RC} \int_0^t v(t')dt' - v_C(0) \tag{5.17}$$

Except for the constant $v_C(0)$, the output signal $v_o(t)$ is thus minus the integral of the input signal $v(t)$. In a similar fashion, the $I - V$ relationships for inductors and capacitors (Eqs. 5.1c, 5.4, 5.6, and 5.7) can be used in Op-Amp circuits to design differentiators as well as integrators.

More examples of this type are discussed in the problem set at the end of the chapter. Combinations of circuits of this general type, along with the pure resistive circuits presented in Chapter 4, can be used to make an "analog computer" which performs mathematical manipulations of a signal or of a combination of signals.

5.6 SUMMARY RESULTS FOR INDUCTORS AND CAPACITORS

The differential and integral $I - V$ relationships for an inductor are

$$v_{ab}(t) = L\frac{di_{ab}(t)}{dt}$$

$$i_{ab}(t) = \frac{1}{L} \int_0^t v_{ab}(t')dt' + i_{ab}(0)$$

and for a capacitor are

$$i_{ab}(t) = C\frac{dv_{ab}(t)}{dt}$$

$$v_{ab}(t) = \frac{1}{C} \int_0^t i_{ab}(t')dt' + v_{ab}(0)$$

For dc circuits, an inductor acts like a short circuit and a capacitor acts like an open circuit. The continuity relationship for an inductor is

$$i_L(t^-) = i_L(t^+)$$

that is, the current through an inductor is a continuous function of time. The analogous relationship for a capacitor is

$$v_C(t^-) = v_C(t^+)$$

The energy stored in an inductor carrying the current i is given by

$$W = \frac{1}{2}Li^2$$

while the energy stored in a capacitor is

$$W = \frac{1}{2}Cv^2$$

Finally, inductors and capacitors in series add as follows:

$$L_E = L_1 + L_2$$

$$C_E = \frac{C_1 C_2}{C_1 + C_2}$$

whereas the corresponding parallel relationships are

$$L_E = \frac{L_1 L_2}{L_1 + L_2}$$

$$C_E = C_1 + C_2$$

5.7   THE DISPLACEMENT CURRENT*

We now reexamine Kirchhoff's current law in the region between the capacitor plates since free electrons cannot pass across that region. We have already argued that $I_1 = I_2$ in the wires which are attached to the plates (Fig. 5.15). In the space *between* the plates, KCL is only satisfied if we include the "displacement current" defined by the expression

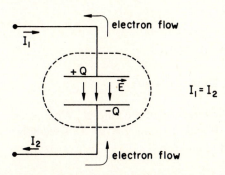

$I_1 = I_2$

FIGURE 5.15.
Illustrating that the current flow into a capacitor equals the current flow out of the capacitor.

* Sections marked with an asterisk are not required for the logical flow of the text.

$$I_D = \left(\frac{\epsilon_o dE}{dt}\right) A$$

where ϵ_o is a constant, E is the electric field inside the capacitor, and A is the cross-sectional area of the capacitor. Defined in this way a current flows even in a vacuum when the electric field changes in time. This concept plays a very important role in electromagnetic field theory, a topic well beyond the scope of the present text. To understand this idea a little better, consider what happens when the switch closes in the circuit shown in Fig. 5.16. The current $i(t)$ starts to flow onto the top of the capacitor and off the bottom. This builds up a positive charge on top and a negative charge below. While $i(t)$ is flowing, the electric field E inside the capacitor increases with time ($dE/dt > 0$) and the displacement current is downward. This current "completes" the circuit even with a vacuum between these two plates. Note also in Fig. 5.16b that the displacement current flows from high to low potential, so the capacitor is a receiver of energy in this case as it charges up. Returning to Fig. 5.11b, we see that when a capacitor is discharging, dE/dt is negative inside the capacitor. The displacement current then flows from low to high potential, and the capacitor is acting as an energy source.

FIGURE 5.16.
A circuit (a) illustrating displacement current flow for $t > 0$. In (b) the electric field direction inside the capacitor is shown in an expanded view.

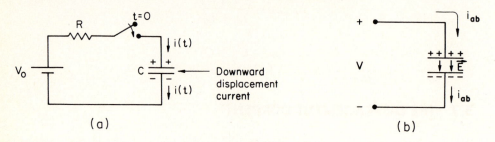

(a)　　　　　　　　　　　　　　(b)

PRACTICE PROBLEMS
AND
ILLUSTRATIVE EXAMPLES

Problem 5.1

The current through an inductance $L = 1.0 \times 10^{-3}$ H is

$$i(t) = 2 + 4\cos(10^5 t) \text{ mA}$$

Find the voltage drop across the inductor in the direction of the current. What is the power being received by the inductor at $t_1 = 0$ and at $t_2 = (\pi \times 10^{-5})/4$ seconds? How much energy is stored in the inductor at t_1 and t_2?

Ans.: $\quad v = -400 \sin(10^5 t)$ mV
$\qquad\quad P = 0$ at t_1

$$P = -1.35 \text{ mW at } t_2 \text{ (inductor is}$$
$$\text{acting as a source)}$$
$$W_T = 1.8 \times 10^{-8} \text{ J at } t_1$$
$$W_T = 1.17 \times 10^{-8} \text{ J at } t_2$$

Problem 5.2

The voltage across a capacitance $C = 1.0 \times 10^{-6}$ F is

$$v(t) = [100 + 2 \sin(10^7 t)] \text{ V}$$

Find the current through the capacitor in the direction of the voltage drop. What is the power, P, being supplied to the capacitor at $t_1 = 0$ and at $t_2 = (\pi \times 10^{-7})/2$ second? How much energy is stored in the capacitor at t_1 and t_2?

$$\text{Ans.:} \quad i = 20 \cos(10^7 t) \text{ A}$$
$$P = 2000 \text{ W at } t_1 \text{ (capacitor is re-}$$
$$\text{ceiving energy)}$$
$$P = 0 \text{ at } t_2$$
$$W_T = 0.5 \times 10^{-2} \text{ J at } t_1$$
$$W_T = 0.52 \times 10^{-2} \text{ J at } t_2$$

Problem 5.3

A 1-μF capacitor is initially uncharged. The current given in Problem 5.1 is applied at $t = 0$ and maintained. Find the voltage across the capacitor at the time $t_1 = \pi \times 10^{-5}$ second and $t_3 = 10\pi \times 10^{-5}$ second. Find the energy stored in the capacitor at t_1 and t_2. Is the capacitor acting as a source or a receiver of energy at t_1 and t_2? Find the power supplied or received at these two times. To what voltage will the capacitor charge as t goes to infinity? Is this realistic?

Problem 5.4

A 1-μH inductor initially carries no current. The voltage given in Problem 5.2 is applied at $t = 0$ and maintained. Find the current through the inductor at the time $t_1 = (\pi/2 \times 10^{-7})$ sec. Find the energy stored in the inductor as well as the power supplied or received then.

Problem 5.5

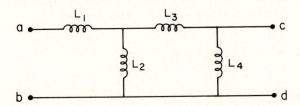

Find the equivalent inductance at the terminal $a-b$. Suppose the terminal $c-d$ is short-circuited. Find the new inductance.

$$\text{Ans.:} \quad L_E = L_1 + \frac{L_2(L_3 + L_4)}{L_2 + L_3 + L_4}$$

$$L_E = L_1 + \frac{L_2 L_3}{L_2 + L_3}$$

Problem 5.6

Find the equivalent inductance at the terminal $c - d$ in the circuit above. Suppose $a - b$ is then short-circuited. Find the new inductance at the terminal $c - d$. Evaluate the latter if all the inductors are 10 mH.

Problem 5.7

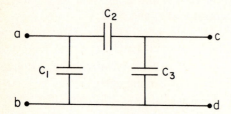

Find the equivalent capacitance at the terminal $a - b$. If $c - d$ is short-circuited, find the new equivalent capacitance.

EXAMPLE 5.1

A 100-V battery, V_T, is applied across the terminal $a-b$ in the circuit of Problem 5.7. Find the voltage V_{cd} and the charge on all the capacitors if $C_1 = 10\mu\text{F}$, $C_2 = 6\mu\text{F}$, and $C_3 = 12\mu\text{F}$.

Solution: Capacitors C_2 and C_3 are in series, so their equivalent capacity is

$$C_E = \frac{C_2 C_3}{C_2 + C_3} = 4\mu\text{F}$$

The total charge on this equivalent capacity is given by

$$Q_E = C_E V_T = 4 \times 10^{-4} \text{ C}$$

Now we have shown in the text that the charges on C_2 and C_3 must be equal to each other and therefore equal to Q_E. The voltage across C_3 must then be

$$V_3 = \frac{Q_3}{C_3} = \frac{Q_E}{C_3} = 33.3 \text{ V}$$

This result is similar to the current divider law since it can be written

$$V_3 = \frac{1}{C_3} Q_E = \frac{C_E}{C_3} V_T = \left(\frac{C_2}{C_2 + C_3} \right) V_T$$

Analogously,

$$V_2 = \left(\frac{C_3}{C_2 + C_3}\right) V_T = 66.7 \text{ V}$$

The respective charges are given by $Q = CV$ and are $Q_1 = 10^{-3}$ C, $Q_2 = 4 \times 10^{-4}$ C, and $Q_3 = 4 \times 10^{-4}$ C. Note $Q_2 = Q_3 = Q_E$, which checks with the argument above.

Problem 5.8

Find the equivalent capacitance at the terminal $c-d$ for the circuit in Problem 5.7 if $C_1 = C_2 = 10$ pF and $C_3 = 1$ pF. Find the charge on each capacitor if the voltage across $c-d$ is 10 V. For capacitor C_3, how many excess electrons are accumulated on one plate of C_3 and how many are missing on the other? (The charge on each electron is 1.6×10^{-19} C.)

Problem 5.9

Find the potential at the upper node and at the node between the 6μF and 2μF capacitors in the circuit below with respect to ground. What is the charge on each capacitor?

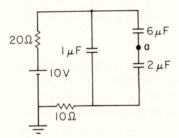

Problem 5.10

In the circuit for Problem 5.9 calculate the equivalent capacitance C_E of the combination 1μF in parallel with the series 6-μF and 2-μF pair. Show that the charge on C_E equals the sum of the charges on the 6-μF capacitor and the 1-μF capacitor. Show that the charge value also equals the sum of the charge on the 1-μF and the 2-μF capacitor.

Problem 5.11

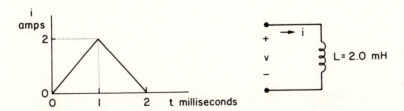

The current $i(t)$ through the inductor L has the time variation shown.

(a) Show that voltage $v(t)$ for $0 < t < 1$ msec. is constant and find its value.   Ans.: $v = 4$ V

(b) Repeat (a) for the time interval $1 \leq t \leq 2$ msec.   Ans.: $v = -4$ V

(c) Find the power being received by the inductor at $t_1 = 0.25$ msec and $t_2 = 1.5$ msec.   Ans.: at $t_1 = 0.25$ msec, $P = 2$ W, at $t_2 = 1.5$ msec, $P = -4$ W

(d) Find the total energy supplied to the inductor between 0 and 2 msec.   Ans.: 0 joules

EXAMPLE 5.2

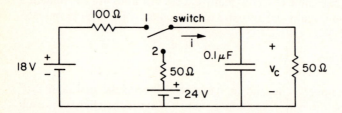

In the circuit above, the switch is moved from position 1 to position 2 at $t = 0$. If the switch had been in position 1 for a long time prior to $t = 0$, find i and v_C: (a) at $t = 0^-$ (just before the switch is moved); (b) at $t = 0^+$ (just after the switch is moved); and (c) a long time after the switch is moved.

Solution: (a) Since the switch has been in position 1 for a long time, we may assume that all the currents and voltages are constant, determined by the 18-V dc source. In Chapter 6 we shall show how to calculate how long "a long time" is for such a circuit. If v_C is constant the current through the capacitance is zero, since $i_C = C(dv_C/dt)$. In other words, a capacitor acts like an open circuit for dc voltages. The circuit for $t = 0^-$ is then

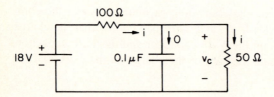

Solving,

$$i(0^-) = \left(\frac{18}{150}\right) = 0.12 \text{ A}$$

$$v_C(0^-) = 50i = 6 \text{ V}$$

(b) The voltage across a capacitance cannot change instantaneously, so

$$v_C(0^+) = v_C(0^-) = 6 \text{ V}$$

To find $i(0^+)$, the circuit is

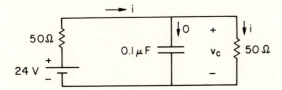

Solving, $i(0^+) = (24 - 6)/50 = 0.36$ A. We may also find $i_C(0^+)$, since

$$i_C(0^+) = i(0^+) - 0.12 = 0.24 \text{ A}$$

and using the relation

$$i_C = C\left|\frac{dv_C}{dt}\right|$$

We also find that

$$\frac{dv_C(0^+)}{dt} = \frac{i_C(0^+)}{C} = 2.4 \times 10^6 \text{ volts/sec}$$

Note that the currents i and i_C may change instantaneously, and that the voltage v_C will start to increase after $t = 0^+$ at an initial rate of 2.4×10^6 volts/sec.

(c) After a long time has passed, the currents and voltages will reach constant values determined by the 24-V dc source. (As we shall see in Chapter 6, a "long time" in this case will be a few hundred microseconds.) After the currents and voltages have become constant, the current through the capacitor will be zero and the circuit becomes

Solving,

$$i = \frac{24}{100} = 0.24 \text{ A}$$

$$v_C = 50i = 12 \text{ V}$$

Thus, the voltage across the capacitor will change with time from its value of 6 volts at $t = 0^+$ to 12 volts. We shall consider the transition of the currents and voltages from one value to another in the following chapters.

Problem 5.12

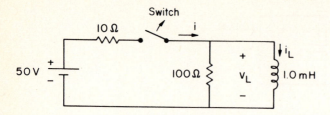

The switch in the circuit shown has been closed for a long time and is opened at $t = 0$. Find i_L, di_L/dt, and v_L for (a) $t = 0^-$ and (b) $t = 0^+$.

$$\text{Ans.: At } t = 0^-: i = i_L = 5 \text{ A}$$
$$di_L/dt = 0, \ v_L = 0$$
$$\text{At } t = 0^+: i = 0, \ i_L = 5 \text{ A}$$
$$v_L = -500 \text{ V}$$
$$di_L/dt = -500 \times 10^3 \text{ A/sec.}$$

Problem 5.13

Find the current i and the voltages v_1 and v_L in the circuit below if the circuit has been established "for a long time."

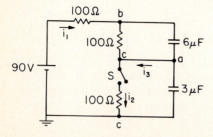

$$R_1 = 2\,\Omega \qquad L = 2 \times 10^{-3} \text{ H} \qquad C = 250 \times 10^{-6} \text{ F}$$
$$R_2 = 2\,\Omega \qquad R_3 = 3\,\Omega \qquad R_4 = 4\,\Omega$$

Suppose at $t = 0$, resistor R_4 suddenly breaks, forming an open circuit. Find the new values of i, v_1 and v_L at $t = 0^+$.

EXAMPLE 5.3

The circuit below has been established for a long time. The switch is closed at $t = 0$. Find the current $i_1(0^+)$, $i_2(0^+)$, and $i_3(0^+)$.

Solution: At $t = 0^-$ no current flows so the voltages at points *a* and *b* are both equal to 90 V. When the switch closes, by the continuity principle the capacitor voltages all remain the same for an instant. The current $i_2(0^+)$ must then equal $90/100 = 0.9$ A. Since $V_b(0^+) = 90$, no current flows to *b* from the battery, $i_1(0^+) = 0$, and also $i_{ab} = 0$. KCL at point *c* shows that $i_3(0^+) = 0.91$ A. The only currents that flow at $t = 0^+$ are around the single mesh at the lower right hand side of the circuit.

Problem 5.14

After a long time the switch in Example 5.3 is opened again at $t = t_1$. Find all three current at $t = t_1^+$.

Problem 5.15

The switch *S* has been in position 1 for a long time and is moved to position 2 at $t = 0$. For $t = 0^+$, find i_L, v_C, di_L/dt, and dv_C/dt.

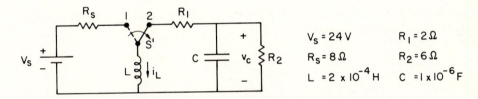

$$V_s = 24 \text{ V} \qquad R_1 = 2\,\Omega$$
$$R_s = 8\,\Omega \qquad R_2 = 6\,\Omega$$
$$L = 2 \times 10^{-4} \text{ H} \qquad C = 1 \times 10^{-6} \text{ F}$$

(Note that the switch symbol S' in this figure is somewhat naive. If the connection at 1 is broken and point 2 is not yet engaged, the system violates the requirement shown in Fig. 1.12 which disallows series connection of a current source [an inductor carrying current] and an open circuit. We require here that the switch is a "make before break" switch.)

Ans.: $i_L(0^+) = 3$ A, $v_C(0^+) = 0$ V

$$\frac{di_L(0^+)}{dt} = -3 \times 10^4 \text{ A/sec}$$

$$\frac{dv_C(0^+)}{dt} = -3 \times 10^6 \text{ V/sec}$$

Problem 5.16

The switch *S* has been in position 1 for a long time and is moved to position 2 at $t = 0$. At $t = 0^+$, find i_L, v_C, di_L/dt, and dv_C/dt.

$$V_s = 24 \text{ V} \qquad R_1 = 6\,\Omega$$
$$R_s = 8\,\Omega \qquad R_2 = 2\,\Omega$$
$$C = 3 \times 10^{-6} \text{ F} \qquad L = 2 \times 10^{-3} \text{ H}$$

EXAMPLE 5.4

Derive the differential equation for the node voltage v_a for $t \geq 0$ (where V_0 is the voltage across the capacitor at $t = 0$ and I_0 is the current through the inductor at $t = 0$).

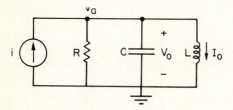

Solution: Using the node voltage method and KCL at node a, we can write

$$i(t) = i_R(t) + i_C(t) + i_L(t)$$

where the terms on the right-hand side are the currents associated with the resistor, capacitor, and inductor, respectively. Then,

$$i_R(t) = \frac{v_a(t)}{R}$$

$$i_C(t) = C\frac{dv_a(t)}{dt}$$

and

$$i_L(t) = \frac{1}{L}\int_0^t v_a(t')dt' + i_L(t = 0)$$

The node voltage equation can then be written

$$i(t) = \frac{v_a(t)}{R} + C\frac{dv_a(t)}{dt} + \left(\frac{1}{L}\right)\int_0^t v_a(t')dt' + I_0$$

To remove the integral, we may take the derivative of each term to yield

$$\frac{di(t)}{dt} = \left(\frac{1}{R}\right)\frac{dv_a(t)}{dt} + \frac{Cd^2v_a(t)}{dt^2} + \left(\frac{1}{L}\right)v_a(t)$$

Problem 5.17

In the circuit below with the switch open, write down the differential equation for the voltage $i(t)$ through the resistor R_2. Find the initial value of this current and of the voltage $v(0^+)$ across resistor R_2 just after the switch opens.

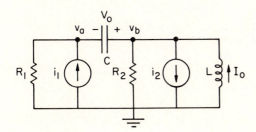

$R_1 = R_2 = 4\,\Omega$

$C = 1\,\mu F$

Ans.: $(1/C)i(t) + (R_1 + R_2)di(t)/dt = 0$

$i(0^+) = 0.5$ A

$v(0^+) = 2$ V

Problem 5.18

Derive the differential equation for the voltage $v_C(t)$ in the circuit of Example 5.2 for $t \geq 0$ with the switch is position 2.

Problem 5.19

Write two differential equations from which the node voltages v_a and v_b can be found for $t \geq 0$. The currents i_1 and i_2 are known functions of time and I_0 and V_0 are, respectively, the current through L and the voltage across C at $t = 0$.

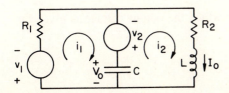

Ans.: $i_1 = v_a/R_1 + C[d(v_a - v_b)/dt]$

$di_2/dt = C[d^2(v_a - v_b)/dt^2] - (1/R_2)dv_b/dt - (1/L)v_b$

Problem 5.20

Derive the differential equation for the current $i_L(t)$ in Problem 5.12 with the switch closed.

Problem 5.21

Write two differential equations from which the loop currents i_1 and i_2 can be found for $t \geq 0$. The voltages v_1 and v_2 are known functions of time and I_0 and V_0 are, respectively, the current through L and the voltage across C at $t = 0$.

Problem 5.22

Derive two coupled differential equations for the currents i and $i_L(t)$ for $t \geq 0$ in Problem 5.13.

Problem 5.23

Derive the second-order differential equation for $v_c(t)$ for $t \geq 0$ in Problem 5.15.

Problem 5.24

With the switch in position 2 derive the second-order differential equation for $v_a(t)$ in Problem 5.16 where node a is the jundtion of R_1, L and R_2.

Problem 5.25

For the circuit shown, write two independent coupled second-order equations from which the two loop currents i_a and i_b may be found for $t \geq 0$. V_0 and I_0 are the voltage and current across the capacitor and the inductor at $t = 0$, respectively.

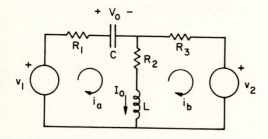

Problem 5.26

Find the output $v_0(t)$ of the circuit below in terms of the derivative of the input function $v(t)$.

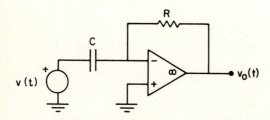

$$\text{Ans.: } v_0(t) = -RC(dv(t)/dt)$$

Problem 5.27

Design a circuit having several Op-Amps for which the output voltage is given by

$$v_0(t) = \frac{dv_1(t)}{dt} + 2\left(\frac{dv_2(t)}{dt}\right)$$

$v_1(t)$ and $v_2(t)$ are given functions of time.

Ans.: A solution (among many possible) is

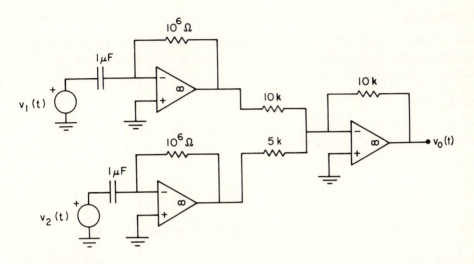

Problem 5.28

By analogy with the capacitor circuits discussed in the text and in the problems above, design ideal Op-Amp circuits using resistors and inductors to perform integration and differentiation of signals.

Ans.:

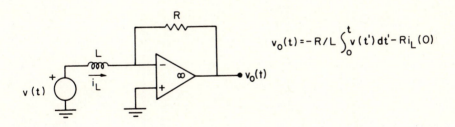

$$v_0(t) = -R/L \int_0^t v(t') \, dt' - R i_L(0)$$

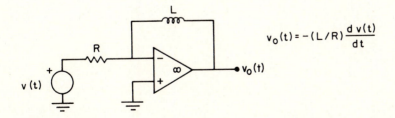

$$v_0(t) = -(L/R) \frac{d\,v(t)}{dt}$$

Problem 5.29

Design a circuit patterned after Problem 5.26 including specification of the source voltage $v(t)$ and the elements R and C for which the output voltage has the form

$$v_0(t) = -(1.0) \cos(\omega t)$$

where $\omega = 10^6$ rad/sec. Do not use any resistors less than 1 kΩ.

Problem 5.30

Let $v(t)$ be the ramp function $v(t) = 10(t/\tau)$ where $\tau = .001$ sec. This function linearly rises from 0 volt to 10 volts in 1 ms. If $R = 1$ kΩ in the differentiation circuit of Problem 5.26, find C such that $v_0 = -1$ V.

Problem 5.31

Use the integration circuit of Fig. 5.14 to design a circuit whose output voltage is the ramp function of Problem 5.30. That is, choose $v(t)$, R, and C and an initial condition, $i_L(0)$, such that

$$v_0(t) = 10t/\tau$$

with $\tau = 1$ msec. Do not use any resistor less than 1 kΩ in the design.

Problem 5.32

The voltage waveform in Problem 5.30 is applied to the circuit in Problem 5.26 with $R = 10$ kΩ and $C = 10^{-7}$ F. Sketch the output voltage as a function of time.

Chapter 6

Circuits with Energy Storage Elements

Inductance and capacitance are the properties of linear circuit elements which have the ability to store energy in magnetic and electric fields, respectively. The voltage–current relationships at the terminals of these elements are time dependent. The equations describing voltage and current in a circuit can still be derived using KVL and KCL, but they now become differential equations. These are, of course, more difficult to solve than the algebraic equations of Chapters 2 and 3. However, there has been considerable study of such equations and straightforward general solutions exist.

In this chapter we first consider the voltages and currents that may exist when no sources are acting, the so-called natural response. The exponential character of the solutions leads to the introduction of the important impedance function. We then study the combined effect of these natural voltages and currents and those forced to exist by sources. This is termed the complete response. At this point we are prepared only to analyze the forced and complete response for dc sources. We return to a more general study of complete response later in the text.

6.1 NATURAL RESPONSE

When a backyard swing or a pendulum is elevated and released, its "natural response" is to oscillate back and forth every few seconds (Fig. 6.1). The motion can be determined from Newton's laws subject to the initial (or boundary) condition that the swing was elevated and hence had some initial stored energy. A complete analysis would include the friction in the support which causes the system oscillation to slowly die out. This friction yields a resistance to the motion and eventually converts the free energy stored in the swing into heat at the pivot point. Resistors

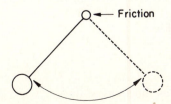

FIGURE 6.1.
The natural response of a pendulum is to oscillate at a frequency determined by its length.

FIGURE 6.2.

A hockey puck set in motion will eventually slow down and come to rest.

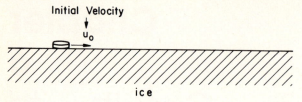

play an analogous role to friction in electrodynamic systems. The characteristic frequency with which the pendulum oscillates is called its natural frequency, and its motion is called the natural response.

Not all mechanical or electrical systems with stored energy oscillate. If a hockey puck is set in motion on a long stretch of ice (Fig. 6.2), it will eventually slow down and stop due to the effect of friction. If we plot the velocity of the hockey puck as a function of time, the curve will look something like the one shown in Fig. 6.3. The curve is one of a variety of exponential forms which we shall meet and which are solutions of first order differential equations arising in mechanical and electrical system analysis. The detailed motion depends upon the initial velocity, the frictional constant of the surface, and the mass of the hockey puck. The time period τ is important in describing this curve since it is the time required for the velocity of the puck to decrease by a factor of e, where e is the base of the natural logarithms. In fact, between any two times t and $t + \tau$, the curve decreases by a factor e. We call τ the time constant.

FIGURE 6.3.

Decay of the velocity with a time constant τ. At time τ the velocity equals U_o/e where $e \simeq 2.718$.

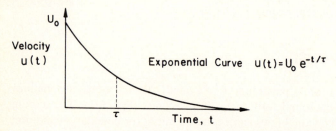

An example of an analogous electrical circuit is shown in Fig. 6.4. With the switch closed, the capacitor will be charged to the battery voltage V_o. When the switch is opened, the net negative charge on the bottom plate of the capacitor flows through the resistor to cancel the net positive charge at the top plate of the capacitor, yielding the conventional current, i, shown in the figure. This current flows until the capacitor is fully discharged and $v = 0$.

Kirchhoff's voltage law for the right-hand loop yields

$$-v + iR = 0$$

where the voltage and the current must be considered as functions of time,

FIGURE 6.4a, b.
Fully charged capacitor (a), and discharge current when the switch is opened (b).

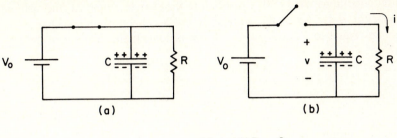

$$-v(t) + i(t)R = 0$$

Substituting Eq. 5.7 for $v(t)$, and being careful about the sign conventions, yields the equation

$$\frac{1}{C}\int_0^t i(t')dt' - v(0) + i(t)R = 0 \qquad (6.1)$$

This equation has only one variable, $i(t)$, but it is an integral rather than a differential equation. We can convert Eq. 6.1 to a differential equation by taking the derivative of each term, which yields

$$\frac{1}{C}i(t) + R\frac{di(t)}{dt} = 0 \qquad (6.2a)$$

or

$$\frac{di(t)}{dt} = \frac{-i(t)}{RC} \qquad (6.2b)$$

Equations of this same form occur in many areas of science and technology. One example arises in the study of radioactive decay in which the number n of radioactive atoms decreases at a rate that is proportional to the number of atoms present at any given time, that is,

$$\frac{dn}{dt} = -\alpha n$$

The solution to equations of this type is well known and is described by the exponential function. Direct substitution for the case of radioactivity shows that the solution for $n(t)$ is

$$n(t) = n_0 e^{-\alpha t}$$

For our case the solution to Eq. 6.2b is of the form

$$i(t) = I_1 e^{-t/RC} \qquad (6.2c)$$

where the constant I_1 must yet be found. This is characteristic of the solution to differential equations. The complete solution to a differential equation must include the initial conditions. At $t = 0^-$, just before the switch is opened, the voltage across the capacitor is V_0. Since the voltage across a capacitor cannot change instantly, it remains equal to V_0 at $t = 0^+$. At that instant, then, the current through the resistor is

$$i(0^+) = \frac{V_0}{R}$$

So from Eq. 6.2c, setting $t = 0$,

$$\frac{V_0}{R} = I_1 e^0 = I_1$$

and the full solution is

$$i(t) = (V_0/R)e^{-t/RC} \tag{6.2d}$$

The current in the above equation is a function of time and has the graphical form shown in Fig. 6.4c. When $t = \tau = RC$, the function $e^{-t/RC} = e^{-1}$ and

$$i(\tau) = \left(\frac{V_0}{R}\right)\left(\frac{1}{e}\right) \tag{6.3a}$$

The time required for the current to decrease by a factor $1/e$ from its initial value $i(0) = V_0/R$ is called the RC time constant,

$$\tau = RC \tag{6.3b}$$

Since R and C values can vary over many orders of magnitude, circuits can be easily constructed with characteristic times, τ, from less than picosecond to tens of minutes or more. Note that a current flows in this circuit even though no forcing term exists for $t > 0$. Such a current is called a natural current.

Since the voltage across a resistor is always given by Ohm's law, we may find the voltage across the capacitor in the circuit of Fig. 6.4 using Eq. 6.2d and $v(t) = i(t)R$

FIGURE 6.4c.
Current flow as a function of time after the switch is opened.

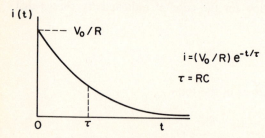

$$v(t) = V_o e^{-t/RC} \tag{6.4}$$

This result is general. The time dependences of the current and the voltage in an electrical circuit exhibiting natural response are identical. The two equations (6.2d) and (6.4) describe the discharge of a capacitor through a resistor and are both characterized by the time constant $\tau = RC$.

TEXT EXAMPLE 6.1 ───────────────────────────────────

For the circuit of Fig. 6.4 suppose $R = 6$ kΩ. Choose C in such a way that the time constant is 10 μsec. If a 12-kΩ resistor is placed in parallel with the 6-kΩ resistor, what will the new time constant be?

Solution: In order to have $\tau = 10\mu$ sec we may solve Eq. 6.3b for C

$$C = \frac{\tau}{R} = \frac{10^{-5}}{6 \times 10^3} = 1.67 \times 10^{-9} F$$

If a 12-kΩ resistor is placed in series with R, the equivalent resistance, R_E, is 4 kΩ. The capacitor will now discharge through a smaller resistance and

$$\tau' = R_E C = 6.7\mu\text{sec}$$

───────────────────────────────────

6.2   GENERAL SOLUTIONS TO THE DIFFERENTIAL EQUATIONS OF CIRCUIT THEORY

We now take a more formal approach to the solution of ordinary differential equations with constant coefficients, that is, equations of the form

$$a_o i + a_1 \frac{di}{dt} + a_2 \frac{d^2 i}{dt^2} + \ldots = f(t) \tag{6.5a}$$

where the a's are all constants. Equation 6.2b is of this form with $f(t) = 0$. When $f(t) = 0$ as in that case, the equation is termed homogeneous. The order of the equation is that of the highest derivative. Since there is only a first derivative in the differential equation that describes the R–C circuit in the example above, it is a first-order equation. It is easy to show that if two solutions $i_1(t)$ and $i_2(t)$ exist for Eq. 6.5a, then their sum $i_3(t) = i_1(t) + i_2(t)$ is also a solution. This shows that the differential equation is linear. In the vernacular of differential equations, then, if $f(t)$ is not equal to zero Eq. 6.5a is a linear, ordinary, first-order inhomogeneous differential equation with constant coefficients. The term ordinarily refers to the type of derivative in the equation. In circuit analysis, ordinary derivatives occur, as opposed to partial derivatives, which would lead to partial differential equations.

Mathematical analysis of equations of this type shows that the solution to a homogeneous equation may always be written in the form

$$i(t) = I_1 e^{s_1 t} + I_2 e^{s_2 t} + \ldots I_n e^{s_n t} \tag{6.5b}$$

where the I's and s's are constants and n is the order of the differential equation.

As an example for applying this general solution, consider the circuit in Fig. 6.5 in which the inductor carries a current at time $t = 0$ and we wish to determine the time history of the current $i(t)$ for $t \geq 0$. The switch S' is again a make before break switch.

FIGURE 6.5.
Example illustrating natural current flow in a series R–L circuit. The switch is thrown at $t = 0$.

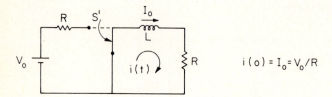

Using KVL around the loop,

$$L\frac{di}{dt} + Ri(t) = 0 \tag{6.6a}$$

which has the same form as Eq. 6.5a. Using the solution suggested by Eq. 6.5b, we assume

$$i(t) = I_1 e^{st}$$

and substitute this into Eq. 6.6a. After taking the derivative, the equation yields

$$LsI_1 e^{st} + RI_1 e^{st} = 0$$

This may be written

$$(sL + R)I_1 e^{st} = 0$$

or

$$(sL + R)i(t) = 0 \tag{6.6b}$$

Now, either $i(t) = 0$ for all time, which is not a very interesting result, or the quantity in the parentheses must vanish. For the latter case,

$$sL + R = 0$$

Solving this equation yields one value for s, namely,

$$s_1 = -\left(\frac{R}{L}\right)$$

and

$$i(t) = I_1 e^{-Rt/L}$$

The initial condition must still be used to find I_1. From the equation directly above, $i(0) = I_1 e^0 = I_1$. But $i(0) = V_0/2R$ in our example, and hence

$$i(t) = \frac{V_0}{2R} e^{-Rt/L}$$

This equation shows that the time constant for an R–L circuit is given by $\tau = L/R$. In the next section we begin to develop the tools which will allow us to solve for the time response of general electrical circuits without having to derive and solve the appropriate differential equation in each individual case.

6.3 THE IMPEDANCE FUNCTION

Equation 6.6b is rewritten here:

$$(R + sL)i(t) = 0 \tag{6.7a}$$

The quantity in the parentheses is very important in electrical engineering and is called the series "impedance function" $Z_s(s)$, for the circuit. With this definition the equation can be written

$$Z_s(s)i(t) = 0 \tag{6.7b}$$

where the subscript s stands for series. The two terms in the parentheses of Eq. 6.7a are the corresponding "impedances" of the resistor and the inductor,

$$Z_R = R \tag{6.8a}$$

$$Z_L = sL \tag{6.8b}$$

Note that setting $Z_s(s)$ equal to zero and solving the resulting algebraic equation for s yields the appropriate value of s for the natural response. That is,

$$Z_s(s) = R + sL = 0$$

has the solution

$$s_1 = -\left(\frac{R}{L}\right)$$

The impedance of a linear circuit element is defined as the ratio of the voltage drop across the element to the current through the element in the direction of the voltage drop, for voltages and currents that vary exponentially with time. Applying this definition we may determine the impedance of each of our circuit elements.

(a) For a resistor R, if $v(t) = V_1 e^{st}$, then $i(t) = v(t)/R = \frac{V}{R} e^{st}$. Thus, the impedance $Z_R = v(t)/i(t) = R$ as in Eq. 6.8a.

(b) If the current through an inductance L is $i(t) = I_0 e^{st}$, the voltage across the impedance is $v(t) = L(di/dt) = sLI_0 e^{st}$. Thus, $Z_L = v(t)/i(t) = sL$ as in Eq. 6.8b.

(c) For a capacitance C, if $v(t) = V_1 e^{st}$, then $i(t) = C(dv/dt) = sCV_1 e^{st}$. Thus, the impedance $Z_C = v(t)/i(t) = 1/sC$.

This yields the final elemental impedance form

$$Z_C = \frac{1}{sC} \tag{6.8c}$$

Out of context, the definition of the impedance function in terms of exponential variations may seem somewhat arbitrary. However, it stems from the fundamental facts that:

(a) Electric circuits containing R, C, and L elements and linear sources are described by linear, ordinary differential equations.

(b) The solution to a linear, ordinary, homogeneous differential equation is always given by the sum of exponential functions.

After substituting exponential solution forms into the differential equation, these differential equations can be manipulated in such a way that the impedance function can be factored out as a polynomial function of s. The advantage gained is that the differential equations that arise out of circuit theory analysis reduce to polynomial algebraic equations for the parameter s, which are, of course, much easier to solve. In Section 6.4 we apply the impedance function to solve circuit problems. Before doing so, some properties of the impedance are derived for future reference.

Suppose that two elements characterized by impedance values Z_1 and Z_2 are connected in the series configuration illustrated on the left-hand side of Fig. 6.6. By definition, $Z_1 = v_1/i_1$ and $Z_2 = v_2/i_2$. Also the equivalent impedance Z_T of the entire network must satisfy $Z_T = v/i$. By KVL,

$$v = v_1 + v_2 = i_1 Z_1 + i_2 Z_2$$

But in a series circuit $i_1 = i_2 = i$, so

$$v = i(Z_1 + Z_2)$$

FIGURE 6.6.
Series and parallel configuration of two impedances.

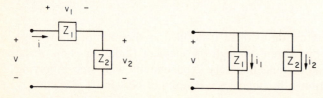

This is a general result. When various elements are connected together in series, the impedance values add like resistors:

$$Z_T = Z_1 + Z_2 \tag{6.9}$$

Note also that from Eq. 6.8a, the impedance must have the same units (ohms) as does resistance. Since Z is a constant for currents and voltages of the form e^{st}, we may use the impedances in exactly the same way we used resistances in the previous chapters. When elements are connected in parallel, as on the right-hand side in Fig. 6.6, KCL requires that

$$i = i_1 + i_2 = \frac{v_1}{Z_1} + \frac{v_2}{Z_2}$$

Since the voltage across the two elements must be the same, $v_1 = v_2 = v$. The total or equivalent impedance is related to i and v by $i = v/Z_T$, so

$$\frac{v}{Z_T} = v\left(\frac{1}{Z_1} + \frac{1}{Z_2}\right)$$

The equivalent impedance is therefore

$$Z_T = \frac{Z_1 Z_2}{Z_1 + Z_2} \tag{6.10}$$

which is identical in form to the result for resistors in parallel.

From the definition of the impedance, $Z = v/i$, we also have the generalized form of Ohm's law valid for currents and voltages which are exponential functions of time,

$$v = iZ \tag{6.11a}$$

or more precisely,

$$v_{ab} = i_{ab}Z \tag{6.11b}$$

where the subscripts have the same meaning as used in Chapter 2 for the current through a resistor and the voltage across the same resistor.

A dc (constant) source may be considered an exponential source with a value of $s = 0$, since $e^{0t} = 1$. For dc, Eqs. 6.8a, 6.8b, and 6.8c yield

$$Z_R = R \tag{6.12a}$$

$$Z_L = sL = 0 \tag{6.12b}$$

$$Z_C = \frac{1}{sC} = \infty \tag{6.12c}$$

These results for dc are consistent with the conclusions in Chapter 5 that an inductance is a short circuit for dc and a capacitance is an open circuit for dc.

TEXT EXAMPLE 6.2 _____

Find the impedance of the parallel $R–C$ circuit shown in Fig. 6.4b. In addition, for the resistor-inductor network on the right-hand side of Fig. 6.7a, consider R_2 and L to be in series with each other and in parallel with R_1. Then find the total equivalent impedance.

Solution: For the parallel $R–C$ network, we apply Eq. 6.10:

$$\frac{R\left(\dfrac{1}{sC}\right)}{R + \dfrac{1}{sC}}$$

This expression may also be written

$$Z_p(s) = \frac{\dfrac{1}{C}}{s + \dfrac{1}{RC}}$$

The latter form of the impedance function is the ratio of two polynomials. In general, an impedance function can always be expressed as the ratio of two polynomials. For the network in Fig. 6.7a, R_2 is in series with the inductor L. Using Eq. 6.9, this portion of the network has the equivalent impedance

$$Z_2(s) = R_2 + sL$$

Z_2 is in parallel with $Z_1 = R_1$, so the total parallel impedance is

$$Z_p(s) = \frac{R_1(R_2 + sL)}{R_1 + R_2 + sL}$$

or equivalently,

$$Z_p(s) = R_1 \frac{s + \dfrac{R_2}{L}}{s + \dfrac{R_1 + R_2}{L}}$$

Again in this last step the function has been written as the ratio of two polynomials.

_____

6.4   NATURAL RESPONSE USING THE IMPEDANCE FUNCTION

We now illustrate how the impedance function may be used to determine the natural response of an electrical system. Earlier we studied $R–C$ and $R–L$ circuits in which currents flowed without a source present in the final configuration. Another example

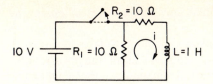

FIGURE 6.7a.
After being closed for a long time, the switch is opened at $t = 0$.

of a system in which currents can flow without a forcing element is given in Fig. 6.7a. Before the switch is thrown, the battery causes currents to flow through the inductor and the two resistors. After the switch is thrown, stored magnetic energy in the inductor forces current to flow through these three elements for a brief time even though no true sources exist. This current flow is what we call the natural response. Our goal here is to determine $i(t)$ for all times, $t \geq 0$. We first derive and solve the differential equations associated with this circuit "the hard way." If the switch has been closed for a long time, we may assume that all the voltages and currents are dc and the voltage across the inductor is zero. Then, just before the switch is opened, at $t = 0^-$, a 1-A current flows through the inductor. Since the current through an inductor cannot change instantaneously, $i(0^+)$ is also equal to 1 A. This result gives the initial condition required to solve the problem completely. The differential equation is found by applying KVL around the loop shown, so that,

$$R_1 i + R_2 i + L\frac{di}{dt} = 0$$

Assuming a solution of the form

$$i = I_1 e^{st}$$

yields

$$(R_1 + R_2 + sL)i_1 e^{st} = (R_1 + R_2 + sL)i(t) = 0$$

If a natural current exists then $i(t)$ is not equal to zero. So, the parentheses must equal zero and,

$$R_1 + R_2 + sL = 0 \qquad (6.13)$$

Solving for s and substituting the specific values of R_1, R_2, and L gives us

$$s_1 = -\frac{R_1 + R_2}{L} = -20 \text{ sec}^{-1}$$

Since from our initial analysis $i(0) = i_1 e^{0t} = i_1 = 1$ A, the final solution is

$$i(t) = 1.0 e^{-20t} \text{ A}$$

Physically, we can argue that the energy stored in the magnetic field of the inductor at $t = 0$ drives a "natural" current flow through the resistors. The current lasts for several hundred milliseconds after the switch opens and has an exponential form.

This is the natural response of the system without any true sources of energy to drive or force the system for $t \geq 0$.

We now solve the problem using the impedance method. Note that Eq. 6.13 corresponds to setting the series impedance function to zero and solving the algebraic equation for s. Without ever setting up the differential equation, we thus could have evaluated the series impedance to current flow through the elements, set it equal to zero,

$$Z_s(s) = R_1 + R_2 + sL = 0$$

and solved the first-order polynomial equation for s to give

$$s_1 = -\frac{R_1 + R_2}{L}$$

Then, knowing that the solution must be of the form $i(t) = I_1 e^{s_1 t}$, we can apply the initial condition to the function

$$i(t) = I_1 e^{-20t}$$

and obtain the same result as before for $i(t)$. Our first application of the impedance function is therefore that we can analyze the circuits which are described by differential equations without ever even deriving the equations! The reason this function is so useful depends upon the mathematical proof alluded to earlier that a function of the form given in Eq. 6.5b always yields solutions to a homogeneous, linear, ordinary differential equation with constant coefficients.

The fundamental generalized Ohm's law relationship between voltage, current, and impedance was given in Eq. 6.11a:

$$v = iZ$$

This equation can be used to give some insight into why setting $Z_s = 0$ yields the appropriate value or values of s in Eq. 6.5b. After the switch is thrown, the circuit of Fig. 6.7a carries the natural current $i_N(t)$, even though no source exists. To illustrate this, we can redraw the circuit, inserting a "phantom" voltage source of magnitude zero volts in series with R_1, R_2, and L (remembering that a zero voltage source is equivalent to a short circuit), as in Fig. 6.7b. Using the fundamental relationship, Eq. 6.11a, we can consider the current $i_N(t)$ to be related to the voltage source v_s by the expression,

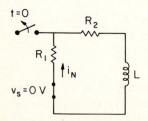

FIGURE 6.7b.
Circuit for determining the natural current $i_N(t)$ due to a phantom voltage source.

$$v_s = i_N(t)Z_s$$

where Z_s is the series impedance function

$$Z_s(s) = R_1 + R_2 + sL$$

This expression would hold if v_s was a true source varying exponentially in time. If $v_s = 0$, as in the present case, we have

$$0 = i_N(t)Z_s(s)$$

The only way this equation can be satisfied for $i_N(t) \neq 0$ is if

$$Z_s(s) = 0$$

In other words, a natural current can only exist in this circuit for the single value of the exponent s which satisfies the relationship

$$R_1 + R_2 + sL = 0$$

or

$$s_1 = -\frac{R_1 + R_2}{L}$$

Since we set $Z_s(s) = 0$, this method is called the impedance zero method.

A natural voltage will also exist across any element carrying a natural current, since $v_n = i_N Z_E$, where Z_E is the impedance of the element. As an alternative to the phantom voltage source, we could consider such a voltage as due to a "phantom" current source of value zero amperes in parallel with the element. For this same example we can draw the circuit in the equivalent configuration shown in Fig. 6.7c and can write

$$i_s = \frac{v_N}{Z_p(s)}$$

where Z_p is the impedance of R_1 in parallel with the series impedance $R_2 + sL$. The subscript p stands for the parallel impedance network, and we have from Text

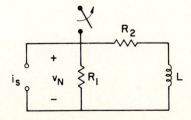

FIGURE 6.7c.
Circuit for determining the natural voltage $v_N(t)$ due to a phantom current source.

Example 6.2,

$$Z_p(s) = \frac{R_1(R_2 + sL)}{R_1 + R_2 + sL}$$

Since $i_s = 0$, we have

$$0 = \frac{v_N}{Z_p(s)} = v_N \frac{1}{Z_p(s)}$$

The only way this can hold for $v_N(t) \neq 0$ is if $Z_p(s) = \infty$ or equivalently if $1/Z_p(s) = 0$. This is called the impedance pole method, because the pole of a function occurs at a value of the variable which makes the function infinite. This will occur if the denominator of $Z_p(s)$ equals zero, so again we have the result

$$s_1 = -\frac{R_1 + R_2}{L}$$

The impedance zero and impedance pole methods must always give the same result, namely, that the natural currents and voltages must all display the same exponential time variation.

TEXT EXAMPLE 6.3 _____

Use both the impedance zero and pole methods to determine the characteristic parameter s after the switch opens in the circuit of Problem 5.17, a diagram of which is reproduced below.

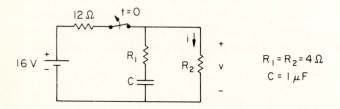

Solution: When the switch is opened, the capacitor C is charged and the natural current i will flow. That current flows through each of the elements C, R_1, and R_2. The series impedance function is

$$Z_s(s) = R_1 + R_2 + \frac{1}{sC}$$

Using the impedance zero method, we set this function equal to zero, yielding

$$s_1 = -\frac{1}{(R_1 + R_2)C}$$

This result means that once the switch is opened, the capacitor discharges through the two series resistors. To understand this method further, we can apply a phantom voltage source as the one shown on the left-hand side of the figure below.

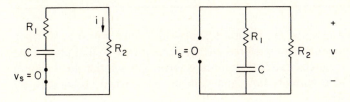

Then, from the generalized Ohm's law, we write

$$v_s = iZ_s = i\left(R_1 + R_2 + \frac{1}{sC}\right)$$

But since $v_s = 0$, it follows that if $i \neq 0$, then it must be the case that $Z_s = 0$. Substituting the actual values for the three elements,

$$s_1 = -1.25 \times 10^5 \text{sec}^{-1}$$

and the solution is of the form,

$$i(t) = I_0 e^{(-1.25 \times 10^5)t}$$

Using the impedance pole method to solve the same problem, we can view the circuit as shown on the right-hand side above. The parallel impedance across the network which remains after the switch is opened is

$$Z_p(s) = \frac{R_2(R_1 + 1/sC)}{R_2 + (R_1 + 1/sC)}$$

$$Z_p(s) = \frac{R_1 R_2 sC + R_2}{(R_1 + R_2)sC + 1}$$

Setting $Z_p(s) = \infty$ or equivalently $1/Z_p(s) = 0$ yields

$$s_1 = -\frac{1}{(R_1 + R_2)C}$$

as found above. To understand the pole zero method we can view the natural voltage as being due to a phantom circuit source as shown in the figure above. Then, from Ohm's law,

$$i_s = \frac{v}{Z_p}$$

Since $i_s = 0$, if a natural voltage exists ($v \neq 0$) it must follow that $1/Z_p = 0$. This argument corresponds to setting $1/Z_p(s) = 0$ or $Z_p(s) = \infty$ which is the pole method used above.

We are now prepared to solve the natural response problems which arise when inductors and/or capacitors containing stored energy remain in a system after some time t, but in which no sources exist. In summary form, the solution to a natural response problem in which currents flow and voltages exist when no sources are present involves three steps:

1. Determine the value or values of the exponential parameter s, which satisfies the relationship

$$Z_s(s) = 0$$

   where Z_s is the series impedance to current flow through any element which carries a natural current in the circuit. An equivalent formulation involves finding the parallel impedance and determining the value or values of s, which yield the poles of $Z_p(s)$,

$$Z_p(s) = \infty$$

   or, equivalently,

$$\frac{1}{Z_p(s)} = 0$$

   Using the pole method, we include any element across which there is a natural voltage.

2. Construct the desired current (or voltage) function in the form

$$i_N(t) = A_1 e^{s_1 t} + \cdots A_j e^{s_j t}$$

3. Determine the A_j by applying j boundary conditions. The number j corresponds to the order of the differential equation, which will be the same as the order of the polynomial in s found in step 1 above.

TEXT EXAMPLE 6.4 _____

Find the natural response solution for the voltage $v_N(t)$ across resistor R_2 for $t \geq 0$ in the circuit illustrated in Problem 5.17 and used again just above in Text Example 6.3.

Solution: We have already performed some of the necessary tasks. The characteristic value of the parameter s we found in Text Example 6.3 to be

$$s_1 = -1.25 \times 10^{+5} \text{sec}^{-1}$$

so the natural voltage must be of the form

$$v_N(t) = V_0 e^{(-1.25 \times 10^5)t}$$

This completes steps one and two above. To carry out step 3, we need to determine V_0. Setting $t = 0$ yields

$$v_N(0) = V_0 e^0 = V_0$$

so we need only find the voltage across R_2 at $t = 0$. Just before the switch is opened, the voltage $v(0^-) = 4$ V, since there is a voltage divider formed by R_2 and the 12-Ω resistor. It is tempting to stop here and claim $v_N(0^+) = 4$ V as well, but this is not the case. Only the voltage across the capacitor is a continuous function of time. That voltage is indeed equal to 4 V since, prior to $t = 0$, no current flows through R_1. However, almost instantaneously the equal resistors R_1 and R_2 divide the capacitor voltage such that $v(0^+) = 2$ V as given in the answers to Problem 5.17. The full solution is

$$v_N(t) = 2e^{(-1.25 \times 10^5)t} \text{ volts}$$

6.5 COMPLETE RESPONSE OF FIRST-ORDER SYSTEMS WITH DC SOURCES

After a system consisting of dc sources and circuit elements has reached an equilibrium, or "steady state," condition, energy may be stored in the inductances and capacitances of the system. When a change is made in a system, there will be a transient state during which the system reaches a new equilibrium. In Section 6.3 we studied the situation that arises when the second state (after the switch is thrown) has no true energy sources. Then, the only currents that flow for $t \geq 0$ are natural currents. In terms of the differential equation describing the system, the term $f(t)$ on the right-hand side of Eq. 6.5a is equal to zero and the equation is homogeneous. If energy sources remain in the circuit after the switch is thrown, or if new sources are present, then the currents that flow for $t \geq 0$ are a combination of the natural response and the forced response due to the sources which remain in the circuit. The differential equation describing such systems corresponds to a nonvanishing function $f(t)$ on the right-hand side of Eq. 6.5a. The analysis technique described below then yields the complete response, which includes both the natural and the forced response.

As an example of complete response, suppose we wish to know the current through the inductor as a function of time if both of the switches in Fig. 6.8a are thrown at $t = 0$ and if the switches have been in their initial position for a long time. By a long time we mean long enough that any natural currents associated with the setup of the initial circuit have died out. The system is then in a steady state at $t = 0$, one in which the currents have steady values determined by the sources. KCL states that for $t > 0$,

$$i_R(t) + i_L(t) = I_0$$

FIGURE 6.8a.
An electrical circuit that displays both a natural response and a forced response after the switches are thrown. S_2' is a make before break switch.

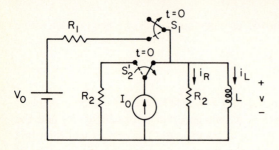

so,

$$\frac{v(t)}{R_2} + i_L(t) = I_0$$

But from Eq. 5.1, the voltage across the inductor is equal to $L\,di_L/dt$, and we can write a single differential equation for $i_L(t)$:

$$\frac{L}{R_2}\frac{di_L}{dt} + i_L(t) = I_0$$

This differential equation is not homogeneous because of the constant term on the right-hand side which corresponds to the source. We need to extend our methods to include such sources.

Since the differential equation is linear, we may use the principle of superposition. If we assume the current on the right-hand side is equal to $I_0 + 0$, we can find the solution for $i_L(t)$ as the sum of two solutions: one for a current source equal to I_0, and the other for a current source equal to zero. In mathematics, the solution using I_0 is called the "particular solution." We will call this solution the "forced response" since it arises from the forcing source. With the right-hand side equal to zero, we are back to the homogeneous equation and the solution to this case is called the "natural" response. The total solution is then the sum of the forced and natural responses. In mathematics the natural response is termed the "characteristic solution" since it is characteristic of the homogeneous differential equation and is independent of the particular forcing function. Note that setting the source term equal to zero corresponds to a phantom current source of value zero.

When these two solutions are found, the initial condition must then be applied to their sum to yield the total solution, which we term the complete response. In this chapter we can only deal with dc sources and therefore limit our analysis to complete response problems involving such sources. Later, we shall include sources and signals with arbitrary time dependence.

The forced response to dc sources is found by applying the principles of Chapters 1 through 3 and the zero frequency (dc) response of all the capacitors and inductors in the final configuration. To do so, we replace all inductors by short circuits and all capacitors by open circuits and solve the resulting resistive circuit. In this example, the forced current through the inductor due to the source will be equal to I_0 since the short circuit corresponding to the inductor will carry the entire source current. The forced response of the system is just $i(t) = I_0$, which the student may show indeed satisfies the inhomogeneous differential equation derived above.

To find the natural response, it is very important to note that we first must suppress any true source or sources which exist in the final configuration. This sets the forcing term on the right-hand side of the differential equation equal to zero. Then, we find at what value of s exponential currents and voltages may still exist in the circuit, in the same way as discussed in Section 6.3. For the present example we can determine the natural voltage v_n as illustrated in Fig. 6.8b with the current source suppressed. To accomplish this, we determine what value of s yields a solution to the equation

$$i = 0 = \frac{v_n}{Z_p(s)} \tag{6.14}$$

where $Z_p(s)$ is the equivalent impedance across which v_n is measured. The zero on the left-hand side corresponds to a phantom current source. In this example, $Z_p(s)$ is the parallel combination of R_2 and L, so

$$Z_p(s) = \frac{R_2 sL}{R_2 + sL} \tag{6.15}$$

As before, the only way v_n can be nonzero in Eq. 6.14 is if the quantity $1/Z_p(s)$ is zero (or equivalently if $Z_p(s)$ is infinite). We seek a value for s such that the ratio of Eq. 6.14 "blows up." This occurs when the denominator of $Z_p(s)$ goes to zero, so

$$R_2 + sL = 0$$

or

$$s_1 = \frac{-R_2}{L}$$

This is the only value of s for which a natural voltage can exist in the circuit when it is not forced by a source.

The natural voltage across the R–L circuit therefore varies as $e^{-R_2 t/L}$. There will also be an associated natural current flow with temporal form $i_{Ln} = I_{1n}e^{-R_2 t/L}$. As $t \to \infty$, this natural current flow will decay to zero, leaving only the forced response.

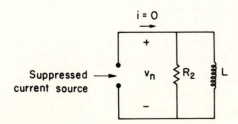

FIGURE 6.8b.
Circuit for analysis of the natural response of the circuit after the switches have been thrown.

The forced response we already showed to be $I = I_0$. The solution of the differential equation is the sum of the natural and forced response:

$$i_L(t) = I_{1n}e^{-R_2t/L} + I_0$$

We still must use the initial condition at $t = 0$ to determine the value of I_{1n}. To accomplish this, we study the original circuit just before the switch is thrown at time $t = 0^-$. Since the system is assumed to be in a steady state, $i_L(0^-) = V_0/R_1$. The current through an inductor cannot change instantly, so at $t = 0^+$ it remains equal to V_0/R_1, and

$$i_L(0) = \frac{V_0}{R_1} = I_{1n}e^0 + I_0$$

Solving for I_{1n},

$$I_{1n} = \frac{V_0}{R_1} - I_0$$

and the complete response for all times greater than zero is

$$i_L(t) = [(V_0/R_1 - I_0)]e^{-R_2t/L}$$

This function has the graphical form shown in Fig. 6.8c. The current through the inductor decays exponentially from its initial value V_0/R_1 to the forced value I_0 after a time equal to several time constants ($\tau = L/R_2$).

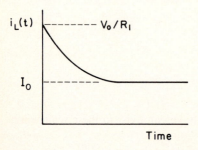

FIGURE 6.8c.
Graph of the current through the inductor versus time.

To summarize, there are four steps to take in solving for the complete response of the systems we study. These steps yield solutions to linear ordinary inhomogeneous differential equations with constant coefficients of order n:

1. Find the solution to the nth-order homogeneous (natural response) equation and thus the n values of s ($s_1, s_2, \ldots, s_n$). In electrical engineering we do this by finding the n solutions to the algebraic equation

$$Z_s(s) = 0 \qquad\qquad (6.16a)$$

where $Z_s(s)$ is the series impedance to the current flow through any essential element of the same circuit with all sources "suppressed," or by finding the n solutions to the algebraic equation

$$Z_p(s) = \infty \qquad (6.16b)$$

or equivalently

$$\frac{1}{Z_p(s)} = 0$$

with all sources suppressed. $Z_p(s)$ is the parallel impedance across any essential element in the circuit. If we are finding the current in a system, the homogeneous or natural solution will have the form

$$i(t) = I_1 e^{s_1 t} + I_2 e^{s_2 t} + \ldots + I_n e^{s_n t}$$

There will still be n unknown coefficients to be determined. An analogous equation exists for $v(t)$ if the voltage is to be found.

2. Find any solution to the particular equation which includes the forcing term. In electrical engineering we often do this by inspection or by analysis of the forced response due to the source.

3. Add the solutions 1 and 2.

4. Apply n boundary conditions to the result of step 3 and hence determine the n coefficients.

We have used the term "essential" since there are some circuits in which one or more impedance values do not enter into the calculation of Z_s or Z_p due to the fact they do not carry any natural current or have any natural voltage drop. An example of such an element is the resistor R_3 in Fig. 6.9 after the switch is opened.

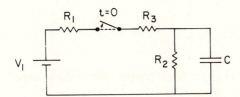

FIGURE 6.9.
Circuit that has an element, resistor R_3, which is not essential after the switch is thrown. The natural response is of the form $e^{-t/R_2 C}$.

TEXT EXAMPLE 6.5 _____

In Text Example 5.1 we found the initial value of the current $i(t)$ in the circuit below. Since there is a source in the circuit after the switch changes position, the example is one involving the complete response to both the 50-V battery and the natural currents due to energy stored in the inductor. Find the complete response $i(t)$ for $t \geq 0$.

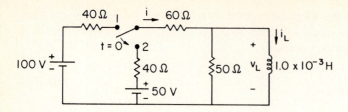

Solution: Following the four steps outlined in the summary above, we first must find the value or values of s. Suppressing the source for $t \geq 0$, we have the following circuit:

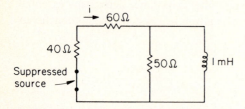

The series impedance method yields

$$Z_s(s) = 100 + \frac{50 \times 10^{-3}s}{50 + 10^{-3}s}$$

$$Z_s(s) = \frac{5000 + 0.1s + 0.05s}{50 + 0.001s}$$

To obtain $Z_s(s) = 0$, we need only set the numerator equal to zero, giving

$$s = \frac{-5000}{0.15} = -3.3 \times 10^4 \text{sec}^{-1}$$

and the natural response is of the form

$$i_N(t) = I_{1n}e^{(-3.3 \times 10^4)t}$$

Step 2 involves finding the forced current due to the 50-V battery. Since the battery is a dc source, the inductor acts like a short circuit. The current is given by

$$i_F = \frac{50}{40 + 60} = 0.5 \text{ A}$$

where the subscript F stands for the forced response. Step 3 merely involves adding these two solutions:

$$i(t) = I_{1n}e^{(-3.3 \times 10^4)t} + 0.5$$

Finally, in step 4 we use the result of Text Example 5.1, in which we showed that $i(0^+) = 0.67$ A,

$$0.67 = I_{1n}e^0 + 0.5$$

Solving yields $I_{1n} = 0.17$, the full solution is

$$i(t) = (0.17e^{(-3.3 \times 10^4)t} + 0.5) \text{ A}$$

6.6   CIRCUIT RESPONSE TO PULSES

One of the important applications of the results discussed in this chapter is the analysis of the circuit response to pulses. Many electrical systems employ single pulses or trains of pulses such as those in Fig. 6.10. A sequence of 5-V pulses such as is shown in Fig. 6.10b is often used as a clock in logic circuits to synchronize the various logic elements.

FIGURE 6.10.
Single pulses and pulse trains such as those used in many digital and analog circuits.

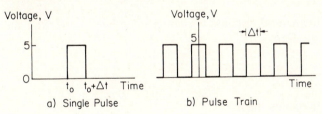

We can analyze the response of linear circuits to a pulse by treating the pulse as the superposition of two voltage steps, as shown in Fig. 6.11. Such pulses and pulse trains are sufficiently important that pulse generators are manufactured and sold as test equipment for modern laboratories. When a voltage source is to be considered as an ideal pulse generator, we shall designate it as shown in Fig. 6.12. An ideal pulse generator has a "rise time," which is infinitesimally small, changing from

FIGURE 6.11.
A 5-volt pulse constructed from two step voltages.

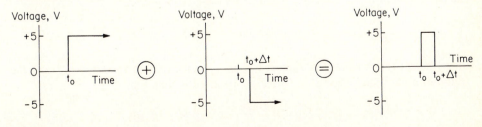

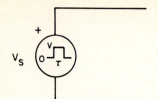

FIGURE 6.12.
Symbol for an ideal pulse generator.

zero to some voltage, V, and returning from that voltage to zero in a very short time. The pulse length τ must be specified in any such source as well as the value of V.

TEXT EXAMPLE 6.6

Find and sketch the voltage output of the system below versus time if the pulse length τ is 10 msec and the initial voltage on the capacitor is zero volts. Find the response if C is reduced to $0.5\mu F$.

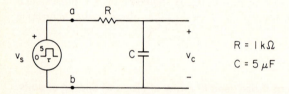

$$R = 1\,k\Omega$$
$$C = 5\,\mu F$$

Solution: We know that the natural response is characterized by $s = -(1/RC)$ and therefore by a function of the form

$$v_{cn}(t) = V_0 e^{-st} = V_0 e^{-200t}$$

If we measure time in milliseconds, then

$$v_{cn}(t) = V_0 e^{-0.2t'}$$

where t' is the time in milliseconds. First, we find the response of the system to a step voltage from 0 V to 5 V at $t = 0$. This is a complete response problem so we need to find the forced response to the step. Since C is an open circuit for a dc voltage,

$$v_{cF} = 5\ V$$

and the full solution is the sum,

$$v_c(t) = V_0 e^{-0.2t'} + 5$$

Since the capacitor voltage is zero at $t' = 0^-$, it must remain so at $t' = 0^+$ and the solution in the time period $t' = 0$ to $t' = 10$ msec is therefore

$$v_c(t') = 5(1 - e^{-0.2t'}) \qquad (0 \le t' \le 10)$$

The effect of the second edge of the pulse may be determined as follows. Let

$$t'' = t' - 10$$

Then, when $t' = 10$, the time $t'' = 0$, and we have our usual equations. At $t'' = 0^-$ the voltage on the capacitor may be determined from the equation above evaluated at $t' = 10$:

$$v_c(t'' = 0) = v_c(t' = 10) = 5(1 - e^{-2}) = 4.32 \text{ V}$$

This shows that the capacitor had charged to 4.32 V just before the pulse ended. This value is used as the initial condition for a new calculation which begins at $t'' = 0$. At this time, the second step function (see Fig. 6.11) brings the voltage at point a to zero. For $t' \geq 10$ the natural response has the same form as for $0 \geq t' \geq 10$ since R and C have not changed:

$$v_{cn}(t'') = V_1 e^{-0.2t''}$$

The new forced response is $v_{cF} = 0$, however, since the second step voltage has returned the voltage to zero volts. The total solution then has the form

$$v_c(t'') = V_1 e^{-0.2t''} + 0$$

To find V_1 we use the initial condition $v_c(t'' = 0) = 4.32$ and

$$v_c(t'' = 0) = 4.32 e^{-0.2t''} \qquad (0 \leq t'' \leq \infty)$$

To express this in terms of t', we substitute $t'' = t' - 10$ and the full solution is

$$v_c(t') = 5(1 - e^{-0.2t'}) \qquad (0 \leq t' \leq 10)$$

$$v_c(t') = 4.32 e^{-0.2(t' - 10)} \qquad (10 \leq t' \leq \infty)$$

A plot of this function is shown below as the solid line in the lower part of the figure.

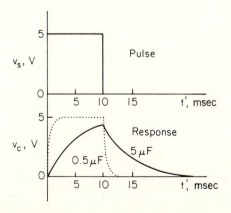

The output voltage across the capacitor does not look very much like the original pulse. If this output signal was sent to some other electronic device which was "expecting" a pulse, the system would most likely not function very well. Suppose now that C is $0.5\mu F$. The only change in the solution is the value for s, which becomes 2000 sec^{-1} rather than 200 sec^{-1}. The mathematical solution becomes

$$v_c(t') = 5(1 - e^{-2t'}) \qquad (0 \le t' \le 10)$$

$$v_c(t') = 4.999 e^{-2(t'-10)} \qquad (10 \le t' \le \infty)$$

which has the solution shown by the dotted line in the figure above. This waveform is much closer to the original pulse.

6.7 NATURAL RESPONSE OF A SECOND-ORDER SYSTEM: OSCILLATORY BEHAVIOR

Thus far we have studied circuits in which the natural currents and voltages decay exponentially in time. It is no accident that systems with only L's or C's have been analyzed, but not circuits containing both. Suppose as shown in Fig. 6.13a, a parallel L–C circuit is initially connected to a battery and the switch is opened at $t = 0$. This situation is somewhat idealized since in any real circuit the inductor would have some small resistance due to the many turns of wire in its construction. However, the response of the idealized circuit is instructive and, as we shall see, corresponds to such mechanical analogies as the frictionless pendulum or loss-free spring. If the switch has been closed for a long time, the current i_n at $t = 0$ is V_0/R and the voltage across the capacitor is zero.

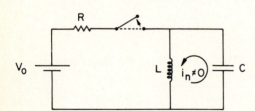

FIGURE 6.13a.
Circuit in which the current oscillates in time after the switch is opened.

Applying KVL around the loop for $t \ge 0$ yields

$$+L\frac{di_n}{dt} + \frac{1}{C}\int_0^t i_n(t')dt' + 0 = 0$$

Taking the derivative of both sides yields

$$L\frac{d^2 i_n}{dt^2} + \frac{1}{C}i_n(t) = 0 \qquad (6.17)$$

This is a second-order, linear, homogeneous, ordinary differential equation with constant coefficients. The examples studied previously in this chapter all involved first-order equations although a number of second-order equations were derived in Chapter 5. Such an equation is identical to the equation of motion of a loss-free spring with spring constant k supported by a frictionless plane and which is loaded with a mass m as shown in Fig. 6.13b.

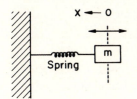

FIGURE 6.13b.
A mass on a spring supported by a frictionless plane will oscillate in time when displaced slightly and released.

From Newton's third law,

$$m\frac{d^2x}{dt^2} = -kx$$

which is identical to Eq. 6.17. This equation also describes a natural response since the mass is not being forced. The solution to this equation is an oscillation about the position $x = 0$, which can be written in the general form

$$x(t) = x_0 \cos (\omega t + \phi)$$

where the frequency ω of oscillation is given by $\omega = \sqrt{k/m}$. The reader should check this solution by substituting into the differential equation above. By analogy with the spring, then, it must be the case that the current $i_n(t)$ in the circuit illustrated in Fig. 6.13a is given by an equation of the form

$$i_n(t) = I_1 \cos (\omega t + \phi) \tag{6.18}$$

Substituting this solution for $i_n(t)$ into Eq. 6.17, the equation is indeed satisfied provided that $\omega = \sqrt{1/LC}$ (analogous to $\omega = \sqrt{k/m}$ for the spring). I_1 and ϕ are the two arbitrary constants required by the fact that the differential equation is of second order. If we plot the function in Eq. 6.18 versus time, it has the form shown in Fig. 6.14. The current oscillates between the maximum values I_1 and $-I_1$ with a period

$$T = f^{-1} = \left(\frac{\omega}{2\pi}\right)^{-1}$$

where f is the frequency in Hz (cycles per second). The symbol ω is called the "radian frequency" and is measured in radians per second. I_1 is the amplitude and ϕ the phase angle, both of which are determined by the initial conditions, that is, exactly how and when the switch is closed, the energy source to which the

FIGURE 6.14.
Plot of the function $I_1 \cos(\omega t + \phi)$.

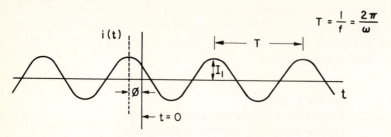

oscillatory circuit was attached, and so on. In our example at $t = 0$, $i = V_0/R$ and $L(di/dt) = 0$. Setting $\phi = 0$ and $i_1 = V_0/R$ satisfies these two conditions, and the full solution is

$$i_n(t) = \frac{V_0}{R} \cos (\omega t)$$

This solution may be used to understand some properties of this second-order system. At $t = 0$ the current is at its maximum value I_1, while the derivative of the current at $t = 0$ is

$$\frac{di}{dt} = -\frac{\omega V_0}{R} \sin (\omega t) = -\frac{\omega V_0}{R} \sin (0) = 0$$

The voltage across the two elements, $v_L = L(di/dt) = v_C$, is therefore zero at $t = 0$. Taken together, these two results show that the capacitor is totally discharged at the same time that the inductor carries the maximum current. The inductor therefore has a magnetic field built up inside while the electric field inside the capacitor vanishes. One quarter period later, when $\omega t = \pi/2$, $i = 0$, but the full voltage has built up across the capacitor. The energy is now stored in the electric field inside the capacitor. The energy is transferred back and forth between the two circuit elements at the frequency ω. If there is no resistor in the circuit, and the capacitor and the inductor are ideal circuit elements, no energy will be dissipated and the oscillation will continue forever. This oscillatory behavior is not only analogous to the spring but to the frictionless pendulum discussed at the beginning of the chapter. In the latter case, energy oscillates between the kinetic energy of the pendulum when it is at the base of its arc and its potential energy when it has reached the highest point. Likewise in the case of the spring, energy oscillates between the kinetic energy of the mass and the potential energy in the compressed spring. All three cases involve natural response since no forcing exists.

Now suppose we analyze this system using the impedance method. The parallel impedance across the L–C circuit is

$$Z_p(s) = \frac{sL/sC}{sL + (1/sC)} = \frac{sL}{s^2LC + 1}$$

To have a nonvanishing "natural" voltage v_n across this network if there is no source attached,

$$i = 0 = \frac{v_n}{Z_p(s)}$$

and $Z_p(s)$ must thus be infinite. The denominator of $Z_p(s)$ vanishes when

$$s^2 LC + 1 = 0$$

which can be written as

$$s^2 = \frac{-1}{LC}$$

And finally we have the two solutions for s:

$$s_1 = +\sqrt{-\frac{1}{LC}} = +\sqrt{-1}\sqrt{\frac{1}{LC}}$$

$$s_2 = -\sqrt{-\frac{1}{LC}} = -\sqrt{-1}\sqrt{\frac{1}{LC}}$$

Both of these values for s include the "imaginary" number $\sqrt{-1}$. Since we expect a solution of the form $i(t) = a_1 e^{s_1 t} + a_2 e^{s_2 t}$, to complete the problem we need to attach a meaning to these expressions when s_1 and s_2 are imaginary. To be consistent, the solution for $i_n(t)$ must be the same as Eq. 6.18. In the next chapter we show that the two solutions are indeed identical and show how imaginary numbers are used to describe oscillatory signals. In Chapter 12 we return to a more complete study of second-order systems, including the effect of resistance.

To summarize the results of this section, we have found that electrical circuits can be constructed which yield oscillating currents and voltages. Second, we have found that these oscillations correspond mathematically to exponential functions with an argument involving $\sqrt{-1}$, or what is called an imaginary argument. Seemingly cumbersome at first, the use of imaginary numbers in mathematically manipulating the parameters of physical systems has proved to be quite advantageous. In the next chapter we review those elements of complex algebra required to analyze electrical circuits that exhibit oscillatory currents and voltages.

PRACTICE PROBLEMS
AND
ILLUSTRATIVE EXAMPLES

EXAMPLE 6.1

The switch in the circuit below has been closed for a long time and is opened at $t = 0$. Find the voltage $v_0(t)$ for $t \geq 0$ if $R = 3\ \Omega$ and $L = 2$ H.

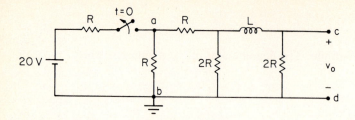

Solution: After the switch is opened, no sources remain in the circuit and the resulting currents and voltages constitute the natural response. Our first task is to find the parameter s which characterizes the natural response. This may be accomplished by either setting the series impedance function equal to zero (zero method) or the parallel impedance (pole method) equal to infinity. Here, we illustrate both techniques which must yield the same value for s.

To determine the series impedance, we may enter anywhere in the circuit. As an illustration consider the phantom voltage source shown below.

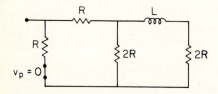

Evaluating the impedance,

$$Z_s = R + R + [2R//(sL + 2R)]$$

$$Z_s = 2R + \frac{2R(sL + 2R)}{sL + 4R} = \frac{12R^2 + 4RsL}{sL + 4R}$$

Setting $Z_s = 0$ is equivalent to setting the numerator of $Z_s = 0$, which yields

$$12R^2 + 4RsL = 0$$

or

$$s = \frac{-3R}{L} = -\frac{9}{2} \ \text{sec}^{-1}$$

Note that this result could have been found as well by analogy with the circuit below.

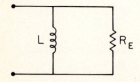

Inspection of the two circuits shows that the equivalent resistance in parallel with L is $3R$.

Using the parallel impedance approach, consider a phantom current source across the output $c-d$,

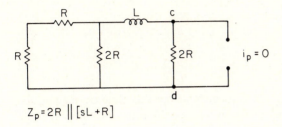

$$Z_p = 2R \,\|\, [sL + R]$$

$$Z_p = \frac{2R(sL + R)}{2R + sL + R} = \frac{2RsL + 2R^2}{3R + sL} = \infty$$

To determine s, we set the denominator equal to zero and

$$s = \frac{-3R}{L}$$

which agrees with the result above. To finish the problem, we write down the form of the natural response function

$$v_0(t) = V_0 e^{st} = V_0 e^{-9t/2}$$

We need yet to determine V_0 from the initial conditions. Before the switch was opened, a current was flowing through the inductor. Since L acts as a short circuit, it is straightforward to show $V_a = 8$ V. The inductor current then is

$$i_L(0^+) = \left(\frac{1}{2}\right)\left(\frac{8}{2R}\right) = \frac{2}{3} \text{ A}$$

where the factor $\frac{1}{2}$ is from the current divider rule. Since this current must be a continuous function of time, the voltage across the terminal $c-d$ at $t = 0^+$ is

$$v_0(0^+) = i_L(0^+)2R = 4 \text{ V}$$

Finally, substituting this result into the functional form above yields the natural response

$$v_0(t) = 4e^{-9t/2}$$

A plot of this function is given below. Note that the voltage decreases by a factor of e in a time $\tau = .22$ sec.

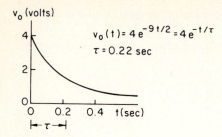

Problem 6.1

Evaluate s in Example 6.1 again by the zero method but this time by placing the phantom voltage source in series with either of the two $2R$ resistors. Show that the same result is attained as found above.

Problem 6.2

Find the natural response current from a to b in the circuit for Example 6.1, that is, $i_{ab}(t)$ for $t \geq 0$. Make a sketch of the temporal function.

Problem 6.3

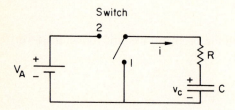

The switch in the circuit shown has been in position 1 for a long time and is moved from position 1 to position 2 at $t = 0$. This is a complete response problem since a source remains in the circuit for $t \geq 0$.

Find the current $i(t)$ for $t \geq 0$.

Find the voltage $v_c(t)$ for $t \geq 0$.

Find i and v_c for $t \gg RC$.

Ans.: $i = (V_A/R)e^{-t/RC}$

Ans.: $v_c = V_A - V_A e^{-t/RC}$

Ans.: $i = 0$; $v_c = V_A$

Problem 6.4

Find the natural responses for the current $i_L(t)$ and voltage $v_L(t)$ in Problem 5.12 for $t \geq 0$.

Problem 6.5

Suppose the resistor between a and b in Example 6.1 breaks at $t = 0.11$ sec (i.e., at $t = \tau/2$). Find the natural response voltage $v_0(t)$ versus time for $t \geq 0.11$ sec. Modify the plot in the example above to show the voltage versus time for $t \geq 0$. *Hint:* Define $t' = (t - 0.11)$ sec and evaluate the new time constant for the modified circuit.

Problem 6.6

The switch in the circuit below is opened at $t = 0$. Find the voltage $v_R(t)$ for $t \geq 0$.

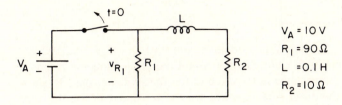

$$V_A = 10 \text{ V}$$
$$R_1 = 90 \, \Omega$$
$$L = 0.1 \text{ H}$$
$$R_2 = 10 \, \Omega$$

Ans.: $v_R(t) = -90e^{-1000t}$ V

Problem 6.7

Find the natural response voltage $v(t)$ and current $i(t)$ in Problem 5.17 for $t \geq 0$.

Problem 6.8

Show that your solution for $i(t)$ in Problem 6.7 satisfies the differential equation given in the answer to Problem 5.17.

Problem 6.9

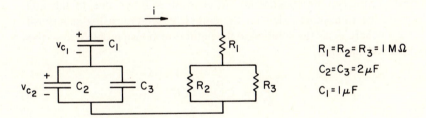

$$R_1 = R_2 = R_3 = 1 \text{ M}\Omega$$
$$C_2 = C_3 = 2\,\mu\text{F}$$
$$C_1 = 1\,\mu\text{F}$$

At $t = 0$, suppose that the capacitors have been charged up in a way such that $v_{c1} = 10$ V and $v_{c2} = 20$ V. Find $i(t)$ for $t \geq 0$.

Ans.: $i(t) = 20e^{-t/1.2}$ microamperes

EXAMPLE 6.2

In Example 5.2 a switch was thrown at $t = 0$ leaving the circuit shown below. In that example we found the values of the currents and voltages at $t = 0^+$ (just after the switch was thrown from position 1 to position 2) and the values of those same quantities after "a long time." Derive the differential equation for the voltage $v_c(t)$ for $t > 0$. Use the pole-zero method to find $v_c(t)$ and $i(t)$ and show that the former does satisfy the differential equation. Quantify what we mean by "a long time" in this circuit.

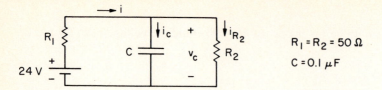

$R_1 = R_2 = 50\ \Omega$

$C = 0.1\ \mu F$

Solution: From Example 5.2 the initial conditions are $v_c(0^+) = 6$ V and $i(0^+) = (24 - 6)/R_1 = 0.36$ A. This is an example of a circuit requiring complete response analysis since a 24-V source remains in the circuit after the switch closes. To show this explicitly in terms of the differential equation for $t \geq 0$ we may derive the node voltage equation for v_c. From KCL,

$$-i + i_c + i_{R2} = 0$$

In terms of v_c, this yields

$$\frac{v_c - 24}{R_1} + C\frac{dv_c}{dt} + \frac{v_c}{R_2} = 0$$

which may also be written as

$$v_c\left(\frac{1}{R_1} + \frac{1}{R_2}\right) + C\frac{dv_c}{dt} = \frac{24}{R_1} \tag{1}$$

This is a linear, ordinary, inhomogeneous differential equation due to the non-vanishing term on the right-hand side. For the complete solution using superposition, $v_c = v_{cn} + v_{cf}$, where v_{cn} is the solution to the natural response problem described by the homogeneous equation

$$v_{cn}\left(\frac{1}{R_1} + \frac{1}{R_2}\right) + C\frac{dv_{cn}}{dt} = 0 \tag{2}$$

and v_{cf} is the particular (forced) solution to the inhomogeneous differential equation

$$v_{cf}\left(\frac{1}{R_1} + \frac{1}{R_2}\right) + C\frac{dv_{cf}}{dt} = \frac{24}{R_1} \tag{3}$$

We now use the pole-zero procedure outlined in the text, then return to Eq. 1 to check the results. For the natural response we shall solve for s using both the impedance zero and pole methods. In usual practice only one method would be used since they yield identical results. With the 24-V source suppressed (set equal to zero), we can consider the natural current to be "driven" by this zero voltage source.

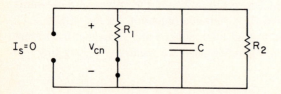

Ohm's law yields

$$0 = i_N Z_s(s) \tag{4}$$

where

$$Z_s(s) = R_1 + \frac{R_2/sC}{R_2 + (1/sC)} = \frac{R_1 R_2 sC + R_1 + R_2}{R_2 sC + 1}$$

Setting this series impedance function equal to zero yields

$$s_1 = -\frac{R_1 + R_2}{R_1 R_2 C} \tag{5}$$

Using the impedance pole method, with the 24-V source suppressed, we can find what natural voltage v_{cn} may exist with a zero current source.

$$0 = \frac{v_{cn}}{Z_p(s)}$$

The only way for v_{cn} not to vanish is if $1/Z_p(s) = 0$ or $Z_p(s) = \infty$. For elements in parallel,

$$\frac{1}{Z_p(s)} = \frac{1}{R_1} + \frac{1}{R_2} + sC = \frac{R_1 + R_2 + sCR_1R_2}{R_1R_2}$$

Setting $1/Z_p(s) = 0$,

$$s_1 = -\frac{R_1 + R_2}{R_1 R_2 C}$$

which is the same result as Eq. 5. $R_1R_2/(R_1 + R_2) = 25 \ \Omega$, so

$$s_1 = \frac{-1}{2.5 \times 10^{-7}} = -4 \times 10^5 \ \text{sec}^{-1}$$

The natural response voltage $v_{cn}(t)$ is therefore

$$v_{cn} = Ae^{-4 \times 10^5 t}$$

where A is yet to be determined. The forced response v_{cf} is just that due to the 24-V dc source. At dc the capacitor is an open circuit, so v_{cf} can be derived from the voltage divider principle $v_{cf} = [R_2/(R_1 + R_2)]24 \ \text{V} = 12 \ \text{V}$. The complete solution is the sum of v_{cn} and v_{cf}:

$$v_c(t) = 12 + Ae^{-4 \times 10^5 t}$$

Now, using the initial condition $v_c(0^+) = 6$,

$$6 = 12 + A; \qquad A = -6$$

Finally, for $t \geq 0$

$$v_c(t) = (12 - 6e^{-4 \times 10^5 t}) \text{ V} \qquad (6)$$

To check this result, we must show that v_c satisfies Eq. 1 as well as the initial conditions. Direct substitution of v_c into Eq. 1 yields

$$\frac{1}{R_p}(12 - 6e^{s_1 t}) - (6s_1 C)e^{s_1 t} = \frac{24}{R_1}$$

where $R_p = R_1 R_2 / (R_1 + R_2) = 25 \ \Omega$. Using $s_1 = -(1/R_p C)$, the two $e^{s_1 t}$ terms cancel, leaving

$$12 = \frac{R_p}{R_1} 24 = \frac{25}{50} 24 = 12$$

which is a valid statement.

The solution also satisfies the initial condition that, at $t = 0$, $v_c = 6$ and that the forced response for $t \gg 1/(4 \times 10^5)$ seconds equals 12 V. Now we see that "after a long time" means after $t \gg R_p C$, where in this case, $R_p C = 2.5$ microseconds. For example, at $t = 25$ microseconds, $v_c = 12 - 6e^{-10} = 12 - .00027$ V, which is very nearly equal to 12 V. Having found $v_c(t)$, we can immediately find $i(t)$.

$$i = \frac{24 - v_c}{R_1} = (0.24 + 0.12e^{-4 \times 10^5 t}) \text{ A} \qquad (7)$$

It is instructive to note that the forced response part of Eq. 7 can be derived directly from the series impedance of Eq. 4 if we replace the suppressed source by the actual source $V_s = 24$ V and set $s = 0$ in $Z_s(s)$. This value for s is appropriate for a dc forcing function, since $s = 0$. For this case

$$24 = i_f Z_s(0) = i_f(R_1 + R_2)$$

$$i_f = 0.24 \text{ A}$$

In Chapter 10 we shall find that a similar approach may be used to find the forced response to oscillatory sources where we must set s equal to its appropriate value for such time variations.

Problem 6.10

Find the complete response solutions for the currents $i_c(t)$ and $i_{R2}(t)$ for $t \geq 0$ in Example 6.2. Show that KCL holds, that is,

$$i(t) = i_c(t) + i_{R2}(t)$$

Problem 6.11

In Problem 6.3 suppose we leave the switch in position 2 for a "long time" and then, at time $t = t_1$, the switch is moved back to position 1.

(a) Find the current $i(t)$ for $t \geq t_1$.   Ans.: $i = -(V_A/R)e^{-(t-t_1)/RC}$

(b) Find the voltage $v_c(t)$ for $t \geq t_1$.   Ans.: $v_c = V_A e^{-(t-t_1)/RC}$

(c) Find the values of i and v_c as $t \to \infty$.   Ans.: $i = 0$; $v_c = 0$

Problem 6.12

In Problem 6.3, $R = 50$ kΩ, $V_A = 10$ V and $C = 0.01 \mu$F. What value must t_1 have for the condition $v_c(t_1) = 9.9$ V to hold? Find the current $i(t)$ at that same time.

Problem 6.13

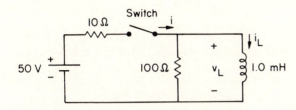

(a) In the circuit shown, the switch has been open for a long time and is closed at $t = 0$. For $t \geq 0$, find i, i_L, and v_L.

Ans.: $i = (5 - 4.55e^{(-9.1 \times 10^3)t})$ A
$i_L = (5 - 5e^{(-9.1 \times 10^3)t})$ A
$v_L = (45.5)e^{(-9.1 \times 10^3)t}$ V

(b) At $t = 1$ sec, the switch is opened. Find i, i_L, and v_L for $t \geq 1$. (Note that this is the circuit of Problem 5.12 except that the switch is opened at $t = 1$ rather than at $t = 0$.)

Ans.: $i = 0$
$i_L = 5e^{-10^5(t-1)}$ A
$v_L = -500e^{-10^5(t-1)}$ V

Problem 6.14

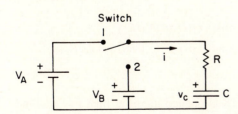

The switch in the circuit shown has been in position 1 for a long time compared to RC. At $t = 0$, it is moved to position 2. Find $i(t)$ and $v_c(t)$ for $t \geq 0$.

Ans.: $i(t) = [(V_B - V_A)/R]e^{-t/RC}$

$v_c(t) = V_B + (V_A - V_B)e^{-t/RC}$

Problem 6.15

In Problem 6.14 let $R = 1$ MΩ, $V_A = 20$ V, $V_B = 10$ V and $C = 0.01\mu$F. At time $t = t_1 = 10$ msec, the switch is returned to position 1. Find $i(t)$ and $v_c(t)$ for $t \geq t_1$.

Problem 6.16

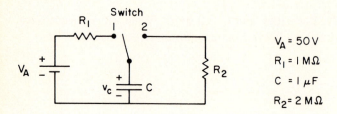

$V_A = 50$ V

$R_1 = 1$ MΩ

$C = 1$ μF

$R_2 = 2$ MΩ

The switch has been in position 1 for a long time compared to R_1C. At $t = 0$, the switch is moved to position 2 and held there for 2 seconds, after which it is moved back to position 1.

(a) Find $v_c(t)$ for $0 \leq t \leq 2$.   Ans.: $v_c = 50e^{-t/2}$ V

(b) Find $v_c(t)$ for $t \geq 2$.   Ans.: $v_c = (50 - 31.6e^{-(t-2)})$ V

Problem 6.17

Find the two times in Problem 6.16 when $v_c(t) = 49$ V.

Problem 6.18

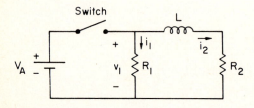

The switch in the circuit shown above has been opened for a long time and is closed at $t = 0$.

Find $i_1(t)$ for $t \geq 0$.   Ans.: $i_1(t) = (V_A/R_1)$

Find $i_2(t)$ for $t \geq 0$.   Ans.: $i_2(t) = (V_A/R_2) - (V_A/R_2)e^{-R_2t/L}$

Problem 6.19

After a long time the switch in Problem 6.18 is opened again at time $t = t_1$. Find $v_1(t)$ for $t \geq t_1$. Let $V_A = 10$ V, $R_1 = 10$ kΩ $= R_2$, and $L = 1$ mH. How large must t_1 be if the criterion for "a long time" is such that $0..9$ mA $\leq i_2(t_1) \leq 1$ mA? (Use the solution to Problem 6.18 for this latter calculation substituting the actual values given here.)

Problem 6.20

In the circuit shown below the switch has been in position 1 for a long time and is moved to position 2 at $t = 0$.

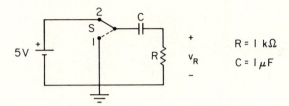

(a) Find the response $v_R(t)$ for $t \geq 0$.

(b) After 10 msec the switch is returned to position 1. Find the voltage response v_R in the time interval 10 msec $\leq t \leq \infty$.

(c) Sketch $v_R(t)$ for -10 msec $\leq t \leq 10$ msec.

Problem 6.21

In the circuit shown, the switch has been in position 1 for a long time and is moved to position 2 at $t = 0$. Find $v(t)$ for $t \geq 0$. (Assume that when the switch is thrown, the current source is turned off so that it is not in series with an open circuit!)

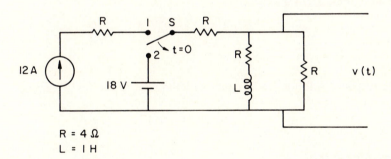

R = 4 Ω
L = 1 H

Ans.: $v(t) = (-9e^{-6t} + 6)$ V

Problem 6.22

In the circuit shown, the switch has been opened for a long time and is closed at $t = 0$. Find $v_c(t)$ for $t \geq 0$.

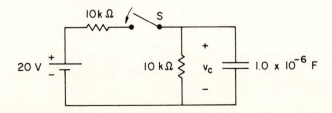

Problem 6.23

For the circuit shown, the switch has been in position 1 for a long time and is moved to position 2 at $t = 0$. Find $i(t)$ for $t \geq 0$.

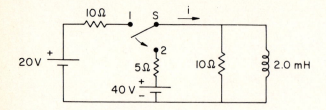

Ans.: $i(t) = (8 - 4e^{(-1.67 \times 10^3)t})$ A

Problem 6.24

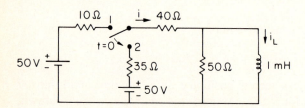

In the circuit shown the switch has been in position 1 for a long time and is moved to position 2 at $t = 0$. Find $i(t)$ for $i \geq 0$.

Ans.: $i(t) = (0.67 + 0.13e^{-3 \times 10^4 t})$ A

Problem 6.25

Find the complete response solution for $i(t)$ and the voltage $v_{ab}(t)$ in Text Example 5.2 for $t \geq 0$.

Problem 6.26

Find the current i_L for $t \geq 0$.

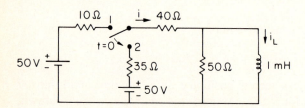

Ans.: $i_L(t) = (2/9 + (10/9)e^{-4.5t})$ A

Problem 6.27

In Example 5.3 of the problem set for Chapter 5, suppose the switch is closed for a long time but at $t = 0$ the $3\mu F$ capacitor "shorts out" to ground. Find the complete response function for $i(t)$ when $t \geq 0$.

Problem 6.28

Find the current $i(t)$ for $t \geq 0$.

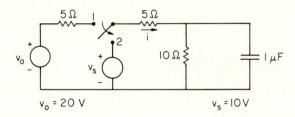

$v_0 = 20$ V $v_s = 10$ V

Ans.: $i(t) = [(2/3) - (2/3)e^{-3 \times 10^5 t}]$ A

Problem 6.29

In the circuit for Problem 6.28 suppose the switch has been in position 2 for a long time. At $t = t_1$ it is returned to position 1. Find $i(t)$ for $t \geq t_1$.

EXAMPLE 6.3

For the circuit shown in (a) below, (1) find the complete response current if a voltage source V_s is applied to the terminals at $t = 0$, and (2) find the complete response current if a current source I_s is applied to the terminals at $t = 0$. Note that the characteristic values for s are not equal! Let the current through the inductor be zero at $t = 0$ in both cases.

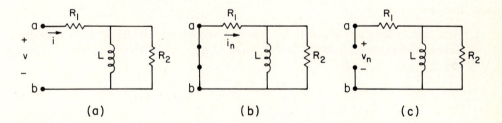

(a)   (b)   (c)

Solution:

1. If a voltage source is applied, the natural response current i_n is the current with the source suppressed as in (b) above. Then, the series impedance is

$$R_1 + \frac{R_2(sL)}{R_2 + sL} = \frac{[(R_1 + R_2)sL + R_1 R_2]}{sL + R_2}$$

Setting the series impedance to zero, $s = \dfrac{R_1 R_2}{(R_1 + R_2)L}$

$$i_n(t) = I_{1n}e^{-R_1 R_2 t/(R_1 + R_2)L}$$

To find the complete response we need to determine the forced response current as t gets very large. Since L acts as a short circuit for dc, $i_F = V_s/R_1$. The solution is then of the form

$$i(t) = i_n(t) + i_F = I_{1n}e^{-R_1R_2t/(R_1+R_2)L} + V_s/R_1$$

Since the initial condition is such that $I_L = 0$, the inductor acts like an open circuit at $t = 0$. Therefore,

$$i(0^+) = \frac{V_s}{R_1 + R_2}$$

Setting $t = 0$ in the expression for $i(t)$, we have

$$\frac{V_s}{R_1 + R_2} = I_{1n} + \frac{V_s}{R_1}$$

Hence,

$$I_{1n} = \frac{-R_2V_s}{R_1(R_1 + R_2)}$$

and

$$i(t) = \frac{-R_2V_s}{R_1(R_1 + R_2)}e^{-R_1R_2t/(R_1+R_2)L} + \frac{V_s}{R_1}$$

2. If a current source is applied, the natural response must be determined with the source suppressed as shown in the figure in (c) above. The impedance "seen" by the open circuit is

$$Z_p = R_1 + \frac{R_2sL}{R_2 + sL} = \frac{R_1(R_2 + sL) + R_2sL}{R_2 + sL}$$

To have $Z_p = \infty$, the denominator must vanish (pole method) so

$$R_2 + sL = 0$$

Solving for s,

$$s = -\frac{R_2}{L} \qquad i_n(t) = I_{1n}e^{-R_2t/L}$$

R_1 does not appear in the solution since it does not carry any natural current (a nonessential element). For the forced response to a current source I_s, L acts as a short circuit and $i_F = I_s$. The form of the solution is

$$i(t) = I_{1n}e^{-R_2t/L} + I_s$$

Since the inductor current vanishes at $t = 0$, the current through R_1 at $t = 0^+$ is equal to I_s. This implies that $I_{1n} = 0$! Hence, the solution is $i(t) = I_s$.

Problem 6.30

Find the complete response current through the inductor in part 2 of Example 6.3 above. Make a sketch of the current through the inductor versus time if $R_2 = 10\ \Omega$ and $L = 10$ H. Sketch the current through the resistor R_2 using the result just found, $i(t) = I_s$, and KCL.

Problem 6.31

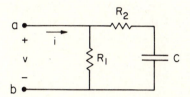

For the circuit above, find (1) the complete response current $i(t)$ if a voltage source V_s is applied to the terminals at $t = 0$, and (2) the complete response current $i(t)$ if a current source I_s is applied to the terminals at $t = 0$. The initial value of the voltage across the capacitor is zero in both cases.

Problem 6.32

Find and sketch the response current $v_R(t)$ in the circuit below if the pulse has $\tau = 10\mu\text{sec}$, $R = 1\ \text{k}\Omega$, $L = 0.3$ mH and the initial inductor current is zero.

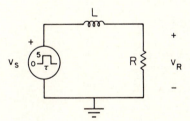

Problem 6.33

Repeat Problem 6.32 for the case where $R = 30\ \text{k}\Omega$, $L = 0.3$ mH and the pulse generator is unchanged.

Problem 6.34

Find and sketch the response voltage v_R in the circuit below if the pulse has $\tau = 10\mu\text{sec}$, $R = 1\ \text{k}\Omega$, $C = 0..001\mu\text{F}$ and if the initial voltage across the capacitor is 0 volt $[v_c(0^-) = 0\text{ V}]$.

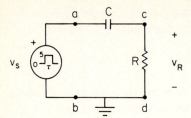

Notice the similarity to Problem 6.11.

Problem 6.35

Find and sketch the response voltage $v_L(t)$ in the circuit below if $\tau = 10$ msec, $R = 10\ \Omega$, $L = 0.01$ H, and the initial current $i_L(0) = 0$ A.

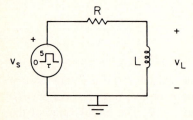

EXAMPLE 6.4

In Problem 5.15, after the switch S' is thrown to position 2, the circuit has both an inductor and a capacitor along with the two resistors R_1 and R_2. Find the series impedance function, set it equal to zero, and solve for the two values of s which characterize the natural response.

Solution: The circuit of interest is reproduced below.

$R_1 = 2\ \Omega$
$R_2 = 6\ \Omega$
$L = 0.2$ mH
$C = 1\ \mu F$

$$Z_s(s) = sL + R_1 + \frac{1}{sC} // R_2$$

$$= sL + R_1 + \frac{R_2/sC}{R_2 + (1/sC)}$$

$$= sL + R_1 + \frac{R_2}{1 + R_2Cs}$$

Using $1 + R_2Cs$ as the common denominator,

$$Z_s(s) = \frac{(sL + R_1)(1 + R_2Cs) + R_2}{1 + R_2Cs}$$

$$= \frac{(R_2LC)s^2 + (L + R_1R_2C)s + (R_1 + R_2)}{(R_2C)s + 1}$$

Both the numerator and the denominator of this expression are polynomials in s. To find the values of s for which $Z_s(s) = 0$, let

$$a = R_2LC$$

$$b = L + R_1R_2C$$

$$c = R_1 + R_2$$

Then, using the quadratic equation

$$s = \frac{-b \pm \sqrt{b^2 - 4ac}}{2a}$$

There are two values for s since the polynomial is of second degree. If

$$b^2 > 4ac$$

that is, if

$$(L + R_1R_2C)^2 > 4R_2LC(R_1 + R_2)$$

then both values of s will be real numbers. However, if $b^2 < 4ac$, then the square root of a negative number must be taken to yield the two solutions for s. In the present case, $b^2 = 4.33 \times 10^{-8}$ whereas $4ac$ is 3.84×10^{-8}. Both values of s are therefore real numbers since $b^2 > 4ac$. The values are

$$s_1 = -5.75 \times 10^4 \text{sec}^{-1}$$

and

$$s_2 = -1.16 \times 10^5 \text{sec}^{-1}$$

This is a "second-order" circuit of the type discussed in detail in Chapter 12.

Problem 6.36

Derive an expression for the characteristic parameters s_1 and s_2 corresponding to the natural response of the second-order system in Example 5.4. (*Hint:* replace the current source with an open circuit and use the pole method on the impedance function for an R, C, and L in parallel.) Under what conditions on R, L, and C will s_1 and s_2 have an imaginary part?

Problem 6.37

Derive a second-order polynomial expression in the variable s whose solution yields the two values of s which characterize the natural response of the circuit in Problem 5.16 after the switch is thrown to position 2. Find the two solutions for s.

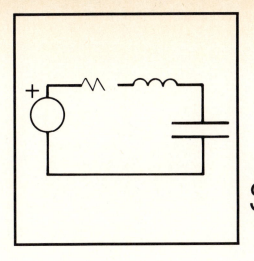

SECTION C

SINGLE-FREQUENCY SYSTEMS AND ELECTRICAL POWER

The impedance formalism has led to solutions involving terms of the form $e^{\pm j\omega t}$. In Chapter 7 we explore the meaning of this mathematical expression, some of the subtleties of complex numbers, and the advantages they hold for analysis of systems that display oscillatory behavior. The forced response to sinusoidal sources is analyzed by evaluating the impedance function $Z(s)$ for the various circuit elements under the condition that $s = j\omega$. Then all of the laws and theorems of circuit analysis hold but with the added complexity that the arithmetic involved deals with complex numbers. This formalism allows for detailed analysis of single-frequency systems such as those used worldwide for the transmission of electrical power. In Chapter 8 we show how to calculate power dissipation in such systems. In Chapter 9 we explore the role which transformers and three-phase systems play in efficiently transmitting power over large distances.

Chapter 7
Complex Numbers and the Phasor Notation

Analysis of oscillatory systems plays a crucial role in electrical engineering as well as in many other branches of science and technology. In part this is due to the near-sinusoidal behavior of many natural systems, as well as the deliberate generation of sinusoidal waveforms such as in 50- or 60-Hz power systems, radio broadcasting, and the like. A second major reason for studying sinusoidal systems stems from Fourier's theorem (discussed in Chapter 10), which proves that an arbitrary function of time can be broken up into the individual contributions of sinusoidal varying functions. This powerful tool thus allows the analysis of linear systems to take place using sinusoidal functions, which is relatively easy; the total solution generated at the end of the process is then obtained by adding up all the sinusoidal solutions. In this chapter we deal with complex algebra and its application to oscillatory systems.

7.1 COMPLEX NUMBER ALGEBRA AND SINUSOIDAL FUNCTIONS

We shall introduce the discussion of complex numbers by equating each complex number with a point in a plane. As shown in Fig. 7.1, each of these points can also be identified with a vector $\mathbf{A}$, as in the figure, which can be expressed in rectangular form in terms of the unit vectors $\hat{a}_x$ and $\hat{a}_y$ along the horizontal axis (x) and the vertical axis (y), respectively:

$$\mathbf{A} = A_x\hat{a}_x + A_y\hat{a}_y$$

FIGURE 7.1.
The complex plane.

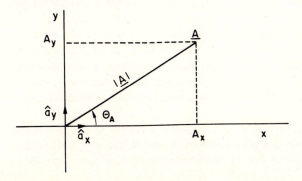

Every point in a plane can therefore be identified with a pair of numbers (A_x, A_y). In this text we denote a complex number by the symbol **A**. There is an important analogy here with the real number system, since every real number can be associated with a point on a number line. For example, every real number can be identified by a point on the x axis in Fig. 7.1. To represent all the complex numbers requires the entire plane, which we call the complex plane.

The vector **A** which links the origin to the point (A_x, A_y) can also be expressed in "polar" notation in terms of its magnitude or length A and the angle θ_A measured counterclockwise from the horizontal axis. We use the notation $\mathbf{A} = A\ \underline{/\theta_A}$ to denote the polar form. Note that the single quantity **A** actually represents two real numbers (A_x, A_y) or equivalently (A, θ_A). In the last section of Chapter 6 we found that the solution to a second-order differential equation involved two parameters, the amplitude and phase of an oscillatory function. As we shall see, since every complex number represents two real numbers, the notation keeps track of both the amplitude and phase of such oscillatory signals. This feature is the single most important reason for introducing complex notation. Inspection shows that the relationships between the rectangular and polar forms are

$$A = \sqrt{A_x^2 + A_y^2}; \qquad \theta_A = \tan^{-1}\left(\frac{A_y}{A_x}\right) \qquad (7.1a)$$

and

$$A_x = A \cos \theta_A; \qquad A_y = A \sin \theta_A \qquad (7.1b)$$

To connect this geometric relationship to algebra, we need an algebraic notation for the complex number **A**. The so-called rectangular form for a complex number could be written

$$\mathbf{A} = A_x(1) + A_y(\sqrt{-1})$$

The two terms are the real and "imaginary parts" of **A**, respectively. The real number (1) is dropped, and defining $j = \sqrt{-1}$ leaves the conventional form we shall use:

$$\mathbf{A} = A_x + jA_y$$

In electrical engineering the symbol j is used for $\sqrt{-1}$ rather than i as is the norm in physics and mathematics. This stems from the pervasive use of the symbol i for current in electrical engineering.

There is, of course, nothing "imaginary" about the component of the complex number **A** along the y axis. However, as we shall see, in order that complex number algebra obey all the same laws as real number algebra, that component must be multiplied by the square root of a negative number. Since the square of all real numbers must be positive, the square root of a negative number is called "imaginary."

The operation of *addition (subtraction)* for complex numbers is defined in the same way as for vectors. For two complex numbers **A** and **B**, the sum (difference) is found by adding (subtracting) their real parts and their imaginary parts separately:

$$\mathbf{C} = \mathbf{A} + \mathbf{B} = (A_x + jA_y) + (B_x + jB_y)$$
$$= (A_x + B_x) + j(A_y + B_y)$$
$$= C_x + jC_y$$

The multiplication operation for complex numbers in the rectangular form follows the normal rules of algebra.

$$\mathbf{C} = \mathbf{A} \times \mathbf{B} = (A_x + jA_y)(B_x + jB_y)$$
$$= A_xB_x + j^2A_yB_y + j(A_yB_x + B_yA_x)$$

Since $j^2 = -1$,

$$\mathbf{C} = C_x + jC_y = (A_xB_x - A_yB_y) + j(A_yB_x + B_yA_x)$$

Defining the multiplication operation in polar form is not quite as straightforward. We need first to find an algebraic form for $A \underline{/\theta}$. In the last chapter we pointed out that the solutions to the differential equations in circuit theory were of the form e^{st}. We have studied many examples wherein s is a real number and one case where we found $s = \pm j\omega$, where $j = \sqrt{-1}$. To relate this result to the complex plane, consider the expression $e^{j\theta}$, where θ is a real number. The function e^x has the series expansion

$$e^x = 1 + x + \frac{x^2}{2!} + \frac{x^3}{3!} + \frac{x^4}{4!} + \cdots$$

Substituting $x = j\theta$ yields

$$e^{j\theta} = 1 + j\theta + \frac{(j\theta)^2}{2!} + \frac{(j\theta)^3}{3!} + \frac{(j\theta)^4}{4!} + \frac{(j\theta)^5}{5!} + \frac{(j\theta)^6}{6!} + \cdots$$

Gathering terms and using the fact that $j^2 = -1$,

$$e^{j\theta} = \left(1 - \frac{\theta^2}{2!} + \frac{\theta^4}{4!} + \cdots\right) + j\left(\theta - \frac{\theta^3}{3!} + \frac{\theta^5}{5!} + \cdots\right)$$

But the two terms in parentheses are just the expansions of $\cos\theta$ and $\sin\theta$, so

$$e^{j\theta} = \cos\theta + j\sin\theta$$

This important equation is called Euler's theorem. We now may relate the location of a complex number in the complex plane to this exponential notation by writing

$$A = A_x + jA_y = A\cos\theta + jA\sin\theta$$

where $A = (A_x^2 + A_y^2)^{1/2}$. Finally,

$$\mathbf{A} = A[\cos\theta + j\sin\theta] = Ae^{j\theta}$$

We now have two algebraic forms for complex numbers: the rectangular and the exponential forms. Note that the polar form $A \underline{/\theta}$ means $Ae^{j\theta}$.

To define multiplication in polar form we now take a clue from the algebra of the exponentials. To multiply exponentials you add the exponents:

$$[Ae^{\alpha_1}] \times [Be^{\alpha_2}] = ABe^{(\alpha_1 + \alpha_2)}$$

while to divide them you subtract the exponents:

$$Ae^{\alpha_1} \div Be^{\alpha_2} = \left(\frac{A}{B}\right) e^{(\alpha_1 - \alpha_2)}$$

Using this result as a guide, the operation of multiplication is defined as follows. If $\mathbf{A} = Ae^{j\theta_A}$ and $\mathbf{B} = Be^{j\theta_B}$,

$$\mathbf{C} = \mathbf{A} \times \mathbf{B} = ABe^{j(\theta_A + \theta_B)}$$

Using the angle notation, $\mathbf{A} = A \underline{/\theta_A}$, $\mathbf{B} = B \underline{/\theta_B}$, and the equivalent result is

$$\mathbf{C} = \mathbf{A} \times \mathbf{B} = AB\underline{/(\theta_A + \theta_B)} = C\underline{/\theta_C} \tag{7.2a}$$

Expressed in words, the product of two complex numbers is another complex number whose amplitude is the product of the amplitudes and whose angle with respect to the x axis (real number line) is the sum of the angles of the two complex numbers. As usual, division is defined as the inverse of multiplication,

$$\mathbf{D} = \frac{\mathbf{A}}{\mathbf{B}} = \left(\frac{A}{B}\right) e^{(j\theta_A - \theta_B)}$$

and in angle notation,

$$\mathbf{D} = \frac{\mathbf{A}}{\mathbf{B}} = \frac{A\underline{/\theta_A}}{B\underline{/\theta_B}} = \left(\frac{A}{B}\right) \underline{/(\theta_A - \theta_B)} = D\underline{/\theta_D} \tag{7.2b}$$

To summarize, the complex number $\mathbf{A}$ can be written in two forms:

$$\mathbf{A} = A_x + jA_y$$

or

$$\mathbf{A} = Ae^{j\theta}$$

The two forms above are referred to as the rectangular and the polar forms, respectively. The latter is also written here as $\mathbf{A} = A \underline{/\theta}$. The relationships between (A_x, A_y) and (A, θ_A) are those given in Eqs. 7.1a and 7.1b.

We now link complex numbers with sinusoidal functions of time. Consider the complex functions of time expressed by $A(t) = Ae^{j\theta(t)}$. When located in the complex plane at some time, the complex number $\mathbf{A}(\theta)$ is always at a distance A from the origin, but its angle varies with time. Suppose $\theta(t)$ has the value ϕ at $t = 0$ and increases linearly in time with angular velocity ω, that is, $\theta(t) = \omega t + \phi$. Then,

$$\mathbf{A}(t) = Ae^{j(\omega t + \phi)} = A\cos(\omega t + \phi) + jA\sin(\omega t + \phi)$$

As indicated by this expression and in the sketch of Fig. 7.2, we have the important result that the real part of $\mathbf{A}(t)$ is a cosine function of time, $A\cos(\omega t + \phi)$. We can write this relationship between $\mathbf{A}(t)$ and this important temporal function as

$$A\cos(\omega t + \phi) = \text{Re}\{\mathbf{A}(t)\} = \text{Re}\{Ae^{j(\omega t + \phi)}\}$$

where Re stands for the expression "the real part of."

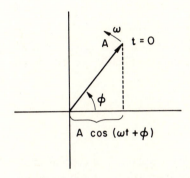

FIGURE 7.2.
The real part of a vector rotating counter-clockwise in the complex plane with frequency ω.

To summarize, $\mathbf{A}(t) = Ae^{j(\omega t + \phi)}$ represents a complex number which "rotates in" the complex plane at a fixed distance from the origin, returning to its initial position $A\,\underline{/\theta}$ every time ωt passes through 2π radians. The real part of this function, $\text{Re}\{\mathbf{A}(t)\}$, is a number that always lies on the real axis at a position defined by the expression $A\cos(\omega t + \phi)$.

Before starting to use complex numbers in electrical engineering, some further tools are needed. It is convenient to define a special complex number, the "complex conjugate," written $\mathbf{A}^*$, such that when $\mathbf{A} = Ae^{j\theta}$,

$$\mathbf{A}^* = Ae^{-j\theta} = A\cos\theta - jA\sin\theta$$

If $\mathbf{A}$ is multiplied by its complex conjugate, the result is the square of the magnitude of $\mathbf{A}$:

$$\mathbf{A} \times \mathbf{A}^* = (Ae^{j\theta}) \times (Ae^{-j\theta}) = A^2$$

We also write this as

$$\mathbf{A} \times \mathbf{A}^* = |\mathbf{A}|^2 \qquad (7.3)$$

where the parallel bar notation stands for the absolute value of the complex number $\mathbf{A}$. Expressed in words, the product of a complex number and its complex conjugate yields the absolute value squared of the complex number. This operation always yields a real number and is equal to the square of the amplitude of the polar form for $\mathbf{A}$. Also, the addition $\mathbf{A}^* + \mathbf{A}$ always gives a real number

$$Ae^{j\theta} + Ae^{-j\theta} = 2A\cos\theta$$

As a special case we have the useful identity

$$\frac{e^{j\theta} + e^{-j\theta}}{2} = \cos\theta \tag{7.4a}$$

For $\theta = \omega t + \phi$, this expression becomes

$$\frac{e^{j(\omega t + \phi)} + e^{-j(\omega t + \phi)}}{2} = \cos(\omega t + \phi) \tag{7.4b}$$

Thus, a cosine function can be thought of as the real part of a rotating vector or as half the sum of two vectors rotating in opposite directions. Finally, as left for an exercise in the problem set, it may be shown that

$$|\mathbf{A} \times \mathbf{B}| = |\mathbf{A}| \times |\mathbf{B}|$$

This result is used several times in later chapters.

The single most compelling reason to use complex numbers in electrical engineering is that their correct use in a calculation keeps track of two numbers simultaneously, the amplitude and phase of an oscillatory signal. The use of the complex plane to represent the complex number system shows this two-dimensionality very clearly.

7.2 THE IMPEDANCE FUNCTION REVISITED

We now start to apply some of these concepts to electrical systems. The three fundamental equations for resistors, inductors, and capacitors are

$$v_{ab} = i_{ab}R$$

$$v_{ab} = \frac{L\,di_{ab}}{dt}$$

$$i_{ab} = \frac{C\,dv_{ab}}{dt}$$

where v_{ab} is the voltage across the element and i_{ab} is the current through it. If i_{ab} has an exponential form, $i_{ab} = I_0 e^{st}$, then each of these three expressions can be solved for $v_{ab}(t)$ to yield the following ratios of voltage to current:

$$\frac{v_{ab}}{i_{ab}} = \begin{cases} R \\ sL \\ \dfrac{1}{sC} \end{cases}$$

These correspond to the respective impedance functions $Z(s)$ derived earlier. For reference, we repeat the definition of impedance: the impedance function of an element, or of a circuit made up of a number of elements, is the ratio of voltage across the element (or circuit) to the current through the element (or circuit) for signals varying exponentially in time.

We may now expand our understanding of $Z(s)$ by recognizing that the parameter s may be a complex number. In fact, since the set of all complex numbers includes the pure real numbers, we can just state that s is a complex number and write it in the rectangular form

$$\mathbf{s} = \alpha + j\omega$$

where α and ω are pure real numbers. This construction associates each complex value for $\mathbf{s}$ with a position in a complex plane (Fig. 7.3) which is identified by a certain distance α along the real axis and a certain distance ω along the imaginary axis. Within this framework there are several different forms that s can take on, each of which is associated with a particular temporal function when expressed as e^{st}:

1. At the origin both the real and imaginary parts of s are zero. Thus, the exponential form $i(t) = I_0 e^{st} = I_0 e^0 = I_0$, which corresponds to a dc current.

2. When s lies on the real axis but is not equal to zero, the waveforms are real exponentials of the form $e^{\pm \alpha t}$, which we found in Chapter 6 to describe the natural response of some circuits.

3. Next, consider the case when s is pure imaginary. In this case the values always come in complex conjugate pairs; that is, $\mathbf{s} = \pm j\omega$. The circuit example studied in Section 6.5 had this property since we found there that setting $Z_s(s) = 0$

FIGURE 7.3.
Complex exponential factors can be represented by their position in a complex plane.

Complex Plane

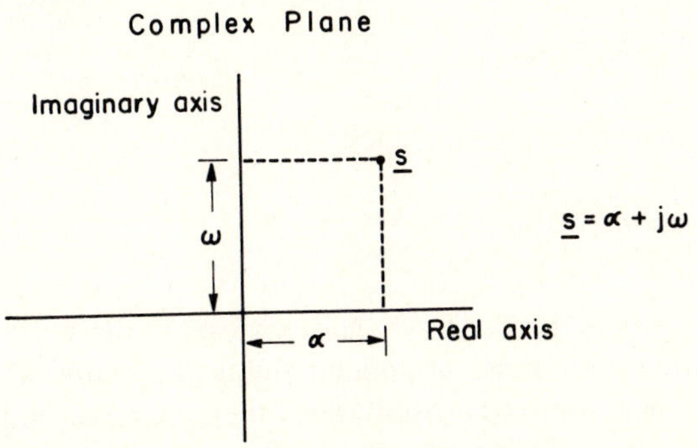

yielded two solutions for s, $s = \pm j \sqrt{(1/LC)}$. When we add the two solutions $I_1 e^{s_1 t}$ and $I_2 e^{s_2 t}$, we get

$$i(t) = I_1 e^{j\omega t} + I_2 e^{-j\omega t} \tag{7.5}$$

Since the physical electrical current $i(t)$ must be a real number, its imaginary part must vanish. That is,

$$\text{Im}\left[i(t) \right] = I_1 \sin(\omega t) + I_2 \sin(-\omega t) = 0$$

But $\sin(-\omega t) = -\sin(\omega t)$, so

$$I_1 \sin(\omega t) - I_2 \sin(\omega t) = 0$$

The requirement that the imaginary part of $i(t)$ vanishes implies $I_1 = I_2$. If we define another constant I_0 such that $I_0 = 2I_1$, then

$$i(t) = \frac{I_0 e^{j\omega t} + I_0 e^{-j\omega t}}{2}$$

and from Eq. 7.4a

$$i(t) = I_0 \cos(\omega t)$$

This mathematical expression represents a pure oscillatory motion in time at the frequency ω. This important result links the formal solution for the series L–C circuit to oscillatory functions of time.

Now there is no reason to require I_1 and I_2 to be real numbers in this derivation. If we substitute constant complex numbers (not functions of time!) into Eq. 7.5, we have

$$i(t) = \mathbf{I}_1 e^{j\omega t} + \mathbf{I}_2 e^{-j\omega t} \tag{7.6}$$

The imaginary part must still vanish, and we leave it as an exercise in the problem set to show that this implies

$$\mathbf{I}_2 = \mathbf{I}_1^*$$

Then, from Eq. 7.4b it follows that the temporal function

$$i(t) = I_0 \cos(\omega t + \phi) \tag{7.7}$$

is the real part of $\mathbf{I}(t) = \mathbf{I}_1' e^{j\omega t}$, where $\mathbf{I}_1' = \frac{1}{2} I_0 e^{j\phi}$ This represents the most general oscillatory function of time with constant angular frequency ω and an "initial" phase angle ϕ at $t = 0$. For sinusoidal oscillations then, $s = j\omega$, and the impedances of the three elemental circuit parameters are

$$\mathbf{Z} = \begin{cases} R \\ j\omega L \\ \dfrac{1}{j\omega C} \end{cases}$$

4. For any position in the complex plane off the axes, s has both real and imaginary parts. To understand this case we need to investigate another second-order system.

Consider the analysis of the following series R–L–C circuit after the switch has been opened (Fig. 7.4). Setting the series impedance function to zero yields the natural response

$$Z_s(s) = R + sL + \frac{1}{sC} = 0$$

which, after rearrangement, can be written

$$s^2 + \left(\frac{R}{L}\right)s + \frac{1}{LC} = 0$$

This second-order (quadratic) equation has the solution

$$s = \frac{-R}{2L} \pm \sqrt{\left(\frac{R}{2L}\right)^2 - \left(\frac{1}{LC}\right)} \qquad (7.8a)$$

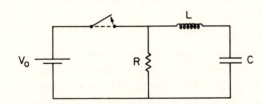

FIGURE 7.4.
Circuit that can lead to a character-istic frequency s with both real and imaginary parts.

If $1/LC > (R/2L)^2$, the square root operation generates an imaginary part to s and we have a result of the form

$$\mathbf{s} = -\alpha \pm j\omega \qquad (7.8b)$$

With s of the form given in Eq. 7.8b, the natural current is of the form

$$i(t) = I_0 e^{-\alpha t} \cos(\omega t + \phi)$$

Since the real part of s is negative (corresponding to exponential decay) and if α is less than ω, a plot of this function looks like that shown in Fig. 7.5. Now we can see why this second-order solution is termed underdamped. Here, because we have taken a $\alpha < \omega$, there are a number of cycles of the oscillation

FIGURE 7.5.
Waveform for a damped oscillatory current.

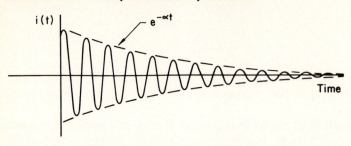

within a time $T = 1/\alpha$ and the sinusoid slowly decays with time. This result is totally analogous to the slow damping of the motion of a pendulum due to friction in the pivot point. In the series R–L–C electrical circuit, the resistor provides the "friction" which eventually dissipates the energy associated with the oscillation. The flow of current between the inductor and capacitor then slowly dies out. In other words, the energy sloshing back and forth between the capacitor and inductor is slowly dissipated in the form of heat in the resistor.

There is considerable literature dealing with electrical and mechanical systems which are of second order and which display natural frequencies of the form given by Eq. 7.8a. For example, if, as in Fig. 7.5, $\alpha < \omega$, the system is called underdamped. If $\alpha = \omega$, the system is critically damped, whereas if $\alpha > \omega$, it is termed overdamped. We study such second-order circuits in detail in Chapter 12. In the series R–L–C electrical circuit described by Eq. 7.8, the relative values of these three elements determine the type of damping for a particular circuit. Note that for any stable system the sign of the real part of s must be negative since otherwise the oscillations would grow with time (e.g., they would be proportional to $e^{+\alpha t}$).

In the next section and the subsequent two chapters we concentrate on pure imaginary values for s and review the elements of complex arithmetic needed to analyze systems exhibiting oscillatory behavior. This corresponds to the study of the forced response of systems to sinusoidal sources which have been in place for "a long time." In Chapter 10 we return to the problem of complete response when the forcing source is sinusoidal.

7.3 APPLICATION OF COMPLEX ALGEBRA TO OSCILLATORY SIGNALS

As we have seen, complex numbers can be written in two forms, the "rectangular" form

$$\mathbf{w} = x + jy$$

or the "polar" form

$$\mathbf{w} = re^{j\theta}$$

which has the equivalent notation

$$\mathbf{w} = r\underline{/\theta}$$

where the symbol $\underline{/\theta}$ means the angle θ measured counterclockwise from the real axis. The meaning of these two forms is illustrated in Fig. 7.6. The two forms are merely two ways of representing the position of a point in the complex plane using either rectangular or polar coordinates.

From a practical viewpoint, certain aspects of complex arithmetic are easier in the rectangular representation whereas some are easier in the polar system. In particular, addition and subtraction are easiest in rectangular form, and multiplication and division are easiest in polar coordinates.

FIGURE 7.6.
Representation of a complex number w in the complex plane.

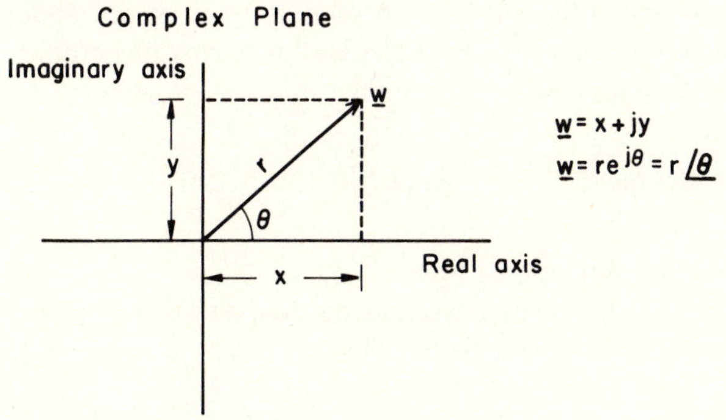

TEXT EXAMPLE 7.1 _____

If $\mathbf{A} = 5 + j5$ and $\mathbf{B} = 10e^{-j\pi/6}$, find $\mathbf{A} + \mathbf{B}$, $\mathbf{A} - \mathbf{B}$, $\mathbf{A} \times \mathbf{B}$, and $\mathbf{A}/\mathbf{B}$.

Solution: Addition and subtraction are much easier in rectangular coordinates, so we must change $\mathbf{B}$ to this form:

$$\mathbf{B} = 10\left[\cos\left(\frac{-\pi}{6}\right) + j\sin\left(\frac{-\pi}{6}\right)\right]$$

$$\mathbf{B} = 10\left[\cos\left(\frac{\pi}{6}\right) - j\sin\left(\frac{\pi}{6}\right)\right]$$

The angle $\pi/6$ radians equals 30° and

$$\mathbf{B} = 8.66 - j5$$

Adding (subtracting) the real and imaginary parts yields the sum (difference) of $\mathbf{A}$ and $\mathbf{B}$, so

$$\mathbf{A} + \mathbf{B} = 13.66$$

$$\mathbf{A} - \mathbf{B} = -3.66 + j10$$

To multiply and divide we change **A** to polar coordinates, $\mathbf{A} = 5\sqrt{2}\,\underline{/+45°}$. Since $\mathbf{B} = 10\,\underline{/-30°}$, we have

$$\mathbf{A} \times \mathbf{B} = 50\sqrt{2}\,\underline{/15°}$$

$$\frac{\mathbf{A}}{\mathbf{B}} = (\sqrt{2}/2)\underline{/75°}$$

The voltages and currents in physical electrical systems must be real numbers. The application of complex numbers to the analysis of such systems comes about because real sinusoidal time functions can also be expressed in terms of complex exponentials. For example, if $i(t) = I_0 \cos(\omega t + \phi)$, then

$$i(t) = Re\{I_0 e^{j(\omega t + \phi)}\} = Re\{\mathbf{i}(t)\}$$

The notation $\mathbf{i}(t)$ refers to a complex function of time.

The time function $\sin(\omega t + \phi)$ can also be expressed as the *real* part of a complex function by expressing the function in its cosine form. This is always possible since

$$\sin(\omega t + \phi) = \cos(\omega t + \phi - \pi/2) = \cos(\omega t + \gamma)$$

where $\gamma = \phi - (\pi/2)$. Thus, $\sin(\omega t + \phi) = Re\{e^{j(\omega t + \phi - \pi/2)}\}$.

As a specific example of the application of complex algebra to the analysis of oscillating electrical signals, consider the circuit shown in Fig. 7.7, for which the applied voltage is $v(t) = V_0 \cos(\omega t)$ and for which we wish to know the current. Since the circuit has been connected for a long time, only the forced response exists. However, the forcing is at a fixed nonzero frequency ω rather than the dc zero frequency forcing of Chapters 1 through 6. In complex notation the voltage is

$$v(t) = Re\{\mathbf{v}\} = Re\{V_0 e^{j\omega t}\}$$

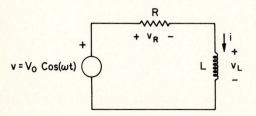

FIGURE 7.7.
Sinusoidal voltage source attached to an R–L circuit.

We do not yet know the current. However, from the definition of the impedance function we know that it is the ratio of the voltage to the current for *exponential time variations*, that is,

$$Z = \frac{v}{i}$$

Since $v(t) = V_0 e^{j\omega t}$ does display an exponential time form, it follows that

$$\mathbf{i} = \frac{\mathbf{v}}{\mathbf{Z}}$$

In this series R–L circuit Z is given by

$$Z(s) = R + sL$$

Now since $\mathbf{s} = j\omega$ for this pure oscillatory system, Z is itself a complex number and

$$\mathbf{i}(t) = \frac{\mathbf{v}(t)}{\mathbf{Z}} = \frac{\mathbf{v}(t)}{R + j\omega L}$$

This division problem is most easily carried out in polar coordinates. In polar form

$$\mathbf{Z} = R + j\omega L = |\mathbf{Z}|\,\underline{/\theta}$$

where

$$|\mathbf{Z}| = \sqrt{R^2 + \omega^2 L^2}$$

and where

$$\theta = \tan^{-1}\left(\frac{\omega L}{R}\right)$$

So the solution for the complex current is

$$\mathbf{i}(t) = \left(\frac{V_0}{|\mathbf{Z}|}\right) e^{j(\omega t - \theta)}$$

which for future reference can also be written with the $e^{j\omega t}$ term factored out as

$$\mathbf{i}(t) = \left(\frac{V_0}{|\mathbf{Z}|}\right) e^{-j\theta} e^{j\omega t} \tag{7.9}$$

When the real parts of the complex voltage $\mathbf{v}(t)$ and of the complex current $\mathbf{i}(t)$ are taken, two time functions are recovered:

$$v(t) = V_0 \cos(\omega t)$$

and

$$i(t) = \left(\frac{V_0}{|\mathbf{Z}|} \right) \cos(\omega t - \theta)$$

These two functions have the time dependent forms plotted in Fig. 7.8. The current has a magnitude given by $V_0/|\mathbf{Z}|$ but displays a phase difference when compared to the voltage. There is no significance to the magnitudes shown since the current and voltage are drawn to different scales.

Figure 7.8 illustrates the effect of the phase difference θ. We see that the current waveform peaks later in time than the voltage, and we say that the current "lags" the voltage. Without the inductor, this would not occur since the impedance function for $L = 0$ is a pure real number (equal to R), and the current would be

$$\mathbf{i}(t) = \left(\frac{V_0}{R} \right) e^{j\omega t}$$

Taking the real part of this function yields

$$i(t) = \left(\frac{V_0}{R} \right) \cos(\omega t)$$

There is no phase shift between $i(t)$ and $v(t)$ for systems in which only ideal resistors and sources are used. This explains why we could use pure real numbers for analysis for such systems. However, once inductors and capacitors are introduced, phase shifts occur due to the energy storage capabilities of these elements. The use of complex numbers allows us to keep track of both the amplitude and phase of $i(t)$ and $v(t)$ since each complex number contains two pieces of information.

FIGURE 7.8.
Voltage and current waveforms for the *R–L* circuit. For convenience the two signals have been given equal amplitude in the plots.

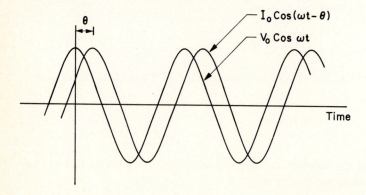

To summarize, when an oscillatory voltage is applied to a circuit consisting of resistors, capacitors, and inductors, the complex current is given by

$$i(t) = \frac{\mathbf{v}(t)}{\mathbf{Z}}$$

where $\mathbf{Z}$ is a complex number found by determining the total impedance of the network. As discussed earlier, the rules for adding $\mathbf{Z}$'s are identical to the rules for resistors. In series,

$$\mathbf{Z}_E = \mathbf{Z}_1 + \mathbf{Z}_2$$

and in parallel,

$$\mathbf{Z}_E = \frac{\mathbf{Z}_1\mathbf{Z}_2}{\mathbf{Z}_1 + \mathbf{Z}_2}$$

Often it is easiest to carry out some of these manipulations using s to represent the frequency, eventually converting to complex numbers by substituting $s = j\omega$.

TEXT EXAMPLE 7.2 _____

Find the current $i(t)$ and the voltage $v_L(t)$ in the circuit illustrated below.

$C = (2\pi)^{-1} \times 10^{-7}$ F
$L = (2\pi)^{-1} \times (10^{-3})$H
$R = 100\ \Omega$
$f = 10^5$ Hz

$v(t) = 100 \cos(\omega t + \frac{\pi}{4})$

Solution: Since $\mathbf{v} = \mathbf{i}\ \mathbf{Z}$, we can find $i(t)$ by taking the real part of the ratio of the complex voltage and the impedance:

$$i(t) = \mathrm{Re}\left\{\frac{\mathbf{v}}{\mathbf{Z}}\right\}$$

By inspection, we can relate the time function $v(t)$ to its complex representation by

$$v(t) = \mathrm{Re}\{100e^{j(\omega t + \pi/4)}\}$$

so

$$\mathbf{v}(t) = 100e^{j(\omega t + \pi/4)}$$

The total impedance is

$$Z = Z_1 + Z_2 = \frac{1}{sC} + \frac{RsL}{R + sL}$$

Setting $s = j\omega$,

$$\mathbf{Z} = \frac{1}{j\omega C} + \frac{j\omega L R}{R + j\omega L}$$

Since $f = 10^5$ Hz, $\omega = 2\pi(10^5)$ and

$$\omega C = 0.01 \text{ ohms}$$

$$\omega L = 100 \text{ ohms}$$

$\mathbf{Z}$ can now be evaluated as follows:

$$\mathbf{Z} = \frac{10^2}{j} + \frac{j\,10^4}{100 + j\,100}$$

Each of these fractions has a complex number in the denominator, which is poor mathematical form. To remove these we can use a trick involving multiplication of the numerator and denominator of each fraction by the complex conjugate of their respective denominators. This operation will *always* convert the denominator into a pure real number.

$$\mathbf{Z} = \frac{10^2}{j} \left(\frac{-j}{-j}\right) + \left(\frac{j\,10^4}{100 + j\,100}\right) \left(\frac{100 - j\,100}{100 - j\,100}\right)$$

Then,

$$\mathbf{Z} = -j\,100 + \frac{(j\,10^4)(100 - j\,100)}{2 \times 10^4} = -j\,100 + (50 + j\,50) = 50 - j\,50$$

For future reference we note that the rectangular and polar representations of the three impedances $\mathbf{Z}_1$, $\mathbf{Z}_2$, and $\mathbf{Z}$ are

$$\mathbf{Z}_1 = -j\,100 = 100\underline{/-\pi/2}$$

$$\mathbf{Z}_2 = 50 + j\,50 = 50\sqrt{2}\underline{/\pi/4}$$

and

$$\mathbf{Z} = 50 - j\,50 = 50\sqrt{2}\underline{/-\pi/4}$$

The complex current is then, in angle notation,

$$\mathbf{i}(t) = \frac{100\underline{/\omega t + \pi/4}}{50\sqrt{2}\underline{/-\pi/4}}$$

Using the exponential form,

$$\mathbf{i}(t) = \sqrt{2}e^{j(\omega t + \pi/2)}$$

Taking the real part,

$$i(t) = 1.414 \cos(\omega t + \pi/2) \ \text{A}$$

For reference this latter function can also be written

$$i(t) = -1.1414 \sin(\omega t) \ \text{A}$$

To find the voltage across the parallel R–L network we use the voltage divider principle:

$$\mathbf{v}_L = \left(\frac{\mathbf{Z}_2}{\mathbf{Z}}\right) \mathbf{v}$$

$$\mathbf{v}_L(t) = \left(\frac{50\sqrt{2}\,\underline{/\pi/4}}{50\sqrt{2}\,\underline{/-\pi/4}}\right) \left[100\underline{/(\omega t + \pi/4)}\right]$$

$$\mathbf{v}_L(t) = 100\underline{/(\omega t + 3\pi/4)} = 100e^{j(\omega t + 3\pi/4)}$$

Finally, the time function is the real part of $\mathbf{v}_L(t)$:

$$v_L(t) = 100 \cos(\omega t + 3\pi/4) \ \text{V}$$

The essential feature of the analysis of single-frequency systems is easy to state in words. The circuit analysis methods of Chapters 1 through 4 may all be applied to these systems if we replace the resistors with impedances and use complex algebra. The voltage divider and current divider laws hold, the superposition principle and Thevenin's theorem are true, and so forth. Once the complex arithmetic is carried out, the real functions of time are extracted by taking the real part of the complex solution. It is crucial to practice these manipulations, however, and a number of examples are given in the problem set.

7.4 PHASOR NOTATION AND MAXIMUM VALUE PHASORS

In Text Example 7.2 the $e^{j\omega t}$ term can be isolated as a multiplicative factor in both the voltage and current functions. In the "phasor notation" this term is suppressed and is "understood" to accompany the phasor, going along for the ride, so to speak, while all complex arithmetic is carried out on the remaining factor. There are two types of phasors: maximum value and "effective" value. Maximum value phasors are presented here, and effective value phasors are presented in Section 7.5. Both types are discussed in the problem set. In subsequent chapters we shall identify which type of phasor is used at the beginning of the chapter and stick to that usage throughout the chapter. In this chapter we affix the subscript m to every maximum value phasor.

A phasor is denoted here by a " ^ " over the quantity. For example, in the analysis associated with the circuit in Fig. 7.7 the temporal voltage is

$$v(t) = V_m \cos (\omega t)$$

the complex voltage is

$$\mathbf{v}(t) = V_m e^{j\omega t}$$

and the maximum value phasor is found by suppressing the $e^{j\omega t}$ term leaving

$$\hat{V}_m = V_m \underline{/0} = V_m$$

In terms of the phasor, the real voltage is $v(t) = Re\{\hat{V}_m e^{j\omega t}\}$. The phasor is in this case a real number equal to the amplitude of the real cosine function, that is, the maximum value of the cosine function. In the same circuit the current phasor is found by suppressing $e^{j\omega t}$ in Eq. 7.9:

$$\hat{I}_m = \left(\frac{V_m}{|\mathbf{Z}|}\right) e^{-j\theta} = I_m e^{-j\theta} = I_m \underline{/-\theta}$$

The corresponding real current is

$$i(t) = Re\{\hat{I}_m e^{j(\omega t)}\} = I_m \cos(\omega t - \theta)$$

In Text Example 7.2 the maximum value phasors are

$$\hat{V}_m = 100 \underline{/\pi/4}$$

$$\hat{I}_m = \sqrt{2} \underline{/\pi/2}$$

and

$$\hat{V}_{Lm} = 100 \underline{/(3\pi/4}$$

Figure 7.9 shows the relationship of a particular phasor $\hat{I}_m = I_m \underline{/-\pi/4}$ to the complex current and to the real sinusoidal current. The complex vector rotates in a circle in the complex plane once every time period $T = 2\pi/\omega$. The angle θ of the phasor gives the *initial angle* at $t = 0$ in this motion, and the amplitude of the phasor yields the radius of the circle. Thus, the phasor $\hat{I}_m$ is a complex number which yields

FIGURE 7.9.
The complex function $i(t)$ rotates once around the complex plane in a time $T = 2\pi/\omega$, where ω is the angular frequency in rad/sec and the initial position is given by the phasor $\emptyset I_m$.

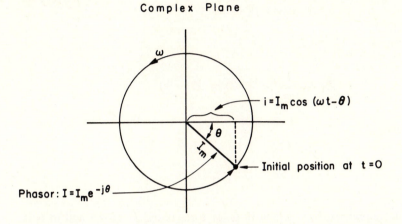

Complex Plane

$i = I_m \cos(\omega t - \theta)$

Initial position at $t = 0$

Phasor: $I = I_m e^{-j\theta}$

the initial value (or position) of a complex function of time $i(t)$ that rotates with an angular velocity ω in the complex plane. The projection of this rotating vector on the real axis is the real temporal current function, $i(t) = I_m \cos(\omega t - \pi/4)$.

To summarize, systems that oscillate at a single frequency ω may be analyzed as follows. All voltages $v(t)$ and currents $i(t)$ are considered as the real parts of the complex function of time, $\mathbf{v}(t)$ and $\mathbf{i}(t)$. These complex functions of time are, in turn, fully characterized by suppressing the $e^{j\omega t}$ term and dealing entirely with the phasors $\hat{I}_m$ and $\hat{V}_m$, which are just complex numbers. A given system is analyzed in terms of complex impedances and the sources that exist. To reconstruct any desired time function $v_k(t)$ or $i_k(t)$ in the system, the corresponding phasor $\hat{V}_{km}$ or $\hat{I}_{km}$ is determined. Then,

$$v_k(t) = Re\{\hat{V}_{km}e^{j\omega t}\}$$

$$i_k(t) = Re\{\hat{I}_{km}e^{j\omega t}\}$$

TEXT EXAMPLE 7.3 _____

In the circuit below find the phasor voltage corresponding to $i(t)$ and $v(t)$. Find the phasors corresponding to i_1 and i_2 and show that KCL holds for those phasors, that is, $\hat{I} = \hat{I}_1 + \hat{I}_2$.

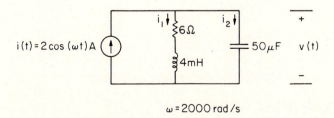

$i(t) = 2\cos(\omega t) A$ i_1 6Ω i_2 $50\mu F$ $v(t)$ $+$ $-$

$4mH$

$\omega = 2000$ rad/s

Solution: First, we evaluate the impedance associated with each leg of the parallel network:

$$\mathbf{Z}_1 = R + j\omega L$$

$$\mathbf{Z}_1 = 6 + j8 = 10\underline{/53.1°}$$

since $\tan^{-1}(8/6) = 53.1°$. Also,

$$\mathbf{Z}_2 = \frac{1}{j\omega C} = -j10 = 10\underline{/-90°}$$

The total impedance is

$$\mathbf{Z}_T = \frac{\mathbf{Z}_1\mathbf{Z}_2}{\mathbf{Z}_1 + \mathbf{Z}_2}$$

The denominator equals $6 - 2j$, which in polar form is $6.3\underline{/-18.4°}$; and in polar form

$$\mathbf{Z}_T = \left(\frac{100}{6.3}\right)\underline{/53.1° + (-90°) - (-18.4°)}$$

$$= 15.8\underline{/-18.4}$$

The phasor for the current source is

$$\hat{I}_m = 2\underline{/0°}$$

In this analysis the current is called the reference phasor since its phase angle is zero. To find the voltage phasor we use

$$\hat{V}_m = \hat{I}_m\mathbf{Z}_T = 31.6\underline{/-18.4°}$$

and the time function is

$$v(t) = 31.6 \cos(\omega t - 18.4°) \text{ V}$$

The two current phasors can be found from $\hat{V}_m$ using $\hat{I}_{1m} = \hat{V}_m/\mathbf{Z}_1$ and $\hat{I}_{2m} = \hat{V}_m/\mathbf{Z}_2$:

$$\hat{I}_{1m} = 3.16\underline{/-71.6°}$$

$$\hat{I}_{2m} = 3.16\underline{/+71.6°}$$

To prove that KCL holds, we can add these in rectangular coordinates:

$$\hat{I}_m = (1 - j3) + (1 + j3) = 2$$

which is indeed equal to the value of the current phasor $\hat{I}_m$. If a calculator is used it is very likely that the round-off procedure yields answers which vary slightly from the answers above. (See the note on page 245 concerning angle units.)

7.5 EFFECTIVE VALUE PHASORS

We turn now to the study of effective value phasors, which are sometimes called rms phasors. As we shall see, these are related to maximum value phasors through the constant factor $\sqrt{2}$, which is a pure real number. This means that all of the complex algebra discussed above remains the same. The effective value phasors (denoted here with a subscript e) are just smaller in absolute value than the corresponding maximum value phasors (denoted here with a subscript m). For example, the relationships between the maximum value phasor and the effective value phasor for the same voltage $\mathbf{v}(t)$ are

$$\hat{V}_e = \left(\frac{1}{\sqrt{2}}\right)\hat{V}_m = \left(\frac{\sqrt{2}}{2}\right)\hat{V}_m = 0.707\,\hat{V}_m \tag{7.10a}$$

and

$$\hat{V}_m = \sqrt{2}\,\hat{V}_e = 1.414\,\hat{V}_e \tag{7.10b}$$

This clumsy subscript notation will only be used in this chapter and will be dropped in subsequent chapters, where all the phasors will either be effective value phasors *or* maximum value phasors throughout a given chapter or section.

Effective value phasors are introduced since they more realistically characterize the effective power which a sinusoidal voltage or current can supply. To be specific, if we have an oscillatory current $i(t) = I_m(\cos \omega t + \theta)$ flowing through a resistor of value R, the average energy dissipated in the resistor during a long time interval T is *not* equal to $I_m^2 R$. Our goal is to find notation such that the phasor current amplitude squared (and multiplied by R) does directly yield the average power dissipated in a resistor. To do this we must first review how to find the average value of a function.

The method by which one can find the average value of a continuous function $i(t)$ in the interval 0 to T can be understood with reference to Fig. 7.10, where

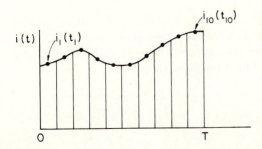

FIGURE 7.10.
Diagram for estimating the average value of $i(t)$ using ten intervals.

we have broken up the time interval into ten discrete segments of length Δt. An approximation to the average value of $i(t)$ is just the sum of the values of the function at the center of each interval divided by the number of intervals:

$$<i(t)> \simeq \frac{i_1(t_1) + i_2(t_2) + \cdots + i_{10}(t_{10})}{10}$$

If we divide the interval into N equal segments, this becomes

$$<i(t)> \simeq \frac{i_1(t_1) + \cdots + i_N(t_N)}{N} = \left(\frac{1}{N}\right) \sum_{k=1}^{N} i_k(t_k)$$

We can multiply and divide this expression by the interval Δt without changing its value, so

$$<i(t)> \simeq \left(\frac{1}{N\Delta t}\right) \sum_{k=1}^{N} i_k(t_k)\Delta t = \left(\frac{1}{T}\right) \sum_{k=1}^{N} i_k(t_k)\Delta t$$

since $N\Delta t = T$. In the limit as Δt goes to zero, the summation becomes an integral, and we have the exact result

$$<i(t)> = \left(\frac{1}{T}\right) \int_0^T i(t)dt = I_{av} \tag{7.11}$$

Obviously, if $i(t)$ has a constant value I_0, then I_{av} equals I_0 and the average value and the dc value are the same. For a general sinusoidal current, $i(t) = I_m \cos(\omega t + \phi)$. Then, the integration Eq. 7.11 yields

$$I_{av} = \frac{I_m}{\omega T}\left[\sin(\omega T + \phi) - \sin \phi\right]$$

where I_m is the maximum value of the oscillatory function. If ωT is an integer multiple of 2π, $\sin(\omega T + \phi) = \sin \phi$ and $I_{av} = 0$. This shows that the average value of a sinusoid integrated over an integer number of cycles is zero. Even if we do not integrate over an integer number of cycles but let T become very large (that is, we integrate over many cycles), I_{av} approaches zero anyway since it is proportional to $1/T$.

We now apply these results to electrical power calculations. Suppose a current $i(t)$ is flowing through a resistance R. The instantaneous power delivered to the resistance is $p(t) = v(t)i(t) = i^2(t)R$. We now *define* the power P delivered to the resistance as the "long time average" of the instantaneous power:

$$P = \lim_{T\to\infty} <p(t)> = \lim_{T\to\infty} \left(\frac{1}{T}\right) \int_0^T v(t)i(t)dt = \lim_{T\to\infty} \left(\frac{1}{T}\right) \int_0^T Ri^2(t)dt \tag{7.12}$$

The power P defined in this way which is supplied by a constant (dc) current, I_0, is simply equal to $I_0^2 R$. However, for a sinusoidal current, $i(t) = I_m \cos(\omega t + \phi)$ and

$$P = \lim_{T \to \infty} \left(\frac{1}{T} \right) \int_0^T I_m^2 R \cos^2(\omega t + \phi) dt$$

Using the trigonometric identity $\cos^2 x = \frac{1}{2} + \frac{1}{2} \cos 2x$, the integral may be evaluated and

$$P = \lim_{T \to \infty} \left[\left(\frac{I_m^2 R}{2T} \right) \int_0^T dt + \left(\frac{I_m^2 R}{2T} \right) \int_0^T \cos(2\omega t + 2\phi) dt \right]$$

Since the second term alone is simply the average value of a sinusoid at twice the frequency, we showed above that in the limit $T \to \infty$ this integral is zero. Evaluating the first integral, which has the value T, yields

$$P = \frac{I_m^2 R}{2} \tag{7.13}$$

To summarize, Eq. 7.13 yields the average power dissipated in a resistor R carrying a sinusoidal current with maximum value I_m.

We now define the effective value of the current I_{eff} in such a way that the average power P derived in Eq. 7.13 is equal to the power delivered to a resistor R by a constant dc current equal to I_{eff}; that is,

$$P = \left(I_{eff} \right)^2 R \tag{7.14}$$

Setting Eq. 7.13 equal to Eq. 7.14 and solving for I_{eff}^2 yields

$$I_{eff}^2 = \frac{I_m^2}{2}$$

so

$$I_{eff} = \frac{I_m}{\sqrt{2}} = \left(\frac{\sqrt{2}}{2} \right) I_m \tag{7.15}$$

Equation 7.15 is related to Eq. 7.10a and shows that the effective value of a sinusoidal current is $\sqrt{2}/2$ times the maximum value I_m. Equation 7.13 shows that the mean power delivered to a resistor is one-half the value of power delivered by a constant current equal to I_m. Note that if we evaluate the power over exactly one cycle of the sinusoidal current, we obtain the same result as found above as T goes to ∞. This is true for any repetitive or "periodic" function of time. Similarly, for an oscillatory voltage we may define the effective value through

$$V_{eff} = \frac{V_m}{\sqrt{2}} = 0.707 \, V_m$$

The effective value is also very often called the root mean square value (I_{rms}) since the effective value is the square *root* of the *mean* value of the *square* of the function $i(t)$. In this text we shall continue to use the term effective value rather than "rms value."

As an example of this terminology, we note that home power systems are described by their effective value. That is, the so-called "120-volt" home voltage actually corresponds to a function of time of the form

$$v(t) = 170 \cos[(2\pi)60t] = \sqrt{2}(120) \cos(377t)$$

The maximum value of the voltage is actually 170 V. It is the effective value that equals 120 V. Suppose this voltage is applied to a lamp with a resistance of 144 ohms. The current through the lamp will be

$$i(t) = \frac{v(t)}{R} = \sqrt{2}\left|\frac{120}{144}\right| \cos(377t) = \sqrt{2} \, [0.833 \cos(377t)]$$

From Eq. 7.14b, the effective values for the voltage and current are thus 120 V and 0.833 A, respectively. The average power can now be found from

$$P = (I_{eff})^2 R$$

which yields 100 watts. The energy supplied if the lamp is on for 1 hour is 100 watt-hours, or 0.1 kW-hr. At an energy charge of 10 cents per kilowatt-hour, the cost will be 1 cent. The instantaneous power to the "100-watt" lamp actually varies between the limits of 200 and 0 watts, since

$$p(t) = i(t)v(t) = 100 + 100 \cos(754t)$$

This expression has a long-time average value of 100 watts since the average value of the cosine term vanishes for large T.

To follow normal practice, in the next two chapters when we describe sinusoidal voltages and currents in phasor notation, the phasor value will correspond to the effective value, except when otherwise specifically stated. For $v(t) = V_m \cos(\omega t + \phi)$ our practice will be to describe v by the phasor $\hat{V}_e = V_0 \,\underline{/\alpha}$, where $V_0 = V_m/\sqrt{2}$. Conversely, given an effective value phasor, say, $\hat{I}_e = I_0 \,\underline{/\phi}$, the temporal function is

$$i(t) = \sqrt{2} \, \text{Re}(\hat{I}e^{j\omega t}) = \sqrt{2} \, I_0 \cos(\omega t + \phi) = I_m \cos(\omega t + \phi)$$

Since for linear operations all the phasors can be multiplied by the same factor

$1/\sqrt{2}$, none of our previous results changes and the phasor manipulations can be done using either maximum value phasors or effective value phasors. In the problem set for this chapter we use both types of phasors to develop some feeling for their interrelationship.

7.6 SUMMARY

In this chapter we have reviewed complex arithmetic and showed how it is used in electrical engineering to describe oscillatory signals. The problem set below is very important since the use of complex numbers requires considerable practice. All of the theorems and the relationships such as the voltage divider principle, which we developed in earlier chapters, hold in complex notation. Many of these tools are used in the problems below.

PRACTICE PROBLEMS
AND
ILLUSTRATIVE EXAMPLES

NOTE: Technically speaking, it is mathematically incorrect to use the notation $5\underline{/30°}$ or $5e^{j30°}$, since the exponent should be dimensionless and the angle should be expressed in radians. We shall, however, interpret $5e^{j30°}$ and $5\underline{/30°}$ as $5e^{j\pi/6}$ whenever necessary because of the convenience of such a representation. Similarly it is customary, but not consistent, to write $\cos(\omega t + 45°)$, where ω is in radians per second and therefore ωt is in radians. It should be understood that $\cos(\omega t + 45°)$ means $\cos(\omega t + \pi/4)$. In the answers below we have sometimes omitted the units for voltage and current for brevity. Also, the student may benefit by reading Sections 8.1 and 8.2 before working on the latter portion of the problem set since some of the material on impedances is reviewed there.

Problem 7.1

Carry out the complex arithmetic and express the answers in both rectangular and polar forms:

(a) $(5 - j12) + 15\underline{/53.1°}$ Ans.: $14 + j0 = 14\underline{/0°}$

(b) $(5 - j12) \times (15\underline{/53.1°})$ Ans.: $195\underline{/-14.3°} = 189 - j48$

(c) $(5 - j12)/(15\underline{/53.1°})$ Ans.: $0.866\underline{/-120.5°} =$
$$-0.44 - j(0.75)$$

(d) $(3 - j5)/(2 + j4)$ Ans.: $-0.7 - j(1.1) = 1.3\underline{/-122.5°}$
$$= 1.3\underline{/+237.5°}$$

(e) $10e^{j\pi/6}$ Ans.: $10\underline{/30°} = 8.66 + j5$

Problem 7.2

Let $\mathbf{A} = 10\underline{/30°}$, $\mathbf{B} = 10\underline{/150°}$, and $\mathbf{C} = 5\sqrt{3} - j5$. Carry out the complex arithmetic and express the answer in both rectangular and polar forms for the operations below.

(a) $\mathbf{A} + \mathbf{B}$

(b) $(\mathbf{A} + \mathbf{B} - \mathbf{C})/\mathbf{A}$

(c) $|\mathbf{A} + \mathbf{C}|$

(d) $(\mathbf{A})(\mathbf{B}/\mathbf{C})$

(e) $\mathbf{C}^2 - \mathbf{A}^2$

(f) $|\mathbf{A}|^{1/2}$

(g) $\mathbf{A}^{1/2}$

(h) Find $\mathbf{X}$ if $\mathbf{AX} = \mathbf{C}$

(i) Find $\mathbf{X}$ if $\mathbf{A} + \mathbf{C} + \mathbf{X} = \mathbf{B}$

Problem 7.3

Given $A = |A|e^{j\theta} = a + jb$, write the following in polar and rectangular forms:

(a) A^* Ans.: $A^* = |A|\underline{/-\theta} = a - jb$

(b) $A + A^*$ Ans.: $2a + j(0) = 2a\underline{/0°}$

(c) $A - A^*$ Ans.: $0 + j(2b) = 2b\underline{/90°}$

(d) A^*A Ans.: $|A|^2\underline{/0°} = a^2 + b^2 + j(0)$

Problem 7.4

Prove that

$$\cos\theta = \frac{e^{j\theta} + e^{-j\theta}}{2}$$

$$\sin\theta = \frac{e^{j\theta} - e^{-j\theta}}{j2}$$

Problem 7.5

Express the following in exponential form:

(a) $3 + j\sqrt{3}$ Ans.: $2\sqrt{3}e^{j(30°)} = 2\sqrt{3}e^{j\pi/6}$

(b) $\cos(\omega t + 60°) + j\sin(\omega t + 60°)$ Ans.: $e^{j(\omega t + 60°)} = e^{j(\omega t + \pi/3)}$

(c) $\sin(\omega t + 60°)$ Ans.: $(1/2j)\{e^{j(\omega t + 60°)} - e^{-j(\omega t + 60°)}\}$
or Im $\{e^{j(\omega t + 60°)}\}$

(d) $5\cos(\omega t + 30°)$ Ans.: $\frac{1}{2}\{5e^{-j(\omega t + 30°)} + 5e^{j(\omega t + 30°)}\}$
or Re$\{5e^{j(\omega t + 30°)}\}$

Problem 7.6

Express the following functions of time in rectangular form:

(a) $5e^{j(\omega t - 60°)}$

(b) $[10e^{j(\omega t - 30°)}]\,[10e^{-j(\omega t - 30°)}]$

(c) $\text{Re}\{\,20e^{j(\omega t - 30°)}\}$

(d) $\text{Re}\{\,5e^{j(\omega t + 90°)}\}$

(e) $20e^{j(\omega t + 60°)}\,/\,5e^{j(\omega t + 30°)}$

Problem 7.7

Find the following:

(a) $\text{Re}\{\,10e^{j\pi/3}\}$ Ans.: 5

(b) $\text{Im}\{\,20e^{-j30°}\}$ Ans.: -10

(c) $\text{Re}\{\,10e^{j60°}\,e^{j100t}\,\}$ Ans.: $10\cos(100t + 60°)$

(d) $\text{Re}\{\,10e^{-20t}\,e^{j60°}\,e^{j100t}\,\}$ Ans.: $10e^{-20t}\cos(100t + 60°)$

Problem 7.8

Find the sum

$$\hat{I}_1 e^{(-2+j100)t} + \hat{I}_1^* e^{(-2-j100)t}$$

where $\hat{I}_1 = 10\,\underline{/30°}$.

$$\text{Ans.: } 20e^{-2t}\cos(100t + 30°)$$

Problem 7.9

Let $\hat{V}_1 = 20\,\underline{/30°}$ and $v_1 = \hat{V}_1 e^{j\omega t}$. Let $\hat{V}_2 = 10\,\underline{/-90°}$ and $v_2 = \hat{V}_2 e^{j\omega t}$. Find:

(a) $\text{Re}(v_1 + v_2)$

(b) $|v_1|$

(c) Show that $|v_1 + v_2| = |\hat{V}_1 + \hat{V}_2|$

(d) Show that $|\hat{V}_1 + \hat{V}_2| \neq |\hat{V}_1| + |\hat{V}_2|$

(e) Show that $|\hat{V}_1\hat{V}_2| = |\hat{V}_1| \cdot |\hat{V}_2|$

Problem 7.10

Show that $i(t) = I_1 e^{j\omega t} + I_2 e^{-j\omega t}$ is a real number if $I_2 = I_1^*$. (*Hint:* Expand $i(t)$ and show that the imaginary part equals zero if $I_2^* = I_1$.)

Problem 7.11

(a) Show that $|A \times B| = |A| \times |B|$ (*Hint:* Use the polar form for **A** and **B**.)

(b) Show that $|A + B| \neq |A| + |B|$ except in special cases.
(*Hint:* Use the rectangular form for **A** and **B**.)

Problem 7.12

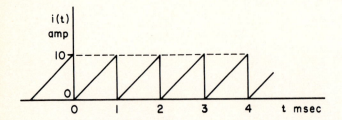

The "sawtooth" current shown above flows through a resistor $R = 10 \, \Omega$. Find the average power dissipated in the resistor. (*Hint:* The function $i(t) = 10^4 t$ for the cycle between $t = 0$ and $t = 1$ milliseconds. Use Eq. 7.12 and the fact quoted in the text that for a periodic function an integral over one full period yields the same result as the $\lim_{T \to \infty}$).

$$\text{Ans.: } P = 333 \text{ W}$$

Problem 7.13

Express the following time functions using maximum value phasors in polar and rectangular forms:

(a) $v(t) = 15 \cos(377t - 60°)$ Ans.: $\hat{V}_m = 15\underline{/-60°} = 7.5 - j(13)$

(b) $i(t) = 10 \sin(10^5 t + \pi/4)$ Ans.: $\hat{I}_m = 10\underline{/-45°} = 7.07 - j(7.07)$

(c) $i(t) = 5 \cos(10^3 t + 30°)$ Ans.: $\hat{I}_m = 5\underline{/30°} = 4.33 + j(2.5)$

Problem 7.14

For the same time functions given in Problem 7.13 find the effective value phasors.

$$\text{Ans.: } \hat{V}_e = (15\sqrt{2}/2)\underline{/-60°} =$$
$$5.3 - j(9.2)$$
$$\hat{I}_e = (10\sqrt{2}/2)\underline{/-45°} = 5 - j(5)$$
$$\hat{I}_e = (5\sqrt{2}/2)\underline{/30°} =$$
$$3.06 + j(1.77)$$

Problem 7.15

Give the effective and maximum value phasors corresponding to the following time functions:

(a) $i(t) = 10 \cos(\omega t + 45°)$
(b) $i(t) = 5\sqrt{2} \cos(100t - 60°)$
(c) $v(t) = 10 \sin(10^5 t + \pi/2)$

Problem 7.16

The following maximum value phasors represent time functions having angular frequency $\omega = 10^3$ rad/sec. Write the time functions:

(a) $\hat{V}_m = 6e^{j\pi/3}$

Ans.: $v(t) = 6\cos(10^3 t + 60°)$

(b) $\hat{I}_m = 5\,\underline{/-20°}$

Ans.: $i(t) = 5\cos(10^3 t - 20°)$

(c) $\hat{V}_m = 5 + j5$

Ans.: $v(t) = 7.07\cos(10^3 t + 45°)$

Problem 7.17

Express the maximum value phasors in the answers to Problem 7.16 as effective value phasors.

Ans.: (a) $\hat{V}_e = (6/\sqrt{2})\,\underline{/60°}$

(b) $\hat{I}_e = (5/\sqrt{2})\,\underline{/-20°}$

(c) $\hat{V}_e = 5\,\underline{/45°}$

Problem 7.18

For each of the following effective value phasors, (a) through (d), write expressions for both the maximum value phasor and the corresponding real temporal function if $\omega = 100$ rad/sec. Also find the impedances given by the ratios in (e) and (f).

(a) $\hat{I}_{ie} = 10\,\underline{/45°}$

(b) $\hat{I}_{2e} = j5\sqrt{2}$

(c) $\hat{I}_{ie} + \hat{I}_{2e}$

(d) $\hat{V}_e = j20$

(e) $\hat{V}_e/\hat{I}_{ie}$ from above values

(f) $\hat{V}_e/\hat{I}_{2e}$ from above values

Problem 7.19

In Problem 7.22 below, three voltage time functions are given in the figure. Find the corresponding maximum and effective value phasors.

EXAMPLE 7.1

In the figure below, three of the currents are given. Find $i_4(t)$.

$i_1(t) = 5\cos(10^4 t - 90°)$

$i_2(t) = 10\cos(10^4 t - 30°)$

$i_3(t) = -10\sin(10^4 t)$

Solution: Given i_1, i_2, and i_3, we can find i_4 since we know from KCL that $i_4 = -(i_1 + i_2 + i_3)$. Using maximum value phasors,

$$\hat{I}_{1m} = 5\underline{/-90°} = -j5$$

$$\hat{I}_{2m} = 10\underline{/-30°} = 8.66 - j5$$

$$\hat{I}_{3m} = 10\underline{/+90°} = j10$$

By KCL, $\hat{I}_{4m} = -(\hat{I}_{1m} + \hat{I}_{2m} + \hat{I}_{3m}) = -8.66 = 8.66\underline{/180°}$ and $i_4(t) = 8.66\cos(10^4t - 180°)$. The corresponding effective value phasors are $\hat{I}_{1e} = (5\sqrt{2}/2)\underline{/-90°}$; $\hat{I}_{2e} = (10\sqrt{2}/2)\underline{/-30°}$; and $\hat{I}_{3e} = (10\sqrt{2}/2)\underline{/90°}$. They must also add up to yield $\hat{I}_{4e} = (8.66\sqrt{2}/2)\underline{/180°}$ since each term in the sum $\hat{I}_{4e} = -(\hat{I}_{1e} + \hat{I}_{2e} + \hat{I}_{3e})$ just differs from the maximum value phasor by the factor $\sqrt{2}/2$.

Problem 7.20

In Example 7.1 suppose the current i_2 flows through some impedance $\mathbf{Z}_2$ to ground. What is the effective value phasor corresponding to the voltage of the node if $\mathbf{Z}_2 = 5\underline{/45°}$? Under these same conditions (e.g., $\mathbf{Z}_2$ is in place and the above voltage is maintained), what impedance $\mathbf{Z}_3$ to ground must the current i_3 flow through?

Problem 7.21

For the situation described in Example 7.1, suppose the maximum value phasor of the node voltage is $50\underline{/+15°}$. Show that the current i_1 cannot be flowing through a pure impedance $\mathbf{Z}_1$ to ground (that is, i_1 must flow to another node). (*Hint:* The real part of any impedance $\mathbf{Z}$ corresponds to the resistance which must be a positive number.)

Problem 7.22

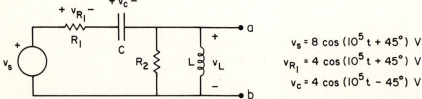

$$v_s = 8\cos(10^5t + 45°)\ V$$
$$v_{R_1} = 4\cos(10^5t + 45°)\ V$$
$$v_c = 4\cos(10^5t - 45°)\ V$$

Using KVL, find $v_L(t)$.

$$\text{Ans.: } v_L(t) = 4\sqrt{2}\cos(10^5t + 90°)V$$
$$= -4\sqrt{2}\sin(10^5t)V$$

EXAMPLE 7.2

Make a phasor diagram using effective value phasors which illustrate that KVL holds in the circuit for Problem 7.22, that is, $\hat{V}_s = \hat{V}_{R1} + \hat{V}_C + \hat{V}_L$.

Solution: The four effective value phasors are shown in the sketch below.

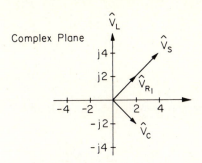

To show that the three phasors add up to $\hat{V}_S$ we place them "tail to tip" in sequence as shown below, which corresponds to addition of the complex numbers. Indeed, $\hat{V}_{R1} + \hat{V}_C + \hat{V}_L = \hat{V}_s$.

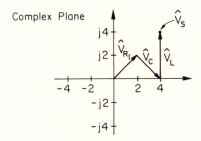

Problem 7.23

In the circuit of Problem 7.22 let $R_2 = 1$ kΩ and L $= 10$ mH. Find the maximum and effective value phasor current supplied by the source. Express the time function for this current.

Problem 7.24

Find the value of R_1 and the capacitor in Problem 7.22. Let $R_2 = 1$ kΩ and L $= 10$ mH, which are the values also used in Problem 7.23.

Problem 7.25

Suppose a voltage $v(t) = 10\sqrt{2} \cos(10^3 t)$ V is applied across an impedance and the current is $i = 4\sqrt{2} \cos(10^3 t + 60°)$ mA. Find the real and imaginary components of the complex impedance **Z**. Find the values of the circuit elements if the impedance consists of two elements in series. Find the two circuit elements if the connection is in parallel.

Problem 7.26

In the circuit below, $R = 2$ MΩ and $C = 1$ pF $(10^{-12}$ F). If $\omega = 10^6$ rad/sec, find the ratio $\hat{V}/\hat{I} = \mathbf{Z}_T$.

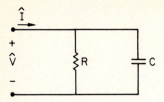

Suppose a series R–C circuit replaces the network above. What values of R and C are required to yield the same $\mathbf{Z}_T$?

$$\text{Ans.: } \mathbf{Z}_T = [0.4 - j(0.8)] \text{ M}\Omega$$
$$R = 4 \times 10^5 \, \Omega = 400 \text{ k}\Omega$$
$$C = 1.25 \text{ pF}$$

Problem 7.27

Repeat Problem 7.26 if $\omega = 2 \times 10^6$ rad/sec.

Problem 7.28

In the circuit below, $\hat{V}_m = 20 \underline{/0°}$, $R = 10 \, \Omega$ and $L = 1$ mH.

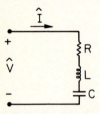

If $\omega = 10^4$ rad/sec, what value must C take on if $\hat{I}_m$ is to be pure real ($\underline{/\hat{I}_m} = 0°$)? With this value for the capacitor, make a phasor diagram showing that

$$\hat{V}_m = \hat{V}_R + \hat{V}_L + \hat{V}_C$$

Maximum value phasors are used.

Problem 7.29

Keeping R, L, and C the same in the circuit of Problem 7.28 (using the value for C found in that problem), find the polar angle of $\hat{I}$ if the source frequency is changed to 10^5 rad/sec. Do the same if $\omega = 10^3$ rad/sec.

Problem 7.30

In the circuit shown below, $i_c(t) = 10\sqrt{2} \cos(10^6 t)$ A, $C = 0.25 \, \mu$F, and $R = 4 \, \Omega$. Find the effective value phasor $\hat{V}_e$. The impedance $\mathbf{Z}$ is unknown.

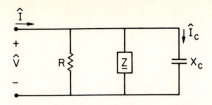

Ans.: $\hat{V}_e = 40 \underline{/-90°}$ V

Problem 7.31

If $\hat{I}_e = [6 - j10]$ A in Problem 7.30, what is the value of the impedance **Z**? Specify values for the elements that make up **Z**.

Problem 7.32

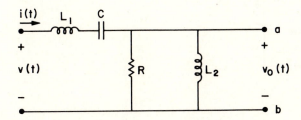

In the figure above, $L_1 = L_2 = 2$ μH, $R = 2$ kΩ and $C = 0.5$ μF. If $v(t) = 20\sqrt{2} \cos(10^3 t + 45°)$ V, find the current $i(t)$.

Problem 7.33

In the circuit for Problem 7.32 above, find $v_0(t)$. Make a plot of the effective value phasors $\hat{V}_0$ and $\hat{I}$ corresponding to $v_0(t)$ and $i(t)$. Show that the phase angle between $\hat{V}_0$ and $\hat{I}$ equals the polar angle of the complex impedance of R and L_2 in parallel.

Problem 7.34

In the circuit below, $R = 2$ kΩ, $L = 0.2$ H, and $i_R = 20\sqrt{2} \cos(10^4 t + 45°)$ mA. It is known that $\hat{V}$ has a phase angle equal to zero.

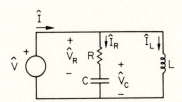

Find the value of C. Find the effective value phasors $\hat{V}$, $\hat{V}_R$, $\hat{V}_C$. Show on a phasor diagram that $\hat{V} = \hat{V}_R + \hat{V}_C$.

Problem 7.35

In the circuit for Problem 7.34, find $\hat{I}$ and $\hat{I}_L$ (effective value).

Problem 7.36

Consider $\hat{V}$ and $\hat{I}$ in the circuit for Problems 7.34 and 7.35. Show that the angle between $\hat{V}$ and $\hat{I}$ is the same as the angle of the equivalent impedance for the network.

Problem 7.37

In the circuit below, $\hat{V}_{1m} = 10\ \underline{/0°}$ V, $\hat{V}_{2m} = 20\ \underline{/45°}$ V, $L = 1$ H, $\omega = 1000$ rad/sec, $R_1 = R_2 = 1$ kΩ, and $C_1 = C_2 = 1\ \mu$F. Use the loop current method to find $\hat{I}_{1m}$ and $\hat{I}_{2m}$.

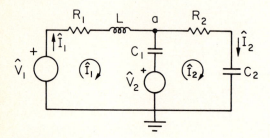

Problem 7.38

In the circuit in Problem 7.37 above, use the node voltage method to find $\hat{V}_{am}$. In addition, show that

$$\hat{V}_{2m} - \hat{V}_{am} = (\hat{I}_{2m} - \hat{I}_{1m})\mathbf{Z}_C$$

where $\mathbf{Z}_C$ is the impedance of the capacitor C_1.

Problem 7.39

Find $\hat{V}_{am}$ using the superposition method in the circuit for Problem 7.37. Explicitly give the values $\hat{V}_{am}'$ and $\hat{V}_{am}''$ corresponding to $\hat{V}_{1m}$ and $\hat{V}_{2m}$, respectively.

Problem 7.40

In the circuit below, $\hat{I}_e = 10\ \underline{/45°}$ mA and $\hat{I}_{Re} = 10\ \underline{/-45°}$ mA. The resistor is 1 kΩ, $L = 1$ mH, and $\omega = 10^6$ rad/sec.

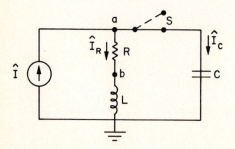

Find the voltage $v_a(t)$. All phasors are effective value. The switch S is closed.

$$\text{Ans.: } v_a(t) = 20 \ \cos(10^6 t) \ V$$

Problem 7.41

Find the value of the capacitor C in Problem 7.40 above.

Problem 7.42

In Chapter 10 we show how to find $v_a(t)$ for $t \geq 0$ after a switch such as the one in Problem 7.40 is opened. This involves the solution of a complete response problem when sinusoidal sources are involved. Here, we ask the student to find the *forced* response part of the solution as t goes to infinity. That is, find the voltage $v_a(t)$ and the current $i_R(t)$ as t goes to infinity if the switch is opened at $t = 0$. By waiting "a long time" any transient response will be dissipated.

Problem 7.43

R=10 Ω

C=1.0 μF

$i(t) = 10 \cos(10^5 t + 30°) \ A$

In the circuit shown, find the equivalent impedance Z seen by the source V_s, the effective value phasor $\hat{I}_e$, and the effective value phasor of the source required to drive the current, $\hat{V}_{se}$. Give also the time function $v_s(t)$.

$$\text{Ans.: } Z = 10 - j10 = 10\sqrt{2}\underline{/-45°}\,\Omega$$
$$\hat{I}_e = (10\sqrt{2}/2)\underline{/30°}\text{A}$$
$$\hat{V}_{se} = 100\underline{/-15°}\text{V}$$
$$v_s(t) = 100\sqrt{2}\ \cos(10^5 t - 15°)\text{V}$$

Problem 7.44

Use the voltage divider principle to find the effective value phasor voltage $\hat{V}_{Ce}$ across the capacitor in the circuit of Problem 7.43,

$$\hat{V}_{Ce} = \left(\frac{Z_C}{Z_C + Z_R}\right)\hat{V}_{se}$$

where Z_C is the capacitor impedance at $\omega = 10^5$ rad/sec and Z_R is the impedance of the resistor. Check also that $\hat{V}_{Ce} = \hat{I}_{Ce} Z_C$, which is just Ohm's law.

Problem 7.45

In the circuit below let $R_1 = R_2 = R_3 = 5\ \Omega$, $L = 5$ mH, and $C = 200\ \mu F$.

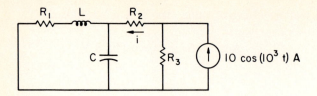

Find the current $i(t)$ using phasors and the current divider method. Either maximum or effective value phasors can be used.

$$\text{Ans.: } i(t) = 3.16 \cos(10^3 t + 18.4°)\text{A}$$

Problem 7.46

Use KCL to find the current through the resistor R_3 in the circuit of Problem 7.45. With this current find the voltage $v(t)$ across the current source. Use the corresponding $\hat{V}$ along with $\hat{I}$ to find the total equivalent impedance connected to the source.

EXAMPLE 7.3

Use the result of Problem 7.43 and the "R_0 method" to find the Thevenin equivalent circuit at the terminals a and b for that circuit. Find the short-circuited current $\hat{I}_{ab} = \hat{I}_N$ and show that

$$\hat{V}_T = \hat{I}_N \mathbf{Z}_0$$

where $\hat{V}_T$ is the Thevenin voltage, $\hat{I}_N$ is the Norton current, and $\mathbf{Z}_0 = \mathbf{Z}_N = \mathbf{Z}_T$ is the Thevenin impedance. All the phasors are effective values and hence the subscript e has been dropped.

Solution: The open-circuited voltage across a–b is equal to $\hat{I}\mathbf{Z}_C$. Using effective value phasors throughout, we have

$$\hat{I} = 5\sqrt{2} \,\underline{/30°}$$

and

$$\mathbf{Z}_c = \frac{1}{j\omega C} = -j10 = 10\underline{/-90°}$$

so

$$\hat{V}_T = 50\sqrt{2} \,\underline{/-60°}\text{ V}$$

To use the "R_0 method" we replace the source with a short circuit and evaluate the impedance seen at the terminals a and b. That is, we find the equivalent impedance of $\mathbf{Z}_R$ in parallel with $\mathbf{Z}_C$.

$$\mathbf{Z}_0 = \mathbf{Z}_T = \frac{10\left[-j10\right]}{10 - j10} = \frac{100\underline{/-90°}}{10\sqrt{2}\underline{/-45°}}$$

$$\mathbf{Z}_T = 5\sqrt{2}\underline{/-45°}\,\Omega$$

The equivalent circuit is

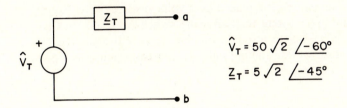

$$\hat{V}_T = 50\sqrt{2}\ \underline{/-60°}$$

$$\underline{Z}_T = 5\sqrt{2}\ \underline{/-45°}$$

The short-circuited current from a to b in the original circuit is

$$\hat{I}_{sc} = \frac{\hat{V}_{se}}{R} = \frac{(100\underline{/-15°})}{10} = 10\underline{/-15°} = \hat{I}_N$$

To check the interrelationship,

$$\hat{I}_N\mathbf{Z}_0 = (10\underline{/-15°})\ (5\sqrt{2}\underline{/-45°})$$

$$= 50\sqrt{2}\underline{/-60°} = \hat{\mathbf{V}}_T$$

which agrees.

Problem 7.47

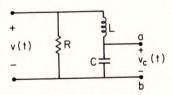

In the circuit above, $v(t) = 10\sqrt{2}\ \cos(10^3 t)$ V, $R = 4\ k\Omega$, $L = 1$H, and $C = 0.5$ μF. Find $v_C(t)$.

$$\text{Ans.: } v_C(t) = 20\sqrt{2}\ \cos(10^3 t)\ V$$

Problem 7.48

Find the Thevenin and Norton equivalent circuits at the port a–b in Problem 7.47 if $v(t) = 10\sqrt{2}\ \cos(10^3 t)$ V. Use effective value phasors to represent $\hat{V}_T$ and $\hat{I}_N$.

Problem 7.49

Find the Thevenin and Norton equivalent circuits at the port a–b in Problem 7.22. Use effective value phasors to represent $\hat{V}_T$ and $\hat{I}_N$ in the circuits. Check using the $\mathbf{Z}_0$ method that $\hat{V}_T = \hat{I}_N \mathbf{Z}_0$, where $\mathbf{Z}_0 = \mathbf{Z}_T = \mathbf{Z}_N$. $R_1 = 500\Omega$, $R_2 = 1k\Omega$, $L = 0.01\text{H}$ and $C = 0.02\mu\text{F}$.

Problem 7.50

In Problem 6.13 suppose the battery is replaced by a voltage source of value $v(t) = 50\cos(10^5 t)$. Find $i(t)$ and $i_L(t)$ a long time after the switch is closed. In Chapter 10 we show how to do the complete response problem for $t \geq 0$ when the source is sinusoidal. In this exercise we are finding only the forced response as t goes to infinity.

Problem 7.51

In Problem 6.22 suppose the battery is replaced by a current source of value $i_s(t) = 4\cos(100t + 30°)$ mA. Find the voltage $v_c(t)$ a long time after the switch is closed.

EXAMPLE 7.4

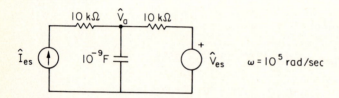

In the circuit above, $\hat{I}_s = 4\,\underline{/0°}$ mA and $\hat{V}_s = 10\,\underline{/30°}$ V. Use superposition to find $v_a(t)$. Effective value phasors are used.

Solution:

$$\mathbf{Z}_C = -j\left(\frac{1}{\omega C}\right) = -j(10^4)$$

Replacing $\hat{V}_s$ with a short circuit,

$$\hat{V}_a' = \hat{I}_s \mathbf{Z}_e$$

where $\mathbf{Z}_e$ is the impedance of the parallel R–C network:

$$\mathbf{Z}_e = \frac{(10^4)[-j(10^4)]}{10^4 - j(10^4)} = \left(\frac{10^4}{\sqrt{2}}\right)\underline{/-45°}$$

This yields

$$\hat{V}_a' = 20\sqrt{2}\,\underline{/-45°} = 20 - j(20)$$

Replacing $\hat{I}_s$ with an open circuit

$$\hat{V}_a'' = \frac{-j(10)}{10 - j(10)} \ (10\underline{/30°})$$

$$\hat{V}_a'' = \frac{(10\underline{/-90°})\,(10\underline{/30°})}{10\sqrt{2}\underline{/-45°}}$$

$$\hat{V}_a'' = 5\sqrt{2}\underline{/-15°} = 6.83 - j(1.8)$$

Adding the two solutions,

$$\hat{V}_a = 26.83 - j(21.8) = 34.6\underline{/-39.1°}$$

Since this is an effective value phasor, the time function is

$$v_a(t) = 48.9 \ \cos(10^5 t - 39.1°) \text{ V}$$

Problem 7.52

Use superposition to find the current $\hat{I}_C$ through the capacitor in Example 7.4. Check that this result is equal to $\hat{V}_a/\mathbf{Z}_C$. Use effective value phasors and give $\hat{I}_C'$ and $\hat{I}_C''$ explicitly

Problem 7.53

Suppose that in addition to the voltage source a current source with the same frequency ($\omega = 10^5$ rad/sec) is attached across a–b in Problem 7.22 with an effective value phasor current

$$\hat{I}_{ba} = 4\underline{/0°} \text{ mA}$$

Use superposition and the result given in that problem answer to find the new value of $v_L(t)$ if the elements are $R_1 = 0.5$ kΩ, $R_2 = 1$ kΩ, $L = 10$ mH, and $C = 0.02$ μF.

Chapter 8

Forced Response to Sinusoidal Signals and Power Computation

The currents and voltages resulting from the application of sources to linear systems are called the "forced response." In general, as we showed in Chapter 6, the currents and voltages will also include a "natural response" component which is the solution of a homogeneous differential equation. The time dependence of the natural response is independent of the source, but in any stable physical system there is some dissipative element and the natural response will decay exponentially to zero. Sometime after the application of a source then, the only response is that due to the source, and the system is said to have reached "steady state." In this context the term "steady state" also includes steady oscillation of the voltages and currents in response to an oscillatory source, the subject of the present chapter.

An important example is the electrical system used for distributing power in which the source voltages and currents vary sinusoidally with time. In the United States, the frequency of the sinusoid is 60 Hz, which corresponds to an angular frequency of $\omega = 2\pi (60)$, or 377 radians per second. In systems used for communication or control, the voltages and currents (the "signals") are often much more complicated, but the analysis may still be carried out on the basis of sinusoidal sources. This is possible because, as discussed in Chapter 10, any reasonably well behaved function of time can be expressed as the *sum* of sinusoids and because, for linear systems, the response can be calculated as the superposition of the response for each of these sinusoids acting alone. Thus, analysis of systems forced by sinusoidal signals has applications far beyond the elementary ones we study here. In the first two sections some of the Chapter 7 material is reviewed, particularly with regard to manipulation of impedances and complex algebra.

In this chapter we deal with the forced response of linear systems to a sinusoidal source and consider power relationships in linear systems. All phasors in this chapter will be *effective value phasors* and no subscript will be used.

8.1 THE IMPEDANCE TRIANGLE AND PHASOR DIAGRAMS

Using the impedance concept and the phasor notation, we can reduce problems dealing with the forced response of a linear system to a solution of a (complex) constant phasor source applied to a (complex) constant impedance. Except for the fact that we deal with complex number operations rather than real number arithmetic, circuits which involve impedances obey all the laws of resistive circuits developed in Chapters 1 through 3.

As discussed earlier, for exponential time functions of the form e^{st}, the ratio of the voltage drop across an R, L, or C element to the current in the direction of the voltage drop defines the impedance Z. For a pure oscillatory function of time, $\mathbf{s} = j\omega$ and the exponential function is $e^{j\omega t}$. For a current $\hat{I}e^{j\omega t}$ and a voltage $\hat{V}e^{j\omega t}$,

$$\mathbf{Z} = \frac{\hat{V}e^{j\omega t}}{\hat{I}e^{j\omega t}} = \frac{\hat{V}}{\hat{I}} \tag{8.1}$$

which is a constant complex number and hence not a function of time.

An arbitrary sinusoidal current $i(t) = \sqrt{2}I_0 \cos(\omega t + \phi) = \mathrm{Re}\{\sqrt{2}\hat{I}e^{j\omega t}\}$ has the effective value phasor $\hat{I} = I_0 e^{j\phi} = I_0\underline{/\phi}$ while a sinusoidal voltage $v(t) = \sqrt{2}V_0 \cos(\omega t + \alpha) = \mathrm{Re}\{\sqrt{2}\hat{V}e^{j\omega t}\}$ has the effective value phasor $\hat{V} = V_0 e^{j\omega t} = V_0\underline{/\alpha}$. If these two phasors represent the current through some element and the voltage across the same element, the associated impedance is the complex constant:

$$\mathbf{Z} = \frac{\hat{V}}{\hat{I}} = \left(\frac{V_0}{I_0}\right)\underline{/\alpha - \phi} = \left(\frac{V_0}{I_0}\right)\underline{/\theta_Z} = |\mathbf{Z}|\underline{/\theta_Z}$$

The *magnitude* of the impedance $\mathbf{Z}$ can be found by simply taking the ratio of the effective value of the associated voltage to the effective value of the associated current. The angle of the impedance θ_Z is the angle by which the voltage phasor leads the current phasor ($\theta_Z = \alpha - \phi$). If we plot the two phasors in a complex plane to yield the so-called "phasor diagram," they look as shown in Fig. 8.1.

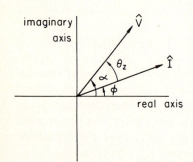

FIGURE 8.1.
Phasors $\hat{V}$ and $\hat{I}$ plotted in a complex plane. Their lengths depend on the scales used for voltage and current.

The complex functions of time $\hat{V}e^{j\omega t}$ and $\hat{I}e^{j\omega t}$ rotate counterclockwise in the complex plane with **v** leading **i** by the angle $\alpha - \phi$. We say that the voltage leads the current, or that the current lags the voltage, in this example. If we plot the impedance in a complex plane, as shown in Fig. 8.2, the complex angle θ_Z is equal to the angle by which the voltage leads the current. This result always holds since $\mathbf{Z} = \hat{V}/\hat{I}$, and complex division corresponds to subtracting the phase angle of the denominator from that of the numerator. (If θ_Z is negative, the voltage lags the current.) The triangle in Fig. 8.2 is called the impedance triangle.

For a resistance, $\mathbf{Z}_R = R$; for an inductance, $\mathbf{Z}_L = j\omega L = \omega L\underline{/+90°}$, and for a capacitance, $\mathbf{Z}_C = (1/j\omega C) = -j(1/\omega C) = (1/\omega C)\underline{/-90°}$. The impedance "triangle"

FIGURE 8.2.
Impedance _Z_ plotted in a complex plane.

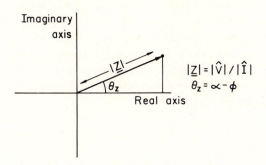

for each of these limiting cases is given in Fig. 8.3. For circuits using only resistors, the voltage is in phase with the current, and we can treat time-varying signals using real number arithmetic. Complex numbers are needed when both the amplitude and phase must be considered. A complex number contains two pieces of information (its amplitude and phase, or its real and imaginary part), and hence the complex notation "keeps track" of both quantities. For a pure inductance, the voltage leads the current (or the current lags the voltage) by 90°; and for a pure capacitance, the voltage lags the current (or the current leads the voltage) by 90°.

FIGURE 8.3.
Impedance "triangle" for the three simplest circuit elements reduces to these three lines.

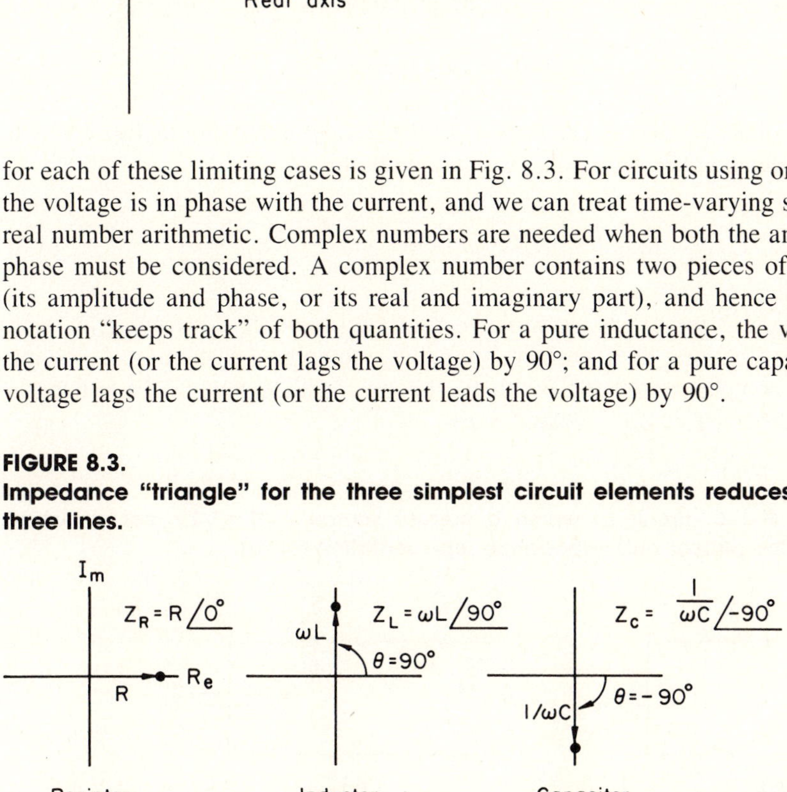

$Z_R = R \underline{/0°}$

$Z_L = \omega L \underline{/90°}$

$Z_C = \dfrac{1}{\omega C} \underline{/-90°}$

Resistor Inductor Capacitor

For the two-element circuit illustrated earlier in Fig. 7.7, which has an inductor and a resistor in series, the impedance triangle is shown in Fig. 8.4. Note that $\mathbf{Z}$ is the sum of $\mathbf{Z}_R$ and $\mathbf{Z}_L$. If a sinusoidal voltage source with phasor value $\hat{V} = V_0 \underline{/0°}$ is applied to this impedance, the current will be $\hat{I} = \hat{V}/\mathbf{Z} = \hat{V}/(|\mathbf{Z}| \underline{/\theta}) = (V_0/|\mathbf{Z}|) \underline{/-\theta}$. When an impedance triangle is located in the upper half of the complex plane,

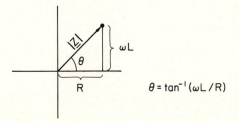

$\theta = \tan^{-1}(\omega L / R)$

FIGURE 8.4.
Impedance triangle for a series _R–L_ circuit.

we say that the impedance is inductive. When it is located in the lower half of the plane, we say it is capacitive. Not surprisingly the series R–L circuit has an inductive impedance triangle.

To further illustrate the usefulness of the impedance concept and phasor notation, consider the analysis of the more complicated series R–L–C circuit in Fig. 8.5. Suppose the current $i_s(t)$ is supplied by a current source such that $i_s(t) = \sqrt{2}I_0 \cos(\omega t + \phi)$ which has been in place for a long time. Then only the forced response at the frequency ω occurs. The complex current source is $\mathbf{i}_s(t) = \sqrt{2}I_0 e^{j(\omega t + \phi)} = \sqrt{2}I_0 e^{j\phi}e^{j\omega t}$. Using phasor notation we suppress the $e^{j\omega t}$ term and have $\hat{I} = I_0\underline{/\phi}$. Applying Ohm's law in phasor notation,

$$\hat{V} = \hat{I}\mathbf{Z}_R + \hat{I}\mathbf{Z}_L + \hat{I}\mathbf{Z}_C \tag{8.2}$$

If the values of R, L, and C are known, then the right-hand side of Eq. 8.2 is a known, complex number. Formally, we may write this as

$$V_0\underline{/\alpha} = RI_0\underline{/\phi} + \omega L I_0\underline{/(\phi + 90°)} + \left(\frac{I_0}{\omega C}\right)\underline{/(\phi - 90°)} \tag{8.3}$$

By solving Eq. 8.3 for V_0 and α, we have converted a problem involving addition of sinusoidal functions to a problem of adding complex constants and solving the resulting complex arithmetic problem. Having found V_0 and α, the voltage is determined to be $v(t) = V_0\sqrt{2}\cos(\omega t + a)$.

FIGURE 8.5.
(a) A series R–L–C circuit to which a current source $i_s(t) = \sqrt{2}\,I_0\cos(\omega t + \phi)$ is applied. (b) The phasor and impedance representation for (a).

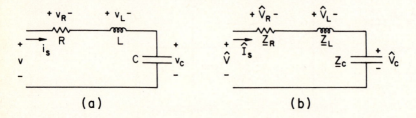

(a)   (b)

It is useful to show the relationship between the three voltage phasors in this example using the phasor diagram of Fig. 8.6. We are usually free to pick one of

FIGURE 8.6.
Phasor diagram (c) for Fig. 8.5 and their sum (d). The three phasors must add up to yield $V_0\underline{/\alpha}$.

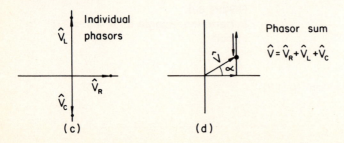

(c)   (d)

the phasors to have a phasor angle equal to zero and call it the reference phasor. If we set $\phi = 0$ in this case, then the current is called the reference phasor. The voltage phasors are shown in the diagram. In this phasor diagram, to be specific we have assumed that the number ωL is greater than $1/\omega C$. If $\omega L = 1/\omega C$, then $|\hat{V}_L| = |\hat{V}_C|$, $\hat{V} = \hat{V}_R$, and $\alpha = 0$; whereas if ωL is less than $1/\omega C$, α will be negative. The voltage may lead the current, be in phase with the current, or lag the current, depending on the relative magnitudes of ωL and $1/\omega C$.

The interrelationships among $\hat{V}_L$, $\hat{V}_R$, $\hat{V}_C$, $\hat{V}$, and $\hat{I}_s$ are independent of the phase angle of the current. If ϕ is not equal to zero, all the phasors shown are merely rotated counterclockwise in the complex plane by the angle ϕ. The phasor diagram just rotates as a rigid body. For a series circuit such as that in Fig. 8.5, it is convenient to assume, as we have done in the example, that the current phasor has zero angle. In the parallel circuit examples we shall do later, it is often convenient to use the voltage as the reference phasor (i.e., assume that the angle of the voltage phasor is zero).

Before leaving this section we define the reciprocal of the impedance, which is called the admittance **Y**, by the equation

$$\mathbf{Y} = \frac{\hat{I}e^{j\omega t}}{\hat{V}e^{j\omega t}} = \frac{\hat{I}}{\hat{V}}$$

For a resistance, $\mathbf{Y}_R = 1/R = G$; for an inductance, $\mathbf{Y}_L = 1/j\omega L = -j(1/\omega L) = (1/\omega L)\underline{/-90\,°}$; and for a capacitance, $\mathbf{Y}_C = j\omega C = \omega C\underline{/+90\,°}$. The admittance function is often used in the analysis of parallel circuits.

TEXT EXAMPLE 8.1

In the circuit below it is known that the current through the resistor R_2 is $i_2(t) = 4\sqrt{2}\cos(\omega t)$ A. Find the source $i_s(t)$. Show that the angle $\underline{/V_2} - \underline{/I_s}$ is equal to the angle of the impedance triangle for the parallel network.

$R_1 = R_2 = 10\,\Omega$
$L = 0.01$ H
$C = 100 \times 10^{-6}$ F
$\omega = 10^3$ rad/sec

Solution: The effective value phasor current corresponding to $i_2(t)$ is $\hat{I}_2 = 4\underline{/0°}$. By Ohm's law,

$$\hat{V}_2 = \hat{I}_2\mathbf{Z}_2$$

where

$$\mathbf{Z}_2 = R_2 + \frac{1}{j\omega C} = 10 - j10 = 10\sqrt{2}\underline{/-45°}$$

and so

$$\hat{V}_2 = 40\sqrt{2}\underline{/-45°}$$

The voltage $\hat{V}_2$ appears across the impedance $\mathbf{Z}_1 = R_1 + j\omega L = 10 + j10 = 10\sqrt{2}\,\underline{/45°}$ as well, so again using Ohm's law,

$$\hat{I}_1 = \frac{\hat{V}_2}{\mathbf{Z}_1} = \frac{40\sqrt{2}\,\underline{/-45°}}{10\sqrt{2}\,\underline{/45°}}$$

$$\hat{I}_1 = 4\,\underline{/-90°}$$

By KCL, $\hat{I}_1 + \hat{I}_2 = \hat{I}_s$. Using rectangular notation for addition,

$$\hat{I}_s = 4 + (-j4) = 4 - j4 = 4\sqrt{2}\,\underline{/-45°}$$

Converting to the temporal function we must multiply the effective value by $\sqrt{2}$ which yields

$$i_s(t) = 8\cos(1000t - 45°)\ A$$

The difference in phase angles of $\hat{V}_2$ and $\hat{I}_2$ is $-45° - (-45°) = 0$. To find the equivalent impedance of the parallel network we can calculate

$$\mathbf{Z}_E = \frac{\mathbf{Z}_1\mathbf{Z}_2}{\mathbf{Z}_1 + \mathbf{Z}_2} = \frac{(10\sqrt{2}\,\underline{/45°})(10\sqrt{2}\,\underline{/-45°})}{20}$$

$$\mathbf{Z}_E = 10\,\underline{/0°}$$

Note that $\mathbf{Z}_E = \hat{V}_2 / \hat{I}_s$.

8.2 REACTANCE

Except for the fact that we are dealing with complex numbers, circuits including impedance obey all the laws of circuits using only resistances. Loop current and node voltage methods are applied in the same way. For impedances in series, the voltage divides in proportion to the impedance. For impedances in parallel, the current divides in inverse proportion to the impedance (or in proportion to the admittances). The principle of superposition holds. Thevenin's and Norton's theorems are obeyed. Many of these applications were explored in the problem set of Chapter 7.

For any combination of elements, not including sources, we can determine the equivalent impedance across a pair of terminals in the same manner with which we found the equivalent resistance for a combination of resistances. For example, consider the circuit of Fig. 8.7, where the input sinusoidal voltage v and current i are represented by the phasors $\hat{V}_1$ and $\hat{I}_1$, respectively. In this figure note that we have used a resistor symbol and then labeled that element $\mathbf{Z}$. This is standard practice since there is no commonly accepted separate symbol for an arbitrary impedance. In this specific case,

FIGURE 8.7.
To find the equivalent impedance of (*a*), we use the series–parallel reductions of (*b*) and (*c*).

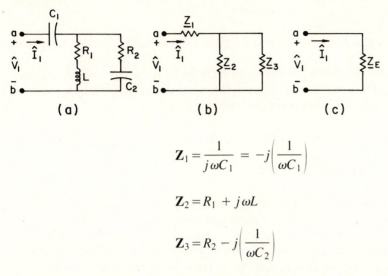

(a) (b) (c)

$$\mathbf{Z}_1 = \frac{1}{j\omega C_1} = -j\left(\frac{1}{\omega C_1}\right)$$

$$\mathbf{Z}_2 = R_1 + j\omega L$$

$$\mathbf{Z}_3 = R_2 - j\left(\frac{1}{\omega C_2}\right)$$

and following through the transformations indicated in parts (b) and (c) of the figure,

$$\mathbf{Z}_E = -j\left(\frac{1}{\omega C_1}\right) - \frac{(R_1 + j\omega L)\left[R_2 - j(1/\omega C_2)\right]}{R_1 + R_2 + j\omega L - (j/\omega C_2)}$$

For any fixed values of R_1, R_2, L, C_1, and C_2, the impedance $\mathbf{Z}_E$ will still depend on the radian frequency of the sinusoidal voltage v_1. This ω will be the same as the radian frequency of the sinusoidal current. With ω fixed, the equivalent impedance $\mathbf{Z}_E$ will be some complex number, $\mathbf{Z}_E = |\mathbf{Z}_E|\underline{/\theta}$, and

$$\mathbf{Z}_E = \frac{\hat{V}_1}{\hat{I}_1} = \frac{V_1\underline{/\alpha}}{I_1\underline{/\theta}} = \left(\frac{V_1}{I_1}\right)\underline{/\alpha - \theta} = |\mathbf{Z}_E|\underline{/\theta}$$

If a voltage $v_1(t) = V_1\sqrt{2}\cos(\omega_1 t + \alpha_1)$ at some other frequency ω_1 is applied to the terminals a–b, we must calculate a *new value* of $\mathbf{Z}_E$ for $\omega = \omega_1$, say,

$$\mathbf{Z}_{E1} = |\mathbf{Z}_{E1}|\underline{/\theta_1}$$

Then the current $i_1(t)$ will be

$$i_1(t) = \frac{V_1\sqrt{2}}{|\mathbf{Z}_{E1}|}\cos(\omega_1 t + \alpha_1 - \theta_1)$$

Since the impedance is a complex number, it can always be written in the rectangular form

$$\mathbf{Z} = |\mathbf{Z}|\underline{/\theta} = |\mathbf{Z}|\cos\theta + j|\mathbf{Z}|\sin\theta$$

We call the real part of $\mathbf{Z}$ the resistive component R and call the imaginary part of $\mathbf{Z}$ the reactive component X, and any impedance can be written in the form

$$\mathbf{Z} = R + jX$$

The values of the real numbers R and X form the sides of the impedance triangle in Fig. 8.4.

The admittance can also be written in its rectangular form. We call the real part of $\mathbf{Y}$ the conductance component G and call the imaginary part of $\mathbf{Y}$ the susceptance component B. Then, any admittance can be written as

$$\mathbf{Y} = G + jB$$

Note that although $\mathbf{Y} = 1/\mathbf{Z}$, with $\mathbf{Z} = R + jX$, it is *not* the case that $G = 1/R$ and $B = 1/X$.

In general, the reactive components of the impedance will depend on the frequency of the source. The resistive component must always be positive, but the reactive component may be either positive (inductive reactance) or negative (capacitance reactance). At any fixed frequency, the equivalent impedance of any set of impedances may be represented by a series combination of a single resistance and either an equivalent inductance ($X = \omega L$) or an equivalent capacitance ($X = -(1/\omega C)$). Similarly, any admittance may be formed of a conductance in parallel with an equivalent inductance ($B = -(1/\omega L)$) or an equivalent capacitance ($B = \omega C$).

As an example, suppose that at a frequency ω of 1000 radians per second the impedance of a given circuit is known to be

$$\mathbf{Z}_{ab} = 10\underline{/30°} = (8.66 + j5) \ \Omega$$

This impedance has an inductive reactance and could be represented by the series network in Fig. 8.8a, where $R = 8.66 \ \Omega$ and $L = 5/\omega = 5$ mH. The same impedance $\mathbf{Z}_{ab}$ can also be realized by a parallel network such as the one given in Fig. 8.8b but with different components R_p and L_p. As an illustration of how to use the admittance function, we note that

$$\mathbf{Y}_{cd} = \mathbf{Y}_R + \mathbf{Y}_L = \frac{1}{R_p} + \frac{1}{j\omega L_p}$$

FIGURE 8.8.
Series and parallel *R–L* networks.

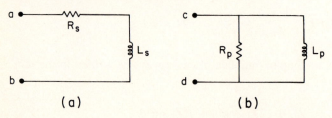

(a)   (b)

To have the same impedance as circuit (a),

$$\mathbf{Y}_{cd} = \mathbf{Z}_{ab}^{-1} = 0.1\underline{/-30°}$$

or

$$\mathbf{Y}_{cd} = 0.0866 - j(0.05) \text{ mho}$$

Thus, $R_p = 0.0866^{-1}$ and $L_p = 1/0.05\omega$, which yield the values $R_p = 11.55\ \Omega$ and $L_p = 20$ mH.

TEXT EXAMPLE 8.2

In the circuit for Text Example 8.1 find the resistive and reactive parts to the equivalent impedance $\mathbf{Z}_E$ for the network. Find the Thevenin and Norton equivalent circuits at the terminals a–b in phasor form.

Solution: The impedance $\mathbf{Z}_1$ and $\mathbf{Z}_2$ are in parallel, so

$$\mathbf{Z}_E = \frac{\mathbf{Z}_1 \mathbf{Z}_2}{\mathbf{Z}_1 + \mathbf{Z}_2}$$

The denominator is best evaluated in rectangular form, $\mathbf{Z}_1 + \mathbf{Z}_2 = (10 + j10) + (10 - j10) = 20\underline{/0°}$. Then,

$$\mathbf{Z}_E = \frac{(10\sqrt{2}\underline{/45°})(10\sqrt{2}\underline{/-45°})}{20\underline{/0°}}$$

$$= 10\underline{/0°}$$

The resistive part of $\mathbf{Z}_E = 10\ \Omega$ and the reactive part is zero. To find the Thevenin equivalent circuit we first note that the voltage $\hat{V}_2 = 40\sqrt{2}\underline{/-45°}$ found in Text Example 8.1 is already the open-circuited voltage phasor which in turn equals the Thevenin voltage phasor $\hat{V}_T$. Suppressing the current source we can use the "R_0 method" to find $\mathbf{Z}_T$.

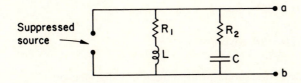

But the parallel impedance which results is just the same as the one found above so $\mathbf{Z}_T = \mathbf{Z}_E = 10\underline{/0°}$, and the Thevenin equivalent circuit is

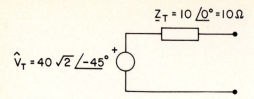

The Norton equivalent circuit is

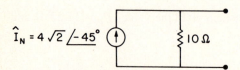

where we have specifically shown that $\mathbf{Z}_N = \mathbf{Z}_T$ is a pure resistance. We can check this result by noting that the Norton current is equal to the short-circuited current and hence equal to the full value of the current source in the original circuit, $I_s = 4\sqrt{2}/{-45°}$.

8.3 POWER COMPUTATION USING PHASORS AND IMPEDANCE

The instantaneous power delivered to any pair of terminals is $p(t) = v(t)i(t)$ where the current is in the direction of the voltage drop. To compute the average power delivered over a period of time (T) requires evaluation of the following integral.

$$<P> = \int_0^T v(t)i(t)dt$$

We are interested here in the long time average of the power, P,

$$P = \lim_{T \to \infty} (1/T) \int_0^T v(t)i(t)dt$$

For sinusoidal sources, we express $v(t)$ and $i(t)$ in the form

$$v(t) = \sqrt{2}V_0 \cos(\omega t + \alpha)$$

$$i(t) = \sqrt{2}I_0 \cos(\omega t + \phi)$$

These two temporal functions then correspond to the effective value phasors

$$\hat{V} = V_0/\alpha$$

$$\hat{I} = I_0/\phi$$

We also *define* the effective value of the voltage to be the absolute value of $|\hat{V}|$, which is the real number $|\hat{V}| = V_0$. The effective value of the current is $|\hat{I}| = I_0$. It is tempting to assume that the average power can be found by multiplying the two effective values. However, since the power is a nonlinear function we must be careful in the application of phasors to the computation of power. Instead, we will use the actual time functions and then in retrospect see how to use the phasors. Inserting these two time functions into the defining equation for P, we see that

$$P = \lim_{T\to\infty} \left(\frac{1}{T}\right) \int_0^T \left[\sqrt{2}V_0 \cos(\omega t + \alpha)\right]\left[\sqrt{2}I_0 \cos(\omega t + \phi)\right] dt$$

Using the identity $\cos(a)\cos(b) = \frac{1}{2}\cos(a+b) + \frac{1}{2}\cos(a-b)$ yields

$$P = \lim_{T\to\infty} \left[\left(\frac{V_0 I_0}{T}\right) \int_0^T \cos(2\omega t + \alpha + \phi)dt + \left(\frac{V_0 I_0}{T}\right) \int_0^T \cos(\alpha - \phi)dt\right]$$

As discussed in Chapter 7, the first term vanishes since it represents the average value of a sinusoidal function. The integral in the second term equals $\cos(\alpha - \phi)T$. The T's cancel and

$$P = |\hat{V}||\hat{I}| \cos(\alpha - \phi) \tag{8.4}$$

This expression shows how to find the power dissipated in some element characterized by phasors $\hat{V}$ and $\hat{I}$. The power equals the product of the effective values of the voltage and of the current multiplied by the cosine of the phase angle difference between $\hat{V}$ and $\hat{I}$. Note that the angle $\alpha - \phi$ is also the angle of the impedance triangle associated with $\mathbf{Z} = \hat{V}/\hat{I}$.

While superposition does not hold for power, it is legitimate to consider only sinusoidal currents and voltages at the same frequency since the long-time average of the product of two sinusoids at different frequencies is zero. The proof of this statement is straightforward. Suppose $v(t) = \sqrt{2}V_0 \cos(\omega_1 t + \alpha_1)$ and $i(t) = \sqrt{2}I_0 \cos(\omega_2 t + \alpha_2)$. Then, the associated average power is

$$P = \lim_{T\to\infty} \left(\frac{1}{T}\right) \int_0^T \left[\sqrt{2}V_0 \cos(\omega_1 t + \alpha)\right]\left[\sqrt{2}I_0 \cos(\omega_2 t + \phi)\right] dt$$

$$= \lim_{T\to\infty} \left(\frac{V_0 I_0}{T}\right) \int_0^T \cos[(\omega_1 + \omega_2)t + \alpha + \phi]dt +$$

$$\lim_{T\to\infty} \left(\frac{V_0 I_0}{T}\right) \int_0^T \cos[(\omega_1 - \omega_2)t + \alpha - \phi]dt$$

If ω_1 is not equal to ω_2, each of the terms of the equation above is the average value of a sinusoid, which we have shown earlier to be zero.

To find the average power supplied by a given sinusoidal voltage source, it is therefore only necessary to find the current through that source at the source frequency. However, if more than one source operates at that frequency, the current used in the calculation is that due to all of the sources acting in unison.

Considering then only a single frequency, from Eq. 8.4 the average power into a pair of terminals a–b, as shown in Fig. 8.9a, is $V_0 I_0 \cos \theta$, where θ is the phase angle difference between $\hat{V}$ and $\hat{I}$. If we use $\hat{V}$ as the reference phasor, a $\alpha = 0$ and $\phi = -\theta$. Since $\cos(-\delta) = \cos(\delta)$, the average power does not depend on whether the current lags or leads the voltage, only by how much it differs. Since $\mathbf{Z} = \hat{V}/\hat{I}$, we note again that the angle of the impedance triangle is equal to the angle θ and, as illustrated in Fig. 8.9b, the average power is the product of the magnitudes of the two effective value phasors (voltage and current) times the cosine of the impedance angle.

FIGURE 8.9.

(a) Time functions and *(b)* phasors for a sinusoidal source.

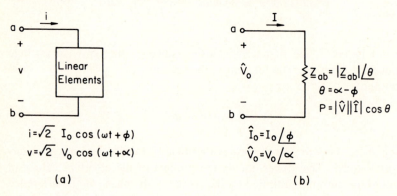

$$i = \sqrt{2}\ I_0 \cos(\omega t + \phi)$$
$$v = \sqrt{2}\ V_0 \cos(\omega t + \alpha)$$

(a)

$$\hat{I}_0 = I_0 \underline{/\phi}$$
$$\hat{V}_0 = V_0 \underline{/\alpha}$$

(b)

Finally, we note that since $|\hat{V}| = |\hat{I}\mathbf{Z}| = |\hat{I}|\,|\mathbf{Z}|$ we can substitute $|\hat{I}|\,|\mathbf{Z}|$ for $|\hat{V}|$ in the expression

$$P = |\hat{V}||\hat{I}|\ \cos\ \theta$$

to yield

$$P = |\hat{I}|^2|\mathbf{Z}|\cos\ \theta$$

but $|\mathbf{Z}|\cos\ \theta = R$, the resistive part of $\mathbf{Z}$. Thus, the power dissipated in any impedance is just the absolute value squared of the current through that impedance times the real part of $\mathbf{Z}$. The total power dissipated in any set of impedances then is the sum of the powers dissipated in each resistor. For a circuit that has k impedances $\mathbf{Z}_k$,

$$P_T = \sum_k |\hat{I}_k|^2 R_k$$

where $R_k = \mathrm{Re}(\mathbf{Z}_k)$. This general result may be applied to the special cases of pure inductance, pure capacitance, and pure resistance as shown in Figs. 8.10a–c.

FIGURE 8.10.
The average power for a pure inductance or capacitance is zero, while for a resistance $P = I_0^2 R = V_0^2 / R$.

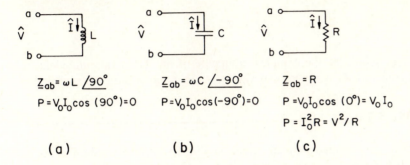

$$Z_{ab} = \omega L \,\underline{/90°}$$
$$P = V_0 I_0 \cos(90°) = 0$$

$$Z_{ab} = \omega C \,\underline{/-90°}$$
$$P = V_0 I_0 \cos(-90°) = 0$$

$$Z_{ab} = R$$
$$P = V_0 I_0 \cos(0°) = V_0 I_0$$
$$P = I_0^2 R = V^2 / R$$

(a)   (b)   (c)

The results above show that for a pure inductance or a pure capacitance there is only an oscillating power with no average component. The average power in any system is supplied only to resistive elements. The total power supplied to a system of resistances, inductances, and capacitances can therefore be found by determining the effective value of the current through each resistor and merely adding together the real numbers $I_{eff}^2 R$ for each resistor. Also, if we reduce some set of impedances to their equivalent impedance at terminals a and b,

$$\mathbf{Z}_{ab} = |\mathbf{Z}_{ab}|\underline{/\theta} = |\mathbf{Z}_{ab}| \cos\theta + j|\mathbf{Z}_{ab}| \sin\theta = R + jX$$

Then, since $|\hat{V}| = |\hat{V}| \, |\mathbf{Z}_{ab}|$,

$$P = V_0 I_0 \cos\theta = I_0^2 |\mathbf{Z}_{ab}| \cos\theta = I_0^2 R$$

and the power $V_0 I_0 \cos\theta$ just equals I_0^2 times the equivalent resistive part of $\mathbf{Z}_{ab}$.

The maximum power transfer theorem is somewhat modified when impedances are involved. As before we replace the circuit attached to the load impedance, $\mathbf{Z}_L$, by the Thevenin equivalent voltage $\hat{V}_T$, in series with the Thevenin equivalent impedance $\mathbf{Z}_T$. The power transferred to the load is then

$$P = |\hat{I}_L|^2 R_L = \frac{|\hat{V}_T|^2}{|\mathbf{Z}_T + \mathbf{Z}_L|^2} R_L$$

$$= \frac{|\hat{V}_T|^2 R_L}{(R_T + R_L)^2 + (X_T + X_L)^2}$$

Now, if $R_L = 0$, $P = 0$ and there is no power transferred at all. Suppose $R_L \neq 0$. Then P will be largest when the denominator is smallest. This occurs when $(X_T + X_L)^2 = 0$, which in turn means that $X_L = -X_T$. Under these conditions, then,

$$P = \frac{|\hat{V}_T|^2 R_L}{R_L + R_T}$$

This is the same form used in the proof of the maximum power transfer theorem in Chapter 3 which maximizes for $R_L = R_T$. Combining these results we find that maximum power is transferred when $R_L = R_T$ and $X_L = -X_T$, or,

$$\mathbf{Z}_L = \mathbf{Z}_T^*$$

that is, the load should be the complex conjugate of the Thevenin equivalent impedance. Of course if $\mathbf{Z}_T$ is a pure real value R_T, the appropriate load is also R_T.

TEXT EXAMPLE 8.3

Find the total power dissipated by the circuit in Text Example 8.1. Show explicitly that this equals the power dissipated in the resistors. Find the maximum power which can be drawn off to a load attached to terminals a–b.

Solution: Since we have found the voltage $\hat{V}_2 = 40\sqrt{2}\underline{/-45^\circ}$ and current $\hat{I}_s = 4\sqrt{2}\underline{/-45^\circ}$ we can use Eq. 8.4, which states that

$$P = |\hat{V}_2|\,|\hat{I}_s|\,\cos(\alpha - \phi)$$

Since there is no phase difference between $\hat{V}_2$ and $\hat{I}_s$,

$$P = (40\sqrt{2})(4\sqrt{2}) = 320 \text{ W}$$

The current dissipated in each resistor is

$$P_1 = |\hat{I}_1|^2 R_1 = 16(10) = 160 \text{ W}$$

$$P_2 = |\hat{I}_2|^2 R_2 = 16(10) = 160 \text{ W}$$

so, indeed,

$$P = \sum_k |\hat{I}_k|^2 R_k = 320 \text{ W}$$

The maximum power supplied to a load corresponds to matching it to the complex conjugate of the Thevenin impedance. In this case, $\mathbf{Z}_T = 10\underline{/0^\circ}$ so we attach a 10-Ω resistor to have a matched load. The equivalent circuit becomes

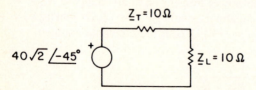

By the voltage divider law, the voltage across the load is $20\sqrt{2}\underline{/-45°}$ and the current is $2\sqrt{2}\underline{/-45°}$. The power dissipated in the load is then

$$P_L = |\hat{I}_L|^2 10 = 80 \text{ W}$$

A common error in power computation is to assume that since $P = |\hat{I}|^2 R$ that it must also hold that $P = |\hat{V}|^2/R$. This is *not* the case except when $\hat{V}$ is the voltage across R alone. Analytically this can be proved as follows. Since $\hat{V} = \hat{I}\mathbf{Z}$ it follows that $|\hat{I}| = |\hat{V}|/|\mathbf{Z}|$ so

$$P = |\hat{I}|^2 R = \frac{|\hat{V}|^2 R}{|\mathbf{Z}|^2}$$

$$= |\hat{V}|^2 \left(\frac{R}{R^2 + X^2} \right)$$

which is equal to $|\hat{V}|^2/R$ *only* when $X = 0$. In Text Example 8.3 note that $|\hat{V}_2|^2/R_2 = 3200/10 = 320$ W, which is not equal to the power dissipated in that impedance $(P_2 = 160 \text{ W})$.

8.4 THE POWER FACTOR AND THE CONCEPT OF COMPLEX POWER

The quantity $\cos\theta$ is called the power factor (p.f.) of the impedance. The power factor of a pure resistance is unity; the power factor of a pure inductance or capacitance is zero. The average power, of course, determines the energy supplied to a load; but from the point of view of the utility supplying the electrical power, the power factor is also of interest. This is the case since for the same average power the *current* supplied to a low power factor load is greater than the current supplied to a unity power factor load by the factor $1/\cos\theta$. As an example, consider the average power supplied to the two loads illustrated in Figs. 8.11a and 8.11b.

FIGURE 8.11a,b.
The average power supplied by the same source to two different loads is 120 kW, but the current to (b) is greater than to (a) by a factor of $\sqrt{2}$.

(a) (b)

In case (a), $P = |\hat{I}||\hat{V}|\cos\theta = 120$ kW. In case (b), $P = |\hat{I}||\hat{V}|\cos\theta = (100\sqrt{2})(1200)(\sqrt{2}/2) = 120$ kW. The average power is the same although the magnitude of the current is 41% higher in case (b). Since the utility must supply a larger current to the circuit of Fig. 8.11b, the generator must have a larger current rating and must also supply the greater I^2R losses in the transmission line. Therefore, a typical utility adds an extra charge for large users (more than 200 kW) when the power factor is less than 0.95. The user may then find it economical to increase the power factor without changing the average power by adding some parallel impedence across the load.

The power factor correction necessary to bring the load of the circuit in Fig. 8.11b to a unity power factor would entail the addition of a parallel capacitive reactance of 12 ohms, as shown in Fig. 8.11c. The corresponding phasor diagram is shown in Fig. 8.11d. In the final configuration the phasor $\hat{I}$ is parallel to $\hat{V}$, the phase angle is zero, and the power factor $\cos\theta = 1$. Since $\cos\theta = \cos(-\theta)$, merely stating the power factor does not specify whether the corresponding load is inductive or capacitive. We say that the power factor is "leading" when the current leads the voltage (capacitive reactance) and that the power factor is lagging when the current lags the voltage (inductive reactance).

The analysis leading to determination of the required parallel load above proceeds as follows. For the circuit of Fig. 8.11c the current in the load, $\hat{I}_L$, is the same as the current with no capacitance (as in Fig. 8.11b) since $\hat{V}$ is unchanged.

$$\hat{I}_L = \frac{\hat{V}}{6 + j6} = 141.4\underline{/-45^\circ} = (100 - j100)\ \text{A}$$

Thus, to assure that $\hat{I}$ is in phase with $\hat{V}$ we must choose C so that

$$\hat{I}_C = +j100 = 100\underline{/90^\circ}$$

hence,

$$\frac{1}{\omega C} = |\mathbf{Z}_C| = \frac{|\hat{V}|}{|\hat{I}_C|} = \frac{1200}{100} = 12\ \Omega$$

FIGURE 8.11c,d.
The power factor is improved by adding a parallel capacitor to an inductive load as shown in (c). The new phasor current is shown in (d) and has the same phase angle as $\hat{V}$.

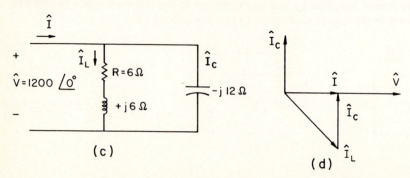

The total current supplied to the parallel combination is now

$$\hat{I} = \hat{I}_L + \hat{I}_C = 100 \underline{/0°}$$

which has a power factor of 1. The average power supplied to the combination is unchanged by the addition of the capacitor and can be again found from either

$$P = |\hat{I}_L|^2 R = (141.4)^2(6) = 120 \text{ kW}$$

or

$$P = |\hat{V}||\hat{I}| \cos \theta = (1200)(100)(1) = 120 \text{ kW}$$

Since the utility must supply whatever current is required by the load, the rating of generators and other equipment (such as the transformers to be discussed later) is based on the total product of the effective values of voltage and current. This product is called the "apparent power" P_A, and the units are volt-amperes (VA). The average power in watts is related to P_A by

$$P = |\hat{V}||\hat{I}| \cos \theta = P_A \cos \theta \qquad \text{where } P_A = |\hat{V}||\hat{I}| \qquad (8.5)$$

This result leads to the introduction of the concept of the complex power defined by

$$\mathbf{P} = \mathbf{V} \cdot \mathbf{I}^*$$

where as before the asterisk denotes the complex conjugate operation. Since $\mathbf{V} = \hat{V}e^{j\omega t}$ and $\mathbf{I} = \hat{I}e^{j\omega t}$, taking the complex conjugate of $\mathbf{I}$ before multiplying removes the time dependence in the product since the two factors cancel each other, $(e^{j\omega t})(e^{-j\omega t}) = 1$. The complex power (not a phasor!) is then

$$\mathbf{P} = \hat{V}(\hat{I}^*) = P + jP_R$$

where P is the true average power and P_R is the so-called reactive power. Since $\hat{V} = \hat{I}\mathbf{Z}$ and since $\hat{I}(\hat{I}^*) = I^2$, we have the important result that the complex power is also given by

$$\mathbf{P} = I^2\mathbf{Z} = I^2R + jI^2X \qquad (8.6)$$

Since $\mathbf{P}$ is proportional to $\mathbf{Z}$, a plot of $\mathbf{P}$ in the complex plane has the same form as the impedance triangle. The quantity I^2X is equal to the reactive power P_R, while I^2R is equal to the true average power P (Fig. 8.12). The absolute value of the complex power is equal to the apparent power defined above, $|\mathbf{P}| = P_A$. To emphasize that the imaginary part of $\mathbf{P}$ does not correspond to power dissipation, instead of calling its unit watts, we use the term volt-ampere-reactive, which is symbolized by (VAr) or (kVAr), for kilovolt-ampere-reactive.

If a number of loads appear in parallel across a given power system as is the usual case for an electrical utility company, for example, the total current is the

FIGURE 8.12.
The power triangle of (b) can be found from the impedance triangle of (a).

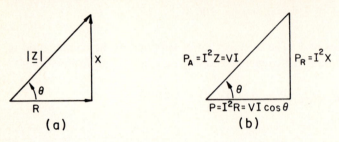

(a)　　　　　　　　　　(b)

complex sum of the current to each load. In the case of the three parallel loads in Fig. 8.13, by KCL, $\hat{I}_T = \hat{I}_1 + \hat{I}_2 + \hat{I}_3$. Thus, since $\hat{V}$ appears across each load, the total complex power $\mathbf{P}_T = (\hat{V})(\hat{I}_T^*)$ is equal to the sum

$$\mathbf{P}_T = \hat{V}\hat{I}_1^* + \hat{V}\hat{I}_2^* + \hat{V}\hat{I}_3^*$$

FIGURE 8.13.
Three loads in parallel across a power line.

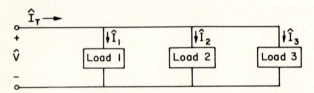

The complex powers add like complex numbers when the loads are in parallel. The total real power is the sum of the real parts of each complex power, and the total reactive power is the sum of the imaginary parts of each complex power:

$$P_T = P_1 + P_2 + P_3$$

$$P_{RT} = P_{R1} + P_{R2} + P_{R3}$$

Referring back to Fig. 8.11, for the inductive load, $\mathbf{P}_L = 120$ kW $+j120$ kVAr; for the capacitive load, $\mathbf{P}_C = 0 - j120$ kVAr and hence the total complex power $\mathbf{P}_T = 120$ kW. The ease with which the complex power can be found when loads are in parallel (the normal case) is the prime reason for introducing the concept.

TEXT EXAMPLE 8.4 _____

In the circuit below, a 60-Hz power generator supplies a load through a transmission line that has a finite resistance $R_T = 2\ \Omega$. The load is found to have a lagging power factor equal to 0.866. A capacitor is placed in parallel with the load as shown, which brings the power factor to unity.

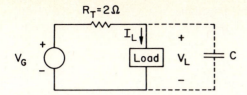

With the capacitor in position the voltage across the load is found to have an effective value of 4 kV and the apparent load power is 400 kVA. Find the value of the load impedance and of the capacitance which was added in parallel with the load to bring the power factor to unity. Find the average power supplied by the generator with and without the capacitor and the value of $\hat{V}_G$.

Solution: We use the concept of complex power to solve this problem. We are free to define one angle, so let $\hat{V}_L = 4000 \underline{/0°} \, \text{V}$ and $\hat{I}_L = I_L \underline{/\alpha} \, \text{A}$. Then the complex power to the load with the capacitor in place is

$$\mathbf{P}_L = \hat{V}_L \hat{I}_L^*$$

and hence

$$|\mathbf{P}_L| = |\hat{V}_L| \, |\hat{I}_L^*|$$

The absolute value of $\mathbf{P}_L$ is the apparent power to the load, which is given as 400 kVA. Likewise, $|\hat{V}_L| = 4000$ V, so $|\hat{I}_L^*| = (4 \times 10^5)/(4 \times 10^3) = 100 \, \text{A} = |\hat{I}_L|$. The load impedance $\mathbf{Z}_L$ can now be determined. Since $\hat{V}_L = \hat{I}_L \mathbf{Z}_L$,

$$|\mathbf{Z}_L| = \frac{4000}{100} = 40 \, \Omega$$

We still need to find the polar angle of $\mathbf{Z}_L$. We do know that the original power factor was 0.866, so

$$\underline{/\mathbf{Z}_L} = \pm 30°$$

The $\pm$ sign is used since we do not yet know the sign of the angle. However, the power factor was quoted as lagging, so the load current lags the load voltage. This implies that $\alpha = -30°$, and that the load is inductive, so

$$\mathbf{Z}_L = 40 \underline{/+30°} = 34.6 + j20$$

The load current is $\hat{I}_L = \hat{V}_L/\mathbf{Z}_L = 100 \underline{/-30°} = 86.6 - j50 \, \text{A}$. There are several ways to proceed now to find the value of C. By brute force we could evaluate the parallel impedance $\mathbf{Z}_L // (jX_C)$, where $X_C = -(1/\omega C)$ and find the value of C which makes this a pure real number. Instead, we shall use the complex power concept. In the final state we want the total complex power $\mathbf{P}$ to be a pure real number. Since complex powers in parallel add as ordinary complex numbers,

$$\mathbf{P}_T = \mathbf{P}_L + \mathbf{P}_C$$

The power triangle has the same shape as the impedance triangle, so $\mathbf{P}_L = 400\underline{/+30°}$ kVA $= 346 + j200$ kVA. To make $\mathbf{P}_T$ pure real, $\mathbf{P}_C = -j200$ kVA, which can be written

$$|\hat{I}_C|^2|X_C| = 2 \times 10^5 \text{ VAr} \tag{1}$$

The load voltage also appears across the capacitor, so $\hat{I}_C = \hat{V}_L/jX_C$, $|\hat{I}_C|^2 = |\hat{V}_L|^2/X_C^2$, and from (1) $|X_C| = (4000)^2/2 \times 10^5 = 80 \ \Omega = 1/\omega C$. At 60 Hz, $\omega = 377$ rad/sec and so $C = 33 \ \mu F$. For reference, $\hat{I}_C = \hat{V}_L/jX_C = +j50$ A and the total final impedance of the load in parallel with the capacitor is $\mathbf{Z}_T = 46.2\underline{/0°}$ which, as expected, is a pure real number.

The total real power consumed is the real part of $\mathbf{P}_T$, which is 346 kW, plus the power lost in the transmission line, $|\hat{I}_T|^2 R_T$:

$$\hat{I}_T = \hat{I}_L + \hat{I}_C = 86.6\underline{/0°} \text{ A}$$

and the total power consumed is 361 kW. To finish the problem we need to find V_G. From the voltage divider principle,

$$\hat{V}_L = \left(\frac{\mathbf{Z}_T}{\mathbf{Z}_T + R_T}\right)\hat{V}_G$$

so $\hat{V}_G = (48.2/46.2)\hat{V}_L = 4173\underline{/0°}$. Using $\hat{V}_G$, we can find the current $\hat{I}'$, which flowed before installing the capacitor:

$$\hat{I}' = \frac{\hat{V}_G}{\mathbf{Z}_L + R_T} = 100\underline{/-28.65°}$$

Note that the magnitude of $\hat{I}'$ is the same as $\hat{I}_L$. Adding the capacitor does not affect the magnitude of the load current at all, and therefore the real power to the load is unchanged, $|\hat{I}'|^2\text{Re}(\mathbf{Z}_L) = 346$ kW. The power dissipated in the transmission line is different (that was the whole idea) since it *was* $|\hat{I}'|^2 R_T = 20$ kW. The total power supplied was 366 kW, and a total of 5 kW has been saved by insertion of the capacitor.

8.5 SUMMARY OF POWER COMPUTATION METHODS

We have developed three basic methods for determining the power dissipated in a network:

1. $P = |\hat{V}_T|\,|\hat{I}_T|\cos\theta$, where $\hat{V}_T$ and $\hat{I}_T$ are the effective value phasor voltage and current delivered to the network and $\theta = \underline{/\hat{V}_T} - \underline{/\hat{I}_T}$. The angle θ is also the angle of the total impedance $\mathbf{Z}_T$.

2. $P = \sum_k {}_k |\hat{I}_k|^2 R_K$, where the summation is over all the resistors R_k in the network and $\hat{I}_k$ is the effective value phasor current through each resistor.

3. $P = \text{Re}(\mathbf{P})$ where $\mathbf{P}$ is the complex power, $\mathbf{P} = \hat{V}_T \hat{I}_T^*$.

8.6 MEASURING INSTRUMENTS

We have previously discussed voltmeters and ammeters in the context of dc measurements. For ac measurements the devices used determine the effective value of the particular parameter of interest, not the amplitude. That is, ac voltmeters measure the root mean square (rms) voltage of a sinusoidal signal and ac ammeters measure the rms current. This is another reason we use effective value phasors since measurement devices determine the effective value of the quantity of interest. An ideal "wattmeter" measures the real power. The symbol for the wattmeter is shown in Fig. 8.14. Note that the wattmeter must be a combination of a voltmeter and an ammeter and reads the average value of their product.

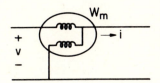

FIGURE 8.14.
The symbol for an ideal wattmeter. The meter reads the average of the product of v and i.

PRACTICE PROBLEMS
AND
ILLUSTRATIVE EXAMPLES

NOTE: All phasors in this problem set are effective value phasors.

Problem 8.1

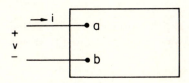

The box contains a linear system with no independent sources.

At the terminals $a–b$, $v = 100 \cos(10^3 t + 45°)$ V, $i = 10 \cos(10^3 t + 15°)$ mA.

(a) Write the phasors $\hat{V}$ and $\hat{I}$.

$$\text{Ans.: } \hat{V} = 50\sqrt{2}\underline{/45°} \text{ V}$$
$$\hat{I} = 5\sqrt{2}\underline{/15°} \text{ mA}$$

(b) Find the input impedance and admittance.

Ans.: $\mathbf{Z} = 10^4 \underline{/30°} \ \Omega$

$\mathbf{Y} = 10^{-4} \underline{/-30°} \ \text{mhos}$

(c) Find R, X, G, and B.

Ans.: $R = 8.66 \ \text{k}\Omega$; $X = 5 \ \text{k}\Omega$

$G = 0.0866 \times 10^{-3} \ \text{mhos}$

$B = -0.05 \times 10^{-3} \ \text{mhos}$

(d) Show two equivalent circuits, each
of which contains only two circuit
elements, that will give the same
impedance at the terminals a–b at
the given frequency.

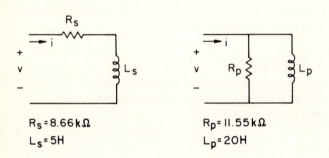

$R_s = 8.66 \ \text{k}\Omega$
$L_s = 5H$

$R_p = 11.55 \ \text{k}\Omega$
$L_p = 20H$

(e) Find the power consumed.

Ans.: $P = 433 \ \text{mW}$

(f) Find the complex power and show
that the real part equals P.

Ans.: $\mathbf{P} = 500 \underline{/30°} = 433 \ W + j250 \ VAr$

Problem 8.2

Show specifically that in Problem 8.1 $|\hat{I}_s|^2 R_s$ and $|\hat{I}_p|^2 R_p$ yield the same power
found in part (e). ($\hat{I}_s$ is the current through R_s, and $\hat{I}_p$ is the current through R_p.)

Problem 8.3

Repeat Problem 8.1 (parts a, b, c, and d) for $v = 100 \cos(10^5 t + 15°) \ V$ and
$i = 5\sqrt{2} \cos(10^5 t + 60°) \ \text{mA}$. In part (d) use the two equivalent circuits shown
below.

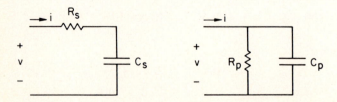

Problem 8.4

In the circuit of Problem 8.3 find the power P and the complex power **P**. Show that $P = \text{Re}(\mathbf{P}) = |\hat{I}_s|^2 R_s = |\hat{I}_p|^2 R_p$, where $\hat{I}_s$ is the circuit through R_s and $\hat{I}_p$ is the current through R_p.

EXAMPLE 8.2

For the circuit shown we desire to produce a current through C such that $i_1 = 2\cos(10^5 t)$ A. What is the required $v_s(t)$? Make phasor diagrams for the voltages and currents. Find the power delivered by V_s and show that it equals $|\hat{I}_1|^2 R_1 + |\hat{I}_2|^2 R_2$.

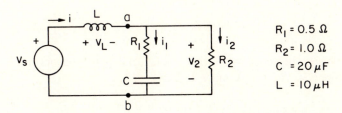

$$R_1 = 0.5\ \Omega$$
$$R_2 = 1.0\ \Omega$$
$$C = 20\ \mu F$$
$$L = 10\ \mu H$$

Solution: Using effective value phasors, $\hat{I}_1 = \sqrt{2}\underline{/0°}$. The impedance $Z_1 = R_1 - j(1/\omega C_1) = 0.5 - j0.5 = 0.5\sqrt{2}\underline{/-45°}$. Then,

$$\hat{V}_2 = \hat{I}_1 Z_1 = 1\underline{/-45°}$$

and

$$\hat{I}_2 = \frac{\hat{V}_2}{R_2} = 1\underline{/-45°} = \frac{\sqrt{2}}{2} - j\left(\frac{\sqrt{2}}{2}\right)$$

The phasor $\hat{I}$ is the sum of $\hat{I}_1$ and $\hat{I}_2$:

$$\hat{I} = \hat{I}_1 + \hat{I}_2 = \sqrt{2} + \frac{\sqrt{2}}{2} - j\left(\frac{\sqrt{2}}{2}\right) = \sqrt{2}(1.5 - j0.5) = 2.24\underline{/-18.4°}$$

and

$$\hat{V}_L = (\hat{I})(j\omega L) = (\hat{I})(j1) = \sqrt{2}(0.5 + j1.5)$$

Finally,

$$\hat{V}_S = \hat{V}_L + \hat{V}_2 = 0.5\sqrt{2} + j1.5\sqrt{2} + 0.5\sqrt{2} - j0.5\sqrt{2} = \sqrt{2} + j\sqrt{2} = 2\underline{/45°}$$

and $v_s(t) = 2\sqrt{2}\cos(10^5 t + 45°)$ V.

The phasor diagram relating all the phasor voltages and all the phasor currents is sketched below. Note that the scales for voltages and current are not the same.

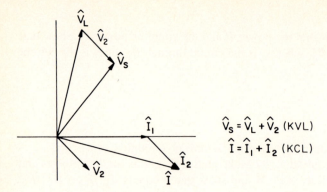

$$\hat{V}_s = \hat{V}_L + \hat{V}_2 \text{ (KVL)}$$
$$\hat{I} = \hat{I}_1 + \hat{I}_2 \text{ (KCL)}$$

The power is given by

$$P = |\hat{V}_s||\hat{I}| \cos \theta$$

where θ is the angular phase difference between $\hat{V}_2$ and $\hat{I}$ and both phasors are effective value phasors. Since $\hat{V}_s = 2\underline{/45°}\,\text{V}$ and $\hat{I} = 2.24\underline{/-18.4°}$,

$$P = 4.48 \cos(63.4°) = 1.99 \text{ W}$$

Checking against the current dissipated in each resistor,

$$P = |\hat{I}_1|^2 R_1 + |\hat{I}_2|^2 R_2$$
$$= 2(0.5) + 1(1) = 2 \text{ W}$$

The difference between the two answers is due to the round-off error in the phase angles and the cosine operation.

Problem 8.5

A source transformation may be used in the circuit of Example 8.2 such that the voltage source $\hat{V}_s$ and impedance $\mathbf{Z}_L$ due to the inductor are replaced by a Norton current source $\hat{I}_N$ in parallel with $\mathbf{Z}_N$.

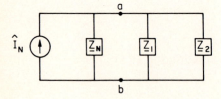

Find $\hat{I}_N$ and $\mathbf{Z}_N$ and show that $\hat{V}_2 = \hat{I}_N \mathbf{Z}_E$ where $\mathbf{Z}_E$ is the equivalent impedance.

Problem 8.6

A voltage $\hat{V} = 150\underline{/0°}\,V$ is applied across a network which has a capacitor in parallel with an impedance $\mathbf{Z}_2$. The latter consists of a resistor in series with an inductor.

The phasor diagram for all the currents and the voltage $\hat{V}$ is given below. $\hat{I}_2$ is the current through $\mathbf{Z}_2$, $\hat{I}_C$ is the capacitor current, and $\hat{I}$ is the total current drawn. Note that the magnitude of the *total* current is equal to the magnitude of the current through *one* of the parallel branches. Find R_2, L_2, and C if the frequency of the source is 60 Hz. Find the total power dissipated.

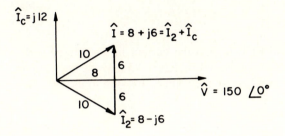

$$\text{Ans.: } R_2 = 12 \ \Omega$$
$$L_2 = 23.9 \ \text{mH}$$
$$C = 212 \ \mu\text{F}$$
$$P = 1.2 \ \text{kW}$$

Problem 8.7

Find the power supplied by the source in Problem 7.32 using $\hat{V}$ and $\hat{I}$. Show that power equals $|\hat{I}_R|^2 R$, where $\hat{I}_R$ is the effective value phasor current through the only resistor in the circuit.

Problem 8.8

Repeat Problem 8.7 for the circuit in Problem 7.34.

Problem 8.9

Find the power P supplied by the current source in Problem 7.40. Show that $P = |\hat{I}_R|^2 R$.

$$\text{Ans.: } P = 100 \ \text{kW}$$

Problem 8.10

Show that the power supplied by each source in Problem 7.37 add up to the powers dissipated in R_1 and R_2.

Problem 8.11

Let $i(t) = 5 \cos(10^4 t - 45°)$V, $R = 100 \ \Omega$, $C = 1 \ \mu\text{F}$, and $L = 10 \ \text{mH}$ in the circuit shown below. Show that the entire current $i(t)$ flows through the resistor. Evaluate the power dissipated by finding $\hat{V}_a$ and using $|\hat{I}| \ |\hat{V}_a| \cos(\text{phase difference})$. Show that this equals the power dissipated in the resistor. Find the current through L and explain how this can exist if the resistor carries "the entire $i(t)$".

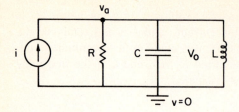

Problem 8.12

Use the answer to Problem 7.45 and KCL to find the current through resistor R_3. Then find the voltage across the current source in Problem 7.45 and use the formula $P = |\hat{V}| \, |\hat{I}| \, \cos \theta$ to find the power supplied by the source. Be sure to use effective value phasors.

Problem 8.13

In the same circuit used in Problems 8.12 and 7.45 show that $P = |\hat{I}_1|^2 R_1 + |\hat{I}_2|^2 R_2 + |\hat{I}_3|^2 R_3$.

Problem 8.14

Find the power supplied by the two sources in Example 7.4. Show that their sum equals the power dissipated in the two resistors.

EXAMPLE 8.3

Find the load impedance which must be attached to the port a–b in Problem 7.43 (and Example 7.3) to maximize the power transfer. Find that power.

Solution: To attain maximum power transfer the load should be matched to the complex conjugate of the Thevenin and Norton equivalent impedance. From the result in Example 7.3,

$$\mathbf{Z}_L = \mathbf{Z}_T^* = 5\sqrt{2}\underline{/+45°} = 5 + j5$$

Since $\hat{V}_T = 50\sqrt{2}\underline{/-60°}$, we can find the power transferred by first evaluating the current in the series circuit:

$$\frac{\hat{V}_T}{Z_T + Z_T^*} = \frac{50\sqrt{2}\underline{/-60°}}{10} = 5\sqrt{2}\underline{/-60°}$$

The power in the *load* is then

$$P_L = |\hat{I}_L|^2 R_L = 250 \text{ W}$$

Problem 8.15

Find the load impedance which must be attached in parallel with the inductor in Problem 8.11 to maximize the power transfer. What is the maximum power transferred to that load?

Problem 8.16

Repeat Problem 8.15 for the circuit in Problem 7.22 (also see Problem 7.49).

Problem 8.17

In Problem 7.30 and 7.31 find the total power P using $\hat{V}$ and $\hat{I}$. Use this result to find the resistive part of **Z**.

$$\text{Ans.: } 400 \text{ W}$$
$$\text{Re}\{\,\mathbf{Z}\,\} = 0$$

EXAMPLE 8.4

Find the complex power in each of the three elements in the circuit studied in Problem 8.17 and in Problems 7.30 and 7.31, as well as the total complex power $(\mathbf{P}_T = \hat{V}\hat{I}^*)$. Show that

$$\mathbf{P}_T = \mathbf{P}_1 + \mathbf{P}_2 + \mathbf{P}_3$$

and that $\text{Re}(\mathbf{P}_T) = 400$ W as found above.

Solution: Since $\hat{V} = 40\underline{/-90°}$ we may find each of the currents:

$$\hat{I}_R = \frac{\hat{V}}{R} = 10\underline{/-90°} = -j10$$

$$\hat{I}_C = \frac{\hat{V}}{jX_C} = \frac{\hat{V}}{-j4} = 10\underline{/0°} = 10$$

From KCL,

$$\hat{I}_Z = \hat{I} - \hat{I}_R - \hat{I}_C$$

$$\hat{I}_Z = (6 - j10) + j10 - 10$$

$$\hat{I}_Z = -4 = 4\underline{/-180}$$

Since $\mathbf{P}_j = \hat{V}_j(\hat{I}_j)^*$,

$$\mathbf{P}_R = 400$$

$$\mathbf{P}_Z = 40j$$

$$\mathbf{P}_C = -40j$$

Adding these complex numbers,

$$\mathbf{P}_T = 400 \text{ W}$$

This is already a pure real number, and hence Re($\mathbf{P}_T$) = 400 W, which agrees with the answer above.

Problem 8.18

Find the complex power **P** in Example 8.2 and show that its real part equals P.

Problem 8.19

Find the complex power in each of the parallel elements in Problem 8.5. Show that they add up to the total complex power $\mathbf{P}_T = \hat{V}_{ab}(\hat{I}_N)^*$, that is,

$$\mathbf{P}_T = \mathbf{P}_N + \mathbf{P}_1 + \mathbf{P}_2$$

Problem 8.20

A circuit is given in Problem 7.22 and is studied in some detail in the subsequent problems. Find the power dissipated in the circuit and the resistance R_1 using only the information in Problem 7.22 plus the fact that $R_2 = 1$ kΩ and $L = 10$ mH. (*Hint:* Find the current through R_1 by evaluating $\hat{V}_L/\mathbf{Z}_P$, where $\mathbf{Z}_P$ is the impedance $R_2//j\omega L$.)

Problem 8.21

Find the complex power in Problem 7.25 from the values given for $\hat{V}$ and $\hat{I}$. Find **Z** from $\mathbf{P}=|\hat{I}|^2\mathbf{Z}$ and compare it with the evaluation of **Z** in Problem 7.25 (that is, with $\mathbf{Z}=\hat{V}/\hat{I}$).

Problem 8.22

In the circuit of Example 8.2 above, find the load impedance seen by the source $\hat{V}_s$ in polar and rectangular forms. Find the complex power and show that it is equal to $|\hat{I}|^2\mathbf{Z}$.

$$\text{Ans.:} \quad \mathbf{Z}=(\hat{V}_s/\hat{I}) = 0.89\underline{/63.43°} =$$
$$0.4 + j0.8 \; \Omega$$
$$\mathbf{P}=2 + j4w$$
$$|\mathbf{I}|^2\mathbf{Z}=5\mathbf{Z} = 2 + j4 = \mathbf{P}$$

Problem 8.23

In the circuit of Example 8.2 it is desired to produce a current $i_2 = 3 \cos(10^5 t)$ A. Find the required v_s. Calculate the total power dissipated and show it equals the power lost in the two resistors.

Problem 8.24

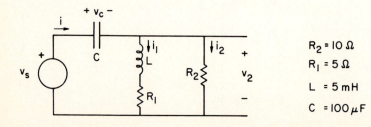

$R_2 = 10 \; \Omega$

$R_1 = 5 \; \Omega$

$L = 5 \, \text{mH}$

$C = 100 \, \mu\text{F}$

In the circuit shown it is desired to produce a voltage $v_2 = 100 \sqrt{2} \cos(10^3 t)$ V. Find all the phasors and the voltage $v_s(t)$.

$$
\begin{aligned}
\text{Ans.: } \hat{V}_R &= 100\underline{/0°}; \\
\hat{I}_2 &= 10\underline{/0°}; \\
\hat{I}_1 &= 10\sqrt{2}\underline{/-45°}; \\
\hat{I} &= 20 - j10; \\
\hat{V}_C &= -100 - j200; \\
\hat{V}_s &= -j200 = 200\underline{/-90°}; \\
v_s &= 200\sqrt{2} \cos(10^3 t - 90°) = \\
&\quad\;\, 200\sqrt{2} \sin(10^3 t)
\end{aligned}
$$

Problem 8.25

Find the total power P supplied by the voltage source in the circuit for Problem 8.24 using $\hat{V}_s$ and $\hat{I}$. Show that

$$P = |\hat{I}_1|^2 R_1 + |\hat{I}_2|^2 R_2$$

Problem 8.26

Find the complex power **P** in Problem 8.24. Use **P** and the fact that $\mathbf{P} = |\hat{I}|^2 \mathbf{Z}$ to find the total impedance **Z**. Show that

$$P = |\hat{I}|^2 R$$

Show also that $|\hat{V}_s|^2/R$ is *not* equal to the power!

EXAMPLE 8.5

A reactive element is to be added in parallel to the network in Problem 8.6 to make $\hat{I}$ pure real without changing the power consumed. Find the type of element and its value. Show that $|\hat{I}|$ is smaller with this element in place than without.

Solution: Since $\hat{I} = 8 + j6$, we need a new element with $\hat{I}_N = -j6$. From Ohm's law,

$$Z_N = \frac{150}{-j6} = j25$$

The element therefore is an inductance such that

$$\omega L = 377L = 25 \ \Omega$$

and hence

$$L = 66.3 \text{ mH}$$

In the original circuit $|\hat{I}| = 10$ A, whereas in the new circuit $|\hat{I}| = 8$ A. The new power is $P = |\hat{V}||\hat{I}| \cos \theta = (150)(8) = 1.2$ kW, which is unchanged.

It is illustrative to solve this problem as well using the complex power idea. In the original circuit

$$P = (150\underline{/0°})(10\underline{/36.87°})^*$$

$$= 1.5 \times 10^3 \underline{/-36.87°}$$

$$P = 1.2 \text{ kW} - j(0.9)|\text{kVAr}|$$

To make P pure real we need a new element in parallel such that $P + P_N$ is pure real, Clearly,

$$P_N = +j(0.9) \text{ kVAr}$$

The element must be inductive and such that

$$\hat{V}\hat{I}_N{}^* = j(0.9) \text{ kVAr}$$

Since $\hat{V} = 150\underline{/0°}$,

$$\hat{I}_N = -j6 \text{ A}.$$

From Ohm's law,

$$Z_N = \frac{\hat{V}}{\hat{I}_N} = j25$$

which agrees with the result found above.

Problem 8.27

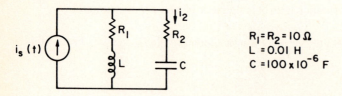

R$_1$ = R$_2$ = 10 Ω
L = 0.01 H
C = 100 × 10^{-6} F

For the steady-state current $i_2 = 4 \cos(1000t)$ A, find the required source $i_s(t)$ and the total complex power. Show that the real part of P is equal to the power dissipated in R_1 and R_2.

$$\text{Ans.:} \quad i_s(t) = 4\sqrt{2} \cos(1000t - 45°) \text{ A}$$

$$\text{Re}(P) = \text{Re}(160 + j0) = 160 \text{ W}$$

$$|\hat{I}_2|^2 R_2 = 80 \text{ W} = |\hat{I}_1|^2 R_1$$

Problem 8.28

In the circuit of Problem 8.27, $i_s(t)$, R_2, and C remain the same but R and L are twice as large. Find the complex power. What is the power factor?

Problem 8.29

Power is supplied to the load illustrated below from an ideal 10-kV, 60-Hz line. The average power to the load is 500-kW at a power factor of 0.5 lagging. The power factor is to be corrected to unity by adding the capacitor C. Find the required value of C.

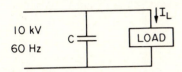

Ans.: 23 μF

Problem 8.30

In a circuit similar to the one shown above in Problem 8.29, the load impedance is $10\underline{/30°}$ and $|\hat{I}_L| = 100$ A with the capacitor out of the circuit. What is the power factor? What value capacitor is needed to correct the power factor to unity? The source is an ideal 60-Hz source.

Problem 8.31

As shown in the figure below, a 60-Hz generator (V_G) supplies a load through a transmission line of resistance R_T.

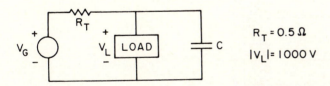

A capacitance C is added in parallel with the load to correct the power factor to unity. With the capacitor in place the load-voltage effective value is 1000 V. The load takes 200 kVA at a lagging power factor of 0.707 (inductive). Find the value of the capacitance and the average power supplied by the generator.

Ans.: 375 μF

151.4 kW

Problem 8.32

Find the generator voltage with the capacitor in place in Problem 8.31. Show that the power supplied by the generator equals 151.4 kW using $\hat{V}_G$ and $\hat{I}_G$.

Problem 8.33

In the circuit for Problem 8.31 suppose the *generator* voltage is $1000\underline{/0°}$. Find the power supplied to the load if the same impedance is used, $C = 375$ μF, and $R_T = 0.5$ Ω as above.

Problem 8.34

As shown in the diagram below, a 60-Hz generator (V_G) supplies a load through a transmission line of resistance R_T. The load voltage $\hat{V}_L$ is $1000\underline{/0°}$ and $R_T = 1.0$ Ω. The load takes an apparent power P_A of 200 kVA at a lagging power factor of 0.5 (inductive load). Find the total power supplied by the generator. What capacitor must be placed in parallel with the load to bring the power factor to unity? What is the new power supplied by the generator?

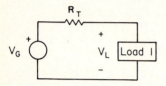

Problem 8.35

An inductance furnace draws 50 kVA at 0.5 power factor lagging from a 200-V, 60-Hz line. The power factor is to be corrected to 0.85 lagging by adding a parallel capacitor C. Find the required value of C.

Ans.: $C = 18.4 \times 10^{-4}$ F

Problem 8.36

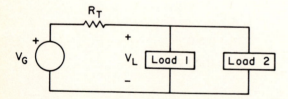

As shown in the diagram above, a 60-Hz generator (V_G) supplies two loads through a transmission line of resistance R_T. The load voltage is 1000 V. $R_T = 0.5$ Ω. Load 1 takes 200 kVA at a lagging power factor of 0.5. Load 2 takes 400 kVA at a leading power factor of 0.5. Find the power supplied by the generator.

Ans.: 360 kW

Problem 8.37

In a circuit similar to that of Problem 8.36, load 1 has an impedance $10\underline{/30°}$ Ω and load 2 has an impedance of $5\underline{/60°}$ Ω. $R_T = 1$ Ω and the measured load voltage is 1000 V. Find the power supplied by V_G.

EXAMPLE 8.6

In the circuit below, $\mathbf{Z}_L = 10\underline{/45°}$ Ω and the voltmeter reads 100 V. Predict the ammeter and wattmeter reading.

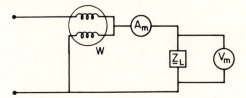

Solution: The complex power is defined by

$$\mathbf{P} = \hat{V}(\hat{I})^*$$

Since $|\hat{A} \times \hat{B}| = |\hat{A}||\hat{B}|$ and $|(\hat{A})^*| = |\hat{A}|$,

$$|\mathbf{P}| = |\hat{V}||\hat{I}|$$

Similarly, since $\hat{V} = \hat{I}\mathbf{Z}$,

$$|\hat{V}| = |\hat{I}||\mathbf{Z}|$$

The voltmeter reads $|\hat{V}| = 100$ V and $|\mathbf{Z}| = 10$ so $|\hat{I}| = 10$ A, which is the required ammeter reading. Since the complex power has the same shape as the impedance triangle,

$$\mathbf{P} = |\mathbf{P}|\underline{/45°}$$

But, since we now know $|\hat{I}| = 10$ A, $|\mathbf{P}| = 1000$ VA. This means in turn that

$$P = \text{Re}\{\mathbf{P}\} = 1000\cos(45°)$$

and finally,

$$P = 707 \text{ W}$$

This is the reading of the wattmeter.

Problem 8.38

For the same circuit as drawn in Example 8.6 suppose $\mathbf{Z}_L = 5\underline{/30°}$ and the ammeter reads 5 A. Predict the reading of the voltmeter and wattmeter.

Problem 8.39

For a load impedance $\mathbf{Z}$, an ammeter in series with $\mathbf{Z}$ reads 10 A, a voltmeter across $\mathbf{Z}$ reads 200 V, and a wattmeter reads 1200 W as the power supplied to $\mathbf{Z}$.

(a) Determine the power, reactive power, apparent power, and power factor.

Ans.: $P = 1200$ W;
$P_X = 1600$ VAr;
$P_A = 2000$ VA; p.f. $= 0.6$

(b) Determine the resistive and reactive components of **Z**. (Note that we do not yet know if **Z** is inductive or capacitive.)

Ans.: $R = 12 \, \Omega$; $|X| = 16 \, \Omega$

(c) When the same voltage is supplied to the series combination of a new pure capacitive reactance and the same **Z**, the ammeter reads 12 A. Determine the new wattmeter reading. Is **Z** capacitive or inductive?

Ans.: $P = 1728$ W;
Z is inductive

Problem 8.40

For the same circuit as drawn in Example 8.6 suppose the load is inductive, the wattmeter reads 1000 W, and the ammeter reads 20 A. Find the real part of **Z**. Can the voltmeter reading be determined? Explain.

Problem 8.41

In the circuit below, $R_1 = 10 \, \Omega$, $R_2 = 20 \, \Omega$, and $X_L = 20 \, \Omega$. If $|\hat{I}_2| = 14.1$ A and $|\hat{I}_C| = 10$ A, predict the reading of the ammeter and wattmeter.

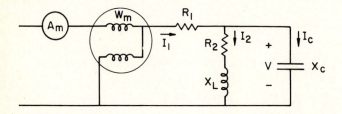

Ans.: 10 A and 5 kW

Problem 8.42

In the circuit of Problem 8.41 the ammeter reads $\sqrt{1200}$ A and the wattmeter reads 33,600 W. If $R_1 = 12 \, \Omega = R_2$, $X_L = 9 \, \Omega$ and $X_C = -60 \, \Omega$, find the voltage $|\hat{V}|$ across the capacitor. Using $\hat{V}$ as the reference phasor ($\underline{/\hat{V}} = 0°$), find $\hat{I}_2$, $\hat{I}_C$, and $\hat{I}_1$.

Chapter
9
Electric Power Systems

Power systems involve the generation, transmission, and distribution of electrical energy. The systems are very large and complicated. For the United States as a whole the maximum use at any one time is about 5×10^{11} watts (500×10^6 kilowatts, or 500,000 megawatts); the electrical energy use per year exceeds 2.5×10^{12} kilowatt-hours. Assuming an average rate of 6 cents per kilowatt-hour the revenue is over 150 billion dollars per year. A single utility system may extend over a wide geographical area, even several states. Numbers of utilities are tied together so that power may be sent from one to another. Many and varied devices ("loads") tap energy from the network. These load devices may range from a few-watt night lamp to a megawatt motor. Since the loads may be turned on and off at random times, the system varies with time. In spite of the ever changing demand the system is required to maintain as far as possible the same voltage levels and a constant sinusoidal voltage with a frequency of 60 Hz.

In this chapter we consider only some basic elements of such a system. We show why different parts of the system operate at different voltage levels and how this is accomplished using transformers. We also consider three-phase systems since it is cost effective to transmit large powers over large distances using three phases and since the instantaneous power supplied by such a system may be designed in such a way that it is independent of time. Most large generators and large motors are wired for three-phase operation. Effective value phasors are used in this chapter.

9.1 VOLTAGE LEVELS

In power systems the electrical energy is routed on transmission lines which may be many kilometers in length. Since the conductors that make up these lines have nonzero resistance there will be energy expended in the lines. We would like to keep these energy losses to a minimum. Consider the two alternatives for delivering the same power to a load shown in Fig. 9.1, where R_T is the resistance of the transmission line, V_s is the source voltage, and R represents the load. The designators V_s, V, and I stand for $|\hat{V}_s|$, $|\hat{V}|$, and $|\hat{I}|$, the absolute value of the effective value phasors.

As shown in Fig. 9.1 the power that must be supplied by the low voltage source (a) to a 12-kW load is 22-kW, whereas the high-voltage source (b) need supply only

FIGURE 9.1.
The same power delivered at different voltages through the same transmission line.

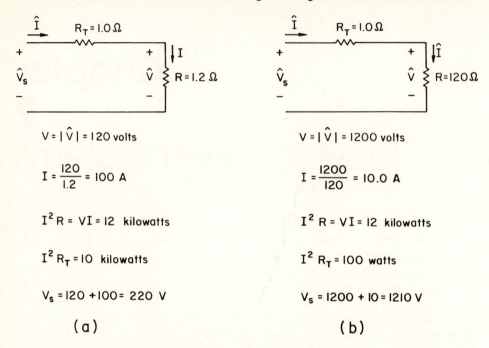

$$V = |\hat{V}| = 120 \text{ volts}$$

$$I = \frac{120}{1.2} = 100 \text{ A}$$

$$I^2 R = VI = 12 \text{ kilowatts}$$

$$I^2 R_T = 10 \text{ kilowatts}$$

$$V_s = 120 + 100 = 220 \text{ V}$$

$$\text{(a)}$$

$$V = |\hat{V}| = 1200 \text{ volts}$$

$$I = \frac{1200}{120} = 10.0 \text{ A}$$

$$I^2 R = VI = 12 \text{ kilowatts}$$

$$I^2 R_T = 100 \text{ watts}$$

$$V_s = 1200 + 10 = 1210 \text{ V}$$

$$\text{(b)}$$

12.1 kW to deliver the same 12 kW to the load. The principle is clear. Since the loss in the transmission line is proportional to the square of the current, we should keep this current as small as possible. Further, since the power supplied to the load is the product of voltage and current, we can keep the current small by raising the voltage while supplying the same power.

Of course we could reduce the loss in (a) by using a transmission line with smaller resistance. To reduce the loss to 100 watts, the resistance would have to be reduced by a factor of 100. Since for the same material the resistance of a wire is proportional to its length divided by its cross-sectional area, the cross-sectional area would have to be increased by a factor of 100 (see Section 1). This would require the use of 100 times as much metal and would increase the cost and the weight of the line by the same factor. The recent discovery of materials that become superconductive (zero resistance) at temperatures well above absolute zero may provide new opportunities for the design of transmission lines in the 21st century.

The example above explains why present power systems transmit power around the country at voltages that now exceed 750 kilovolts. But these voltages are too high for use in distribution systems in cities and of course much too high for safe use in homes. The ability to use different voltage levels, ranging from 120 volts to thousands of kilovolts in different parts of the system, is made possible by the use of transformers.

9.2 THE TRANSFORMER

One reason for the prevalence of alternating current in preference to direct current in power systems is the fact that an alternating voltage can be conveniently transformed in magnitude by means of a "transformer." In its simplest form a transformer consists of two separate coils ("windings") with a common core. The basic construction is shown schematically in Fig. 9.2. When the core is made up of magnetic material such as iron, almost all the magnetic flux produced by the currents threads both coils. While transformers may in general involve nonmagnetic cores, those used in power systems are composed of magnetic material. The windings of a transformer are identified according to the direction of power flow. The "primary" winding is connected to the source of electrical energy; the "secondary" winding is connected to the load.

FIGURE 9.2.
Basic construction of a transformer.

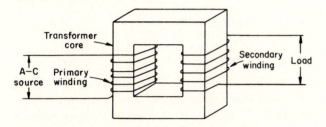

The action of a transformer is based on the same (Faraday's) law discussed in Chapter 5 which provides the voltage–current relationship for inductance:

$$v = \frac{d\Psi}{dt} \tag{9.1}$$

where Ψ is the "flux linkage," the magnetic flux crossing the surface defined by the element. The direction of the magnetic field is related to the voltage drop by the right-hand rule. For a single turn of the coil shown in Fig. 9.3, the flux linkage is the magnetic flux in the core, Φ, around which the coil is wound. For N turns, the flux linkage $\Psi = N\Phi$. The importance of Faraday's law is that the relation given by Eq. 9.1 holds no matter what the source of the magnetic flux. When the only source of the magnetic flux is the current i, then the inductance is the self-inductance,

$$L = \frac{\Psi}{i}$$

and

$$v = \frac{d\Psi}{dt} = L\left(\frac{di}{dt}\right)$$

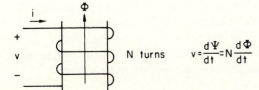

FIGURE 9.3.
Faraday's law for a coil.

as we found in Chapter 5 for an inductor. If the magnetic flux is the same in two different coils, which may have different values for N, as is the case in the arrangement of Fig. 9.4, the voltage across each coil will be proportional to the number of turns.

FIGURE 9.4.
Two sets of coils that are threaded by the same magnetic flux.

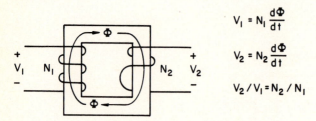

$$V_1 = N_1 \frac{d\Phi}{dt}$$

$$V_2 = N_2 \frac{d\Phi}{dt}$$

$$V_2/V_1 = N_2/N_1$$

Of course, in actual transformers the magnetic flux through the two coils is not exactly the same even when the core is made up of the best ferromagnetic material. Some of the flux paths (the "leakage flux") are completed through the region between the coils. In a detailed study we must also take account of the fact that the coils have some resistance and there are energy losses in the iron. However, in power transformers all of these effects are small. The analysis of transformer operation is simplified if it is first made for an *ideal transformer*. For many purposes actual transformers are close enough to ideal to ignore the difference. Furthermore, the actual effects may be taken into account by combining an ideal transformer with external R, L, and C elements. The definition of an ideal transformer and the circuit symbol used are given in Fig. 9.5. The ratio N_2/N_1 is called the "turns ratio." Only the turns ratio matters for an ideal transformer, which is sometimes represented simply as $1 : a$ above the transformer, where $a = N_2/N_1$. The transformer characteristics may also be defined through the ratio N_1/N_2, in which case the ideal transformer label could be labeled with $a : 1$. The student should note carefully which ratio is used in a given circuit.

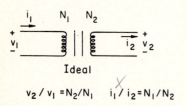

Ideal

$$v_2/v_1 = N_2/N_1 \qquad i_1/i_2 = N_1/N_2$$

FIGURE 9.5.
Circuit model of the ideal transformer.

In an ideal transformer the relationship between the currents shown in Fig. 9.5 is derived from the voltage relationship $N_1v_1 = N_2v_2$ coupled with $v_1i_1 = v_2i_2$, which in turn is based on the assumption that all energy is transmitted through the transformer. Several other properties will be taken as given for an ideal transformer. In an ideal transformer no phase shifts will be introduced by the transformer; if the secondary is open-circuited ($i_2 = 0$) the primary current i_1 is also zero; and the self-inductance of the primary in an ideal transformer is assumed to be infinite.

Using a pair of ideal transformers, the source and load of Fig. 9.1a may now be connected through the same transmission line with greatly reduced losses, as shown in Fig. 9.6. By multiplying the source voltage by a factor of 10, the current in the transmission line is reduced by a factor of 10, and the I^2R_T losses by a factor of 100. In the system of Fig. 9.6, the generator voltage will be only 1.0 V higher than the voltage across the load. The generator supplies 12.1 kW, of which 100W is dissipated in the transmission line resistance R_T, and 12 kW is transmitted to the load.

FIGURE 9.6.
Use of transformers to reduce the losses in a transmission line. All values of current and voltage are magnitudes of the effective value phasors.

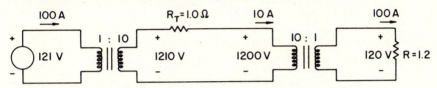

Another consequence of the change of voltage and current levels by an ideal transformer is the change of impedance level. As derived in Fig. 9.7, the impedance $\mathbf{Z}_{ab} = \hat{V}_1/\hat{I}_1$ is not equal to $\mathbf{Z}_L$ but is related to it via the expression $\mathbf{Z}_{ab} = (N_1/N_2)^2\mathbf{Z}_L = a^2\mathbf{Z}_L$. It is important to note in this expression that the turns ratio a is defined as N_1/N_2, the ratio of the number of primary turns to the number of secondary turns. A transformer inserted between the port a–b and the load can thus be used to make the impedance between a and b appear to be larger or smaller than $\mathbf{Z}_L$ depending upon the value of $(N_1/N_2)^2 = a^2$. Again, for the ideal case the power into terminals a–b is equal to the power to the load $\mathbf{Z}_L$ since

$$V_1 I_1 \cos\theta = \left(\frac{V_1}{a}\right)(aI_1)\cos\theta = V_2 I_2 \cos\theta$$

FIGURE 9.7.
The ideal transformer changes the input impedance by a factor of $a^2 = (N_1/N_2)^2$.

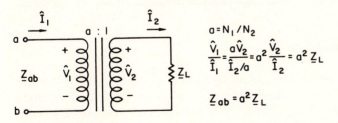

$$a = N_1/N_2$$

$$\frac{\hat{V}_1}{\hat{I}_1} = \frac{a\hat{V}_2}{\hat{I}_2/a} = a^2\frac{\hat{V}_2}{\hat{I}_2} = a^2\underline{Z}_L$$

$$\underline{Z}_{ab} = a^2\underline{Z}_L$$

TEXT EXAMPLE 9.1 _____

The ideal transformer of Fig. 9.7 is connected between a sinusoidal source $v_s = \sqrt{2}(120)\cos(377t)$ and a load whose impedance at 60 Hz is $2 + j2$ ohms. The ideal transformer has a primary to secondary turns ratio of 10. Find the power delivered to the load, the power supplied by the source, and the impedance $\mathbf{Z}_{ab}$. Show that $\mathbf{Z}_{ab} = (N_1/N_2)^2 \mathbf{Z}_L$.

Solution: Since the effective value phasor $\hat{V}_1 = 120\underline{/0°}$ and the turns ratio is 10,

$$\hat{V}_2 = \frac{\hat{V}_1}{10} = 12\underline{/0°}$$

Then,

$$\hat{I}_2 = \frac{\hat{V}_2}{\mathbf{Z}_L} = \frac{12\underline{/0°}}{2 + j2} = 3\sqrt{2}\underline{/-45°}$$

The power P_2 supplied to the load is given by the expression

$$P_2 = V_2 I_2 \cos\theta = (12)(3\sqrt{2})\left(\frac{1}{\sqrt{2}}\right) = 36 \text{ W}$$

To find the power supplied by the source, we use the fact that

$$\hat{I}_1 = \frac{\hat{I}_2}{10} = 0.3\sqrt{2}\underline{/-45°}$$

Then,

$$P_1 = V_1 I_1 \cos\theta = (120)(0.3\sqrt{2})\left(\frac{1}{\sqrt{2}}\right) = 36 \text{ W}$$

as expected, since the ideal transformer is lossless, $P_1 = P_2$.

   The impedance $\mathbf{Z}_{ab} = \hat{V}_1/\hat{I}_1 = 120\underline{/0°}/0.3\sqrt{2}\underline{/-45°} = 200\sqrt{2}\underline{/45°} = 200 + j200$. Finally, $\mathbf{Z}_{ab} = (100)\mathbf{Z}_L = a^2 \mathbf{Z}_L$, as shown in Fig. 9.7.

9.3   GENERATOR ACTION

Most electrical power is obtained by electromechanical energy conversion whereby mechanical energy is converted into electrical energy by means of electrical generators. The mechanical energy may be obtained from fuel which is transformed into heat to produce steam to run a steam turbine or into heat to run an internal combustion engine. Other sources of the mechanical energy are hydropower or wind energy. There are also direct energy-conversion devices, such as fuel cells and photovoltaic generators, that convert chemical or solar energy directly into electrical energy, but these represent a very small part of present power generation.

Mechanical energy is converted into electrical energy by a rotary electrical generator. The fundamental parts of a generator are shown in schematic form in Fig. 9.8. As indicated in the left side of Fig. 9.8, a coil is mounted on a cylinder of magnetic material called the armature which can be rotated in a steady magnetic field produced by field poles. Two views of the system are shown on the left. The action of electric generation is based on the same (Faraday's) law which led to transformer action:

$$v = \frac{d\Psi}{dt}$$

In this case the rate of change of flux linkage occurs because the coil is being mechanically rotated through a magnetic field. Thus, a voltage is produced ("induced") between the ends of the coil. When a load such as a resistor is connected between the ends of the coil through "slip rings," a current flows in the coil. The slip rings allow the motor to rotate while at the same time providing electrical contact. When the rotation is in the direction indicated in Fig. 9.8b, the polarity of the induced voltage is such as to cause current to flow as indicated by the cross and the dot in the conductor cross section. Because the conductors carry current while in a magnetic field, a force is exerted on the conductors in such a direction as to oppose the rotation of the armature. The direction of the force is shown in part (c) of the figure by the arrows labeled F. Therefore, if the armature is to be rotated at the same speed as before the resistor was connected, the mechanical power supplied to the rotor must be increased. The electrical energy supplied to the resistor is provided by the mechanical energy to the shaft.

FIGURE 9.8.
Details of an elementary generator.

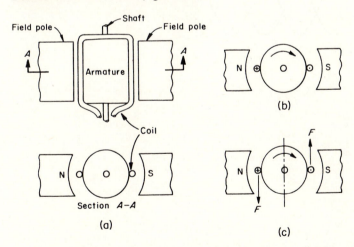

The generators used in power systems are "three-phase synchronous generators." In these machines it is the armature winding that is usually held fixed and the field winding (connected to a dc supply through slip rings) that is rotated. The result is to produce a changing magnetic field in the armature winding. The magnetic flux through the armature winding varies sinusoidally with time, thus inducing a sinusoidal voltage. The rotor is rotated at the speed required to make the frequency of that induced voltage have a value of 60 Hz. The reasons for this type of construction are as follows:

1. The stationary armature is more easily insulated at the high voltage levels for which the synchronous machine is usually designed.
2. Only two relatively small slip rings are required as connections to the rotating dc circuit.

Instead of a single armature coil there are three such identical coils, each taking up one third of the circumference. The magnitude of the voltage induced in each coil is identical; but since the rotating field is changing 120° later in the second coil than in the first, and 120° later in the third coil than in the second, the three induced voltages are not in the same phase. The phase relation among the three voltages is shown in Fig. 9.9a, where AA', BB', and CC' are the three armature coils. In the example shown A', B', and C' are connected to a common terminal N, called the neutral. In practice such a connection of the three phases is called a "wye" connection since it can be drawn as shown in Fig. 9.9b.

FIGURE 9.9.
Phase relationships among the stator voltages are shown in (a). The same connection of the three coils is redrawn in (b) to show why it is called a "*wye*" connection.

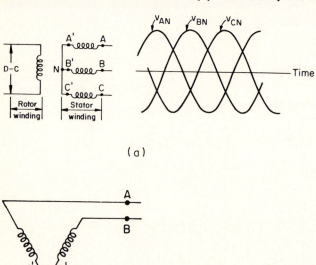

Each of the voltages v_{AN}, v_{BN}, and v_{CN}, are called phase voltages, and the generator is a three-phase synchronous machine. The voltages v_{AB}, v_{BC}, and v_{CA}, are called the "line" voltages, or the "line to line" voltages. As we shall see below, the line voltages are not equal to the phase voltages. The time functions of the three phase voltages are

$$v_{AN}(t) = \sqrt{2}V_p \, \cos(\omega t)$$

$$v_{BN}(t) = \sqrt{2}V_p \, \cos(\omega t - 120°)$$

$$v_{CN}(t) = \sqrt{2}V_p \, \cos(\omega t - 240°) = \sqrt{2}V_p \, \cos(\omega t + 120°)$$

where $\omega = (2\pi)60 = 377$ radians/second. We show below in Text Example 9.2 that $v_{AN}(t) + v_{BN}(t) + v_{CN}(t) = 0$. The same machine may be used as a synchronous motor by supplying electrical energy to the armature, thus reversing the energy conversion. In the motor, electrical energy is converted into mechanical energy.

TEXT EXAMPLE 9.2

Suppose that the three-phase voltages of the synchronous generator shown in Fig. 9.9 are given by the time functions

$$v_{AN}(t) = \sqrt{2}(3600) \cos(377t) \text{ V}$$

$$v_{BN}(t) = \sqrt{2}(3600) \cos(377t - 120°) \text{ V}$$

$$v_{CN}(t) = \sqrt{2}(3600) \cos(377t + 120°) \text{ V}$$

Write the time functions in terms of phasors. Show that $v_{AN}(t) + v_{BN}(t) + v_{CN}(t) = 0$. Find the three voltage phasor differences between the lines, that is, $\hat{V}_{AB} = \hat{V}_A - \hat{V}_B$, $\hat{V}_{BC} = \hat{V}_B - \hat{V}_C$, and $\hat{V}_{CA} = \hat{V}_C - \hat{V}_A$ and show that the corresponding temporal functions also add to zero.

Solution: As usual we may add sinusoidal voltages of the same frequency by using phasors. The three phasors of interest are

$$\hat{V}_{AN} = 3600 \underline{/0°}$$

$$\hat{V}_{BN} = 3600 \underline{/-120°} = -1800 - j1800\sqrt{3}$$

$$\hat{V}_{CN} = 3600 \underline{/+120°} = +1800 - j1800\sqrt{3}$$

Adding these three phasors,

$$\hat{V}_{AN} + \hat{V}_{BN} + \hat{V}_{CN} = 0$$

It therefore follows that the sum of the temporal functions is also equal to zero. Now to find the voltages $v_{AB}(t)$, $v_{BC}(t)$, and $v_{CA}(t)$, we use phasors,

$$\hat{V}_{AB} = \hat{V}_{AN} + \hat{V}_{NB} = 5400 + j1800\sqrt{3} = 3600\sqrt{3}\underline{/30°}$$

$$\hat{V}_{BC} = \hat{V}_{BN} + \hat{V}_{NC} = 0 - j3600\sqrt{3} = 3600\sqrt{3}\underline{/-90°}$$

$$\hat{V}_{CA} = \hat{V}_{CN} + \hat{V}_{NA} = -5400 + j1800\sqrt{3} = 3600\sqrt{3}\underline{/150°}$$

The magnitudes of the line voltages are thus $\sqrt{3}$ times the magnitudes of the phase voltages. Adding the line voltage phasors given above,

$$\hat{V}_{AB} + \hat{V}_{BC} + \hat{V}_{CA} = 0$$

It then follows that the sum of the line voltages is also zero:

$$v_{AB}(t) + v_{BC}(t) + v_{CA}(t) = 0$$

Note that this result could have been deduced immediately by KVL.

9.4 THREE-PHASE SYSTEMS

In a single-phase circuit there may be several voltages available but they all originate from a single sinusoidal source. For example, the three-wire 60-Hz, 240-V system connected to many homes is a single-phase system. The secondary of the transformer feeding the home is center-tapped and grounded. The voltage from each of the other lines to ground is 120 V (effective), whereas the voltage between those lines is 240 V. All the voltages have the *same phase*. (See the figure in Problem 9.21.)

A "polyphase" circuit is an interconnection of a number of single phase circuits, the voltages of which are at the same frequency but differ in phase. The generator shown in Fig. 9.9 yields such a circuit. To modify the voltage levels, a three-phase transformer may be constructed from a combination of three single-phase transformers. Then, as shown in Fig. 9.10, the three transformers may be used to transform the magnitudes of the voltages generated in the system described by Fig. 9.9. In this case both the primaries and secondaries of the transformers are connected in the *wye* configuration. The $\pm$ symbols are used to show that the phase relations in the secondaries are the same as in the primaries.

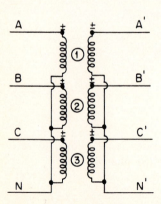

FIGURE 9.10.
Wye–wye connection of three single-phase transformers.

An alternative connection of the three single-phase transformers is the "delta" connection shown in Fig. 9.11. In this case only three wires are needed for each side of the three-phase transformer as there is no neutral reference point. It is also possible to use a *wye* connection on one side of the transformers and a delta connection on the other side. To avoid undue complexity, no transformers are shown in the balance of this section and the primes are dropped in the notation. However, it should be clear that the same change in voltage levels that was described in Section 9.2 can be and is used in three-phase systems.

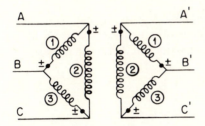

FIGURE 9.11.
Delta–Delta connection of three single-phase transformers.

Henceforth, we shall represent a *wye* connected generator or transformer with the voltage source configuration shown in Fig. 9.12a. The associated voltage phasors are shown in the complex plane in the right-hand side of the figure. The student may wish to refer back to Text Example 9.2 at this point, which deals with just this case. We have dropped the subscript N for brevity here, that is, $\hat{V}_A = \hat{V}_A - \hat{V}_N$ is the phasor voltage with respect to ground. As noted above, $\hat{V}_A$, $\hat{V}_B$, and $\hat{V}_C$ are called the phase voltages.

FIGURE 9.12
(a) Schematic for a *wye*-connected generator or transformer supplying the three lines *A*, *B*, *C*. (b) The phasor diagram on the right shows the relationship between the sources. The vector $\hat{V}_{AB}$ on the right is a graphical representation of the difference in potential between lines *A* and *B*.

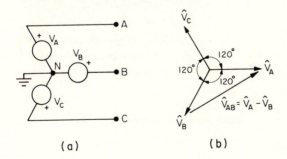

(a) (b)

That the vector labeled $\hat{V}_{AB}$ in Fig. 9.12b is equal to $\hat{V}_A - \hat{V}_B$ may be understood as follows. Two vectors are added graphically by putting the tail of one vector at the tip of the other. Thus, in the figure

$$\hat{V}_B + \hat{V}_{AB} = \hat{V}_A$$

and it follows that

$$\hat{V}_{AB} = \hat{V}_A - \hat{V}_B$$

Note now that if the illustrated vector $\hat{V}_{AB}$ is transposed to the origin, it has a magnitude equal to $\sqrt{3}|\hat{V}_A|$ and makes an angle of $+30°$ with the real axis. This corresponds to the solution in Text Example 9.2,

$$\hat{V}_{AB} = 3600\sqrt{3}\underline{/30°}$$

since each of the sources had effective values of 3600 V.

The corresponding delta configuration at the output of a generator or a set of transformers such as shown in Fig. 9.11 is illustrated in Fig. 9.13. The phasor diagram corresponds to the voltages

$$\hat{V}_{AB} = V_0\underline{/0°}$$

$$\hat{V}_{CA} = V_0\underline{/120°}$$

$$\hat{V}_{BC} = V_0\underline{/-120°}$$

Note that there is no specific point or wire corresponding to ground or zero potential in this configuration which is labeled N in the phasor diagram. In this case the line-to-line voltages are equal to the phase voltages of the sources since they are connected directly between the lines.

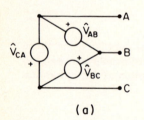
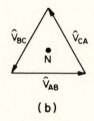

(a)   (b)

FIGURE 9.13.
A delta-connected generator or transformer and the corresponding phasor diagram.

Sources of the two types shown in Figs. 9.12 and 9.13 are typically connected to loads that themselves are either in the *wye* or delta configuration. A *wye–wye* connection is shown in Fig. 9.14. In the *wye–wye* connection shown we are

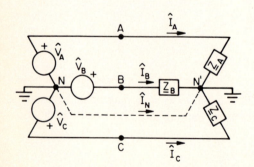

FIGURE 9.14.
Wye–wye connection to loads.

effectively connecting three single-phase sources to three loads but are using only four wires instead of six. The currents $\hat{I}_A$, $\hat{I}_B$, and $\hat{I}_C$ are called the "line" currents. The current $\hat{I}_N$ is the "neutral" current. The currents in the loads $\mathbf{Z}_A$, $\mathbf{Z}_B$, and $\mathbf{Z}_C$ are the "phase" currents and in this connection are identical with the line currents. As noted earlier, the voltages between the lines ($\hat{V}_{AB}$, $\hat{V}_{BC}$, and $\hat{V}_{CA}$) are the "line voltages." The voltages across the loads ($\hat{V}_{AN}$, $\hat{V}_{BN}$, and $\hat{V}_{CN}$) are equal to the phase voltages and are not equal to the line voltages but are related to them by

$$\hat{V}_{AB} = \hat{V}_{AN} - \hat{V}_{BN}$$

$$\hat{V}_{BC} = \hat{V}_{BN} - \hat{V}_{CN}$$

$$\hat{V}_{CA} = \hat{V}_{CN} - \hat{V}_{AN}$$

The currents $\hat{I}_A$, $\hat{I}_B$, and $\hat{I}_C$, may be found from the phase voltages if the point N is connected to N' since then

$$\hat{I}_A = \frac{\hat{V}_A}{\mathbf{Z}_A}$$

$$\hat{I}_B = \frac{\hat{V}_B}{\mathbf{Z}_B}$$

$$\hat{I}_C = \frac{\hat{V}_C}{\mathbf{Z}_C}$$

By KCL, the current in the neutral line is given by

$$\hat{I}_N = -(\hat{I}_A + \hat{I}_B + \hat{I}_C)$$

Note that we have used phasor notation here since all the signals are at the same frequency ω.

If $\mathbf{Z}_A = \mathbf{Z}_B = \mathbf{Z}_C$, as they would be, for example, if the load represented a three-phase motor, the load is called "balanced." The currents $\hat{I}_A$, $\hat{I}_B$, and $\hat{I}_C$ will then have equal magnitudes and the same 120° phase relation as the phase voltages. Thus, the sum of the currents will be zero and no current flows in the neutral return ($\hat{I}_N = 0$). For the case of a balanced load the current is carried by just three wires and the neutral line may be omitted. This provides one practical reason for using three-phase power. Since the neutral line is not required, considerable copper can be saved by using only three wires to transmit the power rather than four. If a neutral line is maintained it may have a small cross section (and thus a higher resistance) since there will be little current through it.

A second important property of three-phase power is that the *instantaneous* power is independent of time for a balanced load. To prove this for equal resistive loads, $\mathbf{Z}_A = \mathbf{Z}_B = \mathbf{Z}_C = R$, we note that the three currents in Fig. 9.14 are given by

$$i_A(t) = \left(\frac{V_0}{R}\right) \cos \omega t$$

$$i_B(t) = \left(\frac{V_0}{R}\right)\cos(\omega t - 120°)$$

$$i_C(t) = \left(\frac{V_0}{R}\right)\cos(\omega t + 120°)$$

The instantaneous power $p(t) = i(t)v(t) = i^2(t)R$ that is supplied by the generator is then given by the expression

$$p(t) = \left(\frac{V_0^2}{R}\right)[\cos^2\omega t + \cos^2(\omega t - 120°) + \cos^2(\omega t + 120°)]$$

Now using the trigonometric identity

$$\cos^2\alpha = \frac{1}{2} + \frac{1}{2}\cos(2\alpha)$$

yields

$$p(t) = \left(\frac{V_0^2}{R}\right)\left[\frac{3}{2} + \frac{1}{2}\cos(2\omega t) + \frac{1}{2}\cos(2\omega t - 240°) + \frac{1}{2}\cos(2\omega t + 240°)\right]$$

Finally, using $\cos(\alpha \pm \beta) = \cos\alpha\cos\beta \pm \sin\alpha\sin\beta$, we have

$$p(t) = 3\left(\frac{V_0^2}{2R}\right)$$

This shows that the power supplied is constant in time and equal to three times the average power in each of the loads. Single-phase power systems on the other hand supply power that pulsates in time. This means that both generators and loads display much smoother operation in polyphase systems than in single-phase systems.

An alternative connection of the sources and loads, the "delta–delta" connection is shown in Fig. 9.15. The currents in the load impedances $\mathbf{Z}_A$, $\mathbf{Z}_B$, and $\mathbf{Z}_C$ in

FIGURE 9.15.
Delta–delta connection to loads.

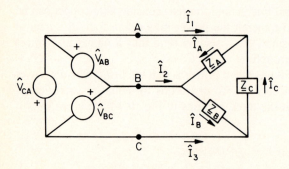

the circuit of Fig. 9.15 will be identical to those in the circuit of Fig. 9.14. These currents $\hat{I}_A$, $\hat{I}_B$, and $\hat{I}_C$ are called the phase currents. In this configuration the line voltages are the same as the phase voltages but the line currents are not the same as the phase currents. The relations are

$$\hat{I}_1 = \hat{I}_A - \hat{I}_C$$

$$\hat{I}_2 = \hat{I}_B - \hat{I}_A$$

$$\hat{I}_3 = \hat{I}_C - \hat{I}_B$$

The computations of the power supplied to a three-phase load may be done most simply by adding the powers in the three-phase impedances $\mathbf{Z}_A$, $\mathbf{Z}_B$, and $\mathbf{Z}_C$:

$$P = I_A^2 R_A + I_B^2 R_B + I_C^2 R_C$$

For $\hat{V}_A = \hat{I}_A \mathbf{Z}_A$, $\hat{V}_B = \hat{I}_B \mathbf{Z}_B$, and $\hat{V}_C = \hat{I}_C \mathbf{Z}_C$, this is equivalent to

$$P = V_A I_A \cos \theta_A + V_B I_B \cos \theta_B + V_C I_C \cos \theta_C$$

where the various symbols I_A, V_A, and so on, are the amplitudes of the corresponding effective value phasors. This result holds for *wye* or delta configurations.

TEXT EXAMPLE 9.3

In the circuit of Fig. 9.15, suppose $\hat{V}_{AB} = 1200 \underline{/0°}$, $\hat{V}_{BC} = 1200 \underline{/-120°}$, and $\hat{V}_{CA} = 1200 \underline{/+120°}$. Let $\mathbf{Z}_A = \mathbf{Z}_B = \mathbf{Z}_C = 120 \underline{/60°} = 60 + j60 \sqrt{3}$. Find the total power delivered by the three-phase source.

Solution:

$$\hat{I}_A = \frac{\hat{V}_{AB}}{\mathbf{Z}_A} = \frac{1200\underline{/0°}}{120\underline{/60°}} = 10\underline{/-60°} = 5 - j5\sqrt{3}$$

$$\hat{I}_B = \frac{\hat{V}_{BC}}{\mathbf{Z}_B} = \frac{1200\underline{/-120°}}{120\underline{/60°}} = 10\underline{/-180°} = -10 + j0$$

$$\hat{I}_C = \frac{\hat{V}_{CA}}{\mathbf{Z}_C} = \frac{1200\underline{/+120°}}{120\underline{/60°}} = 10\underline{/+60°} = 5 + j5\sqrt{3}$$

The total power delivered to the load is

$$P = V_{AB} I_A \cos \theta_A + V_{BC} I_B \cos \theta_B + V_{CA} I_C \cos \theta_C$$

where we use the notation $V_{KL} = |\hat{V}_{KL}|$ and $I_K = |\hat{I}_K|$. Then,

$$V_{AB} = V_{BC} = V_{CA} = 1200$$

$$I_A = I_B = I_C = 10$$

$$\cos \theta_A = \cos \theta_B = \cos \theta_C = 0.5$$

and

$$P = (3)(1200)(10)(0.5) = 18 \text{ kW}$$

To check this computation we may determine the power delivered to the load from

$$P = I_A^2 R_A + I_B^2 R_B + I_C^2 R_C$$

$$P = (10)^2(60) + (10)^2(60) + (10)^2(60)$$

$$P = 3 \times 6000 = 18 \text{ kW}$$

which agrees with the previous calculation.

9.5 RATINGS

The rating of all electrical equipment is based upon the temperature to which various parts are submitted when the equipment is operated.

Transformers are rated in terms of the volt–ampere output under specified conditions of voltage and frequency. The nameplate of a transformer provides the following information:

1. Voltage rating. This is given for both the high-voltage winding and the low-voltage winding. For example, if the transformer has a high-voltage winding rated at 2300 volts and a low-voltage winding rated at 230 volts, the nameplate gives the data as 2300/230 volts. (Note that reference is made to the high- and low-voltage windings, not to primary and secondary windings. Either the high- or the low-voltage winding may be used as the primary.)

2. Kilovolt–ampere rating.
3. Frequency rating.
4. Temperature rise.

The rated currents of the transformer windings are not usually given on the nameplate but can be calculated from the kilovolt–ampere rating of the transformer and the voltage ratings of the windings.

The nameplate of a generator usually contains the rating of power, voltage, speed, and temperature rise. The power rating in kilowatts is the rated full-load output of the generator when the terminal voltage is that specified on the nameplate. A current rating is not usually included but can be calculated from the power voltage rating. The temperature rise specified is usually that to be expected if the machine is operated continuously at rated conditions.

The load on any machine is the power that it delivers. Full load is the rated power output of the machine. The nameplate ratings are full-load ratings, that is, the values to be expected when the machine is operated at rated voltage and is delivering rated power.

The voltage rating of three-phase supplies and loads are specified in terms of effective line voltage values. Thus, a 220-volt three-phase supply is one having 220 volts (effective value) between the lines. Three-phase loads are usually balanced, therefore a single power rating and a single current rating are given for the unit. The current rating is the line current, regardless of the type of load connection. The power rating refers to the total power dissipated in the unit. The power dissipated in each phase is one-third of the total power.

PRACTICE PROBLEMS
AND
ILLUSTRATIVE EXAMPLES

Problem 9.1

(a) A 1-ohm resistor is connected to the low-voltage side of a 440/110-volt transformer. If a single resistor is to replace the transformer and the 1-ohm resistor so as to be an equivalent load on the 440-volt line, what should be the resistance and power ratings of the replacement resistor?

$$\text{Ans.: } 16 \ \Omega, 12.1 \text{ kW}$$

(b) Compare the ratio of the replacement resistor to the 1-ohm resistor with the turns ratio of the transformer.

> Ans.: The resistance ratio of 16 : 1 is equal to the square of the 4 : 1 turns ratio.

EXAMPLE 9.1

For the transformer circuit shown find $\hat{V}_L$.

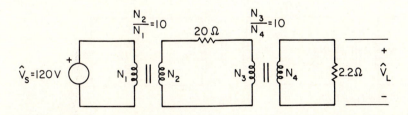

Solution: We will work this two ways. First, we use the basic transformer relationships

$$\hat{V}_S = \left(\frac{N_S}{N_P}\right)\hat{V}_P \tag{1}$$

and

$$\hat{I}_S = \left(\frac{N_P}{N_S}\right)\hat{I}_P \tag{2}$$

where S stands for secondary and P stands for primary. From Eq. 1 the voltage across the secondary of the first transformer is $\hat{V}_2 = 1200$ V. Let $\hat{I}_L$ be the load current. Then, applying Eq. 2 to the second transformer, the current through the 20-Ω resistor is $(1/10)\,\hat{I}_L$ and the load voltage is $2.2(\hat{I}_L)$. The voltage across the primary of the second transformer is 1200 volts minus the voltage drop across the 20-Ω resistor, so

$$\hat{V}_3 = 1200 - 20\left(\frac{\hat{I}_L}{10}\right)$$

But, using the transformer relationship $\hat{V}_3 = 10\hat{V}_L$, this may be written

$$10(2.2\,\hat{I}_L) = 1200 - 2\,\hat{I}_L$$

Solving, $\hat{I}_L = 50$ A, which in turn yields $\hat{V}_L = 110$ V.

Using impedance transformation we could replace the *second* transformer with the impedance $a^2 R_L = 220\,\Omega$ since $a = N_3/N_4 = 10$. The primary voltage at the second transformer is then given by the voltage divider principle:

$$\hat{V}_3 = \frac{220}{220 + 20}(1200) = 1100 \text{ V}$$

Now using the transformer relationship, $\hat{V}_L = \hat{V}_3/10$ and finally $\hat{V}_L = 110$ V, as before. The impedances in this example are all pure resistances, of course.

Problem 9.2

In the 60-Hz circuit below, what is the (effective) value of $|\hat{V}_s|$ necessary to give an (effective) output voltage $|\hat{V}_o|$ equal to 100 volts? What is the power delivered by the source? What percentage of the power goes to the load?

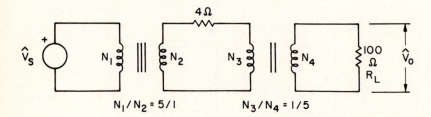

Ans.: 200 V, 200 W, 50%

Problem 9.3

Use the results of Problem 9.2 to deduce the current $\hat{I}_s$ drawn from $\hat{V}_s$. Work from right to left using the impedance transformation $\mathbf{Z}_{ab} = a^2 \mathbf{Z}_L$ to find the equivalent resistance to the source. Note the consistency with the current drawn from $\hat{V}_s$.

$$\text{Ans.: } \hat{I}_s = 1 \text{ A}$$
$$R = 200 \ \Omega$$

Problem 9.4

Repeat Problem 9.2 with $N_1/N_2 = \frac{1}{5}$ and $N_3/N_4 = 5$.

Problem 9.5

Find the equivalent resistance "seen" by the source V_s in Problem 9.4 using impedance transformation. Show consistency with $\hat{V}_s / \hat{I}_s$.

Problem 9.6

Find the power supplied by the generator $\hat{V}_s$ in Example 9.1 using the currents flowing in R_L and in the 20-Ω resistor. Use impedance transformation at the first transformer to find the impedance "seen" by the source $\hat{V}_s$ and show that the power drawn is consistent with your result. The impedance is pure resistive in this case.

Problem 9.7

Suppose the 2.2-Ω load resistor in Example 9.1 is replaced with an impedance $\mathbf{Z}_L = (1 + j3) \ \Omega$. Find the impedance seen by the source and the power supplied by it.

Problem 9.8

Repeat Example 9.1 if $N_1/N_2 = .05$, $N_3/N_4 = .05$, and the load resistor is 2.2 Ω.

Problem 9.9

Find the percentage of the power supplied to the load in the circuits described in Example 9.1 and Problem 9.8. Discuss the result.

Problem 9.10

In Problem 9.8 suppose the load is given by $\mathbf{Z}_L = (2.15 + j2.2) \ \Omega$. Find the load voltage.

Problem 9.11

In Problem 9.10, find the impedance seen by the source and the power supplied by it. Show that this power equals the sum of the load power and the power lost in the 20 Ω resistor.

EXAMPLE 9.2

An audio amplifier has a Thevenin equivalent output resistance of 800 Ω. The amplifier is to be connected to an "8-Ω speaker" (a loudspeaker whose resistance is 8 Ω). The connection may be made directly as shown in (a) or through a transformer as shown in (b). Find the transformer turns ratio a that yields maximum power transfer. Compare the power delivered to the loudspeaker in the two cases.

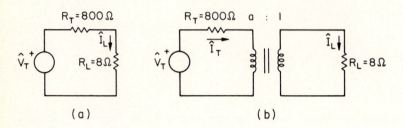

(a)   (b)

Solution: For case (a),

$$P_L = I_L^2 R_L = \frac{V_T^2}{(R_T + R_L)^2} R_L = 1.23 \times 10^{-5}(V_T^2)$$

For case (b), using the result in Fig. 9.7, the resistance presented to the primary of the transformer is $a^2 R_L$. Then,

$$I_T = \frac{V_T}{R_T + a^2 R_L}$$

From the turns ratio,

$$I_L = aI_T = \frac{aV_T}{R_T + a^2 R_L}$$

and

$$P_L = I_L^2 R_L = \frac{V_T^2 a^2}{(R_T + a^2 R_L)^2} R_L$$

Now from Chapter 3 and Chapter 8 we know that maximum power transfer will occur if the entire load, including the transformers, is matched to R_T. (Note that all the impedances in this case are real.) This requires $a^2 R_L = R_T$ and hence that $a = 10$. Substituting into P_L yields

$$P_L = 3.125 \times 10^{-4}(V_T^2)$$

The ratio of P_L in case (b) to P_L in case (a) is 25.4.

By using a transformer to match the resistance of the speaker to the output resistance of the amplifier, we have increased the power delivered to the speaker more than 25 times.

Problem 9.12

If the amplifier in Example 9.2 above is connected to a 32-Ω loudspeaker, find the turns ratio of the transformer required to deliver maximum power to the speaker.

Problem 9.13

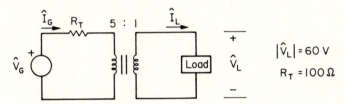

As shown above, a generator ($\hat{V}_G$) supplies 60 volts to a load through a resistance R_T and an ideal transformer with a primary to secondary turns ratio of 5 : 1. The power to the load is 120 W at a power factor of 0.8. Find the power supplied by the generator.

Ans.: 145 W

Problem 9.14

Find the value of the total impedance seen by source $\hat{V}_G$ if the load has an inductive reactance.

Problem 9.15

Repeat Problem 9.13 if $R_T = 200$ Ω, $|\hat{V}_L| = 150$ V, and the turns ratio is 4 : 1. The load power is 240 W with a power factor of 0.8.

Problem 9.16

Find the value of the total impedance seen by the source $\hat{V}_G$ in Problem 9.15 if the load has a capacitive reactance.

Problem 9.17

As shown in the diagram below, a 60-Hz generator (V_G) supplies two 120-volt loads through a transmission line of resistance R_T and an ideal transformer with a turns ratio of 10 : 1.

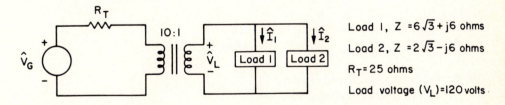

(a) Find the power supplied to loads 1 and 2.

(b) Find the power to R_T and the total power supplied by the generator.

Ans.: (a) $P_1 = 1039$ W; $P_2 = 1039$ W

(b) $P_{R_T} = 100$ W; $P_T = 2178$ W

Problem 9.18

Use impedance transformations to find the total impedance seen by the source $\hat{V}_G$ in Problem 9.17.

Problem 9.19

Find the currents $\hat{I}_1$ and $\hat{I}_2$ in Problem 9.17 (let the phasors be defined such that $\hat{V}_L = 120\underline{/0°}$). Use KCL and the transformer relationship to find the current drawn from the generator. Using this current and the total impedance found in Problem 9.18, calculate the total power supplied and compare it to P_T given in the answer to Problem 9.17.

Problem 9.20

Repeat Problem 9.17 for $\mathbf{Z}_1 = 8 - j8$ Ω and $\mathbf{Z}_L = 8\sqrt{3} - j8$ Ω. R_T and $\hat{V}_L$ remain the same, as does the transformer.

Problem 9.21

A three-wire home power circuit is shown. The various electrical appliances connected are represented by R_1, R_2, and R_3.

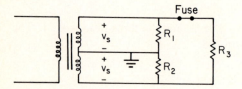

For

$$v_s = 120\sqrt{2}\ \cos\ (377t)$$

and

$$R_1 = 80\ \Omega,\ R_2 = 40\ \Omega,\ R_3 = 60\ \Omega$$

(a) Find the total power delivered to the three loads.

Ans.: 1.5 kW

(b) If the load represented by R_1 is turned off (open-circuited), find the total power delivered to the remaining loads.

Ans.: 1.32 kW

Problem 9.22

A 2 kW oven (pure resistance) is turned on in parallel with R_3 in Problem 9.21. How much steady state effective value current must the fuse carry?

Problem 9.23

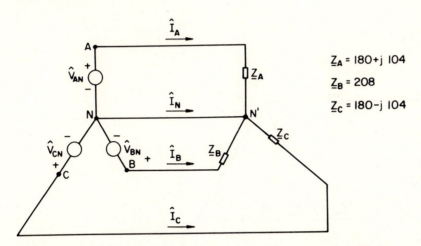

$$Z_A = 180 + j\,104$$

$$Z_B = 208$$

$$Z_C = 180 - j\,104$$

The three sinusoidal voltage sources in the circuit above are connected together as a "three-phase" source in a *wye* configuration, each having the same magnitude and frequency, but different phase angles. The phasor voltages are

$$\hat{V}_{AN} = 120\underline{/0°}\,\text{V} \quad \hat{V}_{BN} = 120\underline{/-120°}\,\text{V} \quad \hat{V}_{CN} = 120\underline{/+120°}\,\text{V}$$

The loads are connected in an unbalanced *wye* configuration.

(a) Find $\hat{I}_A, \hat{I}_B, \hat{I}_C,$ and $\hat{I}_N$.

Ans.: $\hat{I}_A = 0.58\underline{/-30°}\,\text{V}$
$\hat{I}_B = 0.58\underline{/-120°}\,\text{V}$
$\hat{I}_C = 0.58\underline{/+150°}\,\text{V}$
$\hat{I}_N = 0.58\underline{/60°}\,\text{V}$

(b) Find the total power in watts delivered by the three-phase supply.

Ans.: 189 W

Problem 9.24

Repeat Problem 9.23 for the case where the impedances are changed so that $\mathbf{Z}_A = \mathbf{Z}_B = \mathbf{Z}_C = 180 + j\,104\ \Omega$.

Problem 9.25

In the circuit of Problem 9.24 the loads are balanced so that $\hat{I}_N = 0$. Find $\hat{I}_N$ if the impedance $\mathbf{Z}_C$ breaks in such a way that $\mathbf{Z}_C$ becomes an open circuit. That is, find $\hat{I}_N$ in the new configuration.

Problem 9.26

In the circuit for Problem 9.24 suppose the neutral line wire was removed since no current is carried by that wire. In this balanced case the voltage at N' will still be zero (with or without the wire present). Now, let the impedance $\mathbf{Z}_C$ be replaced by an open circuit and find the new *voltage* $\hat{V}_{N'}$, with respect to the neutral reference point N.

Problem 9.27

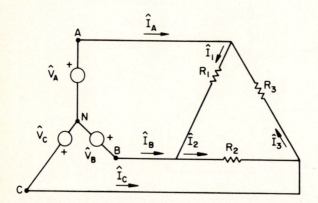

The same three-phase *wye* supply as in Problem 9.23 is connected to the three resistors shown above which are delta-configured.

(a) Find the voltages $\hat{V}_{AB}$, $\hat{V}_{BC}$, and $\hat{V}_{CA}$.

Ans.: $\hat{V}_{AB} = 208 \underline{/30°}$ V
$\hat{V}_{BC} = 208 \underline{/-90°}$ V
$\hat{V}_{CA} = 208 \underline{/150°}$ V

(b) For $R_1 = R_2 = R_3 = 5.2$ ohms, find the currents $\hat{I}_1, \hat{I}_2,$ and $\hat{I}_3$.

Ans.: $\hat{I}_1 = 40\underline{/30°}$ A
$\hat{I}_2 = 40\underline{/-90°}$ A
$\hat{I}_3 = 40\underline{/150°}$ A

(c) Find the total power in watts delivered by the three-phase source.

Ans.: 25 kW

(d) Find the line currents $\hat{I}_A$, $\hat{I}_B$, and $\hat{I}_C$.

Ans.: $\hat{I}_A = 69.3\underline{/0°}$ A
$\hat{I}_B = 69.3\underline{/-120°}$ A
$\hat{I}_C = 69.3\underline{/+120°}$ A

Problem 9.28

Repeat Problem 9.23 for the case where the three impedances are replaced by three impedances $\mathbf{Z}_1 = \mathbf{Z}_2 = \mathbf{Z}_3 = 4.5 + j2.6 \ \Omega$.

Problem 9.29

The three-phase, three-wire line of Problem 9.23 is connected to a balanced load consisting of three 10-ohm resistors connected in a *wye* configuration. This results in a circuit similar to that of Problem 9.23 except that in this case we make no physical connection between the neutrals.

(a) What is the effective value of the voltage, current, and power associated with each 10-ohm resistor? Ans.: 120 V, 12.0 A, 1.44 kW

(b) If resistor R_C should become open-circuited, what would be the voltage, current, and power to each of the two remaining 10-ohm resistors? Ans.: 104 V, 10.4 A, 1.082 kW

Problem 9.30

After the resistor becomes open-circuited in part (b), what is $\hat{V}_{N'}$ with respect to the ground point N?

Problem 9.31

A 440-volt (line to line) three-phase, three-wire *wye*-connected generator supplies power to a *wye*-connected induction motor. The motor delivers 10 horsepowers and operates at an efficiency of 85%. The motor power factor is 0.8. If the motor is a balanced load, what is the current in each line? Note that 1 horsepower is equivalent to 745.7 watts. (*Hint*: The power factor is the same for each of the three coils in the motor.)

Ans.: 14.4 A

Problem 9.32

Repeat Problem 9.31 for a delta-connected motor, everything else being the same.

Problem 9.33

The three-phase source of Problem 9.23 feeds a balanced *wye* load. The total power drawn is 96 kW. Each load is inductive and has a power factor of 0.8. What is the magnitude of each line current (e.g., find $|\hat{I}_A|$)? What is $|\hat{I}_N|$?

Problem 9.34

In Problem 9.33 find the three equal inductive load impedances. Suppose one of the loads in Problem 9.33 burns out and results in an open circuit. Find the phasor current $\hat{I}_N$.

Problem 9.35

Repeat Problem 9.33 with a balanced delta load configuration.

Problem 9.36

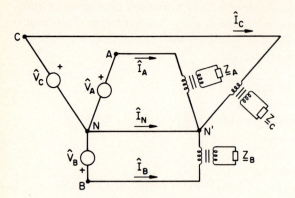

A three-phase, 60-Hz source is connected as shown above to the loads $\mathbf{Z}_A$, $\mathbf{Z}_B$, and $\mathbf{Z}_C$ through identical ideal transformers each having a turns ratio (primary to secondary) of 5 : 1. The phase voltages are $\hat{V}_{AN} = 5000\,\underline{/0°}$, $\hat{V}_{BN} = 5000\,\underline{/-120°}$, and $\hat{V}_{CN} = 5000\,\underline{/+120°}$, whereas $\mathbf{Z}_A = 2 + j2\ \Omega$, $\mathbf{Z}_B = 2\ \Omega$, and $\mathbf{Z}_C = 2\ \Omega$.

(a) Find $\hat{I}_A$, $\hat{I}_B$, $\hat{I}_C$, and $\hat{I}_N$.

(b) Find the total power in watts delivered by the three-phase source.

Problem 9.37

Repeat Problem 9.36 if the load impedances are balanced and equal to $\mathbf{Z}_A = 2 + j2\ \Omega$.

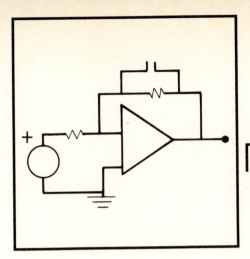

SECTION D

FREQUENCY RESPONSE OF ELECTRICAL SYSTEMS

Thus far we have dealt with the forced response of circuits to single-frequency sources. We now quote the important mathematical result that more complicated signals may be written as a linear superposition of sinusoidal functions each of which itself has a single frequency. Since the systems we study are linear and superposition holds, we may find the response of a system to such a signal by adding together the response of each individual frequency component making up that signal. The crucial mathematical element is called the transfer function and contains all the information needed to calculate how each frequency is transferred through the system. Analysis of this function reveals the frequency response of a system. We develop an analysis method involving so-called "Bode plots" in order to display graphically the frequency response of electrical systems whose poles and zeros are real, and apply the technique to some Op-Amp circuits. We also show that the poles and zeros of the transfer function define both the natural and the forced response and develop a method to find the complete response of systems with sinusoidal sources. Finally, we extend the analysis to second order systems whose transfer functions are characterized by multiple poles or zeros which may be real or complex.

Chapter 10

Transfer Functions, Spectral Synthesis, and Complete Response Revisited

In previous chapters we showed how to analyze the effect of electrical systems on signals that can be characterized by a unique frequency. In this chapter we show that complicated signals can be expressed as a linear superposition of pure sinusoidal ones. In a linear system a single mathematical expression, the transfer function, can be used to characterize the effect of the network on any given frequency. Once each component of the signal is transferred through the system, the output signal can be reconstituted via a linear superposition of the resulting sinusoidal functions of time. We then revisit complete response and show how the transfer function may be used to determine both the natural and the forced response. The student is only expected to understand the concepts of Fourier analysis but not the details of how to evaluate a Fourier integral or a Fourier series expansion. These are left for more advanced courses. Peak value phasors are used in this chapter.

10.1 THE TRANSFER FUNCTION

In the preceding several chapters we have shown how to analyze the response of electrical circuits to signals having a fixed frequency. A typical example might involve finding some output voltage phasor $\hat{V}_o$ when an input source $\hat{V}_i$ is applied to some system (see Fig. 10.1).

In general, the phasor $\hat{V}_o$ and the phasor $\hat{V}_i$ will differ in both amplitude and phase and we can write

$$\hat{V}_o = \mathbf{H}\,\hat{V}_i$$

where $\mathbf{H}$ is a complex constant which determines both the amplitude and phase

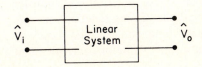

FIGURE 10.1.
Block diagram for a linear system with input $\hat{V}_i$ and output $\hat{V}_o$.

difference between $\hat{V}_o$ and $\hat{V}_i$. If we change the frequency of the source, however, a different value of **H** may be required. That is, the relationship between $\hat{V}_o$ and $\hat{V}_i$ usually depends upon the frequency. The mathematical expression which gives this dependence is called the transfer function since it describes how the input signal is transferred through the network to yield the output signal. The remarkable result is that a single function can be used to describe the entire frequency range.

A transfer function is formally defined as the ratio of any voltage or current in an electrical system to any other voltage or current in that system, when the system is responding to an exponential source of the form e^{st}. The impedance function $Z(s)$ is a particular case of a transfer function since

$$H_1(s) = \frac{V(s)}{I(s)} = Z(s) \tag{10.1a}$$

is the ratio of the voltage across some element or set of elements, to the current through that *same* element or set of elements for exponential variations. Likewise, the function $Y(s) = I(s)/V(s) = 1/Z(s)$ is also a transfer function which is special enough to warrant a name, the admittance function.

A transfer function is often used to relate the output of some two-port system to the input. The ratio of the output voltage V_o to the input voltage V_i of an amplifier,

$$H_2(s) = \frac{V_o(s)}{V_i(s)} \tag{10.1b}$$

is a transfer function. The input $V_i(s)$ is "transferred" through the system and converted into the output $V_o(s)$ via multiplication by $H_2(s)$:

$$V_o(s) = H_2(s)V_i(s)$$

and we say $H_2(s)$ operates on $V_i(s)$ to yield $V_o(s)$. In a linear system, $H(s)$ is a linear operator.

In a system responding to a source varying as e^{st}, all the voltages and currents will also vary as e^{st}. The time dependence in any transfer function will then cancel out and $H(s)$ is independent of time. For example, in Eqs. 10.1a and 10.1b,

$$H_1(s) = \frac{Ve^{st}}{Ie^{st}} = \frac{V}{I}$$

$$H_2(s) = \frac{V_oe^{st}}{V_ie^{st}} = \frac{V_o}{V_i}$$

which are independent of time. We shall use capital letters in defining all transfer functions since they correspond to the complex amplitude of time functions such as $v_o(t) = V_oe^{st}$, $v_i(t) = V_ie^{st}$, and so on. Once a transfer function is defined it may be used to relate the associated voltage and/or current for any value of s provided that the time variations are of the form e^{st}.

When a system is forced to oscillate at a frequency ω, we may set $s = j\omega$ since $e^{j\omega t}$ describes a sinusoidal variation in time. For sinusoidal variations Eq. 10.1b may be written

$$\hat{V}_o(\omega) = \mathbf{H}(j\omega)\,\hat{V}_i(\omega) \tag{10.2}$$

where $\mathbf{H}(j\omega)$ is a complex function of the real variable ω. In general, $\mathbf{H}(j\omega)$ will have a polar amplitude and a phase angle that depend upon ω. This feature allows $\mathbf{H}(j\omega)$ to "keep track" of the amplitude and the phase difference caused by the system which is characterized by the transfer function $H(s)$.

As an example consider the circuit in Fig. 10.2a. If v_b is an exponential function of time, $v_b = V_b e^{st}$, all the currents and voltages will vary as e^{st} and we may replace all elements by their impedances, as shown in Fig. 10.2b. Analysis of the circuit involves only manipulation of complex constants since all the time functions are similar: $i_1 = I_1 e^{st}$, $i_c = I_c e^{st}$, $v_c = V_c e^{st}$, and so on. We may now define various transfer functions, depending on which current or voltage we are trying to determine. If we are interested in v_c, then we may use the voltage divider principle to yield

$$H_1(s) = \frac{V_c}{V_b} = \frac{50\,//(1/sC)}{50 + 50\,//(1/sC)}$$

which can be written as

$$H_1(s) = \frac{50}{2500sC + 100}$$

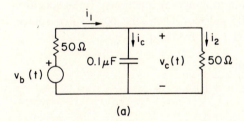

(a)

FIGURE 10.2a.
Circuit used to study transfer functions.

FIGURE 10.2b.
Circuit of Fig. 10.2a with the source $v_b = V_b e^{st}$.

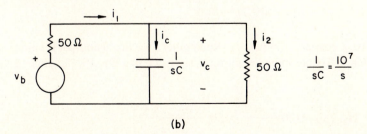

(b)

Including the value of the capacitor and expressing the transfer function in polynomial form,

$$H_1(s) = \frac{2 \times 10^5}{s + 4 \times 10^5} \tag{10.3}$$

Another transfer function of interest is

$$H_2(s) = \frac{V_b}{I_1} = 50 + \frac{50}{50sC + 1} = \frac{2500sC + 100}{50sC + 1}$$

which in polynomial form yields

$$H_2(s) = 50 \left(\frac{s + 4 \times 10^5}{s + 2 \times 10^5} \right) \tag{10.4}$$

The function $H_2(s)$ is simply $Z(s)$, the input impedance to the source v_b.

We may now use the transfer functions $H_1(s)$ and $H_2(s)$ to determine the voltage $v_c(t)$ and current $i_1(t)$ driven by a source $v_b(t)$ *operating at any frequency* ω merely by substituting $s = j\omega$ and evaluating the expressions

$$\hat{V}_c = \mathbf{H}_1(j\omega)\hat{V}_b$$

and

$$\hat{I}_1 = \frac{\hat{V}_b}{\mathbf{H}_2(j\omega)} = \frac{\hat{V}_b}{\mathbf{Z}(j\omega)}$$

and then converting from the phasor notation to temporal functions. To be specific, let $v_b(t) = 24 \cos(\omega t)$, where $\omega = 2 \times 10^5$ rad/sec. Then, the peak value source voltage phasor is $\hat{V}_b = 24 \underline{/0°}$. Evaluating $\mathbf{H}_1(j\omega)$ from Eq. 10.3,

$$\mathbf{H}_1(j\omega) = \frac{2 \times 10^5}{j(2 \times 10^5) + 4 \times 10^5}$$

$$\mathbf{H}_1(j\omega) = \frac{1}{(j + 2)} = \frac{2 - j}{5} = 0.45 \underline{/-26.6°}$$

The output voltage phasor becomes

$$\hat{V}_c = 10.7 \underline{/-26.6°}$$

and finally $v_c(t) = 10.7 \cos(\omega t - 26.6°)$ V. (Remember, we are using peak value phasors in this chapter.) To determine the current, $i_1(t)$, we use Eq. 10.4 evaluated at $s = j(2 \times 10^5)$:

$$\mathbf{H}_2(j\omega) = (50)\frac{j + 2}{j + 1} = 50 \left| \frac{\sqrt{5} \underline{/-26.6°}}{\sqrt{2} \underline{/45°}} \right|$$

$$\mathbf{H}_2(j\omega) = 79.1\underline{/-18.4°} = \mathbf{Z}(j\omega)$$

The current phasor is

$$\hat{I}_1 = 0.30\underline{/18.4°}$$

and $i_1(t) = 0.3 \cos(\omega t + 18.4°)$ A.

If the source were at some other frequency, say, ω', we could find $\hat{V}_c$ and $\hat{I}_1$ at that frequency equally easily using the transfer functions. We just evaluate them at $s = j\omega'$, and then

$$\hat{V}'_c = \mathbf{H}_1(j\omega')\hat{V}'_b$$

and

$$\hat{I}_1 = \frac{\hat{V}'_b}{\mathbf{H}_2(j\omega')}$$

where $\hat{V}'_b$ is the phasor voltage of the source at frequency ω'.

TEXT EXAMPLE 10.1

Evaluate the transfer function $\mathbf{H}(s) = \hat{V}_o / \hat{V}_i$ for the circuit below and apply it to the two signals $v_1(t) = 100 \cos(\omega_1 t)$ and $v_2(t) = 100 \cos(\omega_2 t)$ where $f_1 = 500$ Hz and $f_2 = 5000$ Hz. Comment upon the nature of the filtering realized by the circuit.

$C = (1/100\pi)\ \mu F$
$R = 10^5\ \Omega$

Solution: The voltage divider principle yields

$$\mathbf{H}(s) = \frac{\hat{V}_o}{\hat{V}_i} = \frac{R}{R + (1/sC)} = \frac{RsC}{RsC + 1}$$

For the values of R and C in the circuit above, $RC = 1/1000\pi$ sec. For $s = j\omega_1 = (2\pi f_1)j = (1000\pi)j$,

$$\mathbf{H}(j\omega_1) = \frac{j}{j + 1} = 0.707\underline{/+45°}$$

and for $s = j\omega_2 = j(2\pi f_2) = j(10000\pi)$,

$$\mathbf{H}(j\omega_2) = \frac{j10}{j10 + 1} = 0.995\underline{/+5.7°}$$

The two input phasors are $\hat{V}_1 = 100\underline{/0°}$ at frequency ω_1 and $\hat{V}_2 = 100\underline{/0°}$ at frequency ω_2. The two outputs are

$$\hat{V}_{1o} = \mathbf{H}(j\omega_1)\hat{V}_i = 70.7\underline{/+45°}$$

$$\hat{V}_{2o} = \mathbf{H}(j\omega_2)\hat{V}_i = 99.5\underline{/+5.7°}$$

The output at the lower frequency is attenuated more (71%) and has a greater phase shift ($+45°$) than the higher-frequency signal. This type of network is termed a high pass filter since it passes high-frequency signals with less attentuation.

In the example above the signal at frequency $\omega_2 = 2\pi(5000)$ rad/sec is transferred through the system with an attenuation of 0.5%, while the lower frequency signal, at ω_1, is attenuated by about 29%. This network thus tends to "filter out" low-frequency signals, and we call it a high pass filter. Note also that the phase shift at ω_2 is only 5.7° whereas the phase shift at ω_1 is 45°. Filter networks typically change both the amplitude and the phase of signals that pass through them. The detailed manner by which a given network modifies the amplitude and the phase of signals with different frequencies is called the frequency response of the network. Understanding this concept and the evaluation of the frequency response is one of the major goals of the rest of this section.

In this chapter and the next we spend considerable effort showing how transfer functions can be used to describe the frequency response of electrical circuits. We first point out, quoting Fourier's theorem, that a complicated signal may be broken up into a sum (a linear superposition) of sinusoidal functions and their corresponding phasors. If such a signal is applied to a linear electrical network characterized by a transfer function, that function may be applied to all the individual phasors making up the signal. The output of the network is then found by adding back together all these sinusoidal output phasors.

10.2 FOURIER'S THEOREM

Fourier's theorem states that any piecewise continuous, periodic function which is defined on a finite interval can be written as an infinite sum of sine and cosine functions. For our purposes, this means that any function of time defined over some interval of time $-T/2 \leq t \leq T/2$ can be written in the form

$$v(t) = a_0 + \sum_{n=1}^{\infty} [a_n \cos(n\omega_o t) + b_n \sin(n\omega_o t)] \tag{10.5}$$

where $\omega_0 = 2\pi/T$. The frequency ω_0 is called the fundamental, and its integer multiples are called the harmonics of the frequency ω_0. Except for the constant a_0, each of these terms is an oscillatory function which "fits" into the interval. For example, if $a_1 = a_2 = a_3$, the first three cosine functions are as shown in Fig. 10.3. Expressed slightly differently, Fourier's theorem states that any complicated function of time can be thought of as a sum or superposition of simple pure tones with

FIGURE 10.3.
The fundamental and the first two equal amplitude harmonics of the cosine series on the interval $-T/2 \leq t \leq T/2$.

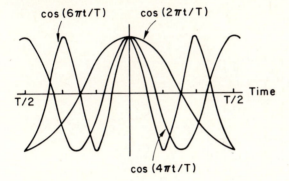

varying amplitudes. The information or complexity of the signal is then contained in the different amplitudes of the various tones.

It is fairly obvious that a set of sinusoids whose frequencies are integer multiples of $\omega_0 = 2\pi/T$ will add up to a time function which is periodic with a period T. The proof follows simply from the fact that each of the terms in the set is periodic; that is,

$$a_n \cos[n\omega_0(t + T)] = a_n \cos[n\omega_0 t + 2\pi n] = a_n \cos(n\omega_0 t) \qquad (10.6a)$$

$$b_n \sin[n\omega_0(t + T)] = b_n \sin[n\omega_0 t + 2\pi n] = b_n \sin(n\omega_0 t) \qquad (10.6b)$$

where we have substituted $T = 2\pi/\omega_0$ in each of the second terms.

The problem of Fourier series *analysis*, however, is the reverse. Given any particular time function with a period T, how do we find the values of the magnitudes $(a_0, a_1, \ldots, a_n$ and $b_1, b_2, \ldots, b_n)$ of the set of sinusoids that will add up to that time function? A detailed discussion of how these coefficients are determined is very important but requires more time and attention to detail than is necessary for the level of this text. Mathematics and higher level engineering courses treat this problem in its proper detail. For our purpose it is sufficient to know that for reasonably well behaved functions such as those which arise in physical systems, a straightforward recipe exists to find these coefficients.

For reference we gave the Fourier series expansions of seven important periodic functions in Table 10.1. The first function represents a constant "dc" signal $v_0(t) = V_0$. Referring to Eq. 10.5 it is clear that $a_1 = V_0$ and all the other coefficients vanish. The second function in the table is a cosine signal for which all the b's in Eq. 10.5 are zero and $a_1 = 1$ is the only a_n which does not vanish. The remaining functions in the table all have more complicated Fourier expansions. To gain some insight into the meaning of the Fourier coefficients we study the third function, the "square wave," in detail here.

The expansion for the square wave can be written very concisely in the form

$$v(t) = \sum_{\substack{n=1 \\ (\text{odd})}}^{\infty} \left(\frac{4}{\pi n}\right) \sin\left(\frac{2\pi n t}{T}\right) \qquad (10.7)$$

Table IO I

The Fourier Series for Several Important Functions

Function Name	Fourier Expansion	Waveform
Constant	$v(t) = V_0$	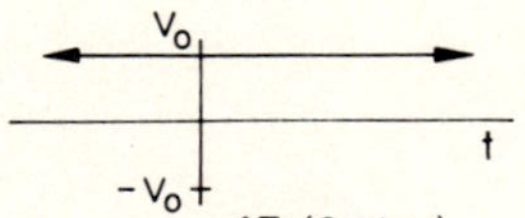
Cosine	$v(t) = V_0 \cos(\omega_0 t)$	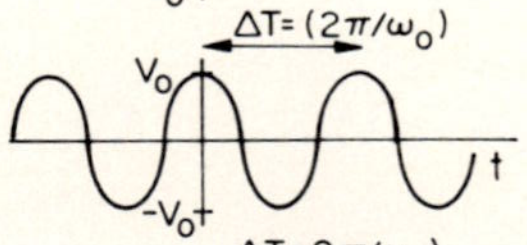 $\Delta T = (2\pi/\omega_0)$
Square Wave	$v(t) = \sum\limits_{n(\text{Odd})} V_0 (4/\pi n) \sin(n\omega_0 t)$	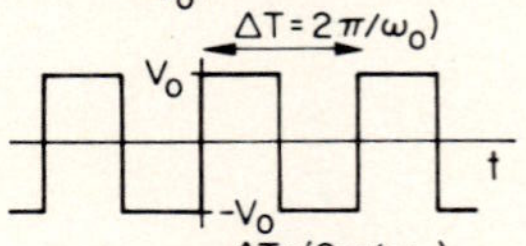 $\Delta T = 2\pi/\omega_0$
Pulse Train	$v(t) = 0.5 V_0 + 0.64 V_0 \sin(\omega_0 t) + 0.21 V_0 \sin(3\omega_0 t)$ $+ 0.13 V_0 \sin(\omega_0 t) + \cdots$	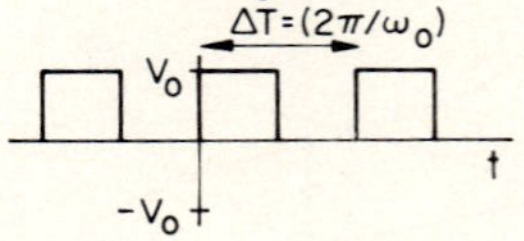 $\Delta T = (2\pi/\omega_0)$
Half – Wave Rectified Sine Wave	$v(t) = (V_0/\pi) + (V_0/2) \sin(\omega_0 t)$ $-(2V_0/3\pi) \cos(2\omega_0 t) - (2V_0/15\pi) \cos(4\omega_0 t)$ $-(2V_0/35\pi) \cos(6\omega_0 t) + \cdots$	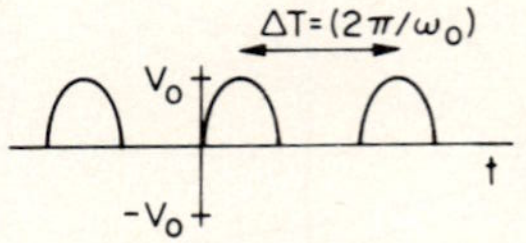 $\Delta T = (2\pi/\omega_0)$
Full – Wave Rectified Sine Wave	$v(t) = (2V_0/\pi) - (4V_0/3\pi) \cos(\omega_0 t)$ $-(4V_0/15\pi) \cos(4\omega_0 t) - (4V_0/35\pi) \cos(6\omega_0 t) + \cdots$	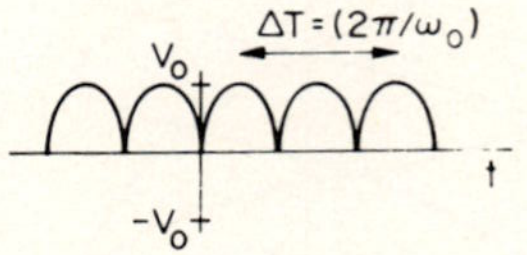 $\Delta T = (2\pi/\omega_0)$

330

The first two terms in this series (corresponding to $n = 1$ and $n = 3$) are plotted in Fig. 10.4a in the interval $-T/2 \leq t \leq +T/2$.

FIGURE 10.4a.
First two terms in the expansion of a square wave.

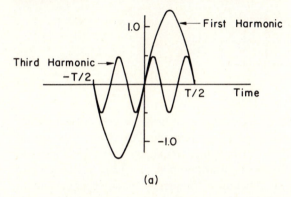

(a)

When we add these two functions together we get the result illustrated in Fig. 10.4b. Note how the second term, $\sin(6\pi t/T)$, "fills in" some of the space near the edges of the square wave where the dominant first term, $\sin(2\pi t/T)$, does not fit very well. Fourier's theorem states that if we are willing to continue to evaluate terms in the expansion (10.7), we can get arbitrarily close to the original square wave.

FIGURE 10.4b.
Sum of the first two terms in the Fourier analysis of a square wave.

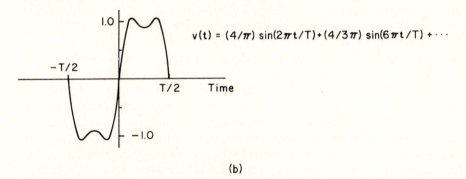

$$v(t) = (4/\pi) \sin(2\pi t/T) + (4/3\pi) \sin(6\pi t/T) + \cdots$$

(b)

Before continuing the discussion it is of interest to point out that outside the interval $-T/2 \leq t \leq T/2$ the expansion can formally still be evaluated by plugging in values for $|t| > T/2$. When this is done the original function is regenerated in every adjoining interval, say, $T/2 \leq t \leq 3T/2$, $-3T/2 \leq t \leq -T/2$, $3T/2 \leq t \leq 5T/2$, and so on. Figure 10.5 shows the extension of the function used above. Such a function is termed a periodic function, since it repeats during every interval period of length T. Turning this result around, we can say that any function that exactly repeats after a definite period T can be evaluated by a Fourier series similar to Eq. 10.7 over the interval T as defined above and the result will be valid for the whole range $-\infty \leq t \leq \infty$. Finally, note that the expansion in Eq. 10.7 corresponds to a unit amplitude square wave. If the square wave amplitude is V_0, say, every term in the expansion should be multiplied by V_0.

FIGURE 10.5.
The expression evaluated for all *t* yields the dashed periodic function.

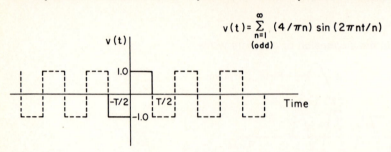

$$v(t) = \sum_{\substack{n=1 \\ (\text{odd})}}^{\infty} (4/\pi n) \sin(2\pi nt/n)$$

Suppose a signal of interest is not periodic; that is, it never repeats its waveform. We can still think about such a function as being "periodic," except that the period T is infinitely long. In such a case the fundamental frequency $\omega_0 = 2\pi/T$ tends to zero as $T \to \infty$. In this case the discrete Fourier components ($n\omega_0$) get very close together and merge into a continuous function of frequency, each separated by an infinitesimal frequency interval $d\omega$. Formally, the relationship between a function $f(t)$ and its frequency representation is then given by the so-called Fourier transform integral

$$f(t) = \left(\frac{1}{\sqrt{2\pi}}\right) \int_{-\infty}^{\infty} g(\omega) e^{j\omega t} d\omega \tag{10.8}$$

Equation 10.8 is analogous to Eq. 10.5 with the summation over a countably infinite number of frequencies ($n\omega_0$) replaced by an integral over all frequencies ($d\omega$) which are noncountable infinite set. The function $g(\omega)$ is analogous to the coefficients a_n and b_n in a Fourier series.

The mathematics and detailed applications associated with Fourier transform integrals is beyond the scope of the present text. The student should be aware that most applications studied in more advanced courses involve the use of Fourier and Laplace transforms (see Section 10.5) rather than Fourier series expansions. For our purpose it is sufficient to realize that an arbitrary function of time $f(t)$ can in principle be converted to its equivalent representation in frequency space, $g(\omega)$. The series expansions for periodic functions are referred to as a discrete Fourier series.

The $a_n \cos(n\omega_0 t)$ and $b_n \sin(n\omega_0 t)$ terms can be combined into a single sinusoidal function with an appropriate phase angle and written in terms of complex coefficients. As we have seen in Chapter 8, the sum of two sinusoids of the same frequency is another of the same frequency. In particular,

$$a_n \cos(\omega_n t) + b_n \sin(\omega_n t) = c_n \cos(\omega_n t + \theta_n)$$

where $c_n = (a_n^2 + b_n^2)^{1/2}$ and $\theta_n = \tan^{-1}(-b_n/a_n)$. We may now express the Fourier series for the $v(t)$ of Eq. 10.5 as

$$v(t) = a_0 + \sum_{n=1}^{\infty} c_n \cos(n\omega_0 t + \theta_n) \tag{10.9}$$

Each of the sinusoids in the Fourier series may be expressed by a peak value phasor

$$\hat{V}_n = |\hat{V}_n|e^{j\theta n} = (a_n^2 + b_n^2)^{1/2}e^{j\theta n}$$

Then,

$$c_n \cos(n\omega_0 t + \theta_n) = \text{Re}(\hat{V}_n e^{jn\omega_0 t})$$

If we now let $\hat{V}_0 = a_0$, we may rewrite Eq. 10.9 as

$$v(t) = \left(\text{Re}(\sum_{n=0}^{\infty} \hat{V}_n e^{jn\omega_0 t}) \right) \tag{10.10}$$

where each of the terms $\hat{V}_n$ in the summation is a phasor corresponding to the frequency $\omega_n = n\omega_0$. Note the similarity to Eq. 10.8.

10.3 SPECTRAL SYNTHESIS

We have shown above that periodic signals which are real functions of time can be expressed as the real part of a sum of complex functions of the form $\hat{V}_n(t) = \hat{V}_n e^{jn\omega_0 t}$, where $\hat{V}_n$ is the phasor at frequency $n\omega_0$. For linear systems we may use the principle of superposition to find the response of the system to each frequency term separately. The output is then the sum of the separate responses.

The block diagram of Fig. 10.6 shows schematically the steps involved in finding the response $v_2(t)$ of a linear system to a particular signal $v_1(t)$. First, the transfer function $H(s)$ relating $\hat{V}_2$ to $\hat{V}_1$ is found. Then, the signal $v_1(t)$ is decomposed into its Fourier components via a Fourier analysis. If $v_1(t)$ is a periodic function, its discrete Fourier series is found. If $v_1(t)$ is not periodic, its Fourier transform is found. The transfer function $H(s)$ is evaluated at each frequency in the spectrum of the signal and a new phasor $\hat{V}_2(\omega) = \mathbf{H}(j\omega)\hat{V}_1(\omega)$ is evaluated at each frequency. Then, the output signal is reconstructed by adding all the phasors $\hat{V}_2$ via a summation if the signal was periodic or by a Fourier integral if the signal was not periodic. In this text all our examples will involve periodic functions and therefore the sum of sinusoids. Fourier integrals will not be referred to again in the text.

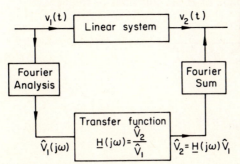

FIGURE 10.6.
The response of a linear system to a signal may be translated to the response to a superposition of sinusoids.

FIGURE 10.7.
Find the response of the *R–L* circuit of (a) to the square wave of (b).

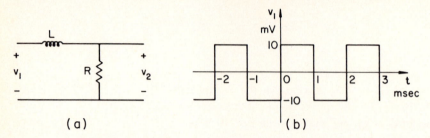

(a)   (b)

As an example of the method diagrammed in Fig. 10.6, consider the response of the network shown in part (a) of Fig. 10.7 to the periodic signal $v_1(t)$ shown in part (b) of the figure. We first write $v_1(t)$ as a sum of the sinusoids using results given above in Table 10.1 for this square wave function of time. Since $T = 2 \times 10^{-3}$ sec and the square wave amplitude is 10 mV, we have

$$v_1 = \left(\frac{40}{\pi}\right) \sin(1000\pi t) + \left(\frac{40}{3\pi}\right) \sin(3000\pi t) + \left(\frac{40}{5\pi}\right) \sin(5000\pi t) + \ldots \text{mV}$$

To find the response of the system to each sinusoid, we first calculate the transfer function

$$H(s) = \frac{V_2}{V_1} = \frac{R}{R + sL}$$

For $s = j\omega$,

$$\mathbf{H}(j\omega) = \frac{R}{R + j\omega L} = \frac{1}{1 + j(\omega L/R)}$$

The algebraic expression for the transfer function may be simplified in form by defining $\omega_p = R/L$. Then,

$$\mathbf{H}(j\omega) = \frac{\underline{/-\tan^{-1}(\omega/\omega_p)}}{[1 + (\omega/\omega_p)^2]^{1/2}} = |\mathbf{H}(j\omega)| \underline{/\theta(\omega)}$$

Both the polar amplitude and the phase angle of **H** are functions of frequency:

$$|\mathbf{H}(j\omega)| = \frac{1}{[1 + (\omega/\omega_p)^2]^{1/2}}$$

and

$$\theta(j\omega) = -\tan^{-1}\left(\frac{\omega}{\omega_p}\right)$$

FIGURE 10.8.
Amplitude and phase of $H(jw) = \hat{V}_2/\hat{V}_1$ as a function of frequency.

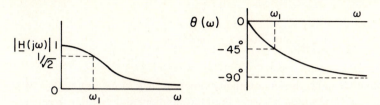

The polar amplitude and the phase angle are sketched as functions of ω in Fig. 10.8.

The result found above is very general, and to evaluate a specific case we must choose values of R and L and hence specify $\omega_p = R/L$. Let $R = 200\pi\ \Omega$ and $L = 0.1$ H. Then, $\omega_p = 2000\pi$ rad/sec. Evaluating $\mathbf{H}(j\omega)$ for the frequency $\omega_1 = 1000\pi$,

$$\mathbf{H}(j1000\pi) = \frac{\hat{V}_2}{\hat{V}_1} = \frac{\underline{/-\tan^{-1}0.5}}{(1 + 0.25)^{1/2}} = 0.89\underline{/-26.6°}$$

For $\omega_3 = 3000\pi$,

$$\mathbf{H}(j3000\pi) = \frac{\hat{V}_2}{\hat{V}_1} = \frac{\underline{/-\tan^{-1}1.5}}{(1 + 2.25)^{1/2}} = 0.55\underline{/-56.3°}$$

For $\omega_5 = 5000\pi$,

$$\mathbf{H}(j5000\pi) = \frac{\hat{V}_2}{\hat{V}_1} = \frac{\underline{/-\tan^{-1}2.5}}{(1 + 6.25)^{1/2}} = 0.37\underline{/-68.2°}$$

For $\omega_7 = 7000\pi$,

$$\mathbf{H}(j7000\pi) = \frac{\hat{V}_2}{\hat{V}_1} = \frac{\underline{/-\tan^{-1}3.5}}{(1 + 12.25)^{1/2}} = 0.27\underline{/-74.1°}$$

and so forth. Note that the polar amplitude of each of these terms decreases as the frequency increases. Multiplying each of the complex numbers above by the appropriate sinusoidal terms in $v_1(t)$, we can now evaluate the first four terms in the Fourier series of the *output* function:

$$v_2(t) = (.89)\left(\frac{40}{\pi}\right)\sin(1000\pi t - 26.6°) +$$

$$(.55)\left(\frac{40}{3\pi}\right)\sin(3000\pi t - 56.3°) +$$

$$(.37)\left(\frac{40}{5\pi}\right)\sin(5000\pi t - 68.2°) +$$

$$(.27)\left(\frac{40}{7\pi}\right)\sin(7000\pi t - 74.1°) + \cdots$$

In the example above we have chosen a transfer function (determined by values of R and L) with an $\omega_p = 2000\pi$ which is larger than the fundamental angular frequency of $v_1(t)$, but which is smaller than the higher harmonics in $v_1(t)$. The result is that the fundamental term ($\omega = \omega_0$) in the output is nearly equal to that same term in the input, but the amplitude of the subsequent terms in the output are greatly reduced. The very high frequency terms in $v_1(t)$ will thus be very small in the output function. Such a transfer function is associated with a "low-pass" filter, since it passes low frequencies but not high ones.

Transfer functions for three *ideal* filters are shown in Fig. 10.9, where only the amplitude of $\mathbf{H}(j\omega)$ is sketched. For the ideal filters of this figure, only the sinusoidal components of the input having frequencies inside the "pass bands" will appear in the output. No real electrical circuit can have these characteristics, but a designer can come close to these forms if desired. Chapter 11 is dedicated to further study of frequency response.

FIGURE 10.9.
Plots of the magnitude of the transfer function of three ideal filters.

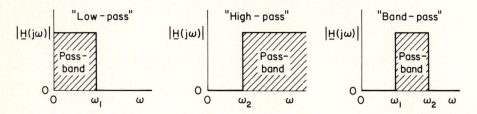

_____

The same signal studied above and illustrated in part (b) of Fig. 10.7 is used as input signal $v_1(t)$ to the high pass filter studied in Text Example 10.1. Find the first four terms of the output function $v_0(t)$.

Solution: The discrete Fourier series expansion of $v_i(t)$ is the same as used above. In terms of phasors the first four terms are

$$\hat{V}_{i1} = \left(\frac{40}{\pi}\right)\angle{-90°} \qquad \text{at } \omega_1 = 1000\pi$$

$$\hat{V}_{i3} = \left(\frac{40}{3\pi}\right)\angle{-90°} \qquad \text{at } \omega_3 = 3000\pi$$

$$\hat{V}_{i5} = \left(\frac{40}{5\pi}\right)\underline{/-90°} \quad \text{at } \omega_5 = 5000\pi$$

$$\hat{V}_{i7} = \left(\frac{40}{7\pi}\right)\underline{/-90°} \quad \text{at } \omega_7 = 7000\pi$$

where we have used the fact that the time function $A \sin(\omega t)$ is represented by the peak value phasor $A\underline{/-90°}$ since $\text{Re}[Ae^{j(\omega t - 90°)}] = \text{Re}[(-j)Ae^{j\omega t}] = A \sin(\omega t)$. Evaluating the transfer function $H(s) = RsC/(RsC + 1)$ at each of these frequencies,

$$\mathbf{H}(j\omega_1) = \frac{j}{j + 1} = 0.707\underline{/+45°}$$

$$\mathbf{H}(j\omega_3) = \frac{j3}{j3 + 1} = 0.95\underline{/+18°}$$

$$\mathbf{H}(j\omega_5) = \frac{j5}{j5 + 1} = 0.98\underline{/+11°}$$

$$\mathbf{H}(j\omega_7) = \frac{j7}{j7 + 1} = 0.99\underline{/+8.1°}$$

Multiplying the respective phasors in these two sets and reconstructing the temporal functions yields

$$v_0(t) = 9.0 \cos(\omega_1 t - 45°) + 4.0 \cos(\omega_3 t - 72°) +$$
$$2.5 \cos(\omega_5 t - 79°) + 1.8 \cos(\omega_7 t - 82°) + \cdots$$

Before leaving the discussion of spectral synthesis in linear systems, we need to investigate power computations when more than one frequency is involved. Since power is determined from the product of two quantities, $p(t) = v(t)i(t)$ or the square of a single quantity, for example, $p(t) = i^2(t)R$, it is not a linear function. This means that the superposition principle does not hold and we must be very careful in power computations.

Consider the case of a current that is composed of two frequency components,

$$i(t) = I_1 \cos(\omega_1 t) + I_2 \cos(\omega_2 t)$$

and a circuit in which $i(t)$ flows through a resistor of value R. The average power is

$$P = \lim_{T \to \infty}\left(\frac{1}{T'}\right)\left\{\int_{-\infty}^{T'} R[I_1^2 \cos^2(\omega_1 t) + I_2^2 \cos(\omega_2 t) + 2I_1 I_2 \cos(\omega_1 t) \cos(\omega_2 t)]dt\right\}$$

The last term, which contains the product of two frequencies, vanishes in the integration and, remembering that the average value of $\cos^2$ is 1/2, we have

$$P = \frac{1}{2}I_1^2 R + \frac{1}{2}I_2^2 R$$

The implication of this result is that we may simply add up the powers associated with frequencies ω_1 and ω_2 as if they were linear quantities. The only requirement is that we include all current or voltage sources at a given frequency in determining the power at that frequency. If effective value phasors are used, $P = |\hat{I}_1|^2 R + |\hat{I}_2|^2 R$ as in Chapter 8.

10.4 COMPLETE RESPONSE REVISITED: SINUSOIDAL FORCING

In Chapter 6 we showed how to determine the complete response when an abrupt transition was made between two states which had either no sources at all or had only dc sources. We now know how to determine the forced response to sinusoidal sources and hence can generalize our previous analysis to include sinusoidal forcing. As a by-product we shall also see that the transfer function can be used to determine both the natural and forced response characteristics. These "new" results actually include the previous work in Chapter 6 since dc is a bona fide frequency ($\omega = 0$) and since the impedance functions $Z_s(s)$ and $Z_p(s)$ we used in the earlier work are themselves transfer functions.

In this section we first work through an example (see Fig. 10.10) illustrating complete response when the imposed source is a sinusoidal function of time using something of a brute force approach. The method is identical to that in Chapter 6 except that the initial condition must include the instantaneous value of the new source at the time of switch closure. Later, we show how the concept of a transfer function unifies the approach to both the complete and natural responses of any linear system.

FIGURE 10.10.
Switch S has been in position 1 for a long time and is moved to position 2 at $t = 0$.

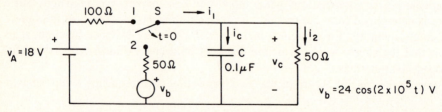

For convenience we shall set the time t_1 at which the switch is thrown to be zero. We shall call $t = 0^-$ the instant just before the change and call $t = 0^+$ the instant just after the change. Then, as shown in Chapter 5 for every current $i_L(t)$ through an inductance, $i_L(0^-) = i_L(0^+)$ and for every voltage $v_c(t)$ across a capacitance, $v_c(0^-) = v_c(0^+)$. In the examples below we assume that the switch has been in its position for $t \le 0$ for a sufficiently long time that we only need to find the forced response to the original circuit since any prior natural response will have decayed to zero by the time the switch is moved from position 1 to position 2 at $t = 0$. Note that unlike previous problems the forcing source for $t \geq 0$ is an oscillatory one. We wish to find the current $i_1(t)$ for $t \geq 0$. After the switch has been thrown, the circuit is identical to the circuit discussed in Section 10.1 (see Fig. 10.2a). With the switch in position 1, we may find the forced response to the initial source by first writing the impedance function:

$$Z_A(s) = 100 + \frac{50(1/sC)}{50 + (1/sC)} = 100 + \frac{50}{50sC + 1}$$

In this case the source is a battery, so setting $s = j0$ yields $Z_A(j0) = 150 \ \Omega$. Then, for $t \leq 0, i_1 = 18/150 = 0.12$ A. This is an "elegant" way to find the forced response by evaluating the impedance function $Z(s) = V(s)/I(s)$ of the circuit for $t \leq 0$, and then setting $s = j0$ since the circuit is dc. Since the source is dc we could have immediately recognized that the capacitance would be an open circuit and that the resulting resistive circuit has $R = 100 \ \Omega + 50 \ \Omega = 150 \ \Omega$. The reader may verify that for $t = 0^-$ the various parameters of interest are as follows:

$$i_1(0^-) = 0.12 \text{ A}$$

$$i_c(0^-) = 0$$

$$i_2(0^-) = 0.12 \text{ A}$$

$$v_c(0^-) = (i_2)(50) = 6 \text{ V}$$

We now draw the circuit for $t = 0^+$, just after the switch is moved to position 2. At that instant the voltage $v_b = 24 \cos(0)$ is 24 V. The capacitor must remain at the same voltage as its value at $t = 0^-$, so $v_c = (0^+) = v_c(0^-) = 6$ V, and we have the circuit of Fig. 10.11. This circuit is also easily solved for the other parameters. At the instant $t = 0^+$, the capacitance acts as a voltage source since its voltage (for that instant) is independent of the rest of the system. The conditions for $t = 0^+$ are

$$v_c(0^+) = 6 \text{ V} \tag{10.11a}$$

$$i_1(0^+) = \frac{24 - 6}{50} = 0.36 \text{ A} \tag{10.11b}$$

$$i_2(0^+) = \frac{6}{50} = 0.12 \text{ A} \tag{10.11c}$$

FIGURE 10.11.
The circuit is drawn for $t = 0^+$. The voltage across the capacitor must be 6 V by continuity and, at $t = 0$, the source voltage $v_b = 24$ V.

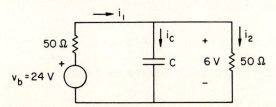

$$i_c(0^+) = i_1 - i_2 = 0.24 \text{ A} \qquad (10.11d)$$

We use these values below to evaluate the complete response.

As in Chapter 6, the natural response can be found by setting the series impedance equal to zero in the configuration which exists at times $t \geq 0$. With the source suppressed, the series impedance function is

$$Z_s(s) = 50 + \frac{50(1/sC)}{50 + (1/sC)} \qquad (10.12)$$

Combining terms and setting $Z_s(s)$ equal to zero gives an equation for s_1:

$$Z_s(s_1) = \frac{2500 s_1 C + 100}{50 s_1 C + 1} = 0$$

Setting the numerator of $Z_s(s_1)$ equal to zero and solving for s_1 yields the natural response exponent:

$$s_1 = -4 \times 10^5 \text{ sec}^{-1}$$

and the natural current is of the form $i_{1n}(t) = A_1 e^{(-4 \times 10^5)t}$.

The *forced* response must be found using the analysis methods of Chapter 8 since the source $v_b(t)$ is sinusoidal at frequency $\omega = 2 \times 10^5$ rad/sec. We have already performed this analysis above in Section 10.1. The impedance is the same as given in Eq. 10.4 provided that we set $s = j\omega$, with $\omega = 2 \times 10^5$ rad/sec. Carrying out the operations in Eq. 10.4 with $s = j\omega$ yields

$$\mathbf{Z} = 79\underline{/-18.4°}$$

and since the phasor $\hat{V}_b = 24\underline{/0°}$,

$$\hat{I}_{1F} = 0.304\underline{/+18.4°}$$

Converting this peak value phasor to the corresponding time function, we have

$$i_{1F}(t) = 0.304 \cos(2 \times 10^5 t + 18.4°)$$

and the total response for $t \geq 0$ is the sum of the natural and forced responses:

$$i_1(t) = A_1 e^{-4 \times 10^5 t} + 0.304 \cos(2 \times 10^5 t + 18.4°)$$

Using the boundary condition from Eq. 10.11b, $i_1(0^+) = 0.36$ A, and setting $t = 0$ in the above expression yields

$$i_1(0) = A_1 + 0.304 \cos(18.4°) = 0.36 \text{ A}$$

Solving gives $A_1 = 0.072$ and, finally, the complete response for $t \geq 0$ is

$$i_1(t) = [0.072 e^{-4 \times 10^5 t} + 0.304 \cos(2 \times 10^5 t + 18.4°)]\text{A}$$

This method follows quite closely the steps used in Chapter 6 for the complete response to dc sources. To determine the current i_1 in this example, the forced response at frequency ω was the solution to the equation

$$\hat{V}_F = \hat{I}_F \mathbf{Z}(j\omega)$$

where $\hat{V}_F$ was the forcing phasor at frequency ω. The values of s characterizing the natural response was found from the equation

$$0 = I_N Z_s(s)$$

which, for $I_N \neq 0$, is only satisfied for one value of s ($s_1 = -4 \times 10^5 \sec^{-1}$ in this case). We may say that the natural current flowed in response to a phantom voltage source of value 0 volts, but that this is only possible at the single "frequency" s_1. Note that both the natural and the forced response were found using the same impedance function. We then applied the boundary condition to the sum of the natural and the forced response.

The impedance function is only one of many possible transfer functions which could be evaluated in this example. Any of these can be used to find both the natural and forced response in this problem or any other problem of this type. In fact, it is most efficient to use the transfer function associated with the desired output parameter. Suppose, for example, we had wised to find the complete response voltage $v_o(t)$ rather than $i_1(t)$ in the above example. We first need the transfer function relating v_c to v_b, which we determined earlier in Eq. 10.3:

$$H_1(s) = \frac{V_c}{V_b} = \frac{2 \times 10^5}{s + 4 \times 10^5} \tag{10.13}$$

This can be written as

$$V_b = \frac{V_c}{H_1(s)} \tag{10.14}$$

To find the natural response we must, as usual, suppress the source. This is equivalent to setting $V_b = 0$. But if $V_b = 0$, the only way that we can have a nonzero voltage V_c in Eq. 10.14 is if $H_1(s)$ is infinitely large, that is,

$$H_1(s) = \infty$$

This equation shows that we must "find the pole" of the function $H_1(s)$ to determine the natural response parameters. From Eq. 10.14, $H_1(s)$ has a pole when $s_1 = -4 \times 10^5 \sec^{-1}$ since the denominator goes to zero there. This value of s is the same as found before using the impedance zero method, a result consistent with the fact that there is only one value for the natural frequency in a system which is described by a first-order differential equation.

The natural response voltage is therefore of the form $v_{cn}(t) = B_1 e^{-4 \times 10^5 t}$, where B_1 is a constant to be determined. In Section 10.1 we have already found the forced response $v_{cf}(t)$ to the sinusoidal source $v_b = 24 \cos(2 \times 10^5 t)$ by using the phasor representation $\hat{V}_b = 24\underline{/0°}$ and substituting the value $s = j\omega = j(2 \times 10^5)$ in the transfer function,

$$\hat{V}_{cF} = \mathbf{H}_1(j\omega)\hat{V}_b$$

which yielded

$$\hat{V}_{cF} = 10.73\underline{/-26.6°}$$

The forced response as a function of time is then

$$v_{cf}(t) = 10.73 \cos(2 \times 10^5 t - 26.6°)V \qquad (10.15)$$

To complete the solution for $v_c(t)$ we add $v_{cn}(t)$ and $v_{cf}(t)$, use the initial condition $v_c(0^+) = 6.0$ V from Eq. 10.11a, and find

$$v_c(t) = [-3.59e^{-4\times 10^5 t} + 10.73 \cos(2 \times 10^5 t - 26.6°)]V$$

We reiterate that the same transfer function $H_1 = V_c/V_b$ may be used to determine both the natural and the forced response.

TEXT EXAMPLE 10.3 ─────────────────────────────────────

In the circuit illustrated below, the switch has been in position 1 for a long time. At $t = 0$ it is thrown to position 2. Find the voltage $v(t)$ for all $t \ge 0$.

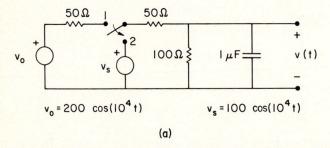

$v_0 = 200 \cos(10^4 t)$ $v_s = 100 \cos(10^4 t)$

(a)

Solution: This problem is somewhat more subtle than the example discussed in the text since the initial and final sources are both sinusoidal. This makes determining the initial conditions more complicated due to the fact that we must first find the forced response to the *initial* sinusoidal source. Note that we cannot merely set $t = 0$ in $v_0(t)$ and solve a dc problem at that time since in general there will be a phase difference between $v(t)$ and $v_0(t)$ which cannot occur in a dc circuit. To do this, we must study the initial circuit in detail:

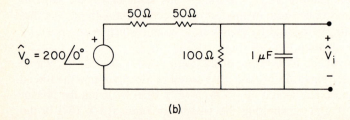

(b)

where we have put a subscript i on the phasor voltage $\hat{V}_i$ to show we are working on the initial conditions. At a frequency $\omega = 10^4$ rad/s, the parallel R and C combination has the impedance

$$Z_p = \frac{R}{RsC + 1} = \frac{100}{10^{-4}s + 1}$$

so for $s = j\omega = j10^4$,

$$\mathbf{Z}_p = 70.7\underline{/-45°} = 50 - j50$$

The voltage $\hat{V}_i$ may be found from the voltage divider principle

$$\hat{V}_i = \left(\frac{\mathbf{Z}_p}{\mathbf{Z}_p + 100} \right) \hat{V}_o$$

$$= \left(\frac{50 - j50}{150 - j50} \right) \hat{V}_o$$

$$\hat{V}_i = (0.45\underline{/-26.6°})(200\underline{/0°}) = 90\underline{/-26.6°}$$

and

$$v_i(t) = 90 \cos(\omega t - 26.6°)$$

Evaluating this function at $t = 0^-$ yields

$$v_i(0^-) = 90 \cos(-26.6°) = 80.5 \text{ V}$$

Since this voltage appears across the capacitor at $t = 0^-$, it must have the same value at $t = 0^+$ and, dropping the subscript i, we have found the initial condition on our desired function

$$v(0^+) = 80.5 \text{ V}$$

We now use the transfer function approach to finding the natural and forced response of the final circuit, which is

(c)

$$H(s) = \frac{V}{V_s} = \frac{Z_p}{Z_p + 50}$$

$$= \frac{100}{150 + 5 \times 10^{-3}s}$$

From the definition of $H(s)$,

$$V = H(s)V_s$$

To find the natural response we have no source so $V_s = 0$ and

$$V = H(s)(0)$$

and the only way we can have a nonzero voltage V is if $H(s) = \infty$. We need to find the pole of $H(s)$ which occurs when the denominator vanishes and

$$s_1 = \frac{-150}{5 \times 10^{-3}} = -3 \times 10^4 \text{ sec}^{-1}$$

The natural response voltage is then of the form

$$v_n(t) = Ae^{-3 \times 10^4 t}$$

To find the forced response we merely evaluate $H(s)$ at $s = j10^4$ and use the same equation $\hat{V}_F = \mathbf{H}\hat{V}_s$, where $\hat{V}_s = 100\underline{/0^\circ}$,

$$\hat{V}_F = \mathbf{H}(j10^4)\hat{V}_s = \left(\frac{100}{150 + j50}\right) 100\underline{/0^\circ}$$

$$= 63.2\underline{/-18.4^\circ}$$

Converting to the time domain and adding the natural and forced response gives

$$v(t) = [Ae^{-3 \times 10^4 t} + 63.2 \cos(\omega t - 18.4^\circ)] \text{ V}$$

From the initial condition we have

$$80.5 = A + 63.2 \cos(-18.4^\circ)$$

which yields $A = 20.5$, and the final answer is

$$v(t) = [20.5e^{-3 \times 10^4 t} + 63.2 \cos(\omega t - 18.4^\circ)] \text{ V}$$

The complete response examples studied in this chapter have all involved transfer functions which are characterized by first-order polynomials. Second-order systems are more complicated; all of Chapter 12 is reserved for this topic.

10.5   A NOTE ON THE LAPLACE TRANSFORM*

Another method for finding the response of systems to time-varying signals is the use of the Laplace transform. The concepts involved are entirely analogous to those used in the Fourier transform. Since you will encounter this method in more advanced study of linear systems it is worth noting that the transfer functions that we have developed are also applicable in the Laplace transform method.

The Laplace transform of a function $v_1(t)$ is given by the expression

$$F(s) = \int_0^\infty v_1(t)e^{-st}dt \qquad (10.16)$$

This expression is analogous to Eq. 10.8 for the Fourier transform. The reverse process of recovering $v_2(t)$ from $V_2(s)$ (analogous to Eq. 10.8) is called the inverse Laplace transform. The mathematical statement is somewhat complicated, but in most engineering problems it is simple to determine $v_2(t)$ using a table of Laplace transform pairs.

One difference between the Laplace and the Fourier transform involves the exponential form. In the latter (see Eq. 10.8) the exponential form involves a purely imaginary argument ($j\omega t$). We can thus consider this as a special case of the Laplace transform in the event that s is purely imaginary. We have seen that $s = j\omega$ applies for purely oscillatory signals. The corresponding functions $\cos \omega t$ and $\sin \omega t$ correspond to oscillations that go on forever both forward and backward in time. A physical interpretation then is that the Fourier series and Fourier transforms are special cases of Laplace transforms which can be used to describe systems which have been in existence for a sufficiently long time that the transient response is finished. Then, the signals can be represented by a linear combination of sinusoidal functions which have been in existence since $t = -\infty$.

If we are interested in the transient response, however, we need to consider the exponent as fully complex, that is, s must be allowed to have both real and imaginary parts. This brings us to the second difference between Eqs. 10.8 and 10.16. In the Laplace transform the integral starts out at $t = 0$. We can thus describe

FIGURE 10.12.
The response of a linear system to a signal may be translated to an algebraic operation on the Laplace-transformed signal.

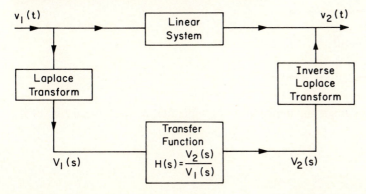

*Sections marked with an asterisk are not required for the logical flow of the text.

a system response to a complicated signal and include the effect of throwing a switch to start it at $t = 0$, by the use of Laplace transforms.

It is important to reiterate that the same transfer function applies to Laplace transforms as to Fourier transforms, that the superposition theorem also holds, and that a similar block diagram applies.

We will not deal any further with Laplace transforms in this text.

10.6 SUMMARY

The same transfer function may be used to find the natural response of a linear electrical system, as well as its response to a sinusoidal source. Furthermore, if the forcing source is a more complicated function of time, we may consider it as a superposition of sinusoidal functions using Fourier analysis methods. The transfer function may then be applied to each Fourier component, evaluated at the appropriate frequency of course, and the output constructed from the sum of all these "transferred" sinusoidal signals. The complete response, natural plus forced, may be determined from the transfer function for the system along with the initial conditions describing how the system was established.

PRACTICE PROBLEMS
AND
ILLUSTRATIVE EXAMPLES

EXAMPLE 10.1

Find the transfer functions $H_1(s) = V_o/V_i$ and $H_2(s) = I_o/V_i$ for the network below. Evaluate the steady-state responses $i_o(t)$ and $v_o(t)$ if $v_i(t) = 100 \cos(\omega t + 45°)$, where $\omega = 10^6$ rad/sec. Evaluate the transfer functions at the frequency $\omega = 0$ (dc) and show that they can also be used to find the response of the network to a dc source by using the analysis techniques of Chapters 1–5.

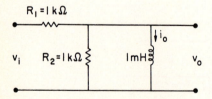

Solution: We first evaluate the impedance of the parallel network consisting of R_2 and L:

$$Z_p = \frac{sLR_2}{R_2 + sL}$$

and the total impedance

$$Z_T = R_1 + Z_p = \frac{R_1(R_2 + sL) + sLR_2}{R_2 + sL}$$

Then, by the voltage divider rule,

$$V_o = \left(\frac{Z_p}{Z_T}\right)V_i$$

and

$$H_1(s) = \frac{Z_p}{Z_T} = \frac{sLR_2}{R_1(R_2 + sL) + sLR_2} = \frac{s}{10^6 + 2s}$$

where in the last step we have substituted the values for R_1, R_2, and L. Likewise, by Ohm's law and the current divider rule,

$$I_o = \left(\frac{R_2}{R_2 + sL}\right)\frac{V_i}{Z_T}$$

and

$$H_2(s) = \frac{R_2}{R_1(R_2 + sL) + sLR_2} = \frac{10^3}{10^6 + 2s}$$

For $\omega = 10^6$ rad/sec,

$$\mathbf{H}_1(j\omega) = \frac{j(10^6)}{10^6 + j(2 \times 10^6)} = \frac{j}{1 + j2} = 0.45\underline{/26.6°}$$

$$\mathbf{H}_2(j\omega) = \frac{10^3}{10^6 + j(2 \times 10^6)} = 4.5 \times 10^{-4}\underline{/-63.4°}$$

The phasors corresponding to the time functions $V_o(t)$ and $I_o(t)$ are then

$$\hat{V}_o = \mathbf{H}_1\hat{V}_i = 45\underline{/71.6°}\,\text{V}$$

$$\hat{I}_o = \mathbf{H}_2\hat{V}_i = 45\underline{/-18.4°}\,\text{mA}$$

and the temporal functions are

$$v_o(t) = 45\cos(10^6 t + 71.6°)\,\text{V}$$

$$i_o(t) = 45\cos(10^6 t - 18.4°)\,\text{mA}$$

At zero frequency $s = j\omega = j(0) = 0$ and $\mathbf{H}_1(0) = 0$. The voltage V_o in this case would be equal to zero, which is correct since an inductor acts as a short circuit at dc. Likewise, $\mathbf{H}_2(0) = 1/R_1$ and $I_o = V_i/R_1$. This corresponds to the fact that if a dc source V_i were placed at the input, the full current will flow through the inductor and that current would be V_i/R_1.

Problem 10.1

Determine the transfer functions $H_1(s) = v_i /i = Z_s(s)$ and $H_2(s) = i_o /v_i$ in the circuit below and find the output signal $i_o(t)$ if $v_i(t) = 100 \cos(\omega t - 45°)$ V, where $\omega = 10^4$ rad/sec. Evaluate the two transfer functions at $\omega = 0$ and use the result to find the dc response currents i and i_o if $v_i = 50$ V.

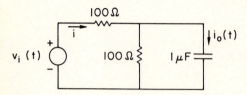

$$\text{Ans.:} H_1(s) = (100)\frac{(s + 2 \times 10^4)}{(s + 10^4)}$$

$$H_2(s) = (0.01)[s/(s + 2 \times 10^4)]$$

$$i_o(t) = 0.45 \cos(\omega t + 18.4°) \text{ A}$$

$$\mathbf{H}_1(0) = 200 \ \Omega; i = 0.25 \text{ A}$$

$$\mathbf{H}_2(0) = 0; i_o = 0$$

Problem 10.2

Find the transfer function

$$H(s) = \frac{v_{cd}}{v_{ab}}$$

in the circuit of Problem 5.7. Evaluate $H(s)$ at $s = j0$ and show that it gives the same result one expects at dc for such a network.

Problem 10.3

Find the transfer function

$$H(s) = \frac{v_c}{v_s}$$

for the circuit in Example 5.2, where v_s represents an arbitrary voltage source in the place of the 24-V battery and the switch is in position 2. Show that the forced response result given in that example for a dc source equal to 24 V may also be derived from $H(s)$. That is, find

$$v_c = \mathbf{H}(j0) \times 24$$

and check against the answer in Example 5.2. Find the forced response if the source

$v_s(t) = 10 \cos(10^5 t + 30°)$ V, that is, find

$$\hat{V}_c = \mathbf{H}(j10^5)\hat{V}_s$$

and convert the phasor back to the time function $v_c(t)$.

Problem 10.4

Find the transfer function

$$H_1(s) = \frac{v_a}{i}$$

in Example 5.4. [$H_1(s)$ is the impedance function.]

$$\text{Ans.:} H_1(s) = \frac{[(1/C)s]}{[s^2 + (s/RC) + (1/LC)]}$$

Problem 10.5

Find the transfer function

$$H_2(s) = \frac{I_o}{i}$$

for the same circuit (Example 5.4) as in Problem 10.4 above. Evaluate $H_2(s)$ and $H_1(s)$ in Problem 10.4 at $s = 0$ and explain the result in terms of the response of the circuit to a dc current source ($s = j0$).

Problem 10.6

Find the transfer function

$$H(s) = \frac{v}{v_s}$$

for the circuit in Problem 5.17 when the switch is in the closed position. Let v_s be an arbitrary source voltage in the same location as the 16-V battery in the derivation. Evaluate $H(s)$ at dc and show that

$$v(0) = H(j0)v_s$$

yields the answer given in that Problem for a 16-V dc source. Find the forced response $v(t)$ if $v_s(t) = 10 \cos(\omega t)$ V, where $\omega = 2.5 \times 10^5$ rad/sec.

Problem 10.7

Evaluate the two transfer functions

$$H_1(s) = \frac{v}{i_s}$$

and

$$H_2(s) = \frac{v}{v_s}$$

for the circuit given in Problem 6.21. $H_1(s)$ is to be evaluated with the switch in position 1 and i_s stands for an arbitrary current source in place of the 12-A dc source. $H_2(s)$ is to be evaluated with the switch in position 2, and v_s stands for an arbitrary voltage source in place of the 18-V dc source. Evaluate $v(0^-)$ and show that

$$v(0^-) = \mathbf{H}_1(j0)12$$

show that the forced response with the switch in position 2 is given by

$$v_F = \mathbf{H}_2(j0)18$$

Problem 10.8

Find the transfer function

$$H_1(s) = \frac{v_C}{v_s}$$

in Problem 6.22 where v_s is an arbitrary source replacing the 20-V battery. Evaluate also

$$H_2(s) = \frac{v_C}{i_s}$$

if a current source replaced the voltage source. Evaluate $\mathbf{H}_1(j0)v_s$ if v_s is a 16V dc source and $\mathbf{H}_2(j0)i_s$ if i_s is a 1 mA dc current source.

Problem 10.9

Find the transfer function

$$H_1(s) = \frac{i_L}{v_4}$$

in the circuit of Problem 6.26, where v_4 stands for an arbitrary voltage source replacing the 4-V dc source. Evaluate H_1 with the switch in position 2. Show that $4H_2(j0) = i_L(0)$ using the answer given for Problem 6.26.

EXAMPLE 10.2

Find the transfer function

$$H(s) = \frac{v_o}{v_i}$$

for the circuit in Problem 10.1, where v_o is the voltage across the capacitor. Evaluate $|\mathbf{H}(j\omega)|$ for the frequencies $\omega = 0$, $\omega = 2 \times 10^4$ rad/sec, and $\omega = \infty$. Make a sketch of $|\mathbf{H}(j\omega)|$ versus ω.

Solution: We use the voltage divider principle to write

$$H(s) = \frac{v_o}{v_i} = \frac{Z_p(s)}{Z_p(s) + 100}$$

where

$$Z_p(s) = \frac{100/sC}{100 + (1/sC)} = \frac{100}{100sC + 1}$$

is the impedance of the 100-Ω resistor and the 10^{-6}-F capacitor in parallel.

$$H(s) = \frac{100}{100 + 10^4 sC + 100} = \frac{100}{10^{-2}s + 200}$$

and finally in polynomial form,

$$H(s) = \frac{10^4}{s + 2 \times 10^4}$$

Evaluating this at several frequencies,

$$\mathbf{H}(j0) = 0.5 = |\mathbf{H}(j0)|$$

$$\mathbf{H}[j(2 \times 10^4)] = \frac{10^4}{j(2 \times 10^4) + 2 \times 10^4} = \left(\frac{1}{2\sqrt{2}}\right)\underline{/-45°}$$

Taking the absolute value of this yields

$$|\mathbf{H}[j(2 \times 10^4)]| = \frac{1}{2\sqrt{2}} = 0.35$$

To find the value of the transfer function at $\omega = \infty$ we take the limit as $\omega \to \infty$:

$$\lim_{\omega \to \infty} \mathbf{H}(j\omega) = \frac{10^4}{j\omega + 2 \times 10^4}$$

As ω becomes very large the $j\omega$ term dominates the denominator and

$$\lim_{\omega \to \infty} \mathbf{H}(j\omega) = \frac{10^4}{j\omega} = \frac{10^4}{\omega}\underline{/-90°}$$

As ω becomes very large $|\mathbf{H}(j\omega)| = 0$. Using these three results the sketch looks as follows:

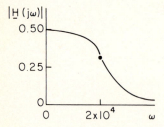

This network is a type of low-pass filter since high frequencies are severely attenuated while low frequencies are only reduced by a factor of two.

Problem 10.10

The transfer function of a certain network is $H(s) = V_o/V_s = 10s/(s + 1000)$. For a steady-state source $v_s(t) = 2 \cos(1000t + 15°)$ V, find the output voltage $v_o(t)$.

$$\text{Ans.: } v_o(t) = 10\sqrt{2} \cos(1000t + 60°) \text{ V}$$

Problem 10.11

The transfer function for a certain network is $H(s) = I_o/I_s = 50(s + 50)/(s + 500)$. For a steady-state source $i_s(t) = 2 \cos(50t)$ mA, find the output current $i_o(t)$. Find the response current i_o if I_s is a dc current source of value $I_s = 2$ mA.

Problem 10.12

Evaluate the absolute value of the transfer function given in Problem 10.10 above at the four different frequencies given below and make a sketch of the results versus ω. That is, find $|\mathbf{H}(j\omega)|$ for $\omega = 0$, $\omega = 1000$ rad/sec, $\omega = 10,000$ rad/sec, and $\omega = \infty$. In the latter case evaluate

$$|\mathbf{H}(j\infty)| = \lim_{\omega \to \infty} |\mathbf{H}(j\omega)|$$

Problem 10.13

The transfer function of a certain amplifier is

$$H(s) = \frac{V_2}{V_1} = 10^9 \left(\frac{s}{(s + 10^3)(s + 10^7)} \right)$$

(a) If $v_1(t) = V_m \cos(10^5 t)$, find the approximate steady-state response $v_2(t)$. (To approximate, set $j10^5 + 10^3 \approx j10^5$ and $j10^5 + 10^7 \approx 10^7$.)

$$\text{Ans.: } v_2(t) = (100V_m) \cos(10^5 t)$$

(b) If $v_1(t) = V_m \cos(10^3 t)$, find the approximate steady-state response $v_2(t)$ using the same approximations as in (a).

Ans.: $v_2(t) =$
$$(100/\sqrt{2})V_m \cos(10^3 t + 45°)$$

(c) If $v_1(t) = V_m \cos(10^7 t)$, find the approximate steady-state response $v_2(t)$.

Ans.: $v_2(t) =$
$$(100/\sqrt{2})V_m \cos(10^7 t - 45°)$$

(d) Make a sketch of the absolute value of the voltage $|\hat{V}_2|$ as a function of frequency.

Ans.:

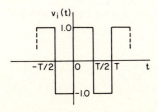

Problem 10.14

Sketch the magnitude of the transfer function $|\mathbf{H}(j\omega)|$ discussed in Problem 10.11 versus ω. (*Hint:* Use $\omega = 0, 50, 500$, and ∞.)

Problem 10.15

Make a sketch of the graph versus ω of the transfer function H_1 given in Problem 10.1. That is, plot $|\mathbf{H}_1(j\omega)|$ versus ω. (*Hint:* Evaluate $|\mathbf{H}_1(j\omega)|$ at $\omega = 0, 10^4, 2 \times 10^4$, and ∞.)

EXAMPLE 10.3

Consider that a square wave signal such as that shown in (a) below is the input to an ideal low-pass filter whose transfer function is $H = v_o/v_i$. The amplitude of the transfer function as a function of the angular frequency is shown in (b), where $\omega_o = 2\pi/T$. Suppose also that for $\omega < 4\omega_o$ the phase angle of this ideal filter is zero. Find the output signal.

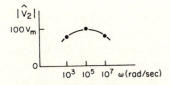
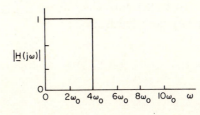

(a)

Solution: We find the mathematical expression for the output voltage and then sketch the resultant waveform. As discussed in the text, $v_i(t)$ may be written as a Fourier series:

$$v_i(t) = \sum_{n=1}^{\infty} \left(\frac{4}{\pi n} \right) \sin(n\omega_o t), \qquad \text{where } n \text{ is odd}$$

The ideal low-pass filter applied to the signal will yield an output given only by the terms in the series with angular frequency less than $4\omega_o$:

$$v_o(t) = \left(\frac{4}{\pi} \right) \sin(\omega_o t) + \left(\frac{4}{3\pi} \right) \sin(3\omega_o t)$$

The resulting signal will appear as shown on the left (below). The higher harmonics are gone and the signal is approximately a square wave with some extra "wiggles" superimposed. If we were to shift the ideal low-pass filter to eliminate all frequencies above $2\omega_o$ even the wiggles will be gone and the resultant output would be the pure sine wave at the fundamental frequency shown on the right (below). The fact that the output signal exceeds the input is not physically possible because an ideal filter of this type cannot be realized even approximately without additional sources.

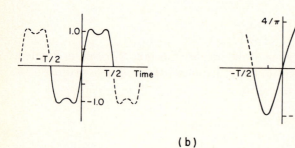

(b)

Problem 10.16

Find the output function $v_o(t)$ if the square wave in Example 10.3 is applied to an ideal high-pass filter with a cut-off at $4\omega_o$. Write down only the largest three terms. Assume that the phase angle of the ideal transfer function is zero for $\omega > 4\omega_o$.

Problem 10.17

As given in Table 10.1, the half-wave rectified sine wave shown below has the Fourier series given by the expression

$$v_i(t) = \frac{1}{\pi} + \left(\frac{1}{2} \right) \sin(\omega_o t) - \left(\frac{2}{3\pi} \right) \cos(2\omega_o t) - \left(\frac{2}{15\pi} \right) \cos(4\omega_o t) - \left(\frac{2}{35\pi} \right) \cos(6\omega_o t) + \dots$$

where $\omega_o = 2\pi/T$.

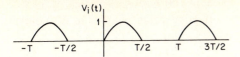

Suppose this signal is passed through a filter with a transfer function such that

$$|\mathbf{H}(j\omega)| = \left|\frac{\hat{V}_o}{\hat{V}_i}\right| = \left[1 + \left(\frac{\omega}{2\omega_o}\right)^2\right]^{-1/2}$$

Evaluate the magnitude of the first five terms in the output signal V_o. (*Hint:* $|\hat{V}_o| = |\hat{\mathbf{H}}\hat{V}_i| = |\hat{\mathbf{H}}||\hat{V}_i|$ at each frequency including $\omega = 0$.)

Ans.: 0.32, 0.45, 0.15, 0.019, 0.006,
all in units of volts

EXAMPLE 10.4

The impedance of a certain network is

$$Z(s) = \frac{V}{I} = 0.1\left(\frac{s + 200}{s + 100}\right)$$

For $v(t) = 2 + 4\cos(100t + 30°)$, find the current $i(t)$ and the power dissipated.

Solution: This is a problem in spectral synthesis since the voltage $v(t)$ has (already) been decomposed into its composite frequencies, namely, zero rad/sec (dc) and 100 rad/sec. The signal at each of these frequencies is transferred through the network via the same transfer function. However, it must be evaluated at the appropriate frequency. At dc,

$$I_{dc} = \frac{V(0)}{Z(0)} = \frac{2}{0.2} = 10 \text{ A}$$

At 100 rad/sec,

$$\hat{I}_{100} = \frac{\hat{V}(j100)}{\mathbf{Z}(j100)} = (40\underline{/30°})\left(\frac{j100 + 100}{j100 + 200}\right)$$

$$= 25.3\underline{/48.4°}$$

To find the total current we reconstruct the signal by adding together the two Fourier components:

$$i(t) = [10 + 25.3\cos(\omega t + 48.4°)] \text{ A}$$

Even though power is a nonlinear quantity, we may add up powers associated with each individual frequency. In this case we have $P_o = VI = (2)(10) = 20$ W at dc and $P_{100} = (4/\sqrt{2})(25.3/\sqrt{2} \cos(-18.4°) = 48$ W at 100 rad/sec. The total is 68 W. Note that we changed peak value phasors to effective value phasors in the course of the power calculations.

Problem 10.18

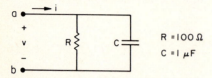

R = 100 Ω
C = 1 μF

(a) For the circuit shown find $Z(s) = v/i$ and evaluate the polar amplitude and angle explicitly as a function of ω.

$$\text{Ans.: } \mathbf{Z} = \frac{100}{[1 + (\omega^2/10^8)]^{1/2}} \underline{/-\tan^{-1}(\omega/10^4)}$$

(b) If the voltage has three Fourier components such that $v(t) = 100 + 200 \cos(10^4 t + 30°) + 50 \cos(2 \times 10^4 t - 45°)]$ V, find $i(t)$.

$$\text{Ans.: } i(t) = [1 + 2.83 \cos(10^4 + 75°) + 1.12 \cos(2 \times 10^4 t + 18.4°)] \text{ A}$$

(c) Find the average power supplied to the terminals a–b.

$$\text{Ans.: } P = 312.5 \text{ W}$$

Problem 10.19

The voltage $v(t)$ given in part (b) of Problem 10.18 is applied to the circuit in Problem 10.1 (which is also discussed in Example 10.2). Find $i_o(t)$.

Problem 10.20

For the circuit below,

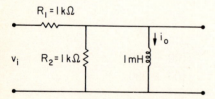

(a) find the transfer function $H(s) = i_o/v_i$.
(b) If $v_i(t) = (10 \cos(10^6 t) + 5)$ V, find $i_o(t)$.

Problem 10.21

The transfer function for a certain system is

$$H(s) = \frac{V_2}{V_1} = 10\left(\frac{s + 1000}{s + 2000}\right)$$

For $v_1(t) = 3 + 4\cos(1000t + 30°) + 6\cos(2000t - 45°)$, find $v_2(t)$. Also, determine the power dissipated in a 10-MΩ resistor if the voltage $v_2(t)$ is applied across that resistor. In this latter portion of the problem you may assume $v_2(t)$ is supplied by an ideal source.

Ans.:

$$v_2(t) = [15 + 25.3\cos(1000t + 48.4°) + 47.4\cos(2000t - 26.6°)] \text{ V}$$
$$P = 22.5 + 32.0 + 112.3 = 166.8 \ \mu\text{W}$$

Problem 10.22

Apply the same composite signal $v_i(t)$ used in Problem 10.21 to the following network,

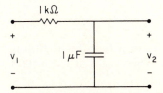

and determine $v_2(t)$. Is this a high-pass or a low-pass filter? If the resulting voltage is applied across a 10-MΩ resistor find the power dissipated. In this latter step you may assume $v_2(t)$ is supplied by an ideal source.

EXAMPLE 10.5

In the circuit below, the switch has been in position 1 for a long time and is moved to position 2 at $t = 0$.

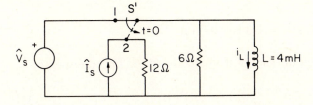

In this circuit,

$$\hat{V}_s = 10\underline{/0°} \text{ V} \qquad \text{and} \qquad \hat{I}_s = \sqrt{2}\underline{/45°} \text{ A}$$

These peak value phasors correspond to signals at a frequency $\omega = 1000$ rad/sec. Find $i_L(t)$ for $t \geq 0$. The switch S' is a "make before break" switch due to the requirement for inductor current continuity.

Solution: Here, the initial and final forced responses are ac signals. The initial condition therefore involves solution of the original circuit using phasor techniques. That *solution* is then evaluated at $t = 0$ to determine the inductor current. It is not correct merely to evaluate $v_s(t)$ at $t = 0$ and then treat that voltage as a dc source.

Following the general approach for complete response problems we first determine the natural response. With the switch in position 2 the transfer function

$$H_1(s) = \frac{I_L}{I_s}$$

is given by the current divider law

$$I_L = \left(\frac{4}{4 + (4 \times 10^{-3})s} \right) I_s = \left(\frac{1000}{s + 1000} \right) I_s$$

and hence

$$H_1(s) = \frac{1000}{s + 1000}$$

Now, since $I_L = H(s)I_s$, the only way that I_L can be nonzero if $I_s = 0$ is if $H(s)$ is
infinite. We therefore seek a pole of the transfer function. This occurs at $s = -1000$
and the natural response current is

$$i_{Ln}(t) = I_o e^{-1000t}$$

The forced response can also be determined from $H_1(s)$ since for $s = j\omega$

$$\hat{I}_{LF} = \mathbf{H}(j\omega)\hat{I}_s$$

$$\hat{I}_{LF} = \frac{(1000)\hat{I}_s}{j1000 + 1000} = \frac{\sqrt{2}\underline{/45°}}{\sqrt{2}\underline{/45°}} = 1\underline{/0°}$$

The forced response is then

$$i_{LF}(t) = 1.0 \cos(10^3 t) \text{ A}$$

The form of the solution is then the sum of the natural and the forced response,

$$i_L(t) = [I_o e^{-1000t} + 1.0 \cos(10^3 t)] \text{ A} \tag{1}$$

To find the initial condition we may use the transfer function

$$H_2(s) = \frac{I_L}{V_s} = \frac{1}{Z_L} = \frac{1}{sL}$$

evaluated with the switch in the original position. Then,

$$\hat{I}_L = \left(\frac{1}{j\omega L} \right) (\hat{V}_s) = (0.25\underline{/-90°})(10\underline{/0°})$$

$$i_2(t) = 2.5 \cos(\omega t - 90°) \text{ A}$$

Evaluating this at $t = 0^-$ yields $i_L(0^-) = 0$. Since the current through an inductor cannot change instantaneously, $i_L(0^-) = i_L(0^+) = 0$ and we have found the necessary initial condition. From Eq. (1) above,

$$i_L(0^+) = 0 = I_o e^o + 1 \cos(0)$$

which implies $I_o = -1$, and finally

$$i_L(t) = 1[\cos(1000t) - e^{-1000t}] \text{ A}$$

Problem 10.23

Repeat Example 10.5 with the exception that the voltage source $\hat{V}_s$ is replaced by a 10-V battery with a 10-Ω resistor in series. The current source remains the same as in the example.

Problem 10.24

Repeat Example 10.5 with the exception that the current source $\hat{I}_s$ is replaced by a 16-V battery with a 4-Ω resistor in series. The 12-Ω resistor remains in the circuit and the voltage source $\hat{V}_s$ remains the same as in the example, that is, a 1000 rad/sec source with a corresponding peak value phasor $10 \underline{/0°}$.

Problem 10.25

In the circuit below, the switch S has been in position 1 for a long time.

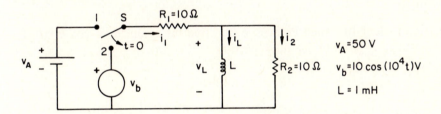

(a) Find the values of i_L, i_1, i_2, and v_L just before 0^- and just after 0^+ when the switch changes position.

$$\text{Ans.: } i_L(0^-) = 5 \text{ A}, \ i_L(0^+) = 5 \text{ A}$$
$$i_1(0^-) = 5 \text{ A}, \ i_1(0^+) = 3 \text{ A}$$
$$i_2(0^-) = 0, \ i_2(0^+) = -2 \text{ A}$$
$$v_L(0^-) = 0, \ v_L(0^+) = -20 \text{ V}$$

(b) Find the functional form of the natural current $i_{in}(t)$ for $t \geq 0$.

$$\text{Ans.: } i_{in}(t) = Ae^{-5000t}$$

(c) Find the forced response current $i_{1F}(t)$.

$$\text{Ans.: } i_{1F}(t) = 0.63 \cos(10^4 t - 18.4°) \text{ A}$$

(d) Find the complete response $i_1(t)$ for $t \geq 0$.

$$\text{Ans.: } i_1(t) = [2.40e^{-5000t} +$$
$$0.63 \cos(10^4 t - 18.4°)] \text{ A}$$

Problem 10.26

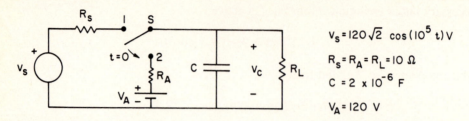

$v_s = 120\sqrt{2}\,\cos(10^5 t)\,\text{V}$

$R_s = R_A = R_L = 10\ \Omega$

$C = 2 \times 10^{-6}\ \text{F}$

$V_A = 120\ \text{V}$

In the circuit shown the switch S has been in position 1 for a long time and is moved to position 2 at $t = 0$.

(a) Find v_c at $t = 0^+$.　　　　Ans.: $v_c = 42.4$ A

(b) Find $v_c(t)$ for $t \geq 0$.　　　Ans.: $v_c = [60 - 17.6e^{-10^5 t}]$ V

Problem 10.27

Returning to Problem 10.3 (and Example 5.2) find the complete response $v_c(t)$ for $t \geq 0$ if the switch is thrown from position 1 to position 2 at $t = 0$. Evaluate $v_c(t)$ for the following two cases:

$$v_s = 24 \text{ V (dc)}$$
$$v_s(t) = 10\,\cos(10^5 t + 30°) \text{ V}$$

Problem 10.28

Returning to Problem 10.7 (and Problem 6.21), find the complete response $v(t)$ if the voltage source is given by the time function

$$v_s(t) = 10\,\cos(4t)$$

where v_s is in the place of the 18-V battery.

Problem 10.29

Find the complete response $v_c(t)$ for the circuit studied in Problems 10.8 and 6.22 in the following four cases: (1) $v_s = 20$ V (dc); (2) $i_s = 1$ mA (dc); (3) $v_s = 20\,\cos(200t)$ V, and (4) $i_s = 1\,\cos(100t)$ mA. In cases 2 and 4 replace the voltage source with the indicated current source.

Problem 10.30

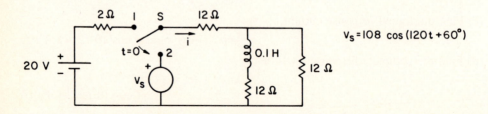

$V_s = 108\,\cos(120t + 60°)$

Switch S has been in position 1 for a long time and is moved to position 2 at $t = 0$.

(a) At $t = 0^+$, find i. Ans.: $i(0^+) = 2.5$ A

(b) Find the impedance $Z(s) = V_s/I$ Ans.: $Z = 24[(s + 180)/(s + 240)]$

(c) Find the natural response current i_n. Ans.: $i_n = Ae^{-180t}$

(d) Find the forced response phasor
current $\hat{I}_F$. Ans.: $\hat{I}_F = 5.58\underline{/52.88°}$ A

(e) Find $i(t)$ for $t \geq 0$. Ans.: $i(t) = [5.58 \cos(120t + 52.88°) - 0.86e^{-180t}]$ A

Problem 10.31

Returning to Problem 10.9 (and Problem 6.26) find $i_L(t)$ for $t \geq 0$ if the 16-V dc source is replaced by the source

$$v_s = 16 \cos(3t) \text{ V}$$

and everything else is the same. Note that though $v_s(t = 0) = 16$ V, the answer is not the same as given in Problem 6.26. Why?

Problem 10.32

Repeat Problem 6.28 if

$$v_o(t) = 20 \cos(10^5 t) \text{ V}$$

that is, find $i(t)$ for $t \geq 0$. (Note again that the answer is different than that given for Problem 6.28, even though $v_o(0) = 20$ V.)

EXAMPLE 10.6

The transfer function of a certain network is

$$H(s) = \frac{V_o}{V_s} = \frac{10s}{s + 1000}$$

It is known that at $t = 0^+$, $v_o = 10$ V. For $v_s = 2 \cos(1000t + 15°)$ V, find the complete response $v_o(t)$ for $t \geq 0$.

Solution: Here, we are given the transfer function without even seeing the circuit and must use it to find both the natural and forced response. Note that an initial condition is also given so we can solve for the complete response. First, to find the natural response consider the fact that

$$V_o = H(s)V_s$$

If we are to have a nonzero output, v_o with no source, that is, if there is to be a natural response with no forcing ($V_s = 0$), then we must satisfy the equation

$$v_{on} = H(s)(0)$$

The only way to have a nonzero V_{on} is if $H(s)$ is infinite. Therefore, we seek that value of s for which H has a pole. This occurs when the denominator vanishes, and thus

$$s_1 = -1000 \ \text{sec}^{-1}$$

is the required value for s_1. The natural response is of the form

$$v_{on} = Ae^{-1000t}$$

where A is unknown. The forced response is

$$\hat{V}_{oF} = [\mathbf{H}(j1000)]2\underline{/15°}$$

$$\hat{V}_{oF} = (10^4\underline{/90°})\frac{2\underline{/15°}}{1000\sqrt{2}\underline{/45°}}$$

$$\hat{V}_{oF} = 14.4\underline{/60°}$$

The sum of the two terms in the time domain is

$$v_o(t) = Ae^{-1000t} + 14.4\cos(1000t + 60°)$$

Note that it is somewhat accidental that in this example $s_1 = \omega$. To complete the problem we use the initial condition at $t = 0$,

$$10 = A + 14.4\cos(60°)$$

$$A = 2.92$$

and finally,

$$v(t) = [2.92e^{-1000t} + 14.4\cos(1000t + 60°)] \ \text{V}$$

Problem 10.33

The transfer function of a certain network is

$$H(s) = \frac{I}{V_s} = \frac{s}{10(s + 10^4)}$$

It is known that at $t = 0^+, i = 2$ A. For $v_s = 100\cos(10^4t + 15°)$ V, find the complete response $i(t)$ for $t \geq 0$.

Ans.:

$$i(t) = [-1.53e^{-10^4t} + 7.07\cos(10^4t + 60°)] \ \text{A}$$

Problem 10.34

The transfer function of a certain network is

$$H(s) = \frac{V_o}{V_s} = \frac{s}{s + 10^4}$$

It is known that at $t = 0^+$, $v_o = 10$ V. For $v_s = 10 \cos(10^4 t + 45°)$ V, find the complete response $v_o(t)$ for $t \geq 0$.

Problem 10.35

In the circuit illustrated below, the switch has been in position 1 for a long time. At $t = 0$ it is thrown to position 2. Find the transfer function $H(s) = V/V_s$ and use it to find the voltage $v(t)$ for all $t \geq 0$. (Use the impedance zero method to check your value for s.)

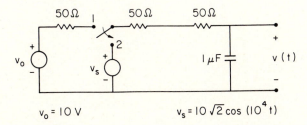

$v_o = 10$ V

$v_s = 10 \sqrt{2} \cos(10^4 t)$

Problem 10.36

In the circuit below, the switch S has been in position 1 for a long time and then is switched to position 2. Find the transfer function $H(s) = V_L / V_b$ and use it to determine voltage $v_L(t)$ for $t \geq 0$.

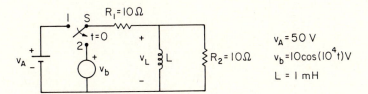

$v_A = 50$ V

$v_b = 10 \cos(10^4 t)$ V

$L = 1$ mH

Problem 10.37

In Example 6.3 arbitrary voltage and current sources were considered. Define $H_1(s) = v_{ab}/i = Z$. Show that

$$H_1(s) = \frac{(R_1 + R_2)sL + R_1 R_2}{sL + R_2}$$

Find the complete response $i(t)$ for $t = 0$ if a voltage source $v_{ab}(t) = 10 \cos(1000t)$ is applied at $t = 0$, if $i_L(0^+) = 1$ mA, and if the elements have the values $R_1 = R_2 = 1 \text{k}\Omega$ and $L = 1$ H.

Problem 10.38

Repeat Problem 10.37 if a current source is applied at $t = 0$ which has the time function $i_s(t) = 1 \cos(1000t)$ mA and everything else is the same. (In this case evaluate $v_{ab}(t)$.)

Problem 10.39

Find the complete response solution to the network described in Problem 10.10 if the source $v_s(t) = 2 \cos(1000t + 15°)$ is applied at $t = 0$ and the initial condition at $t = 0$ is such that $v_o(0^+) = 10$ V.

Problem 10.40

Find the two complete response solutions for the network described in Problem 10.11 if in each case the current source is applied at $t = 0$ and in each case $i_o(0) = 1$ mA.

Chapter 11

Frequency Response— Analysis and Design

Design of an electrical system almost always requires attention to its response to signals at different frequencies. In this chapter we relate the frequency response of a linear system to the poles and zeros of the transfer function. A general formula is presented and then specialized to the case of pure real poles and zeros. This leads to the introduction of Bode plots, which are the logarithmic magnitude and the linear phase angle of the transfer function $\mathbf{H}(j\omega)$ plotted versus the logarithm of the frequency. The Bode plots, particularly in their asymptotic form, are a convenient way to represent $\mathbf{H}(j\omega)$. In addition, we show how to turn the process around and determine the transfer function associated with a given Bode plot. This is a useful design tool since the engineer often has a desired frequency response in mind and must build a system which has the associated transfer function. Some simple filter designs using operational amplifiers are presented. Peak value phasors are used throughout the chapter.

11.1 FREQUENCY RESPONSE—A CASE STUDY

The frequency response of a linear system describes the way in which each frequency component of the input will be modified in amplitude and phase. For a transfer function $H(s)$, the frequency response is $\mathbf{H}(j\omega)$, which has an amplitude and angle both of which are functions of ω. For convenience, we let the amplitude of $\mathbf{H}(j\omega)$ be called A and the angle of $\mathbf{H}(j\omega)$ be called θ. Then,

$$\mathbf{H}(j\omega) = |\mathbf{H}(j\omega)|\,\underline{/\mathbf{H}(j\omega)} = A(\omega)\,\underline{/\theta(\omega)}$$

where both A and θ are functions of ω.

In the circuit of Fig. 11.1a the transfer function may be found using the voltage divider principle. In Chapter 10 we have already sketched the linear amplitude and the phase angle of the frequency response for this circuit as functions of ω. In this section we introduce the decibel notation and apply it to this example of a low-pass filter.

In electrical systems, signals may vary over a very wide range of frequencies. It is common practice therefore to plot the frequency on a logarithmic scale. The

FIGURE 11.1a.
Transfer function and frequency response of an R–L circuit.

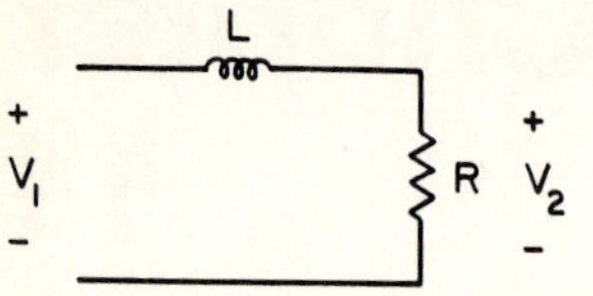

$$H(s) = \frac{V_2}{V_1} = \frac{R}{R + sL}$$

$$\underline{H}(j\omega) = \frac{R}{R + j\omega L} = A \underline{/\theta}$$

$$A = \frac{1}{\sqrt{1 + (\frac{\omega}{\omega_1})^2}} \quad ; \quad \theta = -\tan^{-1}\frac{\omega}{\omega_1} \quad ; \quad \omega_1 = \frac{R}{L}$$

(a)

standard unit for A is also logarithmic in form and uses the decibel, where

$$A_{db} = A \text{ in decibels} = 20 \log_{10} A \tag{11.1a}$$

Since A must be dimensionless in order to take the indicated logarithm, the use of this notation is limited to transfer functions that involve ratios of quantities with the same units. The function $H(s) = V_2/V_1$ in the previous figure is such an example since it is the ratio of two voltages.

The decibel unit was originally defined as the ratios of two *powers*, P_1 and P_2, through the expression

$$\frac{P_1}{P_2} \text{ in decibels} = 10 \log_{10}\left(\frac{P_1}{P_2}\right) \tag{11.1b}$$

Note that the factor 10 appears in (11.*b*) and a factor 20 in (11.*a*). That these two expressions, one involving power and the other using voltage, are equivalent can be seen as follows. If $\hat{V}_1$ and $\hat{V}_2$ are the peak value phasors of the two voltages corresponding to P_1 and P_2, and are impressed across the same resistance value R, then

$$\frac{P_1}{P_2} = \frac{(1/2R)|\hat{V}_1|^2}{(1/2R)|\hat{V}_2|^2} = \frac{|\hat{V}_1|^2}{|\hat{V}_2|^2}$$

Taking the log of both sides and setting $V_1 = |\hat{V}_1|$ and $V_2 = |\hat{V}_2|$,

$$\log_{10}\left(\frac{P_1}{P_2}\right) = 2 \log_{10}\left(\frac{V_1}{V_2}\right)$$

and thus

$$10 \log_{10}\left(\frac{P_1}{P_2}\right) = 20 \log_{10}\left(\frac{V_1}{V_2}\right)$$

Physicists tend to use Eq. 11.1*b* to determine the db relationship whereas engineers use Eq. 11.1*a*. It is important to note that the two expressions yield the same value in decibels, the former (11.1a) employing the ratio of two voltages and the latter (11.1b) the ratio of two powers. In comman usage we ignore the fact that the voltages may not be measured across the same resistance value, and employ an equation such as Eq. 11.1a as the definition of A in db for any dimensionless transfer function.

A db level of $+20$ db means that the power P_1 is 100 times that of P_2 since in that case, $P_1/P_2 = 10^{20/10} = 10^2 = 100$. The corresponding voltage ratio V_1/V_2 is only 10 since power is proportional to the square of a voltage. We reiterate that the db is a *power* unit: $+10$ db is 10 times the power, $+20$ db is 100 times the power, and so on. A number of db values are worked out in Table 11.1. The power ratios and voltage ratios are also given. Note that positive or negative db levels correspond to ratios greater or smaller than unity, respectively. If the power ratio is 2, the db level is $10 \log_{10}(2) = +3.01$ db, whereas if the power ratio is 1/2, the db level is $10 \log_{10}(0.5) = -3.01$ db. We usually round this off and say that the "half-power point" is -3 db. Note that the *voltage* ratio at the half-power point $\sqrt{1/2} = \sqrt{2}/2 = 0.707$. When measuring voltage the "3-db point" is thus the frequency where the voltage V_1 falls to 0.707 times V_2.

We may now evaluate the frequency response of the parameter A for the transfer function of Fig. 11.1a in decibels. First, we note two important properties of logarithms,

$$\log_{10}\left(\frac{1}{a}\right) = -\log_{10}(a)$$

and

$$\log_{10}(a^b) = b \log_{10}(a)$$

Then, from Fig. 11.1a,

$$A(\omega) = \frac{1}{[1 + (\omega/\omega_1)^2]^{1/2}}$$

TABLE 11.1
Some Common db Calculations

Power Ratio P_1/P_2	Voltage Ratio V_1/V_2	db $10 \log_{10}(P_1/P_2) = 20 \log_{10}(V_1/V_2)$
1	1	0
2	$\sqrt{2}$	3.01
4	2	6.02
10	$\sqrt{10}$	10
100	10	20
1000	$\sqrt{1000}$	30
0.5	$\sqrt{2}/2$	-3.01
0.25	1/2	-6.02
0.1	$1/\sqrt{10}$	-10
0.01	0.1	-20
0.001	$1/\sqrt{1000}$	-30

so

$$A_{db}(\omega) = 20 \log_{10}\left(\frac{1}{[1 + (\omega/\omega_1)^2]^{1/2}}\right)$$

$$A_{db}(\omega) = 20 \log_{10}[(1 + (\omega/\omega_1)^2)^{-1/2}$$

and finally,

$$A_{db} = -10 \log_{10}[1 + (\omega/\omega_1)^2]$$

Before sketching A_{db} versus $\log \omega$, let us compute an approximate table of values (to within 0.01 db) as a function of the ratio ω/ω_1:

A_{db}	0.00	0.00	0.00	-0.04	-3.01	-20.04	-40	-60	-80
ω/ω_1	10^{-4}	10^{-3}	10^{-2}	10^{-1}	1	10	10^2	10^3	10^4

The zeros in the table correspond to voltage ratios so close to unity that the db operation yields zero to two significant figures. We sketch A_{db} versus ω (on a logarithmic scale) in Fig. 11.1b. A frequency ratio of 10 to 1 is called a decade, while a frequency ratio of 2 to 1 is called an octave (from the same usage as in the musical scale).

The table of values and the curve in Fig. 11.1b can be approximated by considering three "ranges" for A_{db}, which for reference has the functional form

FIGURE 11.1b.
A_{db} plotted versus ω(log scale) for the transfer function in Fig. 11.1a. The slope is approximately zero for $\omega < \omega_1$ and −20 db/decade for $\omega > \omega_1$.

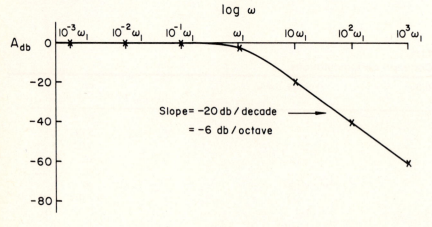

(b)

$$A_{db}(\omega) = -10 \log_{10}\left[1 + \left(\frac{\omega}{\omega_1}\right)^2\right]$$

For $\omega < \omega_1$, the ratio $(\omega/\omega_1)^2$ is small compared to unity, and

$$A_{db} \simeq -10 \log_{10}(1) = 0$$

For $\omega = \omega_1$,

$$A_{db} = -10 \log_{10}(2) \simeq -3$$

For $\omega > \omega_1$, $(\omega/\omega_1)^2$ is large compared to unity and an approximate functional form exists:

$$A_{db}(\omega) = -20 \log_{10}\left(\frac{\omega}{\omega_1}\right)$$

Since the log of a ratio is found by subtraction, $\log_{10}(a/b) = \log_{10}a - \log_{10}b$, we may write this as follows:

$$A_{db}(\omega) = -20 \log_{10}\omega + 20 \log_{10}\omega_1 = -20x + C \qquad (11.2)$$

In the last step we have made the change of variable $x = \log_{10}\omega$ and defined a constant $C = 20 \log_{10}\omega_1$. A plot of the function $-20x + C$ versus x is a straight line of slope -20. Since $x = \log_{10}\omega$ changes by one unit for each decade of ω, we see that, when plotted on a log scale as described here, the line decreases by "20 db per decade." When $\omega = \omega_1$, the two terms in Eq. 11.2 cancel and the expression equals zero. Thus, the straight-line plot corresponding to Eq. 11.2 meets the log ω axis at $x = \log_{10}\omega_1$ where its value equals 0 db.

We can combine the regimes into one plot as shown in Fig. 11.1c. The two straight lines

$$A_{db} = 0 \qquad \omega \leq \omega_1$$

and

$$A_{db} = -20 \log_{10}\omega + 20 \log_{10}\omega_1 \qquad \omega \geq \omega_1$$

meet at $\omega = \omega_1$. Taken together, these lines are called the "asymptotic response" since the actual response is asymptotic to these lines at large and small values of ω. The actual function may be found by smoothly drawing a curve which departs from the asymptotic response and passes through the -3-db point at $\omega = \omega_1$ as shown in Fig. 11.1b and by the dashed curve here.

For this example similar approximations may be made for the angle of $\mathbf{H}(j\omega)$ as a function of ω. From Fig. 11.1a,

$$\theta(\omega) = -\tan^{-1}\left(\frac{\omega}{\omega_1}\right)$$

FIGURE 11.1c.

The asympotic forms for $A_{db}(\omega)$ plotted versus log ω. The dashed curve is the actual frequency response.

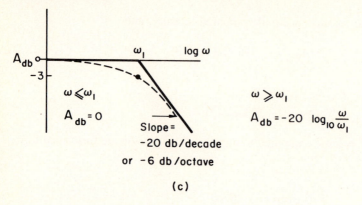

(c)

and, in the same three ranges,

$$\frac{\omega}{\omega_1} << 1 \qquad \theta = 0° \text{(note that for } \omega/\omega_1 = 10^{-1}, \theta = 5.7°)$$

$$\frac{\omega}{\omega_1} = 1 \qquad \theta = -45°$$

$$\frac{\omega}{\omega_1} >> 1 \qquad \theta = -90° \text{ (note that for } \omega/\omega_1 = 10, \theta = 84.3°)$$

As shown in Fig. 11.1d, between $\omega = 0.1\omega_1$ and $\omega = 10\omega$, ω goes from approximately 0° to approximately −90° with an average slope of roughly −45° per decade. This slope does not continue for larger ω since the phase shift approaches the constant value −90° for large ω.

One can understand intuitively why this network acts as a low-pass filter as follows. For dc the inductor acts like a short circuit so the full voltage V appears across the resistor and $V_2/V_1 = 1$. As the frequency is raised, the absolute value of the impedance of the inductor increases ($|Z_L| = \omega L$) and the voltage divider effect

FIGURE 11.1d.

Plot of the phase angle of H versus frequency.

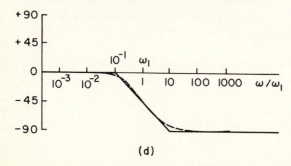

(d)

become more pronounced. This reduces V_2 and a low-pass filter response exists. These results for a simple R–L filter may be generalized to more complicated circuits in the matter described in the next section. Note that in the rest of the discussion we drop the 10 from the $\log_{10}$ expression.

11.2 FREQUENCY RESPONSE—A GENERAL FORMULATION

We have shown from the character of the differential equations which describe linear systems that for voltages and currents of the form e^{st} the ratio of any voltage or current in the system to any other voltage or current in the system is an algebraic function of s. Any $H(s)$ can therefore be written as the ratio of two polynomials times a constant K, and the general form is

$$H(s) = \frac{K(s^n + a_{n-1}s^{n-1} + a_{n-2}s^{n-2} + \cdots a_0)}{s^m + b_{m-1}s^{m-1} + b_{m-2}s^{m-2} + \cdots b_0}$$

Since any polynomial can always in principle be factored into an expression involving its roots, where the number of roots is equal to the degree of the equation, we may with no loss of generality write

$$H(s) = \frac{K[(s - z_1)(s - z_2)(s - z_3)\ldots(s - z_n)]}{(s - p_1)(s - p_2)(s - p_3)\ldots(s - p_m)} \tag{11.3a}$$

The z_1, z_2, z_3, and so forth, in the numerator are the values of s that make the numerator, and thus $H(s)$, equal to zero. These values of s are called the zeros of the transfer function. The p_1, p_2, p_3, and so forth, in the denominator are the values of s that make the denominator equal to zero. They are called the poles of the transfer function since they make $H(s)$ infinite.

As we have seen in Chapter 10, depending upon the circumstances, either the zeros or the poles of the transfer function will yield the values of s which characterize the natural response of the system. If suppressing the sources yields a circuit which displays a natural response when $H(s)$ is equal to zero, then it is the zeros that are used to determine the natural response. If, after suppressing the sources, existence of a natural response is allowed when $H(s)$ is infinite, then it is the poles that are used.

To find the forced response to a sinusoid both the poles and zeros are needed. If we set $s = j\omega$, where ω is the radian frequency of the sinusoid, then, $\mathbf{H}(j\omega)$ is the ratio of the phasors representing the sinusoidal time functions. This, of course, can also be written using the polynomial form in Eq. 11.3a just by setting $s = j\omega$:

$$\mathbf{H}(j\omega) = \frac{\mathbf{K}[(j\omega - \mathbf{z}_1)(j\omega - \mathbf{z}_2)\ldots(j\omega - \mathbf{z}_n)]}{(j\omega - \mathbf{p}_1)(j\omega - \mathbf{p}_2)\ldots(j\omega - \mathbf{p}_m)} \tag{11.3b}$$

Except for the constant $\mathbf{K}$, the forced response is fully determined by the zeros and poles of the transfer function and the frequency of the source. In other words, the poles and zeros of the transfer function contain all the information necessary to define both the natural and the forced responses of the system except for the constant $\mathbf{K}$. In Eq. 11.3b the poles and zeros are written with boldface to show that they are in general complex numbers.

The values of the zeros and poles can be pure real, pure imaginary, or have both a real and imaginary part. For physical systems with no internal energy source the natural response must decay exponentially with time, so the real part of the zeros and poles must always be negative. It can also be shown that if any pole (or zero) has a nonzero imaginary part, there must also exist a pole (or zero) which is the complex conjugate of that pole (or zero). In general, then, the zeros and poles must have the form

$$\mathbf{p}(\text{or } \mathbf{z}) = -a \pm jb$$

Each of the terms in Eq. 11.3b is complex and has a polar form. We may therefore define a set of $A_j(\omega)$ and $\alpha_j(\omega)$ such that

$$j\omega - \mathbf{z}_1 = A_1\underline{/\alpha_1}$$

$$j\omega - \mathbf{z}_2 = A_2\underline{/\alpha_2} \qquad (11.4a)$$

$$.$$

$$.$$

$$.$$

$$j\omega - \mathbf{z}_n = A_n\underline{/\alpha_n}$$

and a set of $B_j(\omega)$ and $\beta_j(\omega)$ such that

$$j\omega - \mathbf{p}_1 = B_1\underline{/\beta_1}$$

$$j\omega - \mathbf{p}_2 = B_2\underline{/\beta_2} \qquad (11.4b)$$

$$.$$

$$.$$

$$.$$

$$j\omega - \mathbf{p}_m = B_m\underline{/\beta_m}$$

Then, Eq. 11.3b may be written in the form

$$\mathbf{H}(j\omega) = \frac{\mathbf{K}[A_1\underline{/\alpha_1} \times A_2\underline{/\alpha_2} \times \ldots A_n\underline{/\alpha_n})}{B_1\underline{/\beta_1} \times B_2\underline{/\beta_2} \times \ldots B_m\underline{/\beta_m}}$$

$$= \frac{\mathbf{K}[(A_1A_2\ldots A_n)\underline{/\alpha_1 + \alpha_2 + \ldots \alpha_n}]}{(B_1B_2\ldots B_m)\underline{/\beta_1 + \beta_2 + \ldots \beta_m}} \qquad (11.5)$$

$\mathbf{H}(j\omega)$ is itself a complex function of the real variable ω which may be written in the form

$$\mathbf{H}(j\omega) = A(\omega) \underline{/\theta(\omega)} \tag{11.6}$$

Combining Eqs. 11.5 and 11.6,

$$A(\omega) = K \left| \frac{A_1 A_2 \ldots A_n}{B_1 B_2 \ldots B_m} \right| \tag{11.7a}$$

and

$$\theta(\omega) = (\alpha_0 + \alpha_1 + \alpha_2 + \cdots \alpha_n) - (\beta_1 + \beta_2 + \cdots \beta_m) \tag{11.7b}$$

where $K = |\mathbf{K}|$ and $\alpha_0 = \underline{/\mathbf{K}}$

Once $\mathbf{H}(j\omega)$ is written in the special form defined by Eq. 11.5, a useful graphical method can be developed. The trick is to identify the functions $A_j \underline{/\alpha_j}$ and $B_j \underline{/\beta_j}$ appropriately with the terms $s - \mathbf{z}_j$ and $s - \mathbf{p}_j$, respectively. First, we plot the poles and zeros in the complex plane. For a system with two real poles and two complex zeros, the plot might look like that in Fig. 11.2a.

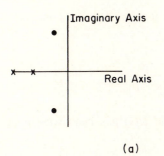

(a)

FIGURE 11.2a.
Pole-zero diagram with two pure real poles and two zeros which have both a real part and an imaginary part. Note that all the real parts are negative and that the zeros are a complex conjugate pair.

Consider one of the poles or zeros, say, $\mathbf{z}_1$, as shown in Fig. 11.2b.

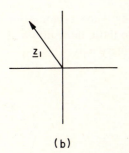

FIGURE 11.2b.
The vector corresponding to the zero z_1.

(b)

We are interested in the value of $H(s)$ when $s = j\omega$. We now show how to evaluate each term in Eq. 11.3b by connecting each zero (or pole) to the point $j\omega$ with a vector $\boldsymbol{\delta}$ as shown in Fig. 11.2c for the zero $\mathbf{z}_1$. We now argue that $\boldsymbol{\delta}$ equals $\mathbf{s} - \mathbf{z}_1$. This follows from the vector relationship

$$\mathbf{z}_1 + \boldsymbol{\delta} = \mathbf{s}$$

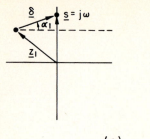

FIGURE 11.2c.

$A_1 = |\underline{\delta}|$ **Vector diagram showing $s = j\omega$, z_1, and**

$\alpha_1 = \angle\underline{\delta}$ $\delta = j\omega - z_1 = A_1 \underline{/\alpha_1}$.

(c)

from which it further follows that

$$\mathbf{s} - \mathbf{z}_1 = \boldsymbol{\delta} = A_1 \underline{/\alpha_1}$$

In the last step we have converted from the vector notation back to complex polar notation. Thus, the factor which enters each of the pole and zero contributions to $|\mathbf{H}(j\omega)|$ in Eq. 11.7a is just the *length* of the vector from the pole to the position $s = j\omega$. Likewise, the α or β angle in Eq. 11.7b is the angle which $\boldsymbol{\delta}$ makes with the real axis.

As an example we evaluate, using the general form for A and θ, the same case studied in Fig. 11.1a. For that circuit,

$$H(s) = \frac{R}{R + sL} = \frac{R/L}{s + (R/L)} = \frac{K}{s - p_1}$$

where p_1 is a negative real number ($p_1 = -R/L$) and $H(s)$ has been written in standard polynomial form. Setting $s = j\omega$,

$$\mathbf{H}(j\omega) = \frac{\mathbf{K}}{j\omega - \mathbf{p}_1} = \frac{\omega_{1p}}{j\omega + \omega_{1p}} = \frac{\omega_{1p}}{B_1 \underline{/\beta_1}}$$

where $\omega_{1p} = R/L$. ($\mathbf{K} = R/L = \omega_{1p}$ in this case.) In Fig. 11.3 the vector δ for three different frequencies is shown by arrows from the pole to those three different frequencies, ω', ω'', and ω''' which are located on the imaginary axis.

FIGURE 11.3.
Graphical determination of $B_1 \underline{/\beta_1}$ as a function of ω.

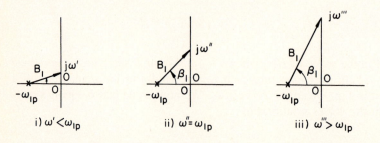

i) $\omega' < \omega_{1p}$ ii) $\omega'' = \omega_{1p}$ iii) $\omega''' > \omega_{1p}$

By inspecting these three figures we may generate Table 11.2.

TABLE 11.2
Graphical Evaluation of $H(j\omega)$ for a One-Pole Low-Pass Filter

ω	B_1	β	$A(\omega) = K/B_1$	$\phi(\omega)$
$\omega << \omega_{1p}$	ω_{1p}	0	1	$0°$
$\omega = \omega_{1p}$	$\sqrt{2}\omega_{1p}$	$45°$	$1/\sqrt{2}$	$-45°$
$\omega >> \omega_{1p}$	ω	$90°$	0	$-90°$

Using these three entries, the frequency response may now be sketched as a function of ω (Fig. 11.4). A circuit with this frequency response is called a one-pole low-pass filter.

FIGURE 11.4.
Sketches of the amplitude and phase of the transfer function for a low-pass R–L circuit.

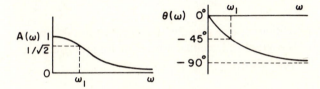

Any system having a single pole will have the same form for its frequency response. For example, the transfer function of the R-C circuit in Fig. 11.5 has a single pole corresponding to the frequency $\omega_{1p} = 1/RC$. The frequency response of the circuit of Fig. 11.5 is exactly the same as that of Fig. 11.1a, with the exception that **K** takes on a different value. They both are one-pole low-pass filters.

FIGURE 11.5.
An R–C circuit whose transfer function has a single pole.

$$\frac{V_2}{V_1} = \frac{1/sC}{R + 1/sC} = \frac{1}{sCR + 1} = \frac{1}{RC}\frac{1}{(s + 1/RC)}$$

TEXT EXAMPLE 11.1 —————————————————————————

For the circuit illustrated below find the transfer function $H(s)$. Evaluate the magnitude and phase angle (that is $A(\omega)$ and $\theta(\omega)$) and sketch them as a function of frequency. Give an intuitive explanation for the fact that the network acts as a high-pass filter.

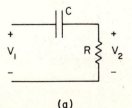

(a)

Solution: The transfer function may be found by using the voltage divider principle:

$$H(s) = \frac{V_2}{V_1} = \frac{R}{R + (1/sC)} = \frac{sCR}{sCR + 1}$$

In polynomial form this is

$$H(s) = \frac{s + 0}{s + (1/RC)}$$

or

$$H(s) = \frac{s - \mathbf{z}_1}{s - \mathbf{p}_1}$$

where $\mathbf{z}_1 = 0$ and $\mathbf{p}_1 = -(1/RC)$, that is, this transfer function has a pole at $s = -(1/RC)$ and a zero at $s = 0$. The graphical method for estimating the frequency response of this circuit is illustrated below.

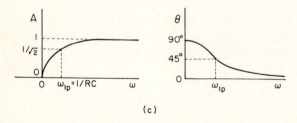

$$A_1 \; \underline{/\alpha_1} = \omega \; \underline{/90°}$$

$$B_1 \; \underline{/\beta_1} = \sqrt{\omega_{1p}^2 + \omega^2} \; \underline{/\tan^{-1} \frac{\omega}{\omega_{1p}}}$$

(b)

For this transfer function, $\mathbf{K} = 1$ since the coefficient in front of the standard polynomial form is unity. Since there is only one pole and one zero, $A = A_1/B_1$ and $\theta = \alpha_1 - \beta_1$.

We first evaluate the magnitude A in three frequency ranges. As $\omega \to 0$, $A_1 \to 0$ and hence $A \to 0$. At $\omega = \omega_{1p}$, $A_1 = \omega_{1p}$ and $B_1 = \sqrt{2}\omega_{1p}$, so $A = 1/\sqrt{2}$. As $\omega \to \infty$, $B_1 = A_1$ and hence $A = 1$. The function $A(\omega)$ is sketched on the left-hand side of the figure below.

(c)

This circuit is a high-pass filter since it allows high-frequency signals through with no change but severely attenuates any component of the signal with $\omega < \omega_{1p}$.

To determine the angle we note that α_1 always equals 90°, so $\theta = 90° - \beta_1$. As

$\omega \to 0$, $\beta_1 \to 0$ and $\theta \to 90°$. At $\omega = \omega_{1p}$, $\beta_1 = 45°$ and $\theta = 45°$. Finally, as $\omega \to \infty$, $\beta_1 \to 90°$ and $\theta = 0°$. The function $\theta(\omega)$ is plotted at the right above.

Note that except for a possible constant multiplier for $A(\omega)$, every system having a zero at $s = 0$ and a single pole at some higher frequency will have the same form for its frequency response. Intuitively we may argue as follows: At dc the capacitor is an open circuit so $V_2 = 0$. At very high frequencies the absolute value of the capacative impedance goes to zero ($|\mathbf{Z}_C| = 1/\omega C$) and the full signal V_1 appears across $R(V_2/V_1 = 1)$. The network acts as a high-pass filter.

11.3 A GENERAL APPROACH TO THE FREQUENCY RESPONSE FOR REAL POLES AND ZEROS—THE BODE PLOTS

We have already anticipated the "Bode plot" several times in this chapter. These are plots of the asymptotic forms of A in decibels and θ in degrees, versus the log of the radian frequency ω. A straightforward analysis applies when all the poles and zeros are real. In such cases, as illustrated in Fig. 11.6 for two poles and two zeros, the constructions which determine the contribution from each pole and each zero are always right triangles. Straightforward trigonometric relationships yield

$$A_1 = (\omega^2 + \omega_{1z}^2)^{1/2}$$

$$\alpha_1 = \tan^{-1}\left(\frac{\omega}{\omega_{1z}}\right)$$

$$B_1 = (\omega^2 + \omega_{1p}^2)^{1/2}$$

$$\beta_1 = \tan^{-1}\left(\frac{\omega}{\omega_{1p}}\right)$$

$$\cdot$$
$$\cdot$$
$$\cdot$$

etc.

from Eqs. 11.6 and 11.7a we may now evaluate the parameters used in constructing Bode plots. To determine the amplitude versus ω we use the equations

$$A_{db}(\omega) = 20\log(A)$$

$$A_{db}(\omega) = 20\log\left|\frac{KA_1A_2\dots}{B_1B_2B_3\dots}\right|$$

The log of a product is the sum of the logs, so we have

$$A_{db}(\omega) = 20\log(K) + 20\log(\omega^2 + \omega_{1z}^2)^{1/2} + \dots - 20\log(\omega^2 + \omega_{1p}^2)^{1/2} - \dots$$

Each term has the same form for real poles and zeros and can be simplified.

FIGURE 11.6.
Figure for determining the frequency response of a transfer function with two real zeros and two real poles.

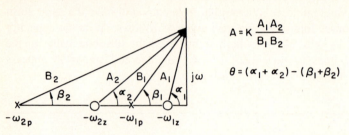

$$A = K \frac{A_1 A_2}{B_1 B_2}$$

$$\theta = (\alpha_1 + \alpha_2) - (\beta_1 + \beta_2)$$

For example,

$$20 \log(\omega^2 + \omega_{1z}^2)^{1/2} = 20 \log[\omega_{1z}(1 + \omega^2/\omega_{1z}^2)^{1/2}]$$

$$= 20 \log(\omega_{1z}) + 10 \log[(1 + (\omega/\omega_{1z})^2]$$

where in the second term the square root has been "moved through" the log operation. The first term is a constant and can be lumped together with the constants from all of the other terms [e.g., $+ 20 \log(\omega_{2z})$... $- 20 \log(\omega_{1p})$...], and with the constant K. This yields a new constant A_0. We may now express the general form of $A(\omega)$ in decibels as

$$A_{db}(\omega) = 20 \log(A_0) + 10 \log\left[1 + \left(\frac{\omega}{\omega_{1z}}\right)^2\right] + 10 \log\left[1 + \left(\frac{\omega}{\omega_{2z}}\right)^2\right] + \cdots$$

$$- 10 \log\left[1 + \left(\frac{\omega}{\omega_{1p}}\right)^2\right] - 10 \log\left[1 + \left(\frac{\omega}{\omega_{2p}}\right)^2\right] - \cdots \quad (11.8)$$

where

$$A_0 = K\left(\frac{\omega_{1z}\omega_{2z}\ldots\omega_{nz}}{\omega_{1p}\omega_{2p}\omega_{3p}\ldots\omega_{mp}}\right) \quad (11.9)$$

Since $\log(1) = 0$, every term but the first in Eq. 11.8 vanishes when $\omega = 0$. This means that

$$A_{db}(0) = 20 \log(A_0)$$

and that the constant A_0 determines the value of $A(\omega)$ at zero frequency.

Except for the first term each of the other terms in Eq. 11.8 has an *identical form*. This fact is the essence of the Bode plot method for real poles and zeros, and hence we examine in detail the common form $10 \log[1 + (\omega^2/\omega_{1z}^2)]$. For $\omega \ll \omega_{1z}$, this term equals $10 \log(1) = 0$ and hence contributes zero to $A_{db}(\omega)$. For $\omega \gg \omega_{1z}$ the term has the form $10 \log(\omega^2/\omega_{1z}^2) = 20 \log(\omega) - 20 \log(\omega_{1z})$. Using the same arguments as given in Section 11.1, this expression represents a straight line when

$\log \omega$ is the variable. In this case the line has a *positive* slope of 20 db per decade. A plot of its contribution to $A_{db}(\omega)$ is shown on the left-hand side of Fig. 11.7. For $\omega = \omega_{1z}$, $A_{db}(\omega_1) = +3$ db. The actual function is shown in the figure as the curved line passing through the point $+3$ db. The straight lines which meet at $x = \log(\omega_{1z})$ are the asymptotic forms. The plot on the right-hand side shows the contribution of a term of the form

$$-10 \log \left[1 + \left(\frac{\omega}{\omega_{1p}} \right)^2 \right]$$

corresponding to a pole (we studied this function in Section 11.1). Such a term contributes zero for $\omega < \omega_{1p}$ and, asymptotically, -20 db per decade for $\omega > \omega_{1p}$. Since the full expression for $A_{db}(\omega)$ is made up of the sum of terms like this, plus $\log(A_0)$, each zero or pole contributes a function such as that shown in (a) or (b) of Fig. 11.7, respectively, to the full response. The sum of all these terms yields the asymptotic form of the function $A_{db}(\omega)$, the Bode plot.

FIGURE 11.7.
Contributions to the frequency response for each zero (a) and each pole (b).

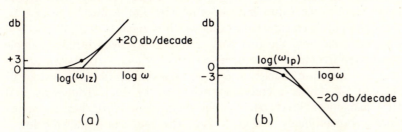

(a)   (b)

Likewise, each angular term is either of the form $+\tan^{-1}(\omega/\omega_{1z})$ or $-\tan^{-1}(\omega/\omega_{1p})$ such that

$$\theta(\omega) = \underline{/K} + \tan^{-1}\left(\frac{\omega}{\omega_{1z}} \right) + \cdots - \tan^{-1}\left(\frac{\omega}{\omega_{1p}} \right) - \cdots$$

The latter have the respective graphical forms shown in Fig. 11.8. Note that these graphs take on the values $\theta = +45°$ for $\omega = \omega_{1z}$ (a zero) and $\theta = -45°$ for $\omega = \omega_{1p}$ (a pole).

FIGURE 11.8.
Contributions of each zero (a) or pole (b) to the total angle $\theta(\omega)$.

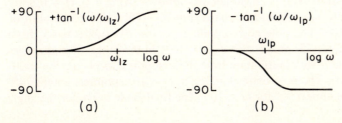

(a)   (b)

A comment on nomenclature is needed. Although in this section the poles and zeros of the transfer function are negative real numbers, we often refer to the corresponding positive real frequencies as the "poles" and "zeros." This is done in the text example below.

TEXT EXAMPLE 11.2 _____

Determine the Bode plots for the transfer function

$$H(s) = \frac{10^9(s + 10)}{(s + 10^4)(s + 10^7)}$$

Solution: This transfer function has a zero at $z_1 = -10$ and poles at $p_1 = -10^4$ and $p_2 = -10^7$. For this case, using Eq. 11.9,

$$A_0 = \frac{K\omega_{1z}}{\omega_{1p}\omega_{2p}} = \frac{10^9(10)}{(10^4)(10^7)} = 0.1$$

and the constant term in $A_{db}(\omega)$ is $20 \log_{10}(0.1) = -20$ db. As mentioned above, this value is just equal to the value in db of $A(\omega) = |H(j\omega)|$ evaluated at $\omega = 0$. The asymptotic Bode plot is constructed as follows. It is equal to -20 db at $\omega = 0$ and remains at that value until the first pole or zero is reached. In this example the lowest pole or zero frequency corresponds to a zero at $\omega_{1z} = 10$ rad/sec, so the asymptotic frequency response changes at $\omega = \omega_{1z}$ to a straight line with a slope of $+20$ db/decade. At 10^4 rad/sec a pole (ω_{1p}) exists, so the slope changes by -20 db/decade. This change just "levels out" the line again. At $\omega = 10^7$ there is another pole (ω_{2p}) which also contributes -20 db/decade. The final plot is shown below.

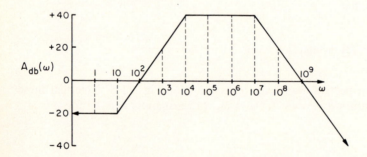

The corresponding plot of $\theta(\omega)$ is constructed as follows. At $\omega = 0$, $\mathbf{H} = 0.1\underline{/0°}$. The zero corresponding to $\omega = 10$ adds a phase shift of $+45°$ and a phase shift of almost $+90°$ for $\omega \geq 10^2$. The pole corresponding to $\omega = 10^4$ subtracts $45°$ from the angle, resulting in a net $45°$ phase shift at $\omega = 10^4$. In the range of frequencies between 10^5 and 10^6, the $-90°$ contribution from the first pole cancels out the $+90°$ from the zero and $\theta = 0°$. The pole corresponding to $\omega = 10^7$ subtracts another $45°$ at $\omega_{2p} = 10^7$ rad/sec and almost $90°$ for $\omega \geq 10^8$. The final Bode curve for the angle is given below.

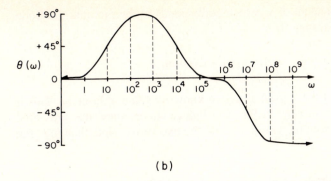

(b)

Before continuing we must discuss the special case which arises when a zero exists at $s = 0$. The high-pass filter studied in Section 10.1 has this property since the corresponding transfer function is of the form

$$H(s) = \frac{s}{s + \omega_{1p}}$$

where $\omega_{1z} = 0$ and $\omega_{1p} = 1/RC$. Analyzing as before yields the equation for the frequency response

$$A_{db}(\omega) = 20 \log \left(\frac{K \cdot 0}{1/RC} \right) - 10 \log \left[1 + \left(\frac{\omega}{\omega_{1p}} \right)^2 \right]$$

Since $\log(0) = -\infty$, however, the first term is undefined and therefore difficult to plot! We do know that the asymptotic plot increases by 20 db/decade due to the zero at $s = 0$ and then "levels out" at $\omega = \omega_{1p}$ due to the pole. The question is, where do we locate this plot relative to the vertical axis? To accomplish this we need only find the value of $A_{db}(\omega)$ at one point since the *form* of the plot is known from the poles and zeros. How this is accomplished is best shown by example.

Consider the transfer function

$$H(s) = 10^3 \left[\frac{s(s + 10^3)}{(s + 10)(s + 10^5)} \right] \tag{11.10}$$

which has zeros at $z_1 = 0$ and $z_2 = -10^3$ and poles at $p_1 = -10$ and $p_2 = -10^5$. The Bode plot will rise from the value $A_{db}(0) = -\infty$ with a slope of $+20$ db/decade. The asymptotic plot will level out at $\omega = 10^3$ rad/sec and then break upward again with a $+20$ db/decade slope at $\omega = 10^7$ rad/sec. Finally, it levels out again at the last pole, $\omega = 10^5$ rad/sec, and stays level as $\omega \to \infty$. Either of these ranges $10 \leq \omega \leq 10^3$ or $10^5 \leq \omega \leq \infty$ may be used to determine the absolute position of the asymptotic plot of $A_{db}(\omega)$. In the limit that s becomes very large, Eq. 11.10 becomes

$$\lim_{s \to \infty} H(s) = \frac{10^3 s^2}{s^2} = 10^3$$

In this range $H(s)$ is independent of s, which corresponds to the fact that the Bode plot has zero slope. In this range then,

$$A_{db}(\omega) = 20 \log(10^3) = 60 \text{ db}$$

We may now make the plot for $A_{db}(\omega)$ since we know its value at large ω. This is shown as the darkest curve in Fig. 11.9a. The lighter curves show the individual straight-line contributions from the two poles and the two zeros. Note that any "flat

FIGURE 11.9a.
Bode plot for $A_{db}(\omega)$ in decibels for $H(s)$ given by Eq. 11.12.

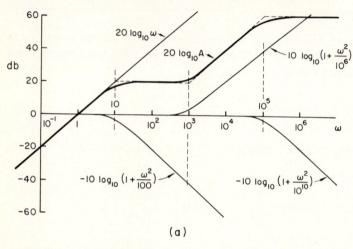

(a)

region" in the curve could be used to determine its position. Considering the range $10 \ll s \ll 10^3$ in Eq. 11.10, we may set $s + 10 \simeq s$, $s + 10^3 \simeq 10^3$, and $s + 10^5 \simeq 10^5$, in which case

$$H(s) \simeq \frac{10^6 s}{10^5 s} = 10$$

This result is independent of s and is such that $20 \log|\mathbf{H}(j\omega)| = 20$, which agrees with the plot in Fig. 11.9a. For reference, the corresponding individual solutions and composite solution for $\theta(\omega)$ are also given for this transfer function in Fig. 11.9b.

As a further example, consider the transfer function

$$H(s) = 10^5 \left[\frac{s}{(s + 100)(s + 10^4)} \right]$$

This has a single zero at $\mathbf{z}_1 = 0$ and poles at $\mathbf{p}_1 = -100$ and $\mathbf{p}_2 = -10^4$. $H(s)$ is independent of s for

$$100 \le s \le 10^4$$

FIGURE 11.9b.
Bode plot of $\theta(\omega)$ for $H(s)$ given by Eq. 11.10.

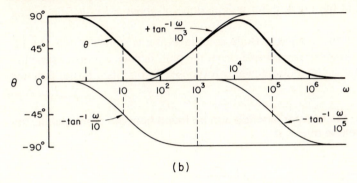

(b)

since in this range

$$H(s) \simeq \frac{10^5 s}{10^4 s} = 10$$

which does not depend on s. In this range

$$A_{db}(\omega) = 20\log(10) = 20 \text{ db}$$

and the asymptotic plot then is flat from 100 to 10^4 at a value of 20 db and rolls off at 20 db/decade above the pole at 10^4. Below $\omega = 100$ the Bode plot rises from $-\infty$ at $+20$ db/decade due to the zero at $s = 0$. The result is the bandpass function shown in Fig. 11.10.

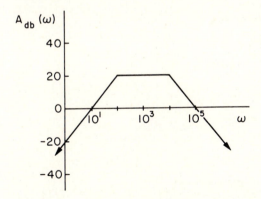

FIGURE 11.10.
Asymptotic Bode plot for $A_{db}(\omega)$ when $H(s) = 10^5\,s/[(s + 100)(s + 10^4)]$.

In summary, we see that an asymptotic Bode plot for real poles and zeros can be sketched simply by knowing the values of the zeros or poles of the function which are called the "break points." In the asymptotic plot for A in decibels, the contribution of each break point is zero until ω is greater than the magnitude of the corresponding zero or pole. If it is a zero, the contribution is a positive slope of 20 db/decade or 6 db/octave. If the break point is at a pole, the contribution is a negative slope of 20 db/decade or 6 db/octave. The contribution of the constant term is simply to shift the whole response up or down but not to change the shape.

In an asymptotic plot for θ the contribution of each zero is $+45°$ at the break point, with a slope of about 45°/decade, going from 0° to 90°. At each pole there is a contribution of $-45°$ at the break point with a slope of about $-45°$/decade, going from 0° to $-90°$.

Another example of an asymptotic Bode amplitude plot is shown in Fig. 11.11. We may now turn the process around and, from the plot, deduce a transfer function that would produce that asymptotic frequency response. To find the transfer function

FIGURE 11.11.
Asymptotic Bode plot. The transfer function can be found from the break points, the slopes, and the magnitude at ω = 0.

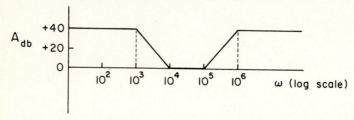

for Fig. 11.11 we note that the slope from $\omega = 10^3$ to $\omega = 10^4$ is -40 db/decade. Therefore, there must be two poles at $\omega = 10^3$. Two zeros are required at $\omega = 10^4$ to level out the curve. Also, the slope from $\omega = 10^5$ to $\omega = 10^6$ is $+40$ db/octave and hence there must be two more zeros at $\omega = 10^5$. Finally, there must be two poles at $\omega = 10^6$. Therefore, the transfer function is of the form

$$H(s) = K\left[\frac{(s + 10^4)^2(s + 10^5)^2}{(s + 10^3)^2(s + 10^6)^2}\right]$$

There are no zeros at $s = 0$ so to find K we use

$$A_0 = K\left[\frac{(10^4)^2(10^5)^2}{(10^3)^2(10^6)^2}\right] = K$$

But from the graph, $A_{db}(0) = +40$, so

$$40 = 20\log A_0 \qquad \text{and} \qquad A_0 = 100$$

We conclude that $K = 100$ and hence

$$H(s) = 100\left[\frac{(s + 10^4)^2(s + 10^5)^2}{(s + 10^3)^2(s + 10^6)^2}\right] \tag{11.11}$$

Eq. 11.11 assumes that $\mathbf{K} = K$, that is $\underline{/\mathbf{K}} = 0°$. From the amplitude plot it is not possible to determine the angle of $\mathbf{K}$.

Of course, the actual frequency response for the transfer function of Eq. 11.11 will not be the asymptotic plot shown in the figure. At $\omega = 10^3$, for example, A_{db} would actually be reduced by 3 db for *each* pole. Since there are two poles, A_{db} would be equal to 34 db. There would be similar rounding at the other values. In summary, given a transfer function we may draw the asymptotic plots. Conversely, given an asymptotic plot, we may derive the transfer function.

TEXT EXAMPLE 11.3 _____

The asymptotic Bode plot for a certain system is shown below. Find the transfer function $H(s)$ assuming **K** is purely real.

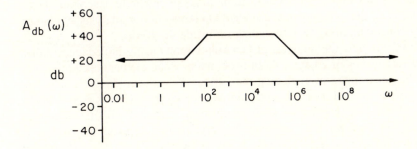

Solution: The slope changes by $+20$ db/decade at $\omega = 10$ rad/sec and by -20 db/decade at $\omega = 10^2$ and $\omega = 10^5$. At $\omega = 10^6$ it again changes by $+20$ db/decade. The transfer function must thus be of the form

$$H(s) = K\left[\frac{(s + 10)(s + 10^6)}{(s + 10^2)(s + 10^5)}\right]$$

The constant K may be found from the value of $A(\omega)$ at $\omega = 0$ since there is no zero at $\omega = 0$. From the plot,

$$A_{db}(0) = 20 = 20\log_{10}(A_0)$$

and $A_0 = 10 = |\mathbf{H}(0)|$. Then,

$$10 = \frac{K(10)(10^6)}{(10^2)(10^5)}$$

and $K = 10$. The required transfer function for **K** purely real is

$$H(s) = 10\left[\frac{(s + 10)(s + 10^6)}{(s + 10^2)(s + 10^5)}\right]$$

_____

11.4 FREQUENCY RESPONSE OF CAPACITOR— RESISTOR OP-AMP CIRCUITS

In this section we apply techniques to operational amplifier circuits similar to those discussed earlier in the text. In circuits using only resistive elements and ideal sources, Ohm's law applies in its elemental form, $v(t) = i(t)R$. This equation, and the circuits it applies to, does not depend upon frequency at all. The transfer function is independent of s and the system responds identically to all frequency components.

In Op-Amp applications, capacitors are often deliberately introduced to control the range of frequency in which the amplifier operates and to match the amplifier circuit to its application. "Stray" capacity also arises inadvertently in circuit design and must be taken into account. In this section we present and analyze operational amplifier circuits using R–C components. The associated transfer functions often have purely real poles and zeros and the Bode analysis methods described above are applicable. Note that we may still use ideal Op-Amps in our analysis and hence not be overly concerned about what is going on inside the device.

We first briefly review the analysis of Op-Amp circuits using impedances rather than the pure resistances of Chapter 4. In brief, each of the elemental circuit types studied may be generalized by replacing the various R_j with $\mathbf{Z}_j$. For example, the inverting amplifier (Fig. 11.12a) has the gain $\mathbf{A}_v = -\mathbf{Z}_2/\mathbf{Z}_1$ which is analogous to the result $A_v = -R_2/R_1$ in Chapter 4. To analyze the circuit from first principles rather than by analogy we use the following arguments. Suppose the source is represented by a phasor $\hat{V}_1$. By the first law of Op-Amps, $V^- = V^+ = 0$ and the current through $\mathbf{Z}_1$ is just $\hat{I}_1 = \hat{V}_1/\mathbf{Z}_1$. by the second law of Op-Amp this current must entirely flow through $\mathbf{Z}_2$. Ohm's law then yields

$$\frac{0 - \hat{V}_0}{\mathbf{Z}_2} = \hat{I}_1 = \frac{\hat{V}_1}{\mathbf{Z}_1}$$

Solving,

$$\mathbf{A}_v = \frac{\hat{V}_0}{\hat{V}_1} = \frac{-\mathbf{Z}_2}{\mathbf{Z}_1}$$

If we now realize that this result is valid for any phasor, then for all frequencies we may define the circuit transfer function by

$$H(s) = \frac{-Z_2}{Z_1}$$

Similarly, the noninverting amplifier of Chapter 4 (Fig. 11.12b) has the gain

$$\mathbf{A}_v = 1 + \frac{\mathbf{Z}_2}{\mathbf{Z}_1} = \frac{\mathbf{Z}_1 + \mathbf{Z}_2}{\mathbf{Z}_1}$$

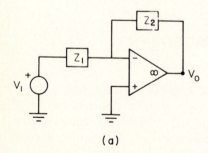

FIGURE 11.12a.
The inverting amplifier.

(a)

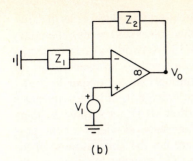

FIGURE 11.12b.
The noninverting amplifier.

(b)

TEXT EXAMPLE 11.4

Derive the transfer functions and describe their properties for the two circuits illustrated below.

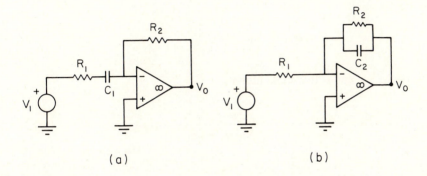

(a)　　　　　　　　　　　(b)

Solution:　Both circuits are inverting amplifiers and hence have transfer functions described by

$$H(s) = -\frac{Z_2}{Z_1}$$

For circuit (a),

$$H_a(s) = -\frac{R_2}{R_1 + (1/sC_1)} = -\frac{R_2}{R_1}\left(\frac{s}{s + \omega_1}\right)$$

where $\omega_1 = 1/(R_1C_1)$. This circuit is a high-pass filter since it has a zero at $s = 0$ and a pole at ω_1. For circuit (b),

$$H_b(s) = -\frac{R_2(1/sC_2)}{R_1[R_2 + (1/sC_2)]} = -\frac{R_2}{R_1}\left(\frac{\omega_2}{s + \omega_2}\right)$$

where $\omega_2 = 1/(R_2C_2)$. This is a low-pass filter with a low-frequency gain of $-(R_2/R_1)$ and a pole at $\omega = \omega_2$.

A major benefit achieved by using Op-Amp circuits to produce a desired frequency response is that it is possible to set individual poles and zeros separately. Consider the responses of the networks in Fig. 11.13. Network (a) can be recognized as a one-pole high-pass filter since the capacitor acts as an open circuit for low frequencies, and the signal V_1 does not "get through" to the output V_2 at dc. Likewise, in network (b) the capacitor shorts out the voltage V_4 at high frequencies and this R–C circuit acts as a one-pole low-pass filter. It is not surprising, then, that when hooked together as in part (c), the resulting network yields a band pass filter. However, the poles of the resulting network are not the same as those of the circuits (a) and (b). This can be shown as follows. The transfer functions for the three circuits are, respectively,

$$H_1(s) = \frac{s}{s + \omega_1}$$

where $\omega_1 = 1/(R_1 C_1)$,

$$H_2(s) = \omega_2\left(\frac{1}{s + \omega_2}\right)$$

where $\omega_2 = 1/(R_2 C_2)$, and (after some algebra)

$$H_3(s) = \frac{\omega_2 s}{s^2 + s\left[\omega_2 + \dfrac{\omega_1(R_1 + R_2)}{R_2}\right] + \omega_1\omega_2}$$

It is simple to show that $H_3(s)$ has two real poles, no matter what values R_1, R_2, C_1, and C_2 take on. This circuit is of second order, a class of circuits we discuss in detail in Chapter 12. The transfer function H_3 has a zero at $\omega = 0$ and two higher-frequency poles but the poles are not equal to the poles of H_1 and H_2. H_3 has this property since the ratio of the voltage across R_1 to the input voltage V_1 depends on R_2 and C_2.

If, as is often the case, we desire to produce a transfer function which has a particular product in the denominator, say,

FIGURE 11.13.
The transfer function of (c) is *not* the product of the transfer functions of (a) and (b).

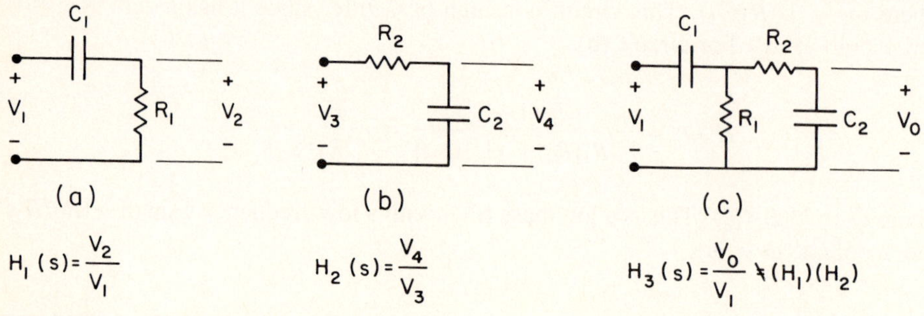

$$(a) \qquad H_1(s) = \frac{V_2}{V_1}$$

$$(b) \qquad H_2(s) = \frac{V_4}{V_3}$$

$$(c) \qquad H_3(s) = \frac{V_0}{V_1} \neq (H_1)(H_2)$$

$$H(s) = K\left[\frac{s}{(s + \omega_1)(s + \omega_2)}\right] \qquad (11.12)$$

where ω_1 and ω_2 are specified, it will take some effort to find the appropriate values of $R_1, C_1, R_2,$ and C_2 in the circuit of Fig. 11.13. On the other hand, it is easy using Op-Amps to achieve the desired result. One possible arrangement is shown in Fig. 11.14. Each of the Op-Amp circuits is an inverting amplifier with a transfer function equal to $-(Z_F/Z_I)$, where Z_F is the feedback impedance and Z_I is the impedance between the negative input and the signal. Then (see Text Example 11.4),

$$H_1(s) = \frac{V_2}{V_1} = -\frac{R_a}{R_1 + (1/sC_1)} = -\frac{R_a}{R_1}\left(\frac{s}{s + \omega_1}\right)$$

where, $\omega_1 = 1/(R_1 C_1)$. Similarly,

$$H_2(s) = \frac{V_0}{V_2} = -\frac{R_2/sC_2}{R_b[R_2 + (1/sC_2)]} = -\frac{R_2}{R_b}\left[\frac{\omega_2}{s + \omega_2}\right]$$

where $\omega_2 = 1/(R_2 C_2)$. But since the first Op-Amp acts as an ideal voltage source,

$$\frac{V_0}{V_1} = \frac{V_2}{V_1} \times \frac{V_0}{V_2}$$

and it follows that

$$H(s) = H_1(s) \times H_2(s)$$

Finally,

$$H(s) = \frac{R_a R_2 \omega_2}{R_1 R_b}\left[\frac{s}{(s + \omega_1)(s + \omega_2)}\right]$$

where $\omega_1 = 1/(R_1 C_1)$ and $\omega_2 = 1/(R_2 C_2)$. We may now separately fix the values of $\omega_1, \omega_2,$ and K to produce the desired response.

FIGURE 11.14.
Two cascaded Op-Amps designed to produce the transfer function $H(s) = V_0/V_1$ of Eq. 11.12.

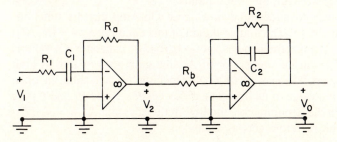

A response in the form of Eq. 11.12 can be achieved with a single Op-Amp, except for the sign, by using the inverting amplifier circuit shown in Fig. 11.15. The transfer function V_0/V_1 for the Op-Amp circuit of Fig. 11.15 is also $-(Z_F/Z_I)$, so

$$H(s) = -\frac{R_2/sC_2}{(R_2 + 1/sC_2)(R_1 + (1/sC_1))}$$

$$H(s) = -\left(\frac{R_2\omega_2}{R_1}\right)\left(\frac{s}{(s + \omega_1)(s + \omega_2)}\right)$$

where $\omega_1 = 1/(R_1C_1)$ and $\omega_2 = 1/(R_2C_2)$. This expression has the same functional form as the one above except that $\mathbf{K} = |\omega_2|(R_2W_2/R_1)\ \underline{/180°}$.

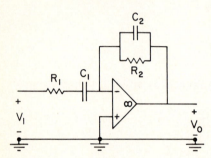

FIGURE 11.15.
The transfer function of Eq. 11.14 may be achieved with a single Op-Amp with the exception that the sign is changed.

If a positive value of $\mathbf{K}$ is required the Op-Amp of Fig. 11.15 may be connected to another inverting amplifier using just resistances. If it is desired to have another pole or zero in the transfer function, this can be achieved by the use of appropriate resistance and capacitance elements in that second Op-Amp circuit. In this way all manner of circuit characteristics can be built up. Some additional examples are given in the problems at the end of this chapter.

If the frequencies ω_1 and ω_2 are close together, we usually refer to the design as a narrow bandpass filter and the circuit only allows frequencies between ω_1 and ω_2 to pass through the system. Such a system might be used to isolate a particular frequency band for study. In a stereo amplifier, the two frequencies would be quite different, ω_1 being several tens of radians per second and ω_2 several thousand radians per second since the human ear responds between about 15 and 15,000 Hz. In this case there is a wide range of frequencies where the response is nearly uniform, with high- and low-frequency cutoffs to match the application. The gain in this frequency range is termed the midband gain. For such a system the *bandwidth* is defined as the difference between the -3db points.

There is an intuitive way to understand how these Op-Amp circuits function. Consider the circuit in Fig. 11.15. The midband gain of this circuit is $\mathbf{A}_v = -(R_2/R_1)$ as studied in Chapter 4. At high frequencies, $\omega > 1/(R_2C_2)$, capacitor C_2 "shorts out" resistor R_2. Since $\mathbf{Z}_2$ is in the numerator of $\mathbf{A}_v$, the gain goes to zero at high frequencies. At low frequencies, $\omega < 1/(R_1C_1)$, capacitor C_1 acts like an open circuit. That is, the absolute value of its impedance exceeds R_1. Since R_1 and C_1 are in series and since $\mathbf{Z}_1$ is in the denominator, the low frequency gain $\mathbf{A}_v = -(\mathbf{Z}_2/\mathbf{Z}_1)$ becomes small. The circuit therefore acts as a bandpass filter.

When amplifiers and filters are used to condition signals before they are processed by a digital computer, they should always be passed through a low-pass filter which limits the frequencies to values lower than one-half the inverse of the sampling time τ of the computer. If this is not done, "aliasing" occurs, an effect whereby high-frequency signals above the sampling frequency are confused with the low-frequency signals of interest. The effect can be understood by analogy with a motion picture of a rotating wagon wheel. Often, the wheel seems to move *backward* when the film is viewed. This occurs when the wheel rotates more than the angular distance between the spokes between the "samples," which are the individual photographs on the movie film. Then, instead of seeming to move forward around the axle, the wheel seems to rotate the other way to the eye. The requirement to prevent or at least minimize the effect in our case is that the signal presented to the digital system must only have frequency components such that

$$\frac{\omega}{2\pi} < 1/2\left(\frac{1}{\tau}\right) \tag{11.13}$$

This is called the Nyquist criterion and is proved in standard texts on digital signal processing.

Filter networks using operational amplifiers play an important role in processing data prior to analysis by digital computers and particularly with regard to the Nyquist criterion. As an example, suppose we wish to construct a low-pass filter with a gain of 10 to condition a signal before it is converted to a digital number, and that the "digitizer" constructs the digital number every millisecond. If the sign of the signal presented to the digitizer is not crucial, an inverting amplifier can be used. For a low-pass filter, the capacitor C_1 in Fig. 11.15 is not needed and the appropriate circuit is as shown in Fig. 11.16a. The transfer function is $V_0/V_1 = -(Z_F/Z_I)$, which yields

$$H(s) = -\frac{R_2\omega_2}{R_1(s + \omega_2)} \tag{11.14}$$

FIGURE 11.16a.
An inverting amplifier with a single pole at $\omega_2 = (R_2C_2)^{-1}$.

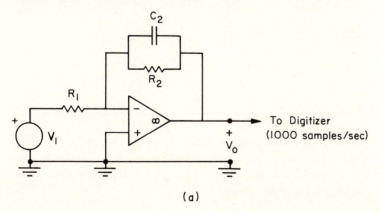

(a)

where $\omega_2 = 1/(R_2 C_2)$. For frequencies $\omega << \omega_2$,

$$H(s) = \frac{-R_2}{R_1}$$

This can be understood from the parallel network in the feedback loop since for low frequencies the capacitor is an open circuit and we have the same result found in Chapter 4 for the pure resistive inverting amplifier. $A_V = -(R_2/R_1)$. To yield a gain of 10, we could take $R_1 = 5$ kΩ and $R_2 = 50$ kΩ. Now with a sampling rate of 1000 sec^{-1} we require a pole at the radian frequency corresponding to

$$\frac{\omega_2}{2\pi} = \frac{1}{2}\left(\frac{1}{0.001}\right)$$

or $\omega_2 = 2\pi(500)$ rad/sec. Since $R_2 = 50$ kΩ,

$$C_2 = [(2\pi)(500)(50,000)]^{-1} = 0.0064 \ \mu F$$

and the required circuit is as shown in Fig. 11.16b. A one-pole low-pass filter may not be sufficient to avoid all aliasing. More poles can be generated with additional inverting Op-Amp circuits such as the one analyzed here or with more sophisticated Op-Amp filter designs which may be found in applications manuals.

FIGURE 11.16b.
An Op-Amp circuit with a one-pole low-pass filter at $f = 500$ Hz and a low frequency gain $A_V = -10$.

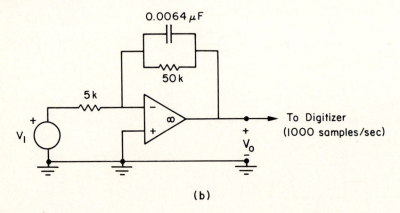

(b)

TEXT EXAMPLE 11.5 _____

Show that the circuit below is a two-pole low-pass filter, which also has a zero. Find values of $R_1, R_2, R_3, C_1,$ and C_2 which give the device a low-frequency gain of 10 and which yield a double pole at 1000 rad/sec. Sketch the asymptotic Bode plot for $A_{db}(\omega)$ and point out the effect of the zero.

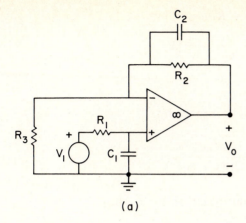

(a)

Solution: By the voltage divider law, the voltage at the positive input is

$$V^+ = \left(\frac{1/sC_1}{R_1 + (1/sC_1)} \right) V_1 = \left(\frac{1}{R_1 sC_1 + 1} \right) V_1 = V^-$$

where the last equality comes from the Op-Amp laws. This expression may be written as

$$V^- = \left(\frac{\omega_1}{s + \omega_1} \right) V_1$$

where $\omega_1 = (R_1 C_1)^{-1}$. The current through the feedback loop is, by Ohm's law,

$$\frac{V_0 - V^-}{Z_F} = \frac{V^-}{R_3}$$

Solving for V_0,

$$V_0 = \left(\frac{R_3 + Z_F}{R_3} \right) V^-$$

where

$$Z_F = \frac{R_2/sC_2}{R_2 + (1/sC_2)} = \frac{R_2}{R_2 sC_2 + 1}$$

This may also be written as

$$Z_F = \frac{R_2 \omega_2}{s + \omega_2}$$

where $\omega_2 = (R_2 C_2)^{-1}$. Substituting into the expression for V_0 gives

$$V_0 = \left(\frac{s + \left(\dfrac{(R_2 + R_3)\omega_2}{R_3} \right)}{s + \omega_2} \right) V^-$$

and finally the transfer function $H(s) = V_0/V_1$ is given by

$$H(s) = \frac{\omega_1(s + \omega')}{(s + \omega_1)(s + \omega_2)}$$

where $\omega' = [(R_2 + R_3)/R_3]\omega_2$. To check this result, let $s = 0$. Then,

$$H(s) = \frac{\omega'}{\omega_2} = \frac{R_2 + R_3}{R_3}$$

which is indeed the gain for the noninverting amplifier shown in Fig. 4.16. For a low-frequency gain of 10 we may choose $R_3 = 3$ kΩ and $R_2 = 27$ kΩ. Then, to have $\omega_2 = 1000$ rad/sec requires that $C_2 = 0.037\,\mu$F. To have $\omega_1 = 1000$ rad/sec we may choose $R_1 = 27$ kΩ as well which yields $C_1 = 0.037\,\mu$F. The zero at ω' will occur in this case at 10,000 rad/sec yielding a Bode plot, which looks as follows:

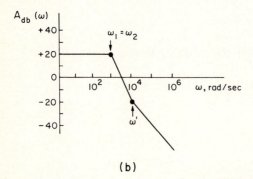

(b)

The zero at $\omega = \omega' = 10^4$ rad/sec reduces the slope of the Bode plot at high frequencies which may or may not be desirable for a given application.

As a final example of resistor–capacitor Op-Amp circuits, we analyze a circuit for which the poles are not pure real numbers. This cannot happen in R–C circuits without active devices, which must have pure real poles and zeros. However, including an Op-Amp changes the possibilities considerably.

TEXT EXAMPLE 11.6 _____

Find the transfer function for the circuit shown below and find its poles and zeros.

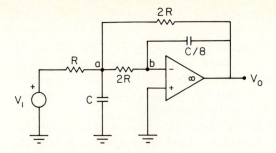

Applying KCL at node a,

$$\frac{V_1 - V_a}{R} = \frac{V_a - V_0}{2R} + \frac{V_a}{1/(sC)} + \frac{V_a}{2R}$$

where we have used the fact that $V^- = V^+ = 0$ V. This equation may also be written

$$2V_1 = (4 + 2RsC)V_a - V_0 \tag{1}$$

Applying KCL at node b and noting that no current flows into the Op-Amp at the negative input,

$$\frac{V_a}{2R} = \frac{-V_0}{8/sC}$$

or

$$V_a = -\left(\frac{RsC}{4}\right)V_0$$

Substitution into Eq. 1 above and solving for the gain V_0/V_1 yields

$$V_0/V_1 = -\frac{4\omega_1^2}{s^2 + 2\omega_1 s + 2\omega_1^2}$$

where $\omega_1 = 1/RC$. Using the binomial equation we find that this expression has poles at

$$s = -\omega_1 \pm j\omega_1$$

The circuit in Text Example 11.5 is described by a second-order equation in the parameter s with complex poles. We have previously touched upon cases of this type but have not seriously undertaken their study. In Chapter 12 we treat this type of circuit in detail including the implication of complex poles and/or zeros in the transfer function.

11.5 FEEDBACK REVISITED*

In Chapter 4 we pointed out that inserting an impedance between the output and the negative input of an Op-Amp feeds back and subtracts a fraction of the output from the input. In this section we review that material and extend it to illustrate several additional features of feedback. In particular we show that negative feedback increases the frequency range over which an Op-Amp works.

One fairly standard model for a feedback system (the same used in Chapter 4) is illustrated in Fig. 11.17. Here, we have generalized the diagram to include the possibility of either positive or negative feedback. To be precise, we require A and F to be positive numbers. For the minus sign (negative feedback),

$$V_0 = A(V_i - FV_0)$$

and

$$A_F = \frac{V_0}{V_i} = \frac{A}{1 + AF}$$

Using the positive sign (positive feedback),

$$A_F = \frac{V_0}{V_i} = \frac{A}{1 - AF}$$

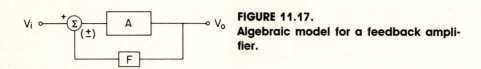

FIGURE 11.17.
Algebraic model for a feedback amplifier.

In Chapter 4 we showed that negative feedback can be used to improve control of a system. Here, we show that negative feedback has the added benefit of improving the bandwidth of an amplifier. Thus far, we have ignored the intrinsic frequency response of operational amplifiers in our design examples. Suppose the gain A of the amplifier has the following transfer function:

$$A(s) = \frac{A_0}{1 + \omega_0 s} = \frac{A_0 \omega_0}{s + \omega_0}$$

which has the graphical form shown in Fig. 11.18. The amplifier thus has a single pole at $s = \omega_0$. The frequency range over which the amplitude has a gain roughly equal to A_0 is called the intrinsic bandwidth of the amplifier, and we denote it by B_0. In this case $B_0 = 1/\tau_0 = \omega_0$.

* Sections marked with an asterisk are not required for the logical flow of the text.

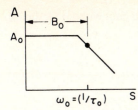

FIGURE 11.18.
Transfer function for an amplifier without feedback.

Now suppose the feedback loop has a constant value for F (that is, F is not a function of s) (see Fig. 11.19).

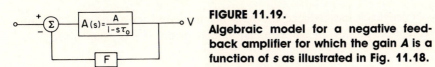

FIGURE 11.19.
Algebraic model for a negative feedback amplifier for which the gain A is a function of s as illustrated in Fig. 11.18.

Carrying out the same algebraic analysis used above,

$$V_0 = A(s)(V_i - FV_0)$$

$$V_0 = \left(\frac{A(s)}{1 + FA(s)} \right) V_i$$

$$V_0 \left(\frac{A_0}{(1 + \tau_0 s)\left(1 + \dfrac{A_0 F}{1 + \tau_0 s}\right)} \right) V_i = \left(\frac{A_0}{(1 + A_0 F + \tau_0 s)} \right) V_i$$

Thus, with feedback,

$$A_F(s) = \frac{A_0/(1 + A_0 F)}{1 + \left(\dfrac{\tau_0}{1 + A_0 F} \right) s}$$

Note that this expression has the same form as

$$A(s) = \frac{A_0}{1 + \tau_0 s}$$

except that

$$A_0 \rightarrow \frac{A_0}{1 + A_0 F} = A_F$$

$$\tau_0 \rightarrow \frac{\tau_0}{1 + A_0 F} = \tau_F$$

The gain at low frequencies ($s \ll (\tau_F)^{-1}$) is reduced from the gain without feedback (as usual for negative feedback):

$$A_F = \left(\frac{1}{1 + A_0 F}\right) A_0$$

whereas the new bandwidth $B_F = 1/\tau_F = \omega_F$ is *increased*:

$$B_F = \frac{1 + A_0 F}{\tau_0} = (1 + A_0 F) B_0$$

The use of feedback has reduced the gain but increased the frequency response. Note also that the product of the gain and the bandwidth is a constant since

$$A_F B_F = A_0 B_0$$

This "gain bandwidth product" is usually given in the specification sheet of an operational amplifier.

The plot shown in Fig. 11.20 can be used to illustrate how the reduction of gain improves the bandwidth of a system. The solid line is a plot of the intrinsic

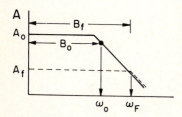

FIGURE 11.20.
Illustration of the way in which the reduction of gain via negative feedback can increase the frequency response of an amplifier.

frequency response of the Op-Amp, with τ_0 and B_0 as indicated. When we reduce the gain to A_F, the new τ_F and hence B_F are determined by the intersection of the dashed line and the solid curve. The gain is much less, but the bandwidth is much greater.

PRACTICE PROBLEMS
AND
ILLUSTRATIVE EXAMPLES

Problem 11.1

For the "high-pass filter" circuit illustrated,

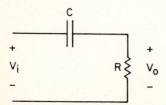

(a) Find the transfer function, $H(s) = V_0/V_i$.

$$\text{Ans.: } H(s) = RsC/(1 + RsC) = \\ s/(s + \omega_1), \text{ where } \omega_1 = 1/RC$$

(b) Suppose the square wave of Example 10.1 (see Table 10.1) is the input to this filter. For the case that $\omega_0 = (1/4)\omega_1$, find the amount in db by which the filter modifies the signal at each of the first three terms. (That is, find $A_{db}(\omega)$ at the frequency of the first three nonvanishing terms, $n = 1, 3, 5$.)

$$\text{Ans.: } -12.3, -4.4, -2.1 \text{ db}$$

Problem 11.2

Graph the asymptotic Bode plot for $A_{db}(\omega)$ of the circuit in Problem 11.1. On the same graph plot the actual value of the three points evaluated in part b above, that is, the points $(1/4)\omega_1, (3/4)\omega_1$, and $(5/4)\omega_1$. (Remember, the Bode plot is logarithmic in both axes.)

EXAMPLE 11.1

Find the transfer function $H(s) = V_2/V_1$ for the circuit shown below and evaluate $|H(j\omega)|$ at $\omega = 0$ and as $\omega \rightarrow \infty$. At what value of ω will $|H(j\omega)|$ be $1/\sqrt{2}$ times the maximum value (the -3db point)?

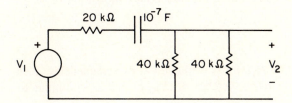

Solution: Let the symbol R stand for the 20k Ω resistor and the symbol C represent the capacitor. Then the two 40-kΩ resistors in parallel equal R and by the voltage divider law,

$$H(s) = \frac{R}{R + [1/(sC + R)]}$$

Substituting values for R and C,

$$H(s) = 0.5\left(\frac{s}{s + 250}\right)$$

At $\omega = 0, |H(j\omega)| = 0$. As $\omega \rightarrow \infty, |H(j\omega)| = 0.5$. The -3-db point should occur at the pole $\omega = 250$ rad/sec. To check this we find

$$H(j250) = 0.5\left(\frac{j250}{j250 + 250}\right) = \frac{0.5}{\sqrt{2}}\underline{/45°}$$

Hence,

$$|\mathbf{H}(j250)| = \frac{0.5}{\sqrt{2}}$$

which shows indeed that the magnitude of the transfer function is reduced by the factor $1/\sqrt{2}$ at $\omega = 250$ rad/sec from its value at very high frequencies.

Problem 11.3

For the circuit of Example 11.1 sketch the magnitude $A_{db}(\omega)$ and the phase $\theta(\omega)$ (in degrees) of $\mathbf{H}(j\omega)$ versus the radian frequency on a log scale.

Ans.:

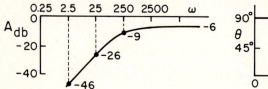 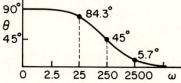

Problem 11.4

Construct the (straight line) asymptotic Bode amplitude plot for Problem 11.3. Compare it to the sketch above by also plotting the four points shown above ($\omega = 2.5, 25, 250, \infty$) on the same graph.

Problem 11.5

In Example 10.1 the transfer function $H_1(s)$ was evaluated at two frequencies. Pick at least four other frequencies and make sketches of $A_{db}(\omega)$ and $\theta(\omega)$ similar to those in Problem 11.3 above.

Problem 11.6

Draw the asymptotic Bode plot for $A_{db}(\omega)$ and $\theta(\omega)$ for the circuit in Problem 11.5 and compare it to the exact value found in Problem 11.5. To do this, plot the exact values for the six points on the same graph as the Bode plot.

Problem 11.7

The transfer function $H(s) = V_2/V_1 = 0.1(s/s + 10^3)$. For $v_1(t) = [10 + 30 \cos(5 \times 10^2 t) + 20 \cos(10^3 t)]$ mV, find $v_2(t)$.

Ans.:
$$v_2(t) = [0 + 1.34 \cos(5 \times 10^2 t + 63.4°) + 1.41 \cos(10^3 t + 45°)] \text{ mV}$$

Problem 11.8

Draw the asymptotic Bode plot for $A_{db}(\omega)$ and $\theta(\omega)$ for the transfer function in Problem 11.7 above. Evaluate these two quantities exactly at the three frequencies used above (0, 500, 1000) and plot the results on the graph of the Bode plot.

Problem 11.9

The transfer function

$$H(s) = 10^9 \left(\frac{s}{(s + 10^3)(s + 10^7)} \right)$$

was studied in Problem 10.13. Sketch the asymptotic Bode plot for $A_{db}(\omega)$ and compare it to the answer for Problem 10.13. To do this, divide values for $|\hat{V}_2(j\omega)|$ in Problem 10.13 by $|\hat{V}_1(j\omega)|$ in the same problem and plot the db levels at the appropriate frequency in the Bode plot found here. Draw a smooth curve through these points which asymptotically approach the Bode plot.

Problem 11.10

Make asymptotic Bode plots of $A_{db}(\omega)$ and $\theta(\omega)$ for the transfer function in Problem 10.10. On the former plot the four points evaluated in Problem 10.12 (e.g., $\omega = 0$, $\omega = 1000$, $\omega = 10^4$, and $\omega = \infty$).

Problem 11.11

Repeat Problem 11.10 for the transfer function in Problem 10.21, that is, make the Bode plot for $A_{db}(\omega)$ and and compare it to the exact values of $\mathbf{H}(j\omega)$ evaluated at $\omega = 0, 1000$, and 2000 rad/sec. Plot the exact values on the same graphs as the Bode plot.

Problem 11.12

The transfer function $H(s) = V_2/V_1 = s/(s + 10^4)$. If $v_1 = 10 + 30 \cos(5 \times 10^3 t) + 20 \cos(10^4 t)$, find v_2.

Ans.:

$$v_2 = 0 + 13.4 \cos(5 \times 10^3 t + 63.4°) + 14.14 \cos(10^4 t + 45°)$$

Problem 11.13

Make sketches of the asymptotic Bode plots $A_{db}(\omega)$ and $\theta(\omega)$ for the transfer function in Problem 11.12 above and compare those to the answers given above at the three frequencies studied.

Problem 11.14

A certain transfer function is

$$H(s) = 100 \left[\frac{s(s + 10^3)}{(s + 10)(s + 10^5)} \right]$$

For the corresponding frequency response, complete the following:

(a) The phase angle at $\omega \to 0$ will be

_____.   Ans.:90°

(b) The phase angle at $\omega \to \infty$ will be

_____.   Ans.:0°

(c) The magnitude at $\omega \to 0$ will be

_____.   Ans.:0

(d) The magnitude at $\omega \to \infty$ will be

_____.   Ans.:100

(e) The magnitude at $\omega = 10^3$ will be

_____.   Ans.: $\sqrt{2}$

(f) The magnitude at $\omega = 10$ will be

_____.   Ans.:$1/\sqrt{2}$

(g) The phase angle at $\omega = 10$ will be

_____.   Ans.:45°

Problem 11.15

Make the asymptotic Bode plot of $A_{db}(\omega)$ for the transfer function in Problem 11.14 above. Compare the plot to the exact magnitudes evaluated at $\omega = 0$, 10, and 10^3 rad/sec and as $\omega \to \infty$ which were given in the answers.

Problem 11.16

The transfer function of a television amplifier is

$$H(s) = 10^9 \left[\frac{s}{(s + 100)(s + 10^7)} \right]$$

Sketch the magnitude of $\mathbf{H}(j\omega)$ as a function of ω using a logarithmic scale for $|\mathbf{H}|$ and a log scale for ω. What is the frequency range (bandwidth) for which $A_{db}(\omega)$ does not differ from its maximum value by more than 3 db? Sketch the amplitude and phase angle using a linear scale versus log ω.

Ans.: Bandwidth $= 10^2 \to 10^7$ rad/sec

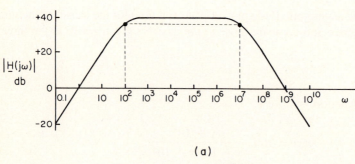

(a)

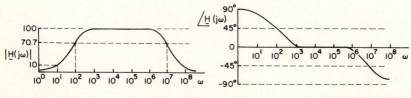

(b)

Problem 11.17

The transfer function of a certain system is

$$H(s) = 0.01 \left[\frac{s(s + 10^6)^2}{(s + 10)(s + 10^4)^2} \right]$$

Sketch the asymptotic Bode plot for the amplitude of $\mathbf{H}(j\omega), A_{db}(\omega)$.

Ans.:

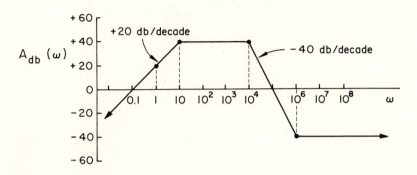

Problem 11.18

For the asymptotic Bode plot shown below find the corresponding transfer function $H(s)$. (Assume real poles and zeros and that $\mathbf{K}$ is real.)

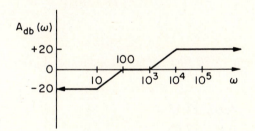

Ans.: $H(s) = 10 \left[\dfrac{(s + 10)(s + 10^3)}{(s + 100)(s + 10^4)} \right]$

Problem 11.19

For the asymptotic Bode plot shown below write the corresponding transfer function $H(s)$, assuming real poles and zeros and that $\mathbf{K}$ is real.

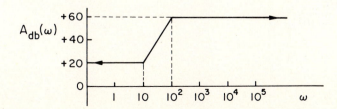

Problem 11.20

The asymptotic Bode plots below show $A_{db}(\omega)$ of ω (on a logarithmic scale). Write the corresponding transfer functions $H(s)$ and determine the constants assuming real poles and zeros and that **K** is real.

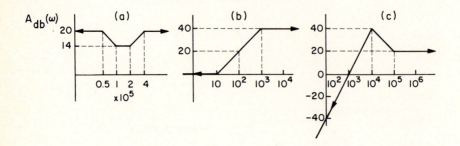

Ans.: (a) $H(s)$

$$= 10\left[\frac{(s + 10^5)(s + 2 \times 10^5)}{(s + 5 \times 10^4)(s + 4 \times 10^5)}\right]$$

$$(b) \ H(s) = 100\left[\frac{s + 10}{s + 10^3}\right]$$

$$(c) \ H(s) = 10\left[\frac{s^2(s + 10^5)}{(s + 10^4)^3}\right]$$

Problem 11.21

For the transfer functions found in Problem 11.20 find the actual values of $A_{db}(\omega)$ in decibels for the following cases:

(a) At $\omega = 4 \times 10^5$ rad/sec in (a).   Ans.: 18.15 db

(b) At $\omega = 10^3$ rad/sec in (b).   Ans.: 37 db

(c) At $\omega = 10^4$ rad/sec in (c).   Ans.: 31 db

Problem 11.22

Sketch the asymptotic Bode plots for the amplitude of the following transfer functions:

$$(a) \ H(s) = 40\left[\frac{s(s + 10^4)}{(s + 10^3)(s + 10^5)}\right]$$

$$(b) \ H(s) = 0.1\left[\frac{s + 10^5}{s + 10^2}\right]$$

Ans.:

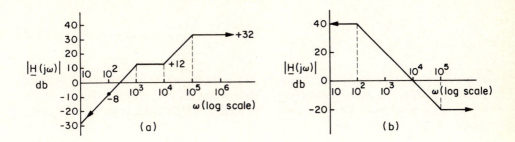

(a)

(b)

Problem 11.23

Find the transfer function $H(s)$ corresponding to the Bode plot for $A_{db}(\omega)$ plotted below assuming real poles and zeros and that **K** is real.

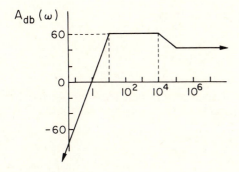

EXAMPLE 11.2

Design a one-pole low-pass filter using Op-Amps with a gain of 100 and a $-3db$ point at 10 kHz. The minimum resistor value should be 10 kΩ.

Solution: There are a number of straightforward solutions. We provide one solution type here and leave another for the problem below. If a change of sign is acceptable we can use the circuit of Fig. 11.16a.

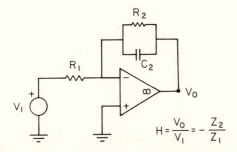

$$H = \frac{V_0}{V_1} = -\frac{Z_2}{Z_1}$$

Substituting for Z_2 and Z_1 yields Eq. 11.14,

$$H(s) = -\frac{R_2}{R_1}\left(\frac{\omega_2}{s + \omega_2}\right)$$

where $\omega_2 = (R_2 C_2)^{-1}$. At high frequencies C_2 shorts out R_2 so the circuit acts as a low-pass filter. At low frequencies the gain is just $\mathbf{A}_v = -(R_2/R_1)$. For $R_1 = 10$ kΩ we must have $R_2 = 100R_1 = 1$ MΩ. To have a -3-db point at 10 kHz we must solve

$$\omega_2 = \frac{1}{R_2 C_2} = 2\pi(10^4) \text{ Hz}$$

For $R_2 = 1$ MΩ,

$$C_2 = [2\pi(10^4)(10^6)]^{-1} \simeq 16 \text{ pF}$$

Problem 11.24

Repeat Example 11.2 using two op-amps in such a way that the sign of $H(s)$ is positive. The minimum resistor should be 10 kΩ.

Problem 11.25

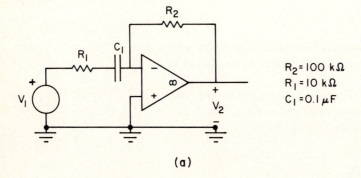

$R_2 = 100$ kΩ
$R_1 = 10$ kΩ
$C_1 = 0.1$ μF

(a)

For the ideal Op-Amp circuit shown above find the transfer function $H(s) = V_2/V_1$ and sketch the asymptotic Bode plot.

Ans.:

$$H(s) = \frac{R_2 s C_1}{R_1 s C_1 + 1} = -\frac{R_2}{R_1}\left(\frac{s}{s + \omega_1}\right), \text{ where } \omega_1 = 1/(R_1 C_1)$$

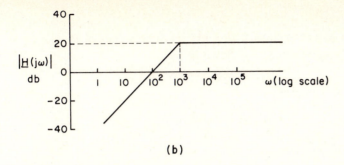

(b)

Problem 11.26

Design a one-pole high-pass filter using an inverting amplifier which has a -3-db point at 2×10^4 rad/sec and a high-frequency gain ($\omega \gg 2 \times 10^4$ rad/sec) of 5. The minimum value resistor should be 10 kΩ.

Problem 11.27

A signal is to be digitized at a rate of 5000 samples/sec. Design a one-pole low-pass filter using an inverting amplifier circuit with unity gain at low frequencies and an asymptotic response that equals -10 db at the Nyquist frequency (2500 Hz). The minimum value resistor should be 10 kΩ.

Problem 11.28

For the ideal noninverting Op-Amp circuit shown below find the transfer function $H(s) = V_0 / V_1$ and sketch the asymptotic Bode plot for $|\mathbf{H}(j\omega)|$.

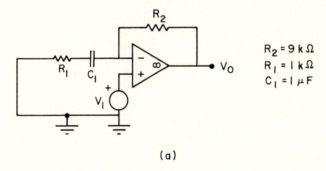

$R_2 = 9 \text{ k}\Omega$
$R_1 = 1 \text{ k}\Omega$
$C_1 = 1 \mu\text{F}$

(a)

$$\text{Ans.:}\ \ H(s) = 10\left(\frac{s + 100}{s + 10^3}\right)$$

The asymptotic Bode plot is shown below.

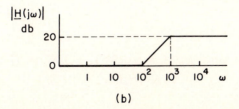

(b)

Problem 11.29

Using the circuit of Problem 11.28 with $R_1 = 1$ kΩ, find the values of R_2 and C_1 that will result in the asymptotic Bode plot shown below.

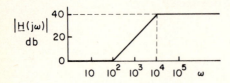

Ans.: $R_2 = 99$ kΩ, $C_1 = 0.1 \mu$F

Problem 11.30

Design a circuit, using two Op-Amps, which has the Bode plot given in Problem 11.16 and is such that $H(s)$ is positive. Since the midband gain is 100, "split" the gain so that each Op-Amp has a gain of 10 in the midband frequency range. The minimum resistor value should be 1 kΩ.

Problem 11.31

Design a circuit using one Op-Amp which has the same Bode plot as shown in Fig. 11.10. The minimum resistor value should be 10 kΩ.

Problem 11.32

(a) and (b) Find the transfer functions for the following two networks and note that $H_b(s)$ is *not* the square of $H_a(s)$.

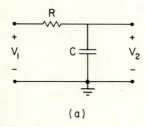

(a)

Ans.: $H_a(s) = 1/(1 + sRC)$

(b)

Ans.: $H_b(s) = 1/[(sRC)^2 + 3sRC + 1]$
$\neq [H_a(s)]^2$

(c) Show that the transfer function for the following network *is* equal to $[H_a(s)]^2$.

$$\text{Ans.: } H_c(s) = [(sRC)^2 + 2sRC + 1]^{-1}$$
$$= H_a^2(s)$$

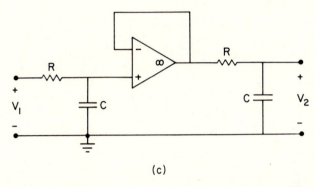

(c)

(Insertion of the "voltage follower" circuit has isolated the two *R–C* networks, and their transfer functions then just multiply. Such an Op-Amp circuit is often termed a buffer since it effectively isolates circuits on either side of the voltage follower.)

Problem 11.33

Suppose the output of circuit (a) in Text Example 11.4 is supplied as the input V_1 of circuit (b). Write down the transfer function for the composite circuit.

Problem 11.34

Let $R_1 = 10 \text{ k}\Omega$ in Problem 11.25. Find R_2 and C_1 such that a highpass filter results with -20db gain at 100 rad/sec and such that the maximum gain of the system is 40 db.

Problem 11.35

Determine the transfer functions $H(s) = V_0/V_i$ for the following network:

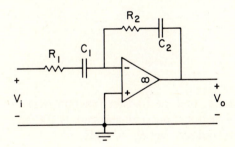

$$\text{Ans.: } H(s) = -\frac{R_2}{R_1}\left(\frac{s + \omega_2}{s + \omega_1}\right),$$
$$\text{where } \omega_1 = 1/(R_1C_1);$$
$$\omega_2 = 1/(R_2C_2)$$

Problem 11.36

Sketch the asymptotic Bode plot for $A_{db}(\omega)$ in the circuit of Problem 11.35 if $\omega_1 > \omega_2$. If $R_1 = 10\ k\Omega$, find R_2, C_1, and C_2 if the high-frequency gain is to be 20 db, the low-frequency gain is to be 0 db, and the pole is to be at 10^4 rad/sec. Where is the zero in this case?

Problem 11.37

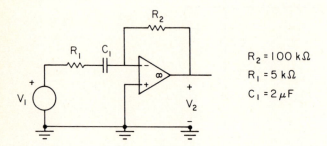

$$R_2 = 100\ k\Omega$$
$$R_1 = 5\ k\Omega$$
$$C_1 = 2\ \mu F$$

For the ideal Op-Amp circuit above find the transfer function $H(s) = V_2/V_1$ and sketch the asymptotic Bode plot. Use the exact transfer function to evaluate $v_2(t)$ if $v_1(t) = 10\cos(100t)$.

Problem 11.38

Using the circuit diagram in Problem 11.37, choose new values of the R's and C's such that the high-frequency gain is -10 and such that the pole occurs at the frequency $f = 15.92$ Hz. Use no resistors less than 10 kΩ.

Problem 11.39

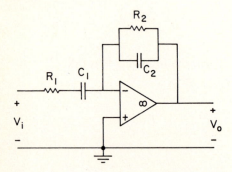

For the ideal Op-Amp circuit shown above, (a) find the transfer function $H(s) = V_0/V_i$. (b) Choose values of the R's and C's such that the midband gain is 100 and the -3-db points occur at 10 rad/sec and 10^5 rad/sec. Set $R_1 = 1\ k\Omega$.

Problem 11.40

For the ideal Op-Amp circuit shown below find the transfer function $H(s) = V_0/V_s$ and sketch the Bode plot for $A_{db}(\omega)$ for the specific values given. The circuit is used to amplify a signal before it is digitized at a sample interval of $\tau = 0.01$ msec.

What is the absolute value of the amplifier gain in the midband and at the Nyquist frequency?

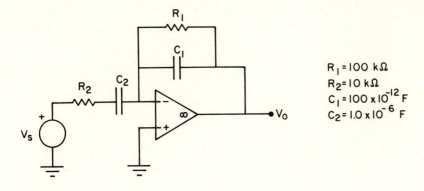

$R_1 = 100 \text{ k}\Omega$

$R_2 = 10 \text{ k}\Omega$

$C_1 = 100 \times 10^{-12} \text{ F}$

$C_2 = 1.0 \times 10^{-6} \text{ F}$

Chapter 12
Second-Order Circuits

Up to this point we have dealt almost entirely with first-order circuits, that is, circuits whose natural response could be described with a single value for s. These, in turn, were circuits characterized by first-order differential equations. The exceptions occurred in Chapter 7, where we showed that the polynomial equation for s in a series $R–L–C$ circuit was of second order, and in Chapter 11, where we found that the circuit of Fig. 11.13c, certain Op-Amp $R–C$ circuits and various transfer functions in the problem set were characterized by higher-order polynomials in s. The detailed study of such systems is the subject of this chapter. Under certain conditions second-order systems display a natural response which is analogous to a pendulum that oscillates in time with a characteristic frequency. If friction is taken into account the oscillation dies out with time. In electrical engineering the type of system is termed underdamped. We study this type of circuit as well as overdamped and critically damped systems. Both the natural and forced responses are investigated and we show that when an underdamped system is forced at its characteristic or resonant frequency, very large amplitude signals can be generated. The use of second-order circuits in filter design is also discussed briefly. Peak value phasors are used.

12.1 CIRCUITS WITH MULTIPLE ENERGY STORAGE ELEMENTS

Circuits with more than one energy storage element usually result in second-order equations. Exceptions occur when the multiple storage elements can be combined into equivalent single elements. Such a case was illustrated in Problem 6.9. An example where such a combination is not possible is shown in Fig. 12.1. To find the natural response of this circuit for $t \geq 0$, the source is suppressed and we set

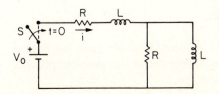

FIGURE 12.1.
Second-order resistor–inductor circuit.

the series impedance function equal to zero:

$$Z_s(s) = R + sL + \frac{RsL}{R + sL} = 0$$

In polynomial form this is

$$Z_s(s) = L\left[\frac{s^2 + (3R/L)s + (R^2/L^2)}{s + R/L}\right] = 0$$

The solutions occur when the numerator vanishes and, using the quadradic formula,

$$s = \frac{1}{2}\left[\frac{-3R}{L} \pm \sqrt{9R^2/L^2 - 4R^2/L^2}\right]$$

Note for this case that no matter what values we choose for R and L, the number inside the square root sign must be positive. This is a general result in the sense that for elementary circuits with resistors and multiple L's *or* multiple C's, no imaginary values of s arise. (As noted in the previous chapter, this result does not necessarily pertain if Op-Amps are used in a circuit, which is why we inserted "elementary" in the previous sentence. The Op-Amp is an active element as opposed to the passive elements in Fig. 12.1.) Carrying out the square root operation,

$$s_\pm = \left(\frac{R}{L}\right)\left(\frac{-3 \pm \sqrt{5}}{2}\right)$$

The natural response current has the functional form

$$i_n(t) = I_1 e^{-2.6Rt/L} + I_2 e^{-0.38Rt/L}$$

with the two unknown coefficients I_1 and I_2 yet to be determined.

TEXT EXAMPLE 12.1

Find the form of the natural response voltage $v_n(t)$ in the circuit below after the switch is opened.

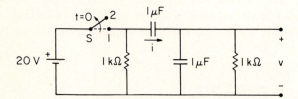

Solution: After the switch is thrown there is no source, so we seek the natural response characteristics of the circuit. We may therefore set the series impedance function equal to zero. Evaluating $Z_s(s)$ we have

$$Z_s(s) = R + \frac{1}{sC} + \frac{R/sC}{R + (1/sC)}$$

where we use the fact that the two resistors and the two capacitors are the same. Simplifying,

$$Z_s(s) = (R)\frac{s^2 + (3/RC)s + (1/R^2C^2)}{s^2 + (s/RC)}$$

This vanishes when the numerator equals zero, which occurs for

$$s_\pm = \frac{-(3/RC) \pm \sqrt{(9/R^2C^2) - (4/R^2C^2)}}{2}$$

and the two solutions are

$$s_\pm = \frac{-3 \pm \sqrt{5}}{2}\left(\frac{1}{RC}\right)$$

or $s_- = -2.6/RC$, $s_+ = -0.38/RC$. Substituting for R and C,

$$v_n(t) = V_1 e^{-2.6 \times 10^3 t} + V_2 e^{-3.8 \times 10^2 t}$$

The two coefficients remain to be determined (see Problem 12.3).

Unlike the first-order case there is no simple procedure for determining the coefficients, and each problem requires some ingenuity. The steps are the same as in previous natural response problems in that the initial conditions are used to determine the coefficients, but two pieces of information are needed. In complete response problems you must add the natural and the forced response and then apply two boundary conditions. These usually involve the value of the function at $t = 0$ and the value of its derivative at that same time. The complete response example given next illustrates the method.

TEXT EXAMPLE 12.2 _____

In the circuit below, the switch is moved from position 1 to position 2 at $t = 0$. Find the complete response current $i(t)$.

(a)

(The switch S' in this circuit must be a "make before break" switch so that the line in series with the inductor is never an open circuit.)

Solution: We use the transfer function $H = I/V_2$. (Note that H is the admittance function $Y = 1/Z_s$.) If the source is suppressed so that $I = (H)(V_2) = (H)(0)$, then H must be infinite to yield a nonvanishing natural current. The series *impedance* function is

$$Z_s(s) = 4s + 16 + \frac{16s}{2s + 8}$$

Since $H = 1/Z_s$ we have

$$H(s) = \frac{2s + 8}{8s^2 + 80s + 128}$$

$H(s)$ is infinite when the denominator is equal to zero, which corresponds to the polynomial equation

$$s^2 + 10s + 16 = 0$$

with the two solutions

$$s_\pm = -5 \pm 3$$

The natural response solution is then

$$i_n(t) = I_1 e^{-8t} + I_2 e^{-2t}$$

Note that setting the series impedance function equal to zero would yield the same values for s. The forced response is found from

$$i_F = \mathbf{H}(0)V_2$$

$$i_F = \left(\frac{8}{128}\right)(8) = 0.5 \text{ A}$$

The form of the solution is then

$$i(t) = I_1 e^{-8t} + I_2 e^{-2t} + 0.5$$

One initial condition is the standard one from the continuity equation for inductors. The initial current through the 4-H inductor is

$$i(0^-) = \frac{16}{16} = 1 \text{ A}$$

which must also equal $i(0^+)$. One equation is therefore

$$1 = I_1 + I_2 + 0.5$$

Another equation can be found from the derivative of $i(t)$. At $t = 0^+$ the current through both inductors is 1A and the current though the 8Ω resistor in parallel with the 2H inductor is zero. The KVL equation at $t = 0^+$ is $8 = (16)(1) + 4di/dt; di/dt$ at 0^+ is therefore equal to -2. The derivative of $i(t)$ is given by

$$di(t)/dt = -8I_1 e^{-8t} - 2I_2 e^{-2t}$$

Evaluating this function at $t = 0^+$ and setting it equal to -2 yields

$$-2 = -8I_1 - 2I_2$$

which is the necessary second equation. Solving the set

$$I_1 + I_2 = 0.5$$

$$8I_1 + 2I_2 = +2$$

yields $I_1 = 1/6$ and $I_2 = 1/3$ and finally

$$i(t) = (1/6)e^{-8t} + (1/3)e^{-2t} + 0.5$$

A plot of this function is shown below. Note that the initial decay follows an e^{-8t} form, but after a short time the current decrease slows down as the e^{-2t} form becomes more important.

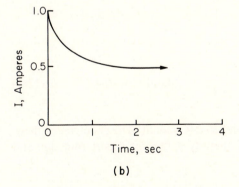

(b)

If the sources V_1 and V_2 in circuits such as the one analyzed above are sinusoidal, the analysis procedure is the same but the algebra is more complicated. For example, the forced response must be determined from an expression of the form

$$\hat{I}_F = \mathbf{H}(j\omega)\hat{V}_2$$

where ω is the frequency of the source $\hat{V}_2$. A problem of this type is worked out as an example in the problem set. The examples thus far involve multiple L's or multiple C's but not both. More interesting circuit responses occur when inductors and capacitors are both present in a circuit.

12.2 NATURAL RESPONSE OF CIRCUITS WITH INDUCTORS AND CAPACITORS

We have already seen in Chapter 7 that the series R–L–C circuit can lead to a natural response characterized by "fully" complex values for s, that is, for values of s with both real and imaginary parts. Another example is presented here and the discussion expanded. Because of the nature of the solutions we find for s, that parameter is sometimes referred to as the "complex frequency."

Consider the parallel R–L–C circuit shown in Fig. 12.2. The natural response current through the inductor for $t \geq 0$ may be determined as follows. The series impedance "seen" by the current i_L is the inductor in series with a parallel R–C combination. The series impedance function is

$$Z_s(s) = sL + \frac{R/sC}{R + (1/sC)}$$

$$Z_s(s) = \frac{RLCs^2 + sL + R}{RsC + 1}$$

Setting the numerator equal to zero yields the polynomial equation

$$s^2 + \left(\frac{1}{RC}\right) s + \frac{1}{LC} = 0$$

which has the two solutions

$$s_{\pm} = -\left(\frac{1}{2RC}\right) \pm \sqrt{\left(\frac{1}{2RC}\right)^2 - \frac{1}{LC}} \qquad \text{(parallel } R\text{–}L\text{–}C) \qquad (12.1)$$

These are the natural frequencies of a parallel R–L–C circuit. The character of these solutions is very much dependent upon the quantity within the square root sign and

FIGURE 12.2.
Parallel R–L–C circuit.

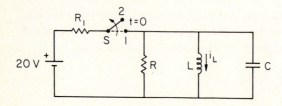

we explore the three different possible situations. For reference, in Eq. 12.2 we also give the two solutions for s in a series $R–L–C$ circuit as derived earlier in Section 7.2.

$$s_{\pm} = -\left(\frac{R}{2L}\right) \pm \sqrt{\left(\frac{R}{2L}\right)^2 - \frac{1}{LC}} \qquad \text{(series } R–L–C\text{)} \qquad (12.2)$$

Underdamped Case

When $4/LC > (1/RC)^2$ in the parallel $R–L–C$ circuit, the quantity inside the square root sign is negative and the solutions are of the form

$$s_{\pm} = -\alpha \pm j\omega_0$$

where

$$\alpha = \frac{1}{2RC}$$

and

$$\omega_0 = \sqrt{\frac{1}{LC} - \left(\frac{1}{2RC}\right)^2}$$

For the series $R–L–C$ circuit, Eq. 12.2, the underdamped case occurs when $1/LC > (R/2L)^2$ and in that case

$$\alpha = -\frac{R}{2L}$$

and

$$\omega_0 = \sqrt{\frac{1}{LC} - \left(\frac{R}{2L}\right)^2}$$

As derived in Chapter 7 and in Section 12.3 below, the natural response current for either underdamped case may be written in the form

$$i_n(t) = I_0 e^{-\alpha t} \cos(\omega_0 t + \phi) \qquad (12.3)$$

That is, if we combine the two solutions of the form $e^{-\alpha t} e^{j\omega_0 t}$ and $e^{-\alpha t} e^{-j\omega_0 t}$ we can write it as given in Eq. 12.3. This time function is a damped oscillator (see Fig. 7.5) and is analogous to the motion of a pendulum which oscillates at some frequency ω_0 with an amplitude which decays in time. In the latter case it is the friction in the pivot that controls the decay rate. In the electrical case the energy lost in the resistance controls the decay rate. The frequency ω_0 is often termed the resonant frequency. It is the frequency at which a second-order underdamped system naturally oscillates. As we shall see, if we drive or force the system at its resonant frequency the response may be very large, larger even than the source that drives the system.

The circuit is termed underdamped since it may require many oscillations before the natural response dies away. If we wish to build an oscillator this is a desirable situation since a good oscillator is one where the amount of energy dissipated in one cycle is much less than the total energy stored in the system. Good mechanical watches have low-friction jeweled settings for this very reason. The quality, or Q, of an oscillator is measured by the ratio

$$Q = 2\pi \frac{\text{maximum energy stored in a cycle}}{\text{energy dissipated per cycle}} = \frac{2\pi E_s}{E_D} \qquad (12.4)$$

when the circuit operates at its resonant frequency. First we determine E_s, the stored energy. We may perform the calculation of Q at any time t and for convenience we choose $t = 0$ when $e^{-\alpha t} = 1$. At the resonant frequency the inductor current at $t = 0$ has the mathematical form

$$i(t) = I_0 \cos(\omega_0 t + \phi)$$

The maximum energy stored in the inductor is therefore $E_s = (\frac{1}{2})Li^2_{max} = (\frac{1}{2})LI_0^2$ (see Chapter 5). The energy dissipated in a single cycle, E_D, equals the average dissipated *power* $(\frac{1}{2})I_0^2 R$ multiplied times the period of the oscillation. (Remember, energy equals power multiplied by time.) The peak value voltage phasor for the parallel circuit in Fig. 12.2 is given by $\hat{V} = (\hat{I})(\mathbf{Z}_L) = (I_0 \underline{/\phi})(j\omega_0 L) = \omega_0 L I_0 \underline{/\phi + 90°}$. The phasor current through the resistor is then

$$\hat{I}_R = \left(\frac{\omega_0 L}{R}\right) I_0 \underline{/\phi + 90°}$$

and the average power dissipated is

$$P_A = \frac{\omega_0^2 L^2 I_0^2}{2R}$$

where the factor of two comes from the use of peak value phasors. The energy dissipated in one cycle is this average power P_A times the total time per cycle, $\tau = 2\pi/\omega_0$, or

$$E_D = P_A \tau = \frac{\pi \omega_0 L^2 I_0^2}{R}$$

Finally, using Eq. 12.4,

$$Q = \frac{R}{\omega_0 L} \qquad (12.5a)$$

In the problem set the student is asked to show that for the series R–L–C circuit

$$Q = \frac{\omega_0 L}{R} \qquad 12.5b)$$

For the parallel R–L–C a high Q occurs when R is large since then very little current flows through the resistor and the energy dissipation is small. An open circuit ($R = \infty$) would yield the highest Q. In a real circuit of this parallel R–L–C type, a so-called tank circuit, if the resistor is replaced with an open circuit, $R = \infty$, the Q is not infinite but is limited by the finite resistance in the wires of the inductor. An example of this type is discussed in the problem set. For the series R–L–C circuit a high Q results when R is small.

For high-Q circuits both equations for the resonant frequency ω_0 are very nearly equal to $(1/LC)^{1/2}$ for the parallel and series R–L–C circuits. We then often ignore the small correction term and set $\omega_0 = (1/LC)^{1/2}$.

TEXT EXAMPLE 12.3 ———————————————————————————

For the circuit in Fig. 12.2, let $R = 1000\ \Omega$, $L = 4$ mH and $C = 0.1\mu$F. Find the resonant frequency and the Q of the circuit. Find ω_0 and Q if the resistor is replace by an open circuit and the L and C are ideal elements.

Solution: For this circuit

$$\omega_0 = \sqrt{\frac{1}{LC} - \left(\frac{1}{2RC}\right)^2}$$

which upon substitution yields an oscillation frequency $\omega_0 = 4.97 \times 10^4$ rad/sec (7.9 kHz). From Eq. 12.5a,

$$Q = \frac{R}{\omega_0 L} = 5$$

For ideal elements and no parallel resistor, $\omega_0 = 5 \times 10^4$ rad/sec and $Q = \infty$.

———————————————————————————————————————

The natural response solution given in Eq. 12.3 has two unknown coefficients I_o and ϕ. As noted above, this is typical of second-order equations. The two constants are determined from two initial conditions as illustrated in the example below.

TEXT EXAMPLE 12.4 ———————————————————————————

Referring to Text Example 12.3 and Fig. 12.2, suppose the switch is opened at $t = 0$. Find the natural response current $i_{Ln}(t)$ for $t \geq 0$ if resistor $R_1 = 1$ kΩ.

Solution: The inductor current has the general form given by Eq. 12.3:

$$i_{Ln}(t) = [I_0 \cos(\omega_0 t + \phi)]e^{-\alpha t}$$

where $\alpha = 1/2RC = 5 \times 10^3$ sec^{-1}. At $t = 0^-$ the inductor current is

$$i_L(0^-) = \frac{20}{10^3} = 20 \text{ mA}$$

By the continuity principle, $i_L(0^+) = 20$ mA. This gives one of the required boundary conditions. For an inductor we also have

$$v_L = \frac{L\,di}{dt}$$

so if we can determine $v_L(0^+)$, we will know the derivative of the current. The capacitor voltage at $t = 0^-$ is zero volts due to the short circuiting of the dc current by the inductor. Since $v_C(0^+) = v_C(0^-)$ and since the capacitor is in parallel with the inductor, $v_L(0^+) = 0$ as well. At $t = 0^+$ the term $e^{-\alpha t} = e^0 = 1$. Our two equations are then

$$20 = I_0 \cos \phi$$

$$0 = -LI_0\omega \sin \phi$$

These two equations are satisfied if $I_0 = 20$ and $\phi = 0$, so our solution is

$$i_L(t) = 20 \cos(\omega_0 t)e^{-5 \times 10^3 t} \text{ mA}$$

Note that R_1 only entered into the initial condition calculation, not the natural response calculation.

Overdamped Case

When the quantities inside the square root sign in Eqs. 12.1 and 12.2 are positive, that is, when

$$\left(\frac{1}{2RC}\right)^2 > \frac{1}{LC} \qquad \text{(parallel } R\text{–}L\text{–}C\text{)}$$

or

$$\left(\frac{R}{2L}\right)^2 > \frac{1}{LC} \qquad \text{(series } R\text{–}L\text{–}C\text{)}$$

then both roots for the natural response parameter s are negative real numbers:

$$s_\pm = -\left(\frac{1}{2RC}\right) \pm \sqrt{\left(\frac{1}{2RC}\right)^2 - \frac{1}{LC}} \quad \text{(parallel } R\text{--}L\text{--}C\text{)}$$

$$s_\pm = -\left(\frac{R}{2L}\right) \pm \sqrt{\left(\frac{R}{2L}\right)^2 - \frac{1}{LC}} \quad \text{(series } R\text{--}L\text{--}C\text{)}$$

This case is identical to the multiple real root examples studied above in Section 12.1.

Critically Damped

A special case occurs when the square root vanishes. Then the solution form $i_n(t) = I_1 e^{s_1 t} + I_2 e^{s_2 t}$ only leads to one function of the form $i_n(t) = I_0 e^{st}$ since $s_1 = s_2$. Since two solutions are needed we must return to study the original differential equations. For reference we derive the differential equation for the series R--L--C circuit in Fig. 12.3. KVL around a counterclockwise loop yields

$$\left(\frac{1}{C}\right)\int_0^t i(t')dt' + v_c(0) + \frac{L\,di}{dt} + iR = 0$$

Taking the derivative of this equation and rearranging terms, we have

$$\frac{di^2}{dt^2} + \left(\frac{R}{L}\right)\left(\frac{di}{dt}\right) + \frac{i}{LC} = 0 \tag{12.6a}$$

From Eq. 12.2 we know that if $R^2 = 4LC$ this system is critically damped. Since only one value of s is determined by our formalism, we need a different form for the solution than used to date. In the problem set the student is asked to show by direct substitution that a function of the form shown in Eq. 12.6b below does satisfy Eq. 12.6a, provided that $s = -R/2L$:

$$i(t) = (I_1 + I_2 t)e^{st} \tag{12.6b}$$

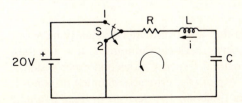

FIGURE 12.3.
A series R--L--C circuit for studying critically damped natural response.

We use this solution form in the example below to illustrate the solution to a critically damped circuit.

TEXT EXAMPLE 12.5 ─────────────────────────────

Find the natural response current associated with the circuit in Fig. 12.3 for $t \geq 0$. Evaluate the expression if $R = 20\Omega$, $L = 0.1$ mH and $C = 1\mu F$.

Solution: We assume a function of the form

$$i(t) = (I_1 + I_2 t)e^{st}$$

with $s = -R/2L$. At $t = 0^-$ the voltage across the capacitor is 20 V and the current through the inductor is zero. At $t = 0$ neither of these values changes, so one equation is $i(0^+) = 0$. Substitution above yields $I_1 = 0$. The derivative of the current may be deduced from the voltage across the inductor at $t = 0$, which is 20 V, since $i(0^+)R = 0$:

$$v_L = \frac{L\,di}{dt} = 20 \text{ V}$$

Since $I_1 = 0$ when the derivative of $i(t)$ is evaluated, it reduces to

$$\frac{di}{dt} = (stI_2 + I_2)e^{st}$$

At $t = 0^+$ this just equals I_2, and we have

$$I_2 = \frac{20}{L}$$

and the solution is thus

$$i(t) = \left(\frac{20t}{L}\right) e^{-Rt/2L}$$

For the particular case $R = 20\ \Omega$, $L = 0.1$ mH and $C = 1\ \mu F$,

$$i(t) = 2 \times 10^5 t e^{-10^5 t} \text{ A}$$

───

Changing the variables in the system so that it is overdamped actually increases the time required to dissipate the stored energy. Critical damping is thus the most desirable condition for rapid decay of the natural response of second-order systems, whether they be electrical or mechanical. A plot of a critically damped response is shown in Fig. 12.4. This graph shows the function $i(t)$ derived in Text Example 12.5 above.

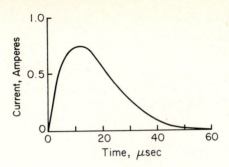

FIGURE 12.4.
The critically damped response from Text Example 12.5.1.

12.3 COMPLETE RESPONSE OF SECOND-ORDER UNDERDAMPED SYSTEMS

When switching occurs in an underdamped system the response may be very complex. In the circuit illustrated below an ac source at frequency $\omega = 10^6$ rad/sec is applied to a circuit which has a resonant frequency very nearly equal to ω. At $t = 0$, $v_c = i_1 = 0$.

FIGURE 12.5.
Circuit used to study complete response with all three (R,L,C) elements in the circuit.

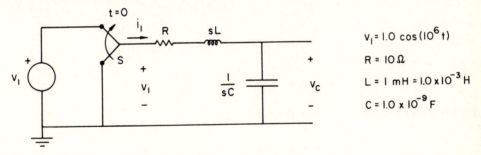

$V_1 = 1.0 \cos(10^6 t)$

$R = 10\,\Omega$

$L = 1\ \text{mH} = 1.0 \times 10^{-3}\ \text{H}$

$C = 1.0 \times 10^{-9}\ \text{F}$

To determine the voltage across the capacitor, $v_c(t)$ for $t \geq 0$, the appropriate transfer function is the ratio V_c/V_1:

$$H(s) = \frac{V_c}{V_1} = \frac{1/sC}{R + sL + (1/sC)} = \frac{1}{s^2LC + sCR + 1}$$

$$= \frac{1/LC}{s^2 + (R/L)s + (1/LC)}$$

Evaluating the expression for the particular case studied here,

$$H(s) = \frac{10^{12}}{s^2 + 10^4 s + 10^{12}}$$

To find the natural response when the voltage source V_1 is suppressed, we need to find what values of s allow a nonvanishing V_c for $V_1 = 0$. Since $V_c = H(s)V_1$, $H(s)$ must be infinite at these values of s. We can find these poles of $H(s)$ by setting the denominator equal to zero, which yields the equation

$$s^2 + 10^4 s + 10^{12} = 0$$

Since this is a series R–L–C circuit we get the same result solving this equation as that given in Eq. 12.2 for the natural response of a series R–L–C, namely,

$$s_{\pm} = -\frac{10^4}{2} \pm \sqrt{\left(\frac{10^4}{2}\right)^2 - 10^{12}} = -5000 \pm \sqrt{25 \times 10^6 - 10^{12}}$$

Since $25 \times 10^6 \ll 10^{12}$, the expression inside the square root sign is negative (underdamped case) and the s values are complex:

$$\mathbf{s}_{\pm} = -5000 \pm j 10^6$$

Note that we have dropped the small deviation of the square root from $10^6 = \sqrt{1/LC}$. This frequency, $\omega_0^2 = 1/LC$, is the frequency which would characterize the natural oscillation if there were no resistor present. The existence of friction shifts the natural frequency by a small amount which we ignore here but which can be quite important. The two natural response solutions are

$$v_{cn1} = \mathbf{A}_1 e^{-5000t} e^{j10^6 t}$$

and

$$v_{cn2} = \mathbf{A}_2 e^{-5000t} e^{-j10^6 t}$$

Since either is a solution, so is their sum, and

$$v_{cn}(t) = v_{cn1} + v_{cn2} = e^{-5000t} \left(\mathbf{A}_1 e^{j10^6 t} + \mathbf{A}_2 e^{-j10^6 t} \right)$$

The voltage v_{cn} must be a real function of time for a physical circuit. As discussed in Section 7.2, this will only be true if $\mathbf{A}_2$ is the complex conjugate of $\mathbf{A}_1$. Let $\mathbf{A}_1 = |\mathbf{A}| e^{j\phi}$ and $\mathbf{A}_2 = \mathbf{A}_1^* = |\mathbf{A}| e^{-j\phi}$. Then,

$$v_{cn}(t) = |\mathbf{A}| e^{-5000t} \left[e^{j(10^6 t + \phi)} + e^{-j(10^6 t + \phi)} \right]$$

$$= |\mathbf{A}| e^{-5000t} 2 \cos(10^6 t + \phi)$$

and finally,

$$v_{cn}(t) = V_{cn} e^{-5000t} \cos(10^6 t + \phi)$$

where a new constant $V_{cn} = 2|\mathbf{A}|$ has been defined. This is the same form as Eq. 12.3.

The natural response of the series R–L–C circuit of Fig. 12.5 is therefore a sinusoidal voltage with an angular frequency of 10^6 rad/sec and with an amplitude that decays exponentially as e^{-5000t}. The period of the sinusoid is $2\pi/10^6 = 6.28 \times 10^{-6}$ sec. The time constant of the decay is $1/5000 = 2 \times 10^{-4}$ sec. This means that there are many periods of the oscillation (about 30) in one time constant. This character of an underdamped response may be related to the Q for a series R–L–C circuit (Eq. 12.5b) as follows. The ratio M of the time constant $\alpha^{-1} = 2L/R$ to the period of the signal at resonance, $\tau = 2\pi/\omega_0$, is given by

$$M = \frac{1}{\alpha\tau} = \frac{\omega_0 L}{\pi R} = \frac{Q}{\pi}$$

For this circuit $Q = 100$ and $M \approx 30$ as found above. A schematic plot of an underdamped oscillator with about five oscillations in one decay time constant ($Q \approx 15$) is shown in Fig. 12.6.

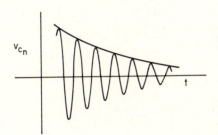

FIGURE 12.6.
Sketch of $|H(j\omega)|$ versus ω for an example of a resonant circuit with a low Q.

To complete the solution we must find the forced response and then determine the two unknown coefficients. The forced response of the circuit in Fig. 12.5 is found using the phasor (peak value) method. For a voltage source $v_1 = V_{1m}\cos(\omega t)$, $\hat{V}_1 = V_1 \underline{/0°}$:

$$\mathbf{H}(j\omega) = \frac{\hat{V}_c}{\hat{V}_1} = 10^{12}\left(\frac{1}{(j\omega)^2 + 10^4(j\omega) + 10^{12}}\right) = 10^{12}\left(\frac{1}{10^{12} - \omega^2 + j10^4\omega}\right)$$

In our example, $v_1 = 1.0 \cos(10^6 t)$, so $\omega = 10^6$ and $\hat{V}_1 = 1 \underline{/0°}$. Then,

$$\mathbf{H}(j\omega) = -j(100) = 100\underline{/-90°}$$

and

$$\hat{V}_c = 100\underline{/-90°}$$

This corresponds to the time function

$$v_{cF}(t) = 100 \cos(10^6 t - 90°)$$

and the full solution is of the form

$$v_c(t) = V_{cn}e^{-5000t} \cos(10^6 t + \phi) + 100 \cos(10^6 t - 90°)$$

To complete the problem we need two initial conditions since there are two constants, V_{cn} and ϕ. The switch was initially grounded and $v_c(0^-) = 0$ and $i_1(0^-) = 0$. The former implies $v_c(0^+) = 0$ and the latter $i_1(0^+) = 0$ since the voltage across a capacitor and the current through an inductor must be continuous functions of time. We can use $v_c(0^+) = 0$ as one of the boundary conditions. The fact that $i_1(0^+) = 0$ implies also that $dv_c(0^+)/dt = 0$ since if there is no current at $t = 0^+$, then the capacitor voltage must not be changing with time. Applying these boundary conditions,

$$v_c(0^+) = V_{cn} \cos \phi = 0$$

$$\frac{dv_c(0^+)}{dt} = -(5000)V_{cn} \cos \phi - (10^6)V_{cn} \sin \phi + 10^8 = 0$$

These two results in turn imply

$$V_{cn} \sin \phi = 100$$

$$V_{cn} \cos \phi = 0$$

These two equations are satisfied if $\phi = 90°$ and $V_{cn} = 100$, so

$$v_c(t) = -100e^{-5000t} \sin(10^6 t) + 100 \sin(10^6 t)$$

where we have used the fact that for any angle α, $\cos(\alpha - 90°) = \sin \alpha$.

As noted above there are about 30 oscillations in the e-folding time of the exponential function (200 μsec). The capacitor voltage starts out at zero, but after 200 μsec it is oscillating with a sine wave amplitude of

$$v_c(t \simeq 200 \ \mu\text{sec}) = (1 - e^{-1})100 \sin(10^6 t)$$

and hence for $t \simeq 200 \ \mu$sec,

$$v_c(t) = 63.2 \sin(10^6 t)$$

After about 100 μsec the natural response has died away entirely, and for $t \geq 1000 \ \mu$sec

$$v_c(t) = 100 \sin(10^6 t)$$

Note that the forced response voltage is much higher that the 1-volt source voltage. This is characteristic of circuits that operate near resonance, such as this one. The result seems unphysical but stems from the fact that there is no energy dissipation in ideal storage elements. Initially during each oscillation of the source, more energy is stored in the L and C combination. This goes on until the energy dissipated in the *resistor* balances the energy added per cycle by the source and the amplitude finally saturates at the value determined by the forced response calculation. This particular circuit is analogous to a backyard swing which, when pushed gently at its resonant frequency, starts at rest and builds up to a large amplitude after some small number of cycles.

If the source voltage is changed to a value very different from the resonant frequency, the response will be much less. This can be seen more clearly by writing $\mathbf{H}(j\omega)$ in terms of its poles:

$$\mathbf{H}(j\omega) = \frac{10^{12}}{(j\omega - \mathbf{s}_+)(j\omega - \mathbf{s}_-)} \tag{12.7}$$

This expression shows that when $j\omega$ equals $j\omega_0$ the term $(j\omega - \mathbf{s}_+) = \alpha$, which is a small number. If the dissipation is zero, $\alpha = 0$ and $\mathbf{H}(j\omega)$ becomes infinite. For a nonzero value of α, $\mathbf{H}(j\omega)$ does not "blow up" but some large finite value of $\mathbf{H}$ will result. In our case we found at the resonant frequency $\mathbf{H}(j\omega) = 100 \; /\!-90°$ and the voltage across the capacitor is 100 times the source voltage. The detailed nature of the frequency response for such circuits is worked out in the next section.

12.4 FREQUENCY RESPONSE OF SECOND-ORDER SYSTEMS — COMPLEX POLES AND ZEROS

We have already treated the frequency response of second-order systems that have pure real poles and zeros in Chapter 11. The overdamped and critically damped cases are thus straightforward and may be analyzed using the results from Chapter 11.

For underdamped systems a very large amplitude response can develop for forcing near the resonant frequency. We anticipate that the transfer function may display a sharp peak in its functional form near the resonant frequency. In the series $R-L-C$ circuit of Fig. 12.5, for example, we found that for a source voltage phasor with a 1-volt amplitude at the resonant frequency the voltage phasor across the capacitor had an amplitude of 100 volts. In this section we investigate the forced response as a function of frequency by studying the transfer function for circuits of this type.

For the circuit of Fig. 12.5,

$$H(s) = \frac{1/LC}{s^2 + (R/L)s + (1/LC)}$$

which may be written in the form

$$H(s) = \frac{\omega_0^2}{(s - \mathbf{s}_+)(s - \mathbf{s}_-)} \tag{12.8}$$

where

$$\mathbf{s}_+ = -\alpha + j\omega_0$$

$$\mathbf{s}_- = -\alpha - j\omega_0$$

In this particular circuit, $\omega_0 = 10^6$ and $\alpha = 5000$.

To review briefly, the natural response may be found from the poles of this transfer function and we obtain a damped oscillator at frequency ω_0. To find the forced response as a function of frequency we investigate $|\mathbf{H}(j\omega)|$, by substituting $s = j\omega$ in Eq. 12.8:

$$\mathbf{H}(j\omega) = \frac{\omega_0^2}{(j\omega - \mathbf{s}_+)(j\omega - \mathbf{s}_-)}$$

The absolute value squared of such a function is found by multiplying it by its complex conjugate $|\mathbf{A}|^2 = \mathbf{A} \cdot \mathbf{A}^*$ (see Chapter 7):

$$|\mathbf{H}(j\omega)|^2 = \frac{\omega_0^4}{(j\omega - \mathbf{s}_+)(j\omega - \mathbf{s}_-)(-j\omega - \mathbf{s}_+^*)(-j\omega - \mathbf{s}_-^*)}$$

Using the results of Problem 12.25 we have

$$|\mathbf{H}(j\omega)| = \frac{\omega_0^2}{[(\omega + \omega_0)^2 + \alpha^2]^{1/2}[(\omega - \omega_0)^2 + \alpha^2]^{1/2}}$$

By evaluating this expression at five important frequencies we can obtain enough information to sketch the function. These results are summarized in Table 12.1, where we have used the fact that $\omega_0 \gg \alpha$.

TABLE 12.1
Evaluation of Series R-L-C Frequency Response

ω	$\|H(j\omega)\|$ (symbolic)	$\|H(j\omega)\|$ (numerical)
0	ω_o^2/ω_o^2	1
$\omega_o - \alpha$	$\omega_o/2\sqrt{2}\alpha$	$100/\sqrt{2}$
ω_o	$\omega_o/2\alpha$	100
$\omega_o + \alpha$	$\omega_o/2\sqrt{2}\alpha$	$100/\sqrt{2}$
∞	ω_o^2/∞^2	0

A linear plot of this function is given in Fig. 12.7, where we have exaggerated the scale somewhat near $\omega = \omega_0$. As mentioned above, the response is very large near the resonant frequency but falls off very quickly for higher and lower frequencies. The bandwidth is defined as the frequency width $\Delta\omega$ for which the response lies within 3 db of the maximum response. Using this definition the bandwidth is $\Delta\omega = 2\alpha$ for the series R–L–C circuit studied here.

There are several relationships of interest. Note that when $\omega = \omega_0$,

$$|\mathbf{H}(j\omega_0)| = \frac{\omega_0}{2\alpha} = Q$$

and hence the peak amplitude equals the Q of the circuit. If there were no dissipation

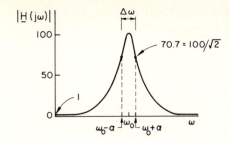

FIGURE 12.7.
Frequency response for the circuit of Fig. 12.5.

at all $Q = \infty$ and the response would also go to infinity. Similarly we can write

$$Q = \frac{\omega_0}{\Delta\omega}$$

since $\Delta\omega = 2\alpha$ and $Q = \omega_0/2\alpha$. This shows that Q of the circuit is also a measure of the width of the frequency response. A high-Q circuit has a very narrow band of frequencies within which it responds ($\Delta\omega$ small compared to ω_0).

Radio and radar transmitters are good examples of systems for which a resonant response is very desirable. The maximum power level possible is usually desired at the transmission frequency. This is accomplished by using resonant or tuned circuits. In the case of a radio transmitter it is easy to understand the importance of the bandwidth of a resonant circuit. For a radio station it can be shown that the transmitted bandwidth must be the order of the frequency range of the original sound source. That is, if some information is to be transmitted on a given "carrier" frequency, the bandwidth of the system must be wide enough to reproduce the information at the receiver.

The formalism developed in Chapter 11 for frequency analysis is applicable in the case of complex poles and zeros but is not as easy to implement analytically. The poles and zeros of the transfer function still contain all of the information but the simple geometric properties associated with real poles and zeros does not hold. In the text example below we show how to evaluate the same transfer function (series R–L–C circuit) as studied above using this method.

TEXT EXAMPLE 12.6 _____

Use the poles and/or zeros of the transfer function for the circuit shown below to sketch the magnitude and phase angle of the transfer function $H(s) = V_2/V_1$.

$$H(s) = \frac{V_2}{V_1} = \frac{1/sC}{R + sL + 1/sC}$$

$$= \frac{1}{LC}\frac{1}{\left(s^2 + \dfrac{R}{L}s + \dfrac{1}{LC}\right)}$$

(a)

Solution: In the transfer function of this circuit (Eq. 12.8), there are two poles—the roots of the quadratic in the denominator. It is convenient to write such quadratics in the form

$$s^2 + 2\alpha s + \omega_n^2$$

where in this case

$$2\alpha = \frac{R}{L}$$

and

$$\omega_n^2 = \frac{1}{LC}$$

Then,

$$H(s) = \frac{\omega_n^2}{(s - p_1)(s - p_2)}$$

where the poles are denoted by p_1 and p_2:

$$p_1 = -\alpha + \sqrt{\alpha^2 - \omega_n^2} \qquad p_2 = -\alpha - \sqrt{\alpha^2 - \omega_n^2}$$

When $\alpha^2 > \omega_n^2$ the square root yields a real number and we have two pure real poles which can be analyzed as discussed in Chapter 11.
In this case $\alpha^2 < \omega_n^2$ and

$$\mathbf{p}_{1,2} = -\alpha \pm j \sqrt{\omega_n^2 - \alpha^2} = -\alpha \pm j\omega_0$$

To a good approximation $\omega_0 = \omega_n$ and the two poles are located at the values $-0.5 \times 10^4 \pm j10^6$.

To determine the frequency response, we cannot use the right triangle approach which led to an analytic general equation for real poles and zeros. However, the *method* is the same; that is, the contribution of each zero and each pole is still determined by the length and angle of the complex vector which links that zero or pole to the position $j\omega$ on the positive imaginary axis. By judiciously choosing a few values of ω we can make a reasonable sketch of the frequency response as follows.

First, $\mathbf{H}(j\omega)$ is written in the standard form (see Eqs. 11.3a and 11.3b)

$$\mathbf{H}(j\omega) = \frac{\hat{V}_2}{\hat{V}_1} = \frac{\omega_n^2}{(j\omega - \mathbf{p}_1)(j\omega - \mathbf{p}_2)}$$

The pole zero diagram is shown in the figure (b) below, with an expanded picture near ω_0 shown in part (c) of the figure.

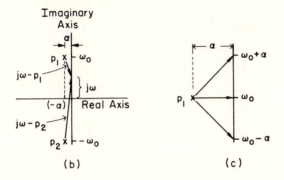

(b) (c)

As before, the frequency response may be determined by examining the reciprocal of the product of vectors drawn from the poles to the position $j\omega$. Setting $\omega_n = \omega_0$, $|\mathbf{H}(j\omega)| = \omega_0^2/(B_1 B_2)$ and $\underline{/\mathbf{H}(j\omega)}x = -\beta_1 - \beta_2$. Five different frequencies (the same ones used in Table 12.1) have been selected and the two contributions to $\mathbf{H}$ evaluated and listed in Table 12.2. For reference remember that $\omega_0^2 = 1/LC$ and $Q = \omega_0/2\alpha$. As an example of how this table is constructed, consider the entries for $\omega = \omega_0 - \alpha$. The vector $j\omega - \mathbf{p}_1$ has a length of $B_1 = \sqrt{2}\alpha$ and makes an angle $\beta_1 = -45°$. The vector $j\omega - \mathbf{p}_2$ has a length B_2 approximately equal to $2\omega_0$ and makes an angle $\beta_2 = +90°$. Then,

$$|\mathbf{H}(j[\omega_0 - \alpha])| = \frac{\omega_0^2}{(\sqrt{2}\alpha)(2\omega_0)}$$

$$|\mathbf{H}| \simeq \frac{\omega_0}{(2\sqrt{2}\alpha)}$$

and

$$\underline{/\mathbf{H}} = -\beta_1 - \beta_2 = -45°$$

TABLE 12.2
Approximate Values for the Contributions of Each Pole to H and the Value of H for Selected Frequencies

ω	$j\omega - p_1$	$j\omega - p_2$	$H(j\omega)$
0	$\omega_o\underline{/-90°}$	$\omega_o\underline{/90°}$	$1\underline{/0°}$
$\omega_o - \alpha$	$\alpha\sqrt{2}\underline{/-45°}$	$2\omega_o\underline{/90°}$	$(\omega_o/2\sqrt{2}\alpha)\underline{/-45°} = Q/\sqrt{2}\underline{/-45°}$
ω_o	$\alpha\underline{/0°}$	$2\omega_o\underline{/90°}$	$(\omega_o/2\alpha)\underline{/-90°} = Q\underline{/-90°}$
$\omega_o + \alpha$	$\alpha\sqrt{2}\underline{/+45°}$	$2\omega_o\underline{/90°}$	$(\omega_o/2\sqrt{2}\alpha)\underline{/-135°} = Q/\sqrt{2}\underline{/-135°}$
∞	$\omega\underline{/90°}$	$\omega\underline{/90°}$	$\lim_{\omega\to\infty}(\omega_o/\omega)^2\underline{/-180°} = 0\underline{/-180°}$

Note that at $\omega = \omega_0$ the amplitude is equal to the Q of the circuit. A sketch of the amplitude and phase response for the transfer function V_2/V_1 is shown below.

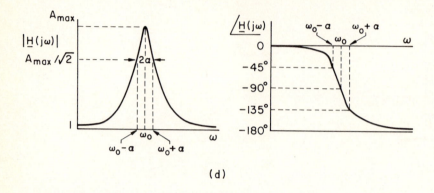

(d)

The graphical pole-zero method is almost as much work as evaluating the expression for $\mathbf{H}(j\omega)$ at a few frequencies. However, Text Example 12.6 does show that the method is the same as that used for real poles and zeros.

12.5 SECOND-ORDER FILTERS

The transfer function derived in Section 12.4 can be considered as a filter. The plot in the text example above shows that the series $R\text{–}L\text{–}C$ circuit has a very peaked response near ω_0. The higher the Q of the circuit, the more narrow the response function. Such a circuit is tunable to a specific frequency, and there are many applications of such circuits.

In general, the design and analysis of filters with R, L, and C elements may proceed as outlined above. The various impedances of the elements are evaluated and a transfer function generated using circuit principles. The character of the transfer function, particularly $|\mathbf{H}(j\omega)|$, then reveals the nature of the filtering achieved by the design. If the poles and zeros are real numbers, then the methods of Chapter 11 may be implemented. For "tuned" or "resonant" circuits the analysis is more challenging but may be worked out as shown in Section 12.4.

PRACTICE PROBLEMS
AND
ILLUSTRATIVE EXAMPLES

Problem 12.1

Find the natural frequencies that characterize the response of the circuit below after the switch is thrown at $t = 0$.

$$R = 10 \text{ k}\Omega$$
$$C = 0.001 \text{ } \mu F$$

$$\text{Ans.:} \quad s_+ = -\left(\frac{3 + \sqrt{5}}{2}\right)\omega_0$$

$$s_- = -\left(\frac{3 - \sqrt{5}}{2}\right)\omega_0$$

$$\omega_0 = 1/RC = 10^5 \text{ rad/sec}$$

EXAMPLE 12.1

Find the natural response function $v_{cn}(t)$ for $t \geq 0$ for the circuit in Problem 12.1. Find the full solution including the coefficients. Assume that the capacitor on the left is not charged for $t \leq 0$.

Solution: From Problem 12.1 we know the form of the natural response solution

$$v_{cn}(t) = Ae^{s_+ t} + Be^{s_- t} \tag{1}$$

We need two boundary or initial conditions to find the coefficients A and B. These are the value of $v_{cn}(t)$ at $t = 0$ and the value of its derivative $dv_{cn}(t)/dt$ at $t = 0$. When the switch is in position 1 the capacitors act like open circuits since the battery is a dc source. The voltage $v_{cn}(0^-)$ is determined by the voltage divider principle and equals 5 V. Since the voltage across a capacitor cannot change instantaneously,

$$v_{cn}(0^+) = v_{cn}(0^-) = 5 \text{ V}$$

To determine the derivative of $v_{cn}(t)$ at $t = 0^+$, recall that for a capacitor $Q = CV$ and that the current is the time derivative of the charge,

$$i = \frac{dQ}{dt} = \frac{C\,dv}{dt}$$

In Chapter 5 we were careful to define the sign conventions as follows:

$$i = \frac{C\,dv}{dt}$$

$$\frac{dv}{dt} = \left(\frac{1}{C}\right)i$$

(a)

This makes physical sense since if a positive charge flows onto the top plate and off of the bottom, v increases with time. In our case at $t = 0^+$ the current flow is as shown in the circuit below:

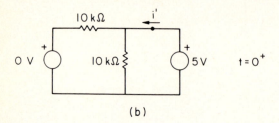

(b)

The 5-V voltage source represents the charged capacitor while the 0-V source represents the uncharged capacitor, both at $t = 0^+$. The current i' is then

$$i' = \frac{5 \text{ V}}{5 \text{ k}\Omega} = 1 \text{ mA}$$

Since $i' = -i(0^+)$ we have

$$\frac{dv_c(0^+)}{dt} = \left(\frac{1}{C}\right) i(0^+) = -\frac{1 \text{ mA}}{10^{-9} \text{ F}}$$

$$= -10^{+6} \text{ V/sec}$$

To apply these results we take the derivative of Eq. 1:

$$\frac{dv_{cn}(t)}{dt} = As_+e^{s_+t} + Bs_-e^{s_-t} \tag{2}$$

Since $s_+ = -2.62 \times 10^5 \text{ sec}^{-1}$ and $s_- = -0.38 \times 10^5 \text{ sec}^{-1}$, evaluating the derivative at $t = 0^+$ and setting it equal to $dv_{cn}(0^+)/dt = -10^6$ V/sec, we have from Eq. 2

$$-10^6 = -(0.262 \times 10^6)A - (0.038 \times 10^6)B$$

or

$$0.262A + 0.038B = 1$$

From Eq. 1 at $t = 0^+$

$$A + B = 5$$

Solving these two simultaneous equations,

$$A = 3.65$$

$$B = 1.35$$

and finally,

$$v_{cn}(t) = 3.65e^{-(2.62\times10^5 t)} + 1.35e^{-(0.38\times10^5)t}$$

Problem 12.2

Repeat the calculations in Problem 12.1 and Example 12.1 if the two resistors R in the circuit are 100 kΩ and nothing else is changed. That is, find s_+ and s_- and the solution $v_{cn}(t)$ for $t \geq 0$.

Problem 12.3

For the second-order circuit below,

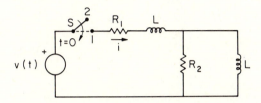

show that for $t \geq 0$ the natural response parameters s_+ and s_- must be pure real numbers no matter what values R_1, R_2, and L take on.

Problem 12.4

For the circuit of Problem 12.3, find the values of the natural response parameters if $L = 1$ mH, $R_1 = 4$ kΩ and $R_2 = 2$ kΩ.

Problem 12.5

Find the full solution to Text Example 12.1, that is, determine the two coefficients V_1 and V_2 in the expression for $v_n(t)$.

Problem 12.6

Find the values of the natural response frequency or frequencies (s_+ and s_-) for the current i_2 in the circuit of Problem 8.27. (Remember that a current source is replaced by an open circuit in determining the natural response of such a system.) Is this circuit underdamped, critically damped, or overdamped?

EXAMPLE 12.2

Find the complete response solution $i(t)$ for $t \geq 0$ in the circuit of Problem 12.3 if the source $v(t)$ is a dc battery with a voltage $v = 10$ V and $R_1 = R_2 = 100$ Ω while $L = 1$ H. Assume that the switch was in position 2 for a long time.

Solution: The natural response component may be found by setting the series impedance function equal to zero. Replacing the voltage source by a short circuit,

$$Z_s(s) = R + sL + \frac{RsL}{R + sL}$$

$$= \frac{(R + sL)^2 + RsL}{R + sL}$$

Setting the numerator equal to zero yields

$$R^2 + 2RsL + s^2L^2 + RsL = 0$$

Dividing by L^2,

$$s^2 + 3\left(\frac{R}{L}\right)s + \left(\frac{R}{L}\right)^2 = 0$$

Defining $\omega_1 = R/L$ leads to the solutions

$$s_+ = -\left(\frac{1}{2}\right)(3 + \sqrt{5})\omega_0 = -262 \text{ rad/sec}$$

$$s_- = -\left(\frac{1}{2}\right)(3 - \sqrt{5})\omega_0 = -38 \text{ rad/sec}$$

The natural response solution is therefore of the form

$$i_n(t) = Ae^{s_+ t} + Be^{s_- t} \tag{1}$$

The forced response is the current i which flows due to the battery as $t \to \infty$. For dc the inductors may be replaced by short circuits. In that case the forced current is

$$i_F = \frac{10 \text{ V}}{100 \text{ } \Omega} = 0.1 \text{ A}$$

The total solution is, then, for $t \geq 0$,

$$i(t) = Ae^{s_+ t} + Be^{s_- t} + 0.1$$

Two boundary or initial conditions are needed. When the circuit is in position 2, no current flows through either inductor, so $i(0^-) = 0$. Since the current through an inductor is a continuous function of time, $i(0^+) = i(0^-) = 0$ and

$$0 = A + B + 0.1$$

To complete the problem we need to evaluate the derivative of the current $i(t)$ at $t = 0$. To accomplish this we consider the following circuit at $t = 0^+$:

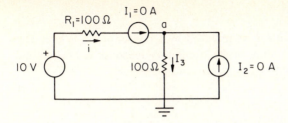

To determine di/dt we use the fact that for an inductor

$$v_L = \frac{L\,di}{dt}$$

or

$$\frac{di_L}{dt} = \left(\frac{1}{L}\right) v_L$$

Since the current $i(t)$ flows through the inductor with which it is in series, we need to know the voltage across that inductor. Referring to the figure above, with $I_1 = I_2 = 0$, the current I_3 must also be zero. Defining ground potential as shown, this implies that $V_a = 0$. Now since $i(0^+) = 0$, there is no potential drop across the 100–Ω resistor labeled R_1. The voltage on each side of the resistor is therefore 10 V and the voltage across I_1 is $10\text{-}V_a = 10 - 0 = 10$ V. Since this current source represents the inductor at $t = 0^+$, we have

$$\frac{di_L}{dt} = \left(\frac{1}{1H}\right)(10 \text{ V}) = 10 \text{ A/sec}$$

By KCL this must also equal di/dt. Taking the derivative of Eq. 1 yields

$$\frac{di}{dt} = As_+ e^{s_+ t} + Bs_- e^{s_- t}$$

Substituting

$$10 = -262A - 38B$$

Solving, we find $A = -0.028$ and $B = -0.072$. The final solution for $t \geq 0$ is then

$$i(t) = -(0.028)e^{-262t} - (0.072)e^{-38t} + 0.1$$

Problem 12.7

Repeat Example 12.2 for the case that $L = 10$ H with everything else unchanged.

Problem 12.8

In Example 12.2 the L in series with R_1 is replaced by a capacitor C. What value of R_1 should be used if critical damping of the oscillatory response is desired and if no other parameters are changed?

Problem 12.9

What single value of s results in the circuit of Problem 12.8 if critical damping is attained?

Problem 12.10

Find the complete response current $i(t)$ in Example 12.2 if $V_0 = 1$ V. Make a plot of this function versus time.

Problem 12.11

Referring to Text Example 12.2, suppose the switch has been in position 2 for a long time and is switched to position 1 at $t = 0$. Find $v(t)$, the voltage across the 2H inductor, for $t \geq 0$. Again, the switch must be a "make before break" switch.

EXAMPLE 12.3

Find the complete response current $i(t)$ in Text Example 12.2 if the source V_2 is not a battery but rather an 8V ac source at frequency $\omega = 4$ rad/sec, that is,

$$v_2(t) = 8\cos(4t)$$

Solution: We already know that the natural response is of the form

$$i_n(t) = I_1 e^{-8t} + I_2 e^{-2t}$$

The forced response is given by the expression

$$\hat{I} = \frac{\hat{V}}{\mathbf{Z}}$$

where $\mathbf{Z}$ is the series impedance function evaluated at $s = j\omega = j4$ and $\hat{V} = 8\ \underline{/0°}$:

$$\mathbf{Z}(s) = 4\left(\frac{s^2 + 10s + 16}{s + 4}\right)$$

so

$$\mathbf{Z}(j\omega) = \frac{4(j40)}{j4 + 4} = 20\sqrt{2}\underline{/45°}$$

and $\hat{I} = 0.2\sqrt{2}\underline{/-45°}$. The total current is then

$$i(t) = I_1 e^{-8t} + I_2 e^{-2t} + (0.2)\sqrt{2}\ \cos(4t - 45°)$$

The initial current through the inductor is still 1 A as in Text Example 12.2 and the time derivative of $i(t)$ is still -2 A/sec. Now the forcing function also has a derivative. The two equations for I_1 and I_2 are,

$$1 = I_1 + I_2 + (0.2)\sqrt{2}\ \cos(-45°)$$

and

$$-2 = -8I_1 - 2I_2 - (0.8)\sqrt{2}\ \sin(-45°)$$

Evaluating the sine and cosine terms,

$$I_1 + I_2 = 0.8$$

$$8I_1 + 2I_2 = 2.8$$

and the solution is

$$i(t) = (0.2)e^{-8t} + (0.6)e^{-2t} + 0.2\sqrt{2}\ \cos(4t - 45°)$$

Problem 12.12

As a final problem involving Text Example 12.2, suppose that the "make before break" switch has been in position 2 for a long time and is switched to position 1 at $t = 0$, but that the source v_1 is an ac signal with $v_1 = 20\ \cos(4t)$. Find the complete response $i(t)$ for $t \geq 0$.

Problem 12.13

In Problem 12.3 let $R_1 = 10$ kΩ, $R_2 = 20$ kΩ, and $L = 10$ mH. Find the complete response current $i(t)$ for $t \geq 0$ if $v(t) = 10\ \cos(10^6 t)$.

Problem 12.14

A 1-volt battery is applied at $t = 0$ by moving the switch from position 2 to position 1 across a parallel R–L–C circuit in a manner similar to the circuit in Fig. 12.2, but with R $= \infty$. If $L = 1$ mH and $C = 10\ \mu$F, what value of R_1 will yield a critically damped response? For that value resistor, find the complete response current $i_L(t)$ for $t \geq 0$ and make a sketch of the waveform.

Problem 12.15

For values of L and C equal to 1.0 mH and 0.1 μF, respectively, in the series R–L–C circuit of Fig. 12.3, find the value of R which yields critical damping. Determine the complete response current $i(t)$ if the switch has been in position 2 for a long time with the capacitor uncharged and is moved to position 1 at $t = 0$. Sketch the current $i(t)$ for $t \geq 0$.

Problem 12.16

Referring to Problem 12.6 (and Problem 8.27), find the complete response solution for $i_2(t)$ if the current source $i_s(t)$ was attached to the network at $t = 0$. Assume that at $t = 0^-$ there was no voltage across the capacitor C and no current through the inductor L.

Problem 12.17

Show that the solution

$$i(t) = (I_1 + I_2 t)e^{st}$$

is a solution to the differential equation

$$\frac{d^2 i}{dt^2} + \left(\frac{R}{L}\right)\frac{di}{dt} + \frac{i}{LC} = 0$$

when $s = -R/2L$ and $L/C = R^2/4$.

Problem 12.18

Find the complex frequencies s_+ and s_- for a series R–L–C circuit if $R = 10\ \Omega$, $L = 10$ mH, and $C = 4\ \mu$F.

$$\text{Ans.:} \quad s_\pm = -500 \pm j(4975)$$

Problem 12.19

The circuit below is a good model for a parallel L–C circuit in which the inductor coils have finite resistance R.

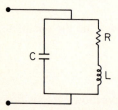

Find an expression for the resonant natural frequency for this network by setting the series impedance equal to zero. Evaluate that frequency for the parameters $L = 4\ \mu$H, $C = 2 \times 10^{-8}$ F, and $R = 10\ \Omega$. Compare that frequency to the resonant frequency for an ideal inductor ($R = 0$).

Problem 12.20

Show that for an underdamped series R–L–C circuit Q is given by

$$Q = \frac{\omega_0 L}{R}$$

EXAMPLE 12.4

The impedance function of a certain second-order network has the form

$$Z(s) = \frac{(s + 1)(s + 2.5)}{(s - s_+)(s - s_-)}$$

where $s_+ = -1 + j2$ and $s_- = -1 - j2$. Write an expression (with undetermined constants) for the open-circuit natural response voltage and the short-circuit natural response current.

Solution: The open-circuited natural response voltage $v_n(t)$ corresponds to the natural response of a circuit which looks as follows:

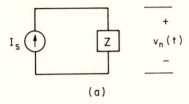

(a)

To find the natural response, I_s is set equal to zero by replacing it with an open circuit:

(b)

From the generalized Ohm's law,

$$v_n = I_s Z(s) = (0)Z(s)$$

The only way to have a nonvanishing voltage (if $I_s = 0$) is if $Z(s) = \infty$. This occurs when the denominator vanishes, that is, when

$$s = -1 + j2 = s_+$$

and

$$s = -1 - j2 = s_-$$

The natural response solution is then of the form

$$v_n(t) = \mathbf{A}e^{s_+t} + \mathbf{B}e^{s_-t}$$

$$\mathbf{v}_n(t) = (\mathbf{A}e^{j2t} + \mathbf{B}e^{-j2t})e^{-t}$$

We have shown previously that, if $v_n(t)$ is to be a real function of time, then $\mathbf{A}$ and $\mathbf{B}$ must be complex conjugates and the solution can be written

$$v_n(t) = Ce^{-t}\cos(2t + \phi)$$

In the second half of the problem we are asked to find the *short*-circuited natural response current. This is equivalent to the natural response of the following circuit:

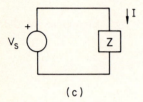

(c)

For such a circuit the natural response is found by replacing V_s with a short circuit:

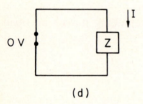

(d)

and evaluating Ohm's law:

$$0 = (I)Z(s)$$

$Z(s) = 0$ when the numerator vanishes, which implies that two solutions for s exist:

$$s_1 = -2.5$$

$$s_2 = -1$$

and finally

$$i_n(t) = De^{-t} + Ee^{-2.5t}$$

The values s_1 and s_2 are the zeros of the impedance function while s_+ and s_- are the poles of that function.

Problem 12.21

For a given network, the open-circuit natural response voltage is $v_n = V_0 e^{-5000t}$ and the short-circuited natural response current is $I_0 e^{-3000t}$.

(a) Find the general form of the impedance function for this network. There will be an arbitrary constant K in your expression.

(b) State the poles and zeros of $Z(s)$.

(c) When a steady voltage of 27 V has been connected to the network for a long time, a steady current of 100 mA flows into the network. We use this information to determine the value of the constant K in the expression for $Z(s)$.

(d) Find the forced response current $i_F(t)$ if a voltage $v(t) = 2 \cos(5000t)$ is applied.

Problem 12.22

For the second-order circuit shown below,

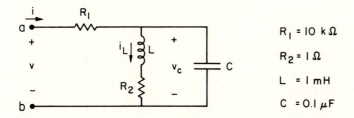

$R_1 = 10 \text{ k}\Omega$

$R_2 = 1 \Omega$

$L = 1 \text{ mH}$

$C = 0.1 \mu\text{F}$

(a) Write the equivalent impedance function at the terminals and find the poles and zeros of the impedance.

(b) Find the forced response $i(t)$ if voltage $v(t) = 20 \cos(10^5 t)$ is applied.

Problem 12.23

Suppose in the circuit for Problem 12.22 there is a short circuit across the terminals a–b. Write an expression for the natural response current $i_{Ln}(t)$ through the inductor. Evaluate the natural response $i_{Ln}(t)$ for $t \geq 0$ if at $t = 0^+$ the voltage across the capacitor is 10 V and the current through the inductor is 0 A.

Problem 12.24

Suppose in the circuit for Problem 12.22 there is an open circuit at the terminals a–b. Write an expression for the natural response voltage $v_{cn}(t)$ across the capacitor. Evaluate the natural response $v_{Cn}(t)$ for $t \geq 0$ if at $t = 0^+$ the voltage across the capacitor is zero and the current through the inductor is 1 mA.

Problem 12.25

Show that $|(j\omega - s_\pm)|^2 = (\omega \pm \omega_0)^2 + \alpha^2$, where $s_+ = -\alpha + j\omega_0$ and $s_- = -\alpha - j\omega_0$. Verify the entries in Table 12.1 using this result.

Problem 12.26

A certain transfer function $H(s) = V_2/V_1 = 2 \times 10^4 \left(\dfrac{s + 50}{s^2 + 50s + 10^6} \right)$.

(a) Find the short-circuited natural response voltage v_{2n} (i.e., for $v_1 = 0$) with undetermined coefficients.

(b) For $v_1 = 0.02 \cos(1000t)$ V, find the forced response $v_{2F}(t)$.

(c) Find the complete response voltage $v_2(t)$ if $v_2(0^+) = 25.32$ V and $dv_2(0^+)/dt = -10^4$ V/sec.

Problem 12.27

Find the complete response voltage $v_L(t)$ in the circuit below if the switch has been in position 1 for a long time and is moved to position 2 and where $i_s(t) = 10 \cos(10^8 t)$ mA.

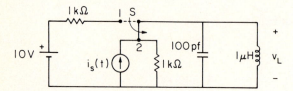

Problem 12.28

Find the Q of the circuit in Problem 12.22 if $R_2 = 0$ and at $t = 0$ a voltage source v_s is applied at a–b. Find the forced response voltage $v_c(t)$ if $v_s(t) = 20 \cos(10^5 t)$. Find the complete response if $v_c(0^+) = 100$ V and $i_L(0^+) = 0$ A.

Problem 12.29

For a transfer function $H(s) = V_2/V_1 = (10^{10})/(s^2 + 10^3 s + 10^{10})$, find the poles and sketch the amplitude of the frequency response $|\mathbf{H}(j\omega)|$ as a function of ω. Show all significant values. Find Q.

$$\text{Ans.: } s_\pm = -500 \pm j10^5$$
$$Q = \omega_0/\Delta\omega = 100$$

Problem 12.30

Sketch the phase angle of $\mathbf{H}$ versus frequency for the circuit in Problem 12.29 above.

Problem 12.31

Consider the circuit of Fig. 12.2 with the switch in position 1. Suppose the voltage source is an arbitrary one of value V_0 and $R_1 = R$. Find the transfer function $H(s) = V_c/V_0$. Then, let $R = 10$ kΩ, $L = 1$ H and $C = 10^{-8}$ F. Sketch $|\mathbf{H}(j\omega)|$ and $\underline{/\mathbf{H}(j\omega)}$. Find Q from $\omega_0/\Delta\omega$.

Problem 12.32

The impedance function of a circuit is

$$Z(s) = 0.01\left(\frac{s^2 + 2 \times 10^3 s + 10^{10}}{s + 2 \times 10^3}\right)$$

(a) Sketch the magnitude and phase of the impedance as a function of the angular frequency ω. Show scales on all axes.
(b) What is the "bandwidth" of the circuit?

Ans.:

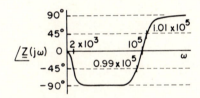

Bandwidth
$= 2 \times 10^3$ rad/sec
$= 318$ Hz

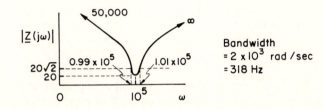

Problem 12.33

The transfer function of a certain network is

$$H(s) = \frac{V_2}{V_1} = \frac{100s}{s^2 + 20s + 10^8}$$

Sketch the magnitude of the transfer function as a function of the angular frequency ω. Show all significant values including the bandwidth of the response.

Ans.:

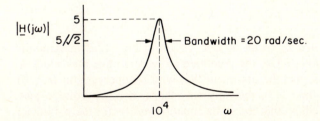

Problem 12.34

With the switch in position 1, consider the source V_0 in Problem 12.1 to be an arbitrary source. Evaluate the transfer function $H(s) = V_c / V_0$ and describe its pole zero properties and filter characteristics.

> Ans.: (a) $H(s) = (1/RC)/[s + (2/RC)]$
>
> (b) pole at $s = -2/RC$
>
> (c) low-pass filter with 3-db point at $\omega_p = 2/RC$

Problem 12.35

Repeat Problem 12.34 above using the circuit in Problem 12.3 where $H(s) = V_{R2} / V$ and V_{R2} is the voltage across the resistor R_2.

Problem 12.36

The transfer function in Problem 12.26 represents a certain filter. Sketch the magnitude of the transfer function $H(s) = V_2 / V_1$ versus frequency. Using the definition $Q = \omega_0 / \Delta\omega$, estimate Q.

Problem 12.37

Consider the transfer function

$$H(s) = \frac{V_2}{V_1} = \frac{(2 \times 10^6)s}{s^2 + 2 \times 10^3 s + 10^{12}}$$

For $v_1(t) = [10 + 0.1 \cos(0.9 \times 10^6 t) + 0.1 \cos(10^6 t)]$ mV, find $v_2(t)$. Note that the signal at $\omega = 10^6$ rad/sec is at the resonant frequency of the system.

Ans.:

$$v_2(t) = \left[0.95 \ \cos(0.9 \times 10^6 t + 90°) + 100 \ \cos(10^6 t)\right] \text{mV}$$

(Note the large gain at the resonant frequency $\omega = 10^6$ rad/sec.)

Problem 12.38

Sketch $|\mathbf{H}(j\omega)|$ for the transfer function in Problem 12.37 above.

Problem 12.39

Evaluate the parallel impedance Z_p of the R–L–C circuit in Fig. 12.2 in symbolic form with the switch in position 2. Use the pole method to determine the values of s for natural response; that is, use the argument that with an open circuit ($I = 0$) attached across the network for $t \geq 0$, the only way that a nonzero voltage may exist ($V = IZ_p$) is if $Z_p = \infty$. Show that the same equation results for the parameter *s* using the impedance zero method by inserting a phantom voltage source of value zero (short circuit) in any of the three parallel legs.

$$\text{Ans.: } Z_p(s) = R\left[\frac{s/RC}{s^2 + (s/RC) + (1/LC)}\right]$$

$$\text{Ans.: } s_\pm = -(1/2RC) \pm$$

$$\frac{1}{2}\left[\left(\frac{1}{RC}\right)^2 - \frac{4}{LC}\right]^{\frac{1}{2}}$$

Problem 12.40

Suppose $R_1 = R = 1000\,\Omega$ in the circuit for Fig. 12.2 and the battery is replaced by an arbitrary voltage source V_s. The other values are the same as in Text Example 12.3. Find the transfer function

$$H(s) = \frac{V_c}{V_s}$$

if the switch is in position 1, where V_c is the voltage across the capacitor and hence across the entire parallel network. Find the complex natural response frequencies for the system. Find Q and explain why adding the resistor R_1 decreases Q over the result given in Text Example 12.3 for the same parallel R–L–C network.

Problem 12.41

Sketch the magnitude of the transfer function $H(s) = V_c/V_s$ found in Problem 12.40 versus frequency. Compare the Q found above with $Q = \omega_0/\Delta\omega$.

Problem 12.42

Sketch the magnitude of the transfer function $H(s) = V_c/V$ for the circuit of Problem 12.22 versus frequency.

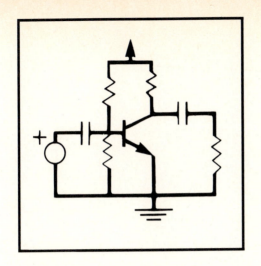

SECTION E

ELEMENTARY ELECTRONIC DEVICES

All of the circuit elements used thus far display a linear relationship between current and voltage. Much of modern electronic design, however, depends upon nonlinear devices such as diodes and transistors. We introduce these circuit elements in this section and show some of their important applications. In order to proceed we need a working knowledge of semiconductor materials. A rigorous study of electronic devices is beyond the scope of this text, however, and the student should look to advanced courses for more details. Semiconductor diodes are discussed in Chapter 13. We show how to include such nonlinear devices in circuits and discuss their important application in power supply design. This design example uses material from many previous chapters, for example, transformers, complete response, transfer functions, and filters. We also explain how to create linear models of electrical systems that are valid for small signal analysis. The latter methodology is particularly important in the primary transistor application we pursue in this text, the small signal amplifier. We study two types of transistors in this book: the *npn* bipolar junction transistor, and the *n*-channel enhancement MOSFET. The idea is to give the student some experience with semiconductor devices without attempting to survey all possible transistor designs and circuit applications. We again use many of the previous developments in the text including transfer functions, Bode plots, and frequency synthesis. Realistic small signal *R–C* amplifiers using these two transistors are treated in some detail using the concepts of dependent sources developed in Chapter 4.

Chapter 13
Semiconductor Diodes and Diode Circuits

Thus far we have considered only circuits containing elements for which the current–voltage relationship is linear. Not all devices behave in this manner. The rapid development of electrical communications was initiated by the invention of the vacuum tube, a nonlinear device. In this chapter we introduce the first nonlinear element to our repertoire—the semiconductor diode. Invention of diodes and transistors has revolutionized electronic circuit design and has made possible most of the advances in electrical engineering in the past several decades. We shall not present an exhaustive study of diodes nor of the transistors and transistor circuits we treat in the rest of the text. Our goal is rather to give a basic understanding of how such devices operate and to illustrate some ways in which they are used in electronic circuits. The subtleties of circuit design and the details of device construction are left for more advanced courses. A common feature to our approach is the construction of linear small signal models for each device introduced. This important simplification allows us to treat these nonlinear devices as linear elements under certain restrictions. All of the laws and theorems presented earlier in the text, including the principle of Fourier synthesis, are then applicable in the linear circuit model.

13.1 SEMICONDUCTOR MATERIALS

Semiconductor diodes are constructed from the group IV elements such as silicon (Si) or from the group III–V compounds such as GaAs. In pure form these materials are relatively poor electrical conductors. This stems from the fact that the electrons are quite tightly bound to the nuclei and that, therefore, there are relatively few free electrons to carry current. At absolute zero temperature the electrons would all be bound and no current would flow at all. As the temperature increases thermal motions will result in broken bonds and in an increasing number of free electrons. There will also be an equal number of "holes" left behind where these electrons have vacated, leaving behind a positive ion which is tightly bound to the crystal lattice. The conductivity will correspondingly increase with temperature as more electrons are free to carry the current. When a free electron randomly encounters a hole, it may "recombine" with it to complete the local valence bond. However, at a given temperature the number of holes and free electrons on the average remains constant even though a continuous process of formation and recombination of free electron and ion–hole pairs takes place.

The Ohm's law derivation described in the last section of Chapter 1 pertains to Si and Ge crystals as well as to metallic conductors such as silver and copper. However, since the free electron density is low, we must also consider the motion of holes through the material. This effect occurs when a broken valence bond (hole) is filled not by a free electron, as in recombination, but by a bound valence electron from a neighboring atom. As illustrated in Fig. 13.1, when an electron jumps from right to left at time t, the hole moves from left to right. This progression of charge from left to right constitutes a conventional current in that same direction. Without an applied voltage, hole motions of this type would be random and no net current would result. However, as shown in the figure, when a voltage is applied the associated electric field slightly distorts the atoms and creates a more favorable condition for hole progression to the right (parallel to **E**). Since the holes move more slowly than the free electrons, one can think about holes as heavy positively charged particles "drifting" in the direction parallel to **E**. Both of these sources of charge flow carry conventional current from high to low potential through the material, but in pure Si or Ge the hole conductivity is much smaller than the electron conductivity.

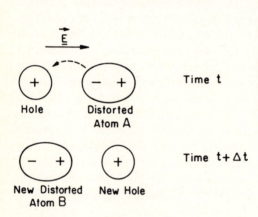

FIGURE 13.1.
Progression of a hole to the right. At time t the electric field due to an applied voltage distorts the atoms, causing enhanced probability that an electron will jump to neutralize the hole next to it. This creates a new hole displaced to the right.

The conductivity of Si or Ge can be considerably increased by a process called doping. For example, if we add (or "dope") some atoms to pure Si or Ge which have five valence electrons (such as antimony, phosphorous, or arsenic), the extra electron bond will be easily broken and more free electrons will be available to carry an electron current. These added "donor" atoms increase the number of *negative* charge carriers and the resulting material is designated as an n-type semiconductor. Similarly, when elements from the third column of the periodic table are added to Si or Ge, they add more holes to the material and hence increase the hole conductivity. Such an "acceptor doped" semiconductor material is termed p-type since more *positive* holes are available for conduction. In fact, sufficient acceptors can be added so that the hole conductivity of a p-type material can exceed that of the free electrons in that same piece of material. The holes are then the majority carriers in a p-type material, whereas the free electrons are the majority carriers in an n-type material.

Before discussing semiconductor devices one more way in which these materials can conduct electricity must be addressed. Consider a p-type material in which there is a gradient in the number of acceptor atoms as illustrated in Fig. 13.2. As

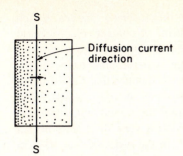

FIGURE 13.2.
Cross-sectional schematic diagram showing nonuniform density of holes with more to the left of the plane *SS* than to the right. Holes will diffuse across that surface carrying a positive current.

the electron–hole pairs are created and destroyed due to thermal agitation, more will move across the plane surface, labeled *S–S*, to the right than will move to the left since there are more holes initially located to the left of *S–S*. This is typical of any "diffusion" process driven by a concentration gradient. The flow of oxygen across a biological cell wall and the spread of perfume atoms across a room from a newly opened bottle are examples from other physical systems in which diffusion is important. In our case the flow of holes also carries an electrical current with it just as the perfume, which is a minor constituent in air, carries its characteristic smell.

In the applications described below the gradients in *p* or *n* doping are not smooth, as illustrated above, but are very sharp. These are called step gradients. An extreme example occurs when a uniform *p*-type material is affixed directly to a uniform *n*-type material. This yields a very sharp gradient in both holes and free electrons, and the resulting device is termed a semiconductor diode.

13.2 SEMICONDUCTOR DIODES

By themselves, the materials discussed above would not be as important as they have become in electrical engineering applications. However, when *p*- and *n*-type materials are joined together, their interaction produces some very useful characteristics. The interface between two such materials illustrated in Fig. 13.3 is called a *pn* junction. At such an interface a number of currents can flow. First, consider the diffusion current. Since more free electrons exist in the *n*-type material, and more holes in the *p*-type, electrons will diffuse from right to left and holes will diffuse from left to right. Both of these diffusion processes create a net conventional current to the right. With no external connections to draw off the diffusion current, it will deposit a net negative charge in the *p*-type material and a net positive charge

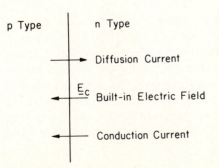

FIGURE 13.3.
Currents and the electric field near a *pn* junction.

in the *n*-type material. An electric field will build up due to these charges, which points from right to left. This field is called the built-in, or contact, electric field $\mathbf{E}_c$. In the presence of this electric field a conduction current will begin to flow from right to left, carried by electrons in the *p*-type material and holes in the *n*-type material. Since no external connections exist, KCL dictates that the diffusion current and conduction current must be equal and opposite. In other words, after a *pn* junction is formed, an electric field builds up until the conduction current is exactly equal and opposite to the diffusion current. After a very short time no net current will flow. The transition region is very small, in the order of 5×10^{-7} m (0.5 micron). The potential difference across the region due to the electric field is called the contact, or threshold potential and is only a few tenths of a volt, typically 0.7 V for Si diodes and 0.3 V for Ge diodes. This voltage difference is an important factor in the detailed operation of a semiconductor diode since it forms a potential barrier which the conventional current carriers must overcome to travel from left to right across the *pn* junction.

In the next paragraphs we describe the current–voltage relationship for a semiconductor diode. For a resistor, the *I–V* curve is a straight line which corresponds to the fact that it is a linear element. A typical diode curve is illustrated in Fig. 13.4, and it is far from being a straight line! The measurement sense of I_D and V_D are shown in the sketch on the right-hand side of the figure. (The symbol for a diode is also shown in that sketch.)

FIGURE 13.4.

Typical *I–V* curve for a germanium diode. The symbol for a diode is shown, as are the direction in which I_D is measured and the polarity by which V_D is defined for the curve.

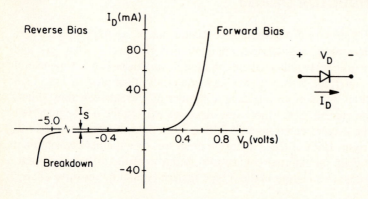

To understand this curve, suppose we attach a battery across the diode as illustrated in Fig. 13.5. The potential difference creates an electric field $\mathbf{E}_A$ which is in the same direction as $\mathbf{E}_c$. This sum of these two fields entirely suppresses the flow of electrons from left to right across the junction by increasing the potential barrier across the boundary. The result is that a very small current flows to the left which is carried entirely by the minority carriers, that is, by the holes in the *n*-type material and by the free electrons in the *p*-type material since only holes can "cross" the boundary. This is the small current I_s in the "reverse bias" portion of the *I–V* curve plotted in Fig. 13.4.

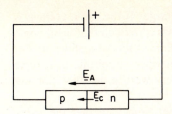

FIGURE 13.5.
Reverse bias hookup for a diode. The applied E_A and the contract electric fields E_C are in the same direction.

At very large negative voltages, the electrons gain so much energy *between* collisions with the lattice that these collisions break apart some of the bonds and release more carriers. These in turn collide with the lattice and release even more carriers. Such an avalanche effect is termed reverse bias breakdown and the diode conducts electricity very well. This regime of voltage corresponds to the very left-hand portion of the curve. Note that the voltage scale is broken since the reverse bias breakdown occurs at several volts of negative potential.

When the diode is "forward biased," the contact potential tends to be canceled by the applied potential. For very small positive forward voltages, the barrier is merely reduced and some small number of free electrons from the n-type material cross the boundary and fill a hole in the p-type material to carry current in the forward direction. When the applied voltage considerably exceeds the contact, or threshold, potential V_T, the device enters an ohmic regime where large numbers of free electrons cross the junction and the I–V curve becomes a straight line just like a resistor. The threshold voltage V_T is different for different diodes and is temperature dependent since the diffusion current depends upon temperature.

13.3 · OPERATION OF SEMICONDUCTOR DIODES

As discussed above the characteristic I–V curve for a typical diode has the schematic form shown in Fig. 13.6. This curve is far from the straight line plotted in Chapter 2 for a resistor and hence we should not expect the device to behave in the same fashion. We can gain some insight into its operation, however, by drawing analogies to the resistor. Suppose the voltage across the diode is considerably larger than V_T.

FIGURE 13.6.
Schematic I–V curve for a diode.

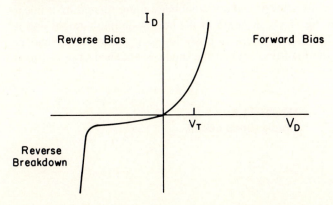

The curve shows that the forward current is then very large. If the same voltage is applied in the reverse direction, a very small current flows (provided that breakdown does not occur). The diode thus acts like a small resistance (almost a short circuit) in the forward direction and a large resistance (almost an open circuit) in the reverse direction. In other words, the diode conducts current very well in one direction but not so well in the other.

The diode controls the flow of current in much the same way that the valve shown in Fig. 13.7 controls the flow of water. If the water pressure is high enough to overcome the spring tension water will flow to the right, the forward biased direction. If the pump is reversed, water cannot flow to the left since the flap will close. In the diode when the voltage V_D across a diode is positive and larger than V_T ("forward biased"), current easily passes through it in the forward direction. When the voltage polarity is reversed, the diode "shuts off" and a weak current flows in the reverse direction. The single most important property of the semiconductor diode is its valvelike character since it affords the engineer a simple device which can be used to control the flow direction of electrical current without any mechanical parts.

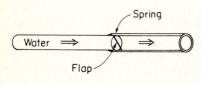

FIGURE 13.7.
An analogy to the diode. When the water pressure is high enough the flap opens and water flows to the right. The pressure required to open the orifice corresponds to the threshold voltage V_T. Flow to the left is shut off when the flap closes against the inner diameter of the smaller pipe.

Our next task is to determine the voltage across and the current through a diode when it is inserted into a circuit. Like a resistor, the diode is a passive device and does not by itself generate any voltage or current. The particular point Q on the curve where the diode operates is determined by properties of the rest of the circuit to which it is attached. The elemental diode circuit is illustrated in Fig. 13.8. When the diode is attached at the points a and b, some current I_D will flow and there will be some voltage V_D across the diode. These two numbers must by definition fall on the I–V curve for the diode. As discussed in Chapter 3 and in Text Example 13.1 below, I_{ab} and V_{ab} for the resistor–battery combination are related by a linear equation which is given in Fig. 13.9. A graph of the corresponding straight line is also plotted in the figure. This line includes all possible I_{ab}–V_{ab} pairs for the battery and resistor and is called the load-line. When the diode is attached at a and b, it

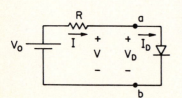

FIGURE 13.8.
Simple diode circuit.

FIGURE 13.9.
I–V curve for the output of the resistor-battery pair in the circuit of Fig. 13.8.

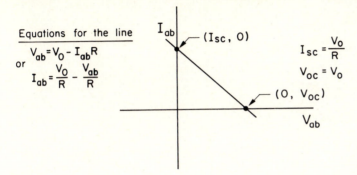

Equations for the line

$$V_{ab} = V_0 - I_{ab}R$$

or

$$I_{ab} = \frac{V_0}{R} - \frac{V_{ab}}{R}$$

$$I_{sc} = \frac{V_0}{R}$$

$$V_{oc} = V_0$$

must be the case that $I_{ab} = I_D$ and $V_{ab} = V_D$. In turn, this requires that the circuit, operating point Q must lie on both curves (Figs. 13.6 and 13.9) simultaneously. This can only occur at the intersection point illustrated in Fig. 13.10, and the solution is found.

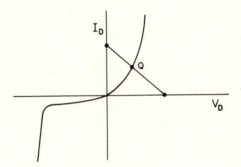

FIGURE 13.10.
Superposition of the I–V curves for the resistor-battery pair and for the diode.

TEXT EXAMPLE 13.1 —————————————————————————

Use the graphical method to find the operating points (I, V) of (1) the 25-Ω resistor in the circuit shown on the left below and (2) of the diode in the circuit shown on the right. The diode used has the characteristics shown in Fig. 13.4. Use Ohm's law to check the solution for the resistor.

(a)

Solution: For both cases we must construct the set of all (I, V) pairs which can characterize the output of the circuit shown below:

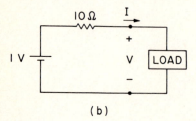

(b)

Since the battery and resistor are linear elements, the (I, V) pairs must fall on a straight line. Two points determine a line, so we are free to choose two particularly simple conditions — a short circuit and an open circuit,

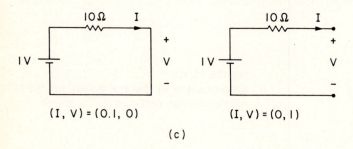

(c)

The load-line is found by joining the two points $(0.1, 0)$ and $(0,1)$, as shown below:

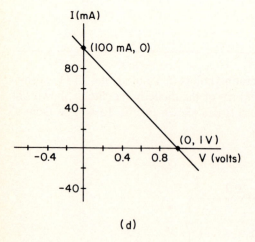

(d)

The next step is to superimpose this curve upon the (I, V) curves for the 25-Ω resistor and for the diode and to find their intersection:

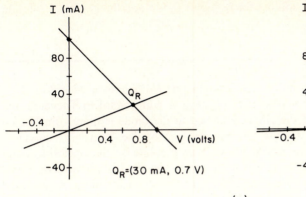 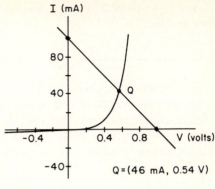

(e)

We may check the resistor solution analytically since mathematical expressions exist for both curves. The output curve for the 1–V, 10-Ω pair is from Ohm's law:

$$I = \frac{1 - V}{10} \qquad (a)$$

or

$$I = -0.1V + 0.1$$

The equation for the 25-Ω resistor curve is

$$V = 25I$$

or

$$I = 0.04V$$

Setting these two equal we can solve for V,

$$0.04V = -0.1V + 0.1$$

$$V = 0.71 \text{ volts}$$

and thence for I using either equation ($I = 0.029$ A). These results agree with the graphical solution within the accuracy of that method. Although we do not have a mathematical expression for this diode curve, we can check for consistency with the voltage drop across the 10-Ω resistor due to the current I_D. That is, from Eq. (a) it should be the case that

$$V_D = 1 - 10I_D$$

where $I_D = 0.046$ A. Indeed, solving for I_D yields $0.54V$ and the graphical solution is consistent.

The parameters which determined the straight-line I–V curves used in Fig. 13.9 were the same as those discussed earlier in the text with regard to Thevenin and Norton equivalent circuits. If the circuit attached to the diode is more complicated than the one in Fig. 13.8, it must first be replaced by the Thevenin equivalent circuit before analysis. As an example, we find graphically the operating point of the circuit shown in Fig. 13.11a when the diode has the I–V relationship given by the solid curve in Fig. 13.11b. (The diode curve in this example is exaggerated for ease of display.) To proceed, we find the Thevenin equivalent circuit to the left of the nodes a and b. That circuit has $V_{Th} = V_{oc} = 3$ V and $R_T = 30\ \Omega$. It follows then that $I_{sc} = 0.10$ A. The load-line is found by connecting the two (I, V) pairs, $(0, V_{oc})$ and $(I_{sc}, 0)$, as shown by the dashed line in Fig. 13.11b. The two curves intersect at Q where $(I_Q, V_Q) = (0.05$ A, 1.5 V$)$. Note that we use V_{Th} to represent the Thevenin equivalent voltage to avoid confusion with V_T, the threshold voltage.

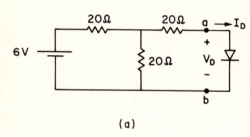

(a)

FIGURE 13.11a.
Find the operating point for this circuit if the diode has the *I–V* curve illustrated in 13.11*b*.

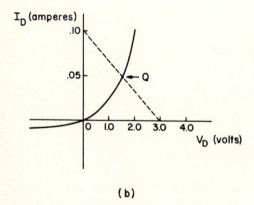

(b)

FIGURE 13.11b.
I–V curve for the diode (solid curve) and for the Thevenin equivalent circuit (dashed line) of the network to which it is attached.

TEXT EXAMPLE 13.2 _____

Find the operating point for the diode in the circuit below if the diode used is identical to the germanium diode whose *I–V* curve is illustrated in Fig. 13.4.

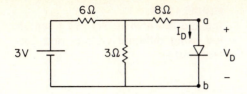

Solution: The open-circuited voltage at *a–b* is [3/(3+6)](3 V) = 1 V, which equals the Thevenin equivalent voltage V_{Th}. The Thevenin resistance can be found from the R_o method after suppressing the battery (short circuit). $R_o = R_T = 8\ \Omega + (3\ \Omega\ //6\ \Omega) = 10\ \Omega$. This is the same circuit analyzed above in Text Example 13.1 and hence has the same solution, $(I_D, V_D) = (46\ \text{mA}, 0.54\ \text{V})$.

13.4   DIODE MODELS WITH APPLICATION TO RECTIFIER CIRCUITS

Engineers often use mathematical models of electronic elements which approximate the properties of the device. The more exact the requirement on the analysis, the more complicated the model must be. The simplest model for a diode corresponds to the approximation of the diode characteristic shown in Fig. 13.12.

FIGURE 13.12.
I–V curve for the simplest diode model, a short circuit for V > 0 and an open circuit for V < 0.

For positive voltages this diode model has zero resistance (short circuit/infinite slope) whereas for negative voltages it has infinite resistance (open circuit/zero slope). In this approximation the diode is a perfect valve, freely allowing current flow for $V > 0$ and stopping all current for $V < 0$. In terms of symbols we denote this "ideal diode" as shown in Fig. 13.13.

FIGURE 13.13.
Symbol for an ideal diode which has the
I–V curve plotted in Fig. 13.12.

Such a model is quite adequate when V_T is small compared to the other voltages in the circuit. For example, suppose we wish to use the germanium diode of Fig. 13.4 in a circuit characterized by tens of volts rather than a few volts. In such an application a more appropriate way to plot the same diode curve is illustrated in Fig. 13.14. On such scales the diode curve is quite close to the ideal diode curve plotted in Fig. 13.12 (except for the reverse bias breakdown region).

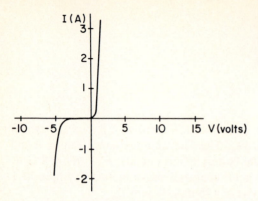

FIGURE 13.14.
The diode curve in Fig. 13.4 replotted using different voltage and current scales.

A slightly more realistic diode model is illustrated in Fig. 13.15. This plot includes the threshold voltage V_T. The I–V curve corresponds to a short circuit for $V > V_T$ and to an open circuit for $V < V_T$. We can make a circuit model for the device with an ideal diode plus a series battery with a voltage equal to V_T, which is oriented in such a way as to oppose the forward current flow (See Fig. 13.16).

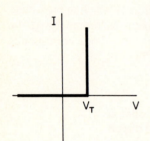

FIGURE 13.15.
I–V curve for a diode model including the effect of a threshold voltage.

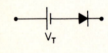

FIGURE 13.16.
Representation of a diode model which includes the threshold voltage in series with a resistor.

Finally, a more precise circuit model and its corresponding I–V graph are given in Fig. 13.17. This model includes an ideal diode, a battery voltage V_T, and a small series resistor equal to the inverse slope of the straight line for $V > V_T$. A method for

FIGURE 13.17.
I–V curve and circuit elements for a slightly more realistic model of a diode including a finite but constant dynamic resistance.

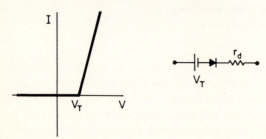

determining r_d is shown in the text example below. We return to this construction in Section 13.6 for a more rigorous mathematical analysis.

TEXT EXAMPLE 13.3

Construct a circuit model of the form shown in Fig. 13.17 which approximates the operation of a certain germanium diode. The diode curve is illustrated in Fig. 13.4.

Solution: First, a line is constructed tangent to the curve for voltages larger than V_T as shown below.

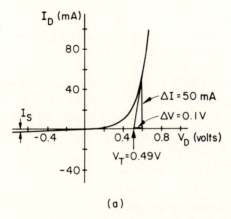

(a)

This line intersects the V axis at $V_T = 0.49$ V. The inverse slope of that line can be estimated from the triangle shown, which yields

$$r_d = \left(\frac{\Delta I}{\Delta V}\right)^{-1} = 2\ \Omega$$

The diode model is

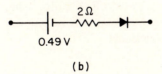

(b)

One of the most important applications of diodes is in the construction of voltage "rectifiers." As we noted in Chapter 8, power systems usually operate with oscillatory signals such as the one illustrated in Fig. 13.18. The voltage spends equal amounts of time at positive and negative values and thus has a long-time average value of zero volts. Many electronic and mechanical applications require a steady or dc voltage to operate, which implies an average voltage that is nonzero. Rectifier circuits made out of diodes provide a method to convert an oscillatory signal into one with a nonzero average. For example, consider the simple circuit shown in Fig.

FIGURE 13.18.
Sinusoidal waveform corresponding to the voltage in the power system in the United States, $v(t) = V_0 \sin \omega t = 170 \sin(377t)$. The "radian frequency" is $\omega = 2\pi/T$, where T is the period of the sine wave.

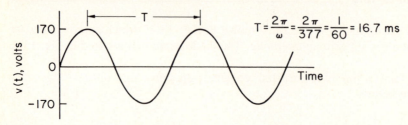

13.19a, where $v(t)$ is the sinusoidal signal of Fig. 13.18 which satisfies the mathematical relationship $v(t) = 170 \sin(377t)$ V, corresponding to the "house" voltage in the United States and Canada.

Now suppose we use the simplest model for a diode to analyze the current which flows through the series resistor and diode, that is, the model in Fig. 13.12. When $v(t)$ has a positive value the diode acts as a perfect short circuit and the current flow is clockwise with magnitude

$$i(t) = \frac{v(t)}{R} = 1.0 \sin(377t) \text{ A} \qquad v(t) > 0$$

When $v(t)$ is negative, the diode acts like an open circuit and

$$i(t) = 0 \qquad v(t) < 0$$

A plot of $i(t)$ versus time for this case is given in the top panel of Fig. 13.19b. This waveform is termed a half-wave rectified sine wave since only half of the sinusoidal signal appears in the current waveform. If more accuracy is required, we could analyze the circuit using the diode model illustrated in Fig. 13.15. In this case the current does not flow at all until $v(t)$ exceeds the voltage V_T and the current waveform looks like that plotted in the middle panel of Fig. 13.19b. The maximum current in this model would be equal to $(V_o - V_T)/R$. Finally, using the more sophisticated model of Fig. 13.17, the maximum current will be reduced further by the finite dynamic resistance r_d of the diode such that

$$i_{max} = (V_o - V_T)(R + r_d)$$

If r_d is small compared to R and if $V_T \gg V_o$, it hardly matters which of the models is used in the analysis. In the next section we show how diodes may be used to

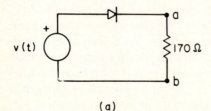

FIGURE 13.19a.
Simple "half-wave" rectifier circuit.

(a)

FIGURE 13.19*b*.
Half-wave rectifier waveform for progressively more sophisticated models of the diode. The variation is exaggerated to emphasize the effect.

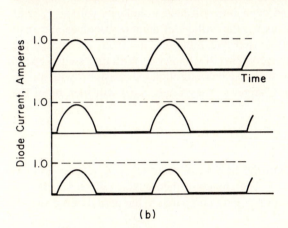

(b)

create full-wave rectified sine signals for which the "deadtime" in the curves of Fig. 13.19*b* is filled in with identical positive voltage excursions.

We have accomplished part of the task required to convert a sinusoidal voltage source into a constant dc voltage source. The curves in Fig. 13.19*b* at least have an average value greater than zero.

13.5   ELEMENTS OF POWER SUPPLY DESIGN

The most common source of electrical energy is an ac power network. However, many transistor based systems require dc voltage levels to function, and it is important to design efficient ac-to-dc power converters. As a source of dc power the circuit in Fig. 13.19*a* may be improved considerably by adding a capacitor in parallel with *R*. In such a circuit the capacitor is charged up on the positive swing of the power source. When the source swings negative the charged capacitor behaves as a storage tank for energy and can continue to supply current to a load, for example, even when the source is at a negative voltage and the diode is reverse biased. A somewhat more flexible power supply design is illustrated in Fig. 13.20. Here, a transformer is used to adjust the voltage v_s in such a way that the dc component of $v(t)$ is the desired power supply voltage.

FIGURE 13.20.
Simple power supply using a transformer, an ideal diode, and an *R–C* network.

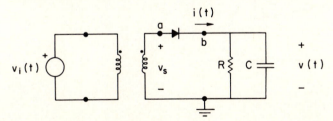

A quantitative analysis of this simple power supply circuit proceeds as follows. When the diode is forward-biased the capacitor charges up to the peak value V_m of the voltage v_s. When $v_s(t)$ begins to decrease on the left-hand side of the diode (point a) in Fig. 13.20, the charged capacitor will tend to hold its voltage v_b near to the value V_m. However, as soon as v_a, the voltage at point a, becomes less than v_b, the ideal diode becomes reverse-biased and shuts off. In effect, it acts like the switch in one of the natural response problems of Chapter 6. With the "switch open" (diode reverse-biased) the capacitor will discharge through the resistor R with a time constant $\tau = (RC)^{-1}$. As found in Chapter 6, the voltage $v_b(t)$ will then be of the form

$$v_b(t) = V_m e^{-t/RC}$$

This time function will be valid until the half-wave rectifier cycle repeats and the diode turns on again ($v_a > v_b + V_T$). The capacitor then charges up to V_m and the process repeats. If the RC time constant is long compared to the period of the signal, that is,

$$\tau = RC > \frac{2\pi}{\omega}$$

then the voltage will not discharge very much before the signal $v_i(t)$ reverses polarity again and charges the capacitor back up. The result is a waveform similar to that illustrated in Fig. 13.21.

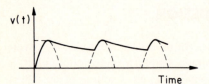

FIGURE 13.21.
Voltage response of a half-wave rectifier with a simple $R-C$ filter.

The solid curve in Fig. 13.21 has an average value that is positive. The deviation of this signal from a straight line (constant) value is called the ripple of the power supply, V_r. There are various quantitative definitions of the ripple voltage in the literature. Here, we take it to be one-half of the peak-to-peak ac voltage which is superimposed upon the dc level, V_L. The various voltages are defined in Fig. 13.22.

FIGURE 13.22.
Waveforms and definitions for a half-wave rectifier-based power supply.

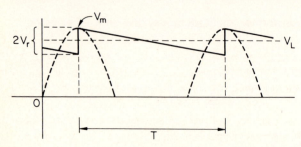

From the figure, $V_m = V_L + V_r$. A good power supply will have a small ripple, that is, $V_r << V_L$ and we measure the quality of the supply by expressing the ratio V_r/V_L as a percentage, the "percent" ripple.

Clearly, the capacitor should be large enough that it does not discharge very much in the time $T = 2\pi/\omega$, a rough measure of the time when it is in the discharge state. As a first guess, then, we need $RC >> T$. If that inequality holds, for $t \leq T$ the function

$$v_b(t) = V_m e^{-t/RC} \tag{13.2}$$

can be approximated by the function

$$v'_b(t) \simeq V_M\left(1 - \frac{t}{RC}\right) \tag{13.3}$$

This holds since the power series expansion of the function e^x is given by

$$e^x = 1 + x + \frac{x^2}{2!} + \cdots$$

For small values of x ($x = -t/RC$ in this case), $e^x \simeq 1 + x$ and Eq. 13.3 is a good approximation to Eq. 13.2. If we approximate the true response by Eq. 13.3 we have the solid straight line segments shows in Fig. 13.22. The ripple magnitude is then given by the equations

$$2V_r = v'_b(0) - v'_b(T)$$

$$2V_r = V_m - \left(1 - \frac{T}{RC}\right)V_m$$

$$2V_r = \left(\frac{2\pi}{\omega RC}\right)V_m$$

and hence

$$\left(\frac{V_r}{V_m}\right)_{hw} = \frac{\pi}{\omega RC} = (2fRC)^{-1} \tag{13.4a}$$

where hw stands for half-wave rectification. Now the expression 13.4a is not exactly the same as the ripple ratio defined earlier, V_r/V_L. However, if the ripple is small then $V_L \approx V_m$ and the two ratios are almost equal. In the remainder of this chapter we shall use the ratio V_r/V_m as the measure of the ripple in a power supply.

TEXT EXAMPLE 13.4 _____

Consider the simple rectifier circuit using an ideal diode shown in the figure below.

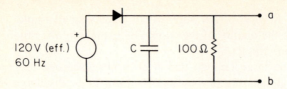

Find the minimum value of C required to yield a ripple voltage which is less than 10% of the dc voltage. What is the dc voltage V_{ab}? How much does the ripple increase if a 500 Ω load is added across $a-b$ and the capacitor remains fixed at the same value?

Solution: From the formula in Eq. 13.4a we have

$$0.1 = \frac{1}{(2)(60)(100)C}$$

and hence

$$C = 833 \ \mu\text{F}$$

is the minimum value of C which will yield a 10% ripple. The dc voltage V_L is $(0.9)(170) = 153$ V. If we add a 500-Ω resistor in parallel, the effective parallel resistance is 83.3 Ω. Since the ripple is inversely proportional to the total resistance parallel to C, the new ripple percentage is $10(100/83.3) = 12\%$.

The ripple can be improved by a factor of 2 if a full-wave rectifier is used since there is half as much time for the voltage to decay before the next charging cycle begins. The ripple ratio is then

$$\left.\left(\frac{V_r}{V_m}\right)\right|_{fw} = (4fRC)^{-1} \tag{13.4b}$$

Two simple full-wave rectifier circuits are illustrated here. In Fig. 13.23 a center-tapped transformer is used and only two diodes are needed. In Fig. 13.24 a less expensive transformer is used but four diodes are required. The transformers allow designs for virtually any desired dc voltage since the turns ratio may be used to

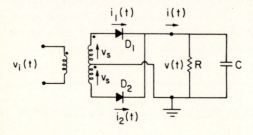

FIGURE 13.23.
Full-wave rectifier circuit using two diodes.

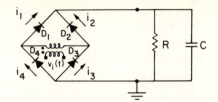

FIGURE 13.24.
Full-wave rectifier using four diodes.

adjust the value of V_m. An illustrative example using a full-wave rectifier is worked out in the problem set.

The reader may wonder why we have not used Fourier analysis methods in the analysis of rectifier circuits. The problem is that a Fourier synthesis approach is only applicable to linear systems. Since the diode is a nonlinear device, extreme care must be used in any such approach. In the circuits of Text Example 13.4 and in Figs. 13.23 and 13.24, for example, once the $R–C$ networks are attached, the half-wave or full-wave rectified waveforms do not exist as a "signal" anywhere in the system.

To apply Fourier methods we could apply the output of a diode rectifier circuit to a small resistance R_s as shown in (a) of Fig. 13.25. Similarly, the circuit shown in (b) uses an Op-Amp voltage follower to yield a half-wave rectified signal $v_1(t)$ which does not depend upon R and which is suitable for linear analysis. The latter is of little use as a power supply since the Op-Amp itself requires dc supplies to function. Op-Amp circuits using rectifiers do have applications for signal processing networks, however. In any case in the circuits studied below using Fourier methods we shall assume that $v_1(t)$ is either a half-wave or a full-wave rectified signal which behaves as an ideal source similar to the output of the circuit in Fig. 13.25b.

The Fourier series expansion of the half-wave and full-wave rectified sine waves were given in Chapter 10 (Table 10.1) and are reproduced below.

$$v_{hw}(t) = \frac{V_o}{\pi} + \left(\frac{V_o}{2}\right)\sin(\omega_o t) - \left(\frac{2V_o}{3\pi}\right)\cos(2\omega_o t) - \left(\frac{2V_o}{15\pi}\right)\cos(4\omega_o t) - \left(\frac{2V_o}{35\pi}\right)(6\omega_o t) + \cdots$$
$$(13.5a)$$

$$v_{fw}(t) = \left(\frac{2V_o}{\pi}\right) - \left(\frac{4V_o}{3\pi}\right)\cos(2\omega_o t) - \left(\frac{4V_0}{15\pi}\right)\cos(4\omega_o t) - \left(\frac{4V_o}{35\pi}\right)\cos(6\omega_o t) + \cdots \quad (13.5b)$$

FIGURE 13.25.
Two circuits which yield approximate ideal half-wave rectified sources.

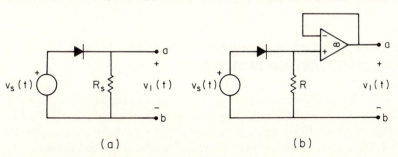

(a)   (b)

The first term of the Fourier series is equal to the average value since all the other terms are sine or cosine functions which have an average value equal to zero. Our objective is to pass this signal through a network that suppresses the high-frequency components of these waveforms and leaves the dc component unchanged.

Suppose we pass these two signals through an ideal low-pass filter, which has a transfer function $H = V_2/V_1$ whose magnitude is plotted in Fig. 13.26a. The only term in either Fourier series which will survive this filter is the first. The result is that the output signals are characterized by constant dc values (Fig. 13.26b).

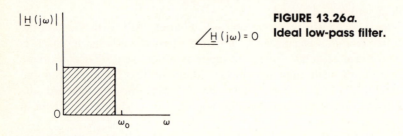

FIGURE 13.26a.
Ideal low-pass filter.

FIGURE 13.26b.
Schematic diagram showing the effect of an ideal low-pass filter on half-wave and full-wave rectified sine waves. The output in each case is a positive dc voltage.

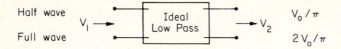

Fourier formalism is used to analyze the so-called choke input power supply filter (see Fig. 13.27) in the problem set. Conceptually the system transfer function is applied to the Fourier series representation of either a half-wave or full-wave rectified "ideal source" signal $v_1(t)$. In each case a dc component survives in $v_2(t)$ along with additional attenuated components at the other frequencies. Most of the "ripple" component will be at the first harmonic of the source ω_0 for a half-wave rectifier and at the second harmonic $2\omega_0$ for a full-wave rectifier.

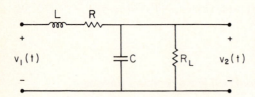

FIGURE 13.27.
Choke input filter for power supply. The resistance in series with the inductor represents the finite resistance of the wires inside the inductor.

13.6 THE RESPONSE OF DIODES TO SMALL TIME-VARYING SIGNALS

Returning now to the elemental diode circuit in Fig. 13.8, suppose in addition to the battery there is also a small time-varying voltage $v_1(t)$ in series with V_0 (see Fig. 13.28). The current which flows through the diode will now have a time variation related to the voltage $v_1(t)$ in addition to the steady current we found earlier.

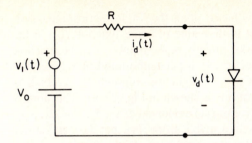

FIGURE 13.28.
Addition of a small time-varying signal due to the elemental diode circuit.

When the applied voltage varies about the value V_o due to $v_1(t)$, the current varies in such a way that the instantaneous operating point moves along the diode curve. Suppose we are using a diode with the characteristic curve illustrated in Fig. 13.29, and that $V_o = 3$ V, $R = 30$ Ω and $v_1(t) = 0.5 \sin(\omega t)$ V. Then, when $t = 0$, $v_1(t) = 0$ and the operating point and the load line will be as shown by the dashed line. The values I_Q and V_Q are 0.05 A and 1.5 V, respectively. As a function of time, the total voltage $V_o + v_1(t)$ will vary in the range 2.5 V to 3.5 V due to the sinusoidal signal. At each extreme of this voltage swing we can freeze the system and construct the two load lines drawn in solid in Fig. 13.29. The three lines are parallel since the resistor R does not change and since it determines the slope of the load line. The current through the diode will vary since the intersection point with the diode curve moves along that curve. The arrows at the left show the values of the current at the voltage extremes. Note that even though the voltage change is symmetric about V_o (± 0.5 V), the current through the diode changes more for $V_o + 0.5$ than it does for $V_o - 0.5$. This feature is characteristic of nonlinear systems. In this case the diode current varies in the range 0.04 A $\leq I_D(t) \leq$ 0.062 A as the applied voltage varies from 2.5 V to 3.5 V.

We now turn to the development of a linear model for the forward-biased diode. Linear models play an important role in the analysis of many systems that are inherently nonlinear. One reason they are developed is to allow use of analytic tools such as Fourier and Laplace methods which rely upon the linearity of the equations. Another is that for small signals, a linear model does indeed yield accurate solutions.

FIGURE 13.29.
The "dynamic" operating point moves along the load line as the signal varies in time.

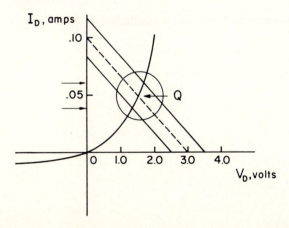

The mathematical approach we use is based upon the Taylor series expansion of a function. Suppose we are interested in the value of some function $f(x)$ in the neighborhood of the point $x = x_o$. Of course, directly at the point $x = x_o$ the function takes on the exact value $f(x_o)$. For values of x in the neighborhood of x_o, we may find $f(x)$ by approximating the function by a straight line through $f(x_o)$ which is tangent to the curve at $x = x_o$. This straight line is shown in Fig. 13.30. The slope S of the tangent line is given by the derivative of $f(x)$ evaluated at x_o, $S = (df/dx)_{x=x_o}$. Very close to Q the straight line is indistinguishable from $f(x)$, but as x gets away from x_o, the straight line deviates from $f(x)$ more and more. In the problem set it is shown that the straight line in the figure has the equation

$$g(x) = f(x_o) + S(x - x_o) \qquad (13.6)$$

This function is a linear approximation to $f(x)$ near $x = x_o$. Equation 13.6 contains the first two terms of the so-called Taylor series expansion of the function. The full Taylor series expansion of a function is

$$f(x) = f(x_o) + \left(\frac{df}{dx}\right)_{x_o} \Delta x + \left(\frac{1}{2}\right)\left(\frac{d^2f}{dx^2}\right)_{x_o} (\Delta x)^2 + \left(\frac{1}{6}\right)\left(\frac{d^3f}{dx^3}\right)_{x_o} (\Delta x)^3 + \cdots$$

If we use enough terms the expansion will yield a value that is nearly equal to $f(x)$ no matter how large Δx becomes. But if we only retain the first two terms in this expansion, the resulting function is a linear approximation to the actual function which is valid in the neighborhood of $x = x_o$. For small signals this holds, and we may use a linear approximation.

To apply this mathematical result to a nonlinear electronic device such as a diode we consider the diode current i_d to be a function of the diode voltage, v_d:

$$i_d = i_d(v_d)$$

We are interested in the diode current in the neighborhood of some operating point which is characterized by the dc values I_Q and V_Q. For the linear approximation to hold, we require that the diode voltage change only by a small amount. Turning to

FIGURE 13.30.
An arbitrary function $f(x)$ may be evaluated near some particular point Q using a Taylor series expansion.

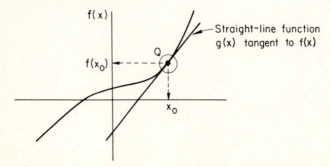

the circuit given in Fig. 13.28, this condition holds provided that the "signal" $v_1(t)$ has an amplitude which at its maximum value is much smaller than the battery voltage V_o. Then, the diode will always be in the forward-biased regime and, furthermore, the circuit is appropriate for linearization. To proceed, we make a Taylor series expansion about the operating point Q:

$$i_d(t) = I_Q + \left(\frac{di_d}{dv_d}\right)_Q (v_d(t) - V_Q) + \left(\frac{1}{2}\right) \left(\frac{d^2 i_d}{dv_d^2}\right)_Q (v_d(t) - V_Q)^2 + \cdots$$

For small values of $v_d(t) - V_Q$, only the first two terms are needed and

$$i_d(t) = I_Q + \left(\frac{di_d}{dv_d}\right)_Q (v_d(t) - V_Q) \tag{13.7}$$

When the derivative is evaluated at Q, the result is simply a number which has the units of inverse resistance. If we define a resistor value r_d such that

$$\left(\frac{di_d}{dv_d}\right)_Q = (r_d)^{-1} \tag{13.8a}$$

then substituting into Eq. 13.7 yields

$$i_d(t) = \frac{v_d(t)}{r_d} + \left(I_Q - \frac{V_Q}{r_d}\right)$$

This equation may also be written as

$$i_d(t) = \frac{v_d(t) - V_T}{r_d} \tag{13.8b}$$

where

$$V_T = V_Q - r_d I_Q \tag{13.8c}$$

Equations 13.8a through 13.8c represent a linear *mathematical* model for the behavior of a diode in the neighborhood of an operating point Q. The *circuit* model for this set of equations is as shown in Fig. 13.31, since the current through the resistor r_d in such a circuit is, by Ohm's law, given by Eq. 13.8b. This linear circuit model may replace the diode in any circuit for which the linearity conditions hold.

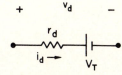

FIGURE 13.31.
Linear small signal circuit model for a forward-biased diode. V_T is the forward bias voltage and r_d the dynamic resistance.

That is, for any circuit in which the diode current and voltage change by amounts much smaller than I_Q and V_Q.

We have yet to interpret the voltage V_T in terms of the original diode curve. From Eq. 13.8*b* we see that $i_d(t) = 0$ when $v_d(t) = V_T$. The voltage V_T is thus the intercept of the tangent to the diode curve with the horizontal (voltage) axis. For a steep diode curve we may identify V_T with the threshold voltage of the diode discussed earlier in this chapter.

For the present example, the diode curve in Fig. 13.28 may be analyzed as illustrated in Fig. 13.32. A straight line tangent to the curve is constructed through the point Q. Its intercept on the V axis is $V_T = 0.8$ V. The derivative may be estimated using the triangle construction shown, which yields $r_d = 17\ \Omega$. As a check on these results we can compare them with the graphical solution for V_Q and I_Q.

FIGURE 13.32.
Illustration of the method for finding the values of r_d and V_T.

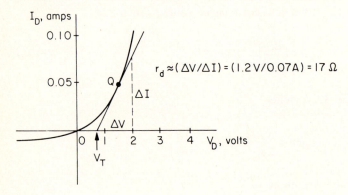

If we replace the diode by its linear model and suppress the small ac signal in the circuit to study the dc response it becomes the circuit shown in Fig. 13.33. For this circuit,

$$I_Q = \frac{3 - 0.8}{47} = 0.047 \text{ A}$$

and

$$V_Q = 0.8 + 17I_D = 1.6 \text{ V}$$

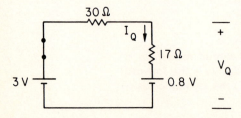

FIGURE 13.33.
DC circuit diagram with the small signal suppressed.

Within the accuracy of the methods used, these results agree well with the numbers quoted above $[(I_Q, V_Q) = (0.05 \text{ A}, 1.5 \text{ V})]$.

Turning now to the ac response we suppress both dc supplies, the real 3-V supply and the model V_T supply. We may do this now, since by creating a linear model the principle of superposition holds (Fig. 13.34). The total current through the diode will now be the sum of the dc current at the operating point plus the time-varying current $i(t)$ calculated from this circuit. Using the linear model and $v_1(t) = 0.5 \sin(\omega t)$,

$$i(t) = \frac{v_1(t)}{(R + r_d)} = \frac{v_1(t)}{47} = 10.6 \sin(\omega t) \text{ mA}$$

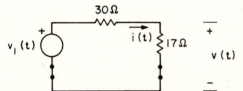

30 Ω

i (t) 17 Ω

$v_1(t)$

v (t)

FIGURE 13.34.
Linear model for the diode circuit with a small series signal and with the dc voltages suppressed.

The total diode current is the sum of the dc current at the operating point and this time-varying current. The amplitude of the time-varying voltage across the diode is given by the voltage divider ratio $r_d/(R + r_d)$ or 17/47 times the amplitude of $v_1(t)$, which yields $0.18 \sin(\omega t)$ V. The peak-to-peak variation of the current (twice the amplitude) calculated from this model is 21.2 mA. This differs slightly from the value found graphically above using Fig. 13.29 which was 22 mA. This is due in part to the fact that the signal voltage in this example, $0.5 \sin(\omega t)$ V, is not "small" compared to the 3-V dc battery, and that the linear approximation is therefore somewhat suspect. Viewed in this light, the result is surprisingly good.

To summarize, as long as the excursions in the diode voltage and current, $v_d(t)$ and $i_d(t)$, are not very large compared to V_Q and I_Q, the dc values at the operating point, the approximation of a constant r_d will yield an accurate prediction of the ac response. If the signal is too large, however, the actual voltage and current levels must be read from the graph and the signal $i_d(t)$ will not necessarily be proportional to the signal $v_1(t)$. In such a large signal case, linear analysis breaks down and we need to use the fully nonlinear curve of the diode. The linear model for a diode is not as important or as useful as the linear models we present for transistors in the next chapter, but its study serves as an introduction to that topic.

TEXT EXAMPLE 13.5

Find the dc current I_Q and the dc voltage V_Q at the operating point of the diode in the left-hand figure below if that diode has the I–V characteristic shown in the right-hand figure. Also, if the small signal $v_i = 21 \cos(\omega t)$ mV and the diode is represented by the diode model shown, find the time-varying voltage across the diode.

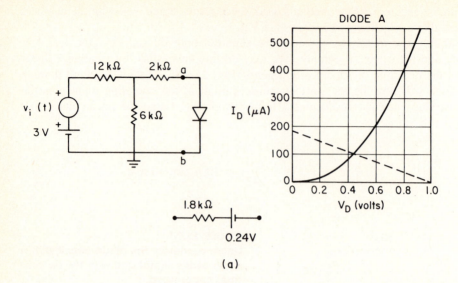

(a)

Solution: The dc response may be found by suppressing the small signal. Since the diode curve refers to the diode terminals, in order to use the load line method we need to "Thevenize" the network to which it is attached. The Thevenin equivalent dc voltage associated with the network is $V_{Th} = V_{oc} = (6/18)3 = 1$ V. The Thevenin equivalent resistance is 2 kΩ + (6 kΩ //12 kΩ) = 6 kΩ. The load line, then, is the dashed line in the figure which links the short-circuit current (167 μA) to the open-circuit voltage (1 V). The operating point is $(I_Q, V_Q) = (100 \ \mu\text{A}, 0.42 \text{ V})$. To determine the ac response we replace the diode by its linear model and suppress the dc sources. The resulting circuit is as shown on the left below. The Thevenin equivalent circuit for the network to the left of a–b is shown on the right below and may be used to analyze the small signal response.

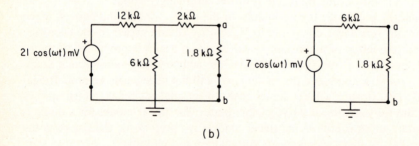

(b)

Note that the same Thevenin resistance applies in this small-signal case. Also, the same fraction of the original signal appears as the Thevenized signal, $6/(12 + 6)v_i = (1/3)v_i$. The time-varying voltage across a–b is, then,

$$v(t) = 1.6 \ \cos(\omega t) \text{ mV}$$

PRACTICE PROBLEMS
AND
ILLUSTRATIVE EXAMPLES

EXAMPLE 13.1

The $I–V$ curve for a 10-Ω resistor as illustrated in Fig. 1.6 is reproduced here:

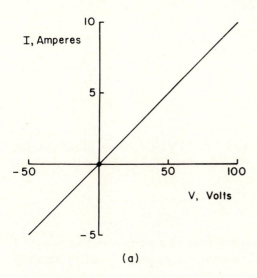

(a)

Use this curve and the load line method to find the operating point of the following circuit:

(b)

when $V_o = 50$ V and R $= 5\ \Omega$.

Solution: The solution is found from the intersection of the $I–V$ curve for the 10 Ω resistor to the right of $a–b$ with the line which corresponds to the equation for the circuit to the left of $a–b$:

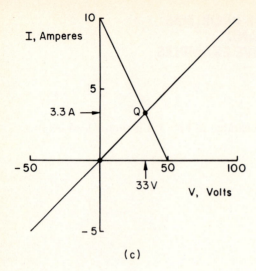

(c)

The operating point Q occurs at $(I_Q, V_Q) = (3.3 \text{ A}, 33 \text{ V})$. Of course, in this case the solution could have been found directly from the circuit using Ohm's law. (Check this yourself.)

Problem 13.1

Use the load line method to find the operating point of the circuit of Example 13.1 if $V_o = 100$ V and R $= 20$ Ω. (The 10-Ω resistor remains in the circuit.) Check this with Ohm's law.

$$\text{Ans.: } (I_Q, V_Q) = (3.3 \text{ A}, 33 \text{ V})$$

EXAMPLE 13.2

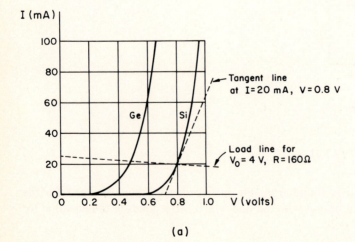

(a)

The measured forward bias volt–ampere characteristics of two *pn* junction diodes are shown above, one made with germanium (Ge) and one made with silicon (Si). Consider the following circuit:

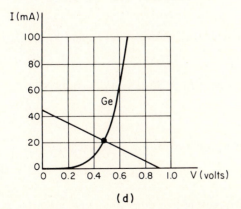

(b)

Find I_D and V_D for the germanium diode.

Solution: The load line is determined from the Thevenin equivalent current "seen" by the diode. The Thevenin equivalent voltage is the open circuit voltage

$$V_{Th} = V_{oc} = \left(\frac{25}{125}\right)(4.5) = 0.9 \text{ V}$$

The Thevenin resistance can be found using the R_o method by suppressing the 4.5-V battery. Then, $R_o = 25//100 = 20 \ \Omega$ and the equivalent circuit is

20 Ω

0.9 V

(c)

The load line is found by connecting the (I, V) pairs $(0, V_{oc})$ and $(I_{sc}, 0)$ which are $(0, 0.9 \text{ V})$ and $(45 \text{ mA}, 0)$, respectively. This load line is shown below along with the diode curve:

I (mA)

100

80

60

Ge

40

20

0

0   0.2   0.4   0.6   0.8   1.0   V (volts)

(d)

The operating point is

$$Q = (22 \text{ mA}, 0.47 \text{ V})$$

Problem 13.2

Find the operating point for the silicon diode if it is inserted in place of the germanium diode in the same circuit as used in Example 13.2.

Problem 13.3

Suppose a 2-volt battery is used in the circuit of Example 13.2 and the resistors R_1 and R_2 are equal. Find their value if the germanium diode is used and if the operating point is $(I_Q, V_Q = 35 \text{ mA}, 0.54 \text{ V})$.

> Ans.: $R_1 = R_2 = 28 \ \Omega$
> (Note: Your answer may be somewhat different due to the graphical analysis.)

Problem 13.4

Repeat Problem 13.3 for the silicon diode if the operating point is $(I_Q, V_Q) = (20 \text{ mA}, 0.8 \text{ V})$, that is, find R_1 and R_2 if $R_1 = R_2$ and the battery voltage V is 2 volts.

Problem 13.5

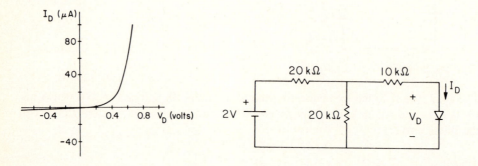

The diode in the circuit above has the characteristics shown. Find I_D and V_D, at the operating point.

> Ans.: $I_D = 26 \ \mu\text{A}, \ V_D = 0.48 \text{ V}$

Problem 13.6

Repeat Problem 13.5 if the 10-kΩ resistor is replaced with a 20-kΩ resistor.

Problem 13.7

For future reference in this problem set, we present two other diodelike curves:

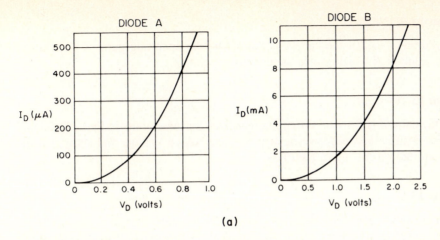

(a)

The diode in the circuit below has the characteristics of diode A above. Find I_D and V_D.

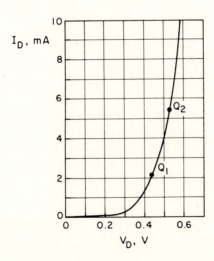

(b)

Ans.: $I_D = 200 \ \mu A$, $V_D = 0.6$ V

Problem 13.8

Design a circuit of the form given in Problem 13.5 which uses a diode whose curve is shown below, which uses 3 equal resistors of value R, which has a battery of value 1.2 V and which has the operating point Q_1 given in the figure below. (In effect, find R.)

Problem 13.9

Repeat Problem 13.8 if the battery is 2.4 V and the operating point is Q_2, that is, find the value of R which yields the operating point Q_2.

Problem 13.10

Sketch the current paths with different color lines for the circuits in Figs. 13.23 and 13.24 for positive and negative polarities of the source voltage and show that the circuits are indeed full-wave rectifiers.

Problem 13.11

The circuit shown below is used as a "full-wave" rectifier. If $v(t) = 170 \sin(377t)$, sketch the current waveform $i(t)$ using the ideal diode model. If instead the diode model of Fig. 13.17 is used and the V_T and r_d of each diode are 1 V, and 1 Ω, respectively, what would be the maximum value of the current?

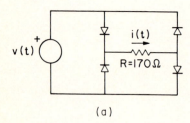

(a)

Ans.:

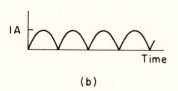

(b)

Ans.: $I_{max} = 0.977$ A

EXAMPLE 13.3

A power supply is to be designed using an R–C filter, a step-down transformer, and the rectifier circuit of Fig. 13.24 to yield a 12-V dc level and a percent ripple with no load which is less than 5%. If $R = 1$ kΩ, find the minimum value of C and the turns ratio N. What is the new percent ripple if that value of C is used and a 10-kΩ load is attached? The ac source is a standard 120-V, 60-Hz supply.

Solution: Using Eq. 13.4b,

$$C = [(0.05)(4)(60)(1000)]^{-1} = 83 \ \mu F$$

To obtain a 12-V dc supply voltage, the turns ratio must be

$$N = \frac{(0.95)(170)}{12} = 13.45$$

If a 10-kΩ load is attached, the new parallel resistance is 917 Ω and the ripple will be 5.5%.

Problem 13.12

Find the turns ratio N of the step-down transformer and the minimum value of the required capacitor C in Fig. 13.20 to yield a 12-V dc power supply and a percent ripple of 5% if $R = 2000\ \Omega$. Assume $v_i(t)$ is a 120-V (effective value!), 60-Hz source. (Why is the answer to this problem the same as that given in Example 13.3?)

$$\text{Ans.: } C = 83\ \mu F$$
$$N = 13.45$$

Problem 13.13

What is the new percent ripple in the circuit in Problem 13.12 if it is further loaded by a 5-kΩ resistor?

Problem 13.14

Find the percent ripple for the circuit of Problem 13.11 if the R–C combination of Text Example 13.4 is used in place of the 170-Ω resistor. What is the new percent ripple if the circuit is further loaded by 1 kΩ?

Problem 13.15

Design a full-wave rectifier with a simple R–C filter using the circuit of Fig. 13.23 which uses a 120-V, 60-Hz source and provides 25-V dc to the resistor which has a value of 4 kΩ. Find the minimum value of C and the step-down turns ratio N if the percent ripple is to be less than 1%. What is the ripple if that value of C is used and an additional load is attached which draws 2.25 mA? You may assume that the new load does not lower the 25-V dc level significantly.

Problem 13.16

In the circuit of Fig. 13.23 let $v_i(t) = 20\cos(1000t)$ V. Each transformer steps the voltage down and has a turns ratio of 5. If $C = 30\ \mu F$ and $R = 10\ k\Omega$, find the dc current through R. What is the percent ripple? Find the percent ripple if a 60-μF capacitor is added in parallel to C.

Problem 13.17

Find the transfer function $H = v_2/v_1$ for the "choke input" filter circuit in Fig. 13.27.

$$\text{Ans.:} \quad \frac{1}{(LC)s^2 + (RC + L/R_L)s + (1 + R/R_L)}$$

Problem 13.18

Suppose a full-wave rectified signal $v_1(t)$ is generated from a source $v_s(t) = 10\cos(1000t)$ which has the characteristics of a true source (e.g., analogous to the circuit of Fig. 13.25b). Let $v_1(t)$ be applied to the inductor choke–input filter of Problem 13.17. What is the dc output voltage if $R = 50\ \Omega$ and $R_L = 1000\ \Omega$? What is the percent ripple due to the first nonzero component of the spectrum (2000 rad/sec) if $C = 10\ \mu F$ and $L = 20\ H$? (*Hint:* Evaluate the transfer function at $s = j(2000)$ and multiply it by the amplitude of the $(2\omega_o)$ term in the Fourier expansion from Eq. 13.5b. Use peak value phasors.)

$$\text{Ans.:} \quad V_{dc} = 6.1\ V;$$
$$\text{percent ripple} = 0.087\%$$

Problem 13.19

Find the change in V_{dc} and the percent ripple if another 1-kΩ load is added in parallel to R_L in Problem 13.18.

Problem 13.20

Repeat Problem 13.18 with the same applied filter but using a half-wave rectified version of the same source voltage $v_s(t)$. Find the percent ripple at the fundamental(1000 rad/sec) and the dc voltage of the supply.

Problem 13.21

Using the choke input filter design of Problem 13.17 find the voltage V_{dc} and the percent ripple (at the fundamental) for a 50-V (peak value), 60-Hz supply which is half-wave rectified to supply $v_1(t)$ if $L = 10\ H$, $R = 20\ \Omega$, $R_L = 300\ \Omega$, and $C = 100\ \mu F$.

Problem 13.22

Repeat Problem 13.21 if the same filter is applied to a full-wave rectified signal derived from the same 50-V (peak value), 60-Hz supply. *Hint:* evaluate the ripple at 120 Hz ($2\omega_0$).

EXAMPLE 13.4

The diode whose curve is shown below is attached to an external circuit such that the operating point is at Q. Find the parameters for the linear model.

Solution: The values may be found from the construction shown

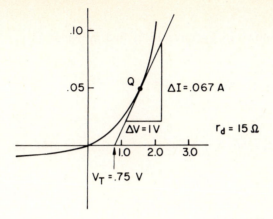

Ans.: $r_d = 15\Omega; \quad V_T = 0.75$ V

Problem 13.23

Find the dynamic resistance r_d and the threshold voltage V_T for the diode below at the operating point Q shown in the figure, where these parameters correspond to the linear model.

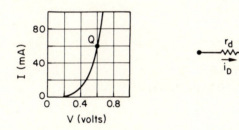

Problem 13.24

Find the linear model parameters for the two operating points Q_1 and Q_2 associated with the diode curve given in Problem 13.8.

Problem 13.25

Find the dynamic resistance and threshold voltage (the linear model parameters) for diode A given in Problem 13.7 above if the operating point has a diode current of 200 μA.

Problem 13.26

Using diode B of Problem 13.7, construct a circuit which has the load line shown. Find the parameters of the linear model at the operating point shown.

Problem 13.27

For the silicon diode characteristic shown in Example 13.2, find the value of r_d when $I = 80$ mA.

$$\text{Ans.: } r_d = 4.5 \ \Omega$$

Problem 13.28

For the germanium diode characteristic shown in Example 13.2, find the value of r_d when $I = 40$ mA.

Problem 13.29

For high forward bias voltages the two diodes whose curves are illustrated in Example 13.2 enter an ohmic regime where the curves are asymptotic to straight lines. Find the dynamic resistance associated with this portion of the curves for both diodes.

EXAMPLE 13.5

Suppose a small time-varying voltage $v_i(t)$ is inserted in series with V in the circuit below. Find $i_d(t)$ and $v_d(t)$ if the silicon diode of Example 13.2 is used.

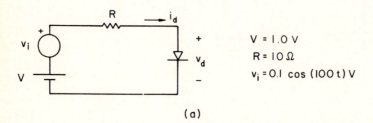

$$V = 1.0 \text{ V}$$
$$R = 10 \ \Omega$$
$$v_i = 0.1 \cos (100 \ t) \text{ V}$$

(a)

Solution: Suppressing the small signal, we may determine the operating point by connecting the point $(I, V) = (100 \text{ mA}, 0)$ with the point $(0, 1 \text{ V})$. This line crosses the Si diode line at the coordinates $(I, V) = (20 \text{ mA}, 0.8 \text{ V})$. The result from the dashed line in Example 13.2 yields $r_d = 4.1 \ \Omega$. Since the signal v_i is small, the diode is always forward-biased and may be modeled as shown below.

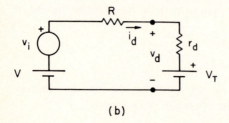

(b)

Now if we suppress the dc sources to determine the time-varying signals, we replace them with short circuits

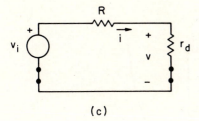

(c)

and can immediately write

$$i(t) = \frac{v_i(t)}{R + r_d} = 7.1 \cos(100t) \text{ mA}$$

$$v(t) = \left(\frac{r_d}{R + r_d}\right)v_i(t) = 29 \cos(100t) \text{ mV}$$

where the voltage divider rule was used in the last step. The total current and voltage are then

$$i_d(t) = [20 + 7.1 \cos(100t)] \text{ mA}$$

$$v_d(t) = [800 + 29 \cos(100t)] \text{ mV}$$

Problem 13.30

The diode in the circuit may be represented by the model shown. Find $v_d(t)$.

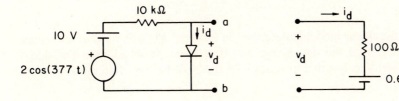

Ans.: $v_d(t) = [0.69 + 0.02 \cos(377t)]$ V

Problem 13.31

Find $i_d(t)$ for the diode in the circuit of Problem 13.30.

Problem 13.32

Find the operating point for the diode-like curve in the circuit shown below as well as the parameters of a linear model which may be used to characterize the diode at that point.

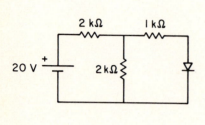

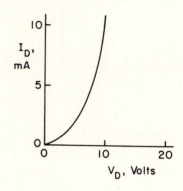

Ans.: $(I_Q, V_Q) = (3.5 \text{ mA}, 6.5 \text{ V})$

$(r_d, V_T) = (1250 \ \Omega, 3.4 \text{ V})$

Problem 13.33

If a small-signal voltage $v_i(t) = 0.5 \cos(\omega t)$ V is inserted in series with the 20-V battery in the circuit for Problem 13.32, find the total voltage across the diode and the total current through the diode.

Problem 13.34

For diode A above, find the values in the linear model that will be a close approximation of the actual diode for any voltage $V_D \geq 0.6$ V.

Ans.: $V_T \simeq 0.4$ V, $r_d \simeq 1000 \ \Omega$

Problem 13.35

If the diode whose linear model is described in Problem 13.34 is used in the circuit of Problem 13.32, find the dc voltage and dc current. If in addition the signal $v_i(t) = 0.5 \cos(\omega t)$ V is inserted in series with the 20-V battery, find the ac voltage and current.

Problem 13.36

In the circuit of Problem 13.7 a voltage $v_s = 10 \cos(500t)$ mV is inserted in series with the battery. Find $i_d(t)$.

Ans.: $i_d(t) = [200 + 5 \cos(500t)] \ \mu A$

Problem 13.37

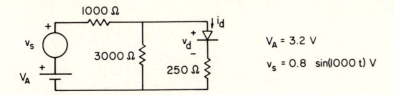

$V_A = 3.2$ V

$v_s = 0.8 \sin(1000 \, t)$ V

The diode in the circuit shown has the characteristic of diode B above. Find the diode current $i_d(t)$ and the diode voltage $v_d(t)$ (include both the dc and ac components in your answer).

Chapter 14

Transistors and Elementary Transistor Amplifiers

In Chapter 13 we introduced the diode, a nonlinear device, into our study, discussed its operation, and presented some applications. We also showed how, under certain conditions, a linear model can be substituted for a diode in circuit analysis. Like the resistor, the diode is a two-terminal, one-port element. In this chapter nonlinear three-terminal elements are discussed. As examples of such devices we introduce two particular types of transistors and subsequently use these same devices in the remainder of the text. The transistor presentation is not intended to be exhaustive but rather to provide concrete examples of methods which have general application. Linear models are derived from the two transistor types and simple amplifier circuits are analyzed using these models. Since a transistor has three terminals, it is a two-port element and may be treated as discussed in Chapter 4. More realistic amplifier circuits are described in Chapter 15. Peak valve phasors are used in this chapter.

14.1 MOSFET TRANSISTORS

In this section we treat the "n channel enhancement MOSFET" transistor. The term MOSFET stands for *m*etal *o*xide *s*emiconductor *f*ield *e*ffect *t*ransistor. As a circuit element, the device is symbolized as shown in Fig. 14.1. The three terminals are called the gate (G), drain (D), and source (S). There are three currents and three voltage differences associated with this (or any other three-terminal) device. Referring to Fig. 14.1 we may define the following sets of voltages and currents for the MOSFET

$$(V_{GS}, V_{DS}, V_{GD})$$

$$(I_G, I_D, I_S)$$

where, as usual, V_{AB} means $V_A - V_B$ and the I_j are as noted in the figure. These six variables may be reduced to just four by application of KVL and KCL. One way to do this is to note that, by KVL,

$$V_{GD} = V_{GS} - V_{DS}$$

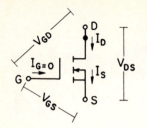

FIGURE 14.1.
Symbol for an *n* channel enhancement MOSFET and definition of the voltages and currents.

and, by KCL,

$$I_S = I_G + I_D$$

so that V_{GD} and I_S can always be found if we know the other four values.

Even with this simplification the description of a three-terminal device is clearly much more involved than is the description of a two-terminal element. For example, although a diode is a nonlinear device, it may still be fully described by a single curve of current versus voltage which may in turn be plotted on normal two-dimensional graph paper. Description of a three-terminal transistor requires four parameters. The interrelationship between four such quantities cannot be easily represented in two dimensions and requires considerable discussion. In the next paragraphs the construction and operation of the *n* channel enhancement MOSFET are outlined, as are the typical interrelationships found between the currents and voltages associated with this device. The linear model for a MOSFET is derived in Section 14.2.

Before discussing this particular transistor type we note that there is a whole family of field effect transistors. We have chosen to study the *n* channel enhancement metal oxide semiconductor FET in this text since it is a very common device and since it lends itself well to circuit examples. Such a device is also sometimes called an enhancement NMOS transistor. In the remainder of the text the term MOSFET will refer to an *n* channel enhancement MOSFET. Study of this device complements the bipolar junction transistor (BJT) analysis to be presented later since the MOSFET may be modeled by a voltage-controlled current source whereas the linear BJT model employs a current-controlled current source. Also, the fact that very little current flows into the gate of a MOSFET makes it a candidate device for the first transistor stage of an operational amplifier, a device which was introduced in Chapter 4 and which has been used throughout the remainder of the text.

As the name implies, in FET devices an electric field controls its operation. The device construction is shown schematically in Fig. 14.2. The p-type material is called the substrate, or bulk, and in most applications, including our own, is electrically connected to the source (*S*) inside the device. To the outside world this makes the unit a three-terminal device. The source and drain (*D*) are both attached to heavily doped *n*-type semiconductors. As discussed in Chapter 13, a *p*-type material conducts primarily by the migration of positive holes whereas an *n*-type material conducts primarily by free electrons.

The gate point of the MOSFET is attached to a metal plate which is in turn insulated from the substrate by a very thin insulating layer of silicon dioxide. This structure is similar to a parallel plate capacitor except that one of the plates is made of a *p*-type semiconductor (silicon). Unlike a metal plate, the *p*-type substrate material can have fixed charges (acceptors and donors) as well as mobile charges

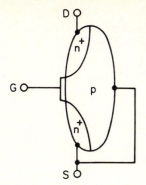

FIGURE 14.2.
Schematic diagram showing the construction of an *n* channel enhancement MOSFET.

(electron and holes). The proper polarities of these charges for an *n* channel MOSFET are shown schematically in Fig. 14.3 if a positive voltage is applied to the gate. In this schematic the device is "ON" and the gate voltage induces an electric field in the insulator which points from the metal plate toward the substrate. This field penetrates a small distance into the substrate and forces the mobile positive charges (holes) away from the insulator but does not affect the fixed charges. In addition mobile negative charges (electrons) supplied by the n^+ material are attracted to the insulator–substrate interface. A large positive gate bias would thus form a continuous sheet of minority carrier charges (electrons) in a channel running all the way from the source to the drain. A zero or negative voltage on the gate would induce no field or a field of the opposite direction into the substrate, and hence holes would either remain, or their density would be enhanced, at the insulator–substrate interface. These two states correspond to the on- and off-states of the MOSFET respectively.

The excess electrons attracted to the region near the gate in the "on" condition (positive gate voltage) form a channel in which a current can flow from drain to source. This channel is shown schematically in Fig. 14.4. The current is carried by electrons moving up through the channel and is supplied by the n^+ material attached to the source and "drained" off through the n^+ material attached to the drain. Note that a steady current through the channel does *not* change the electric field across the channel nor the number of free electrons in the channel since, by KCL, as many electrons must flow out of the channel to the drain as flow into the channel at the source.

The key to the device operation is that the amount of current which can flow through the channel is determined by the number of free electrons there, since for a constant value of V_{GS} the width of the channel remains constant. The number of

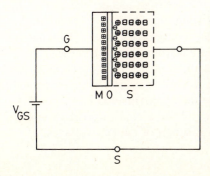

FIGURE 14.3.
Circuit analogy for the gate and substrate of the *n* channel MOSFET. Mobile and fixed charges in the substrate have been denoted by round and square symbols, respectively. The negative mobile carriers are electrons and the positive mobile carriers are holes. The fixed charges in the metal are lattice atoms.

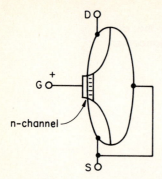

FIGURE 14.4.
Schematic diagram of the *n*-channel created near the gate by an applied positive voltage at the gate.

current carriers is in turn determined by the voltage applied at the gate since that voltage attracts the negative carriers to the channel. The device therefore acts as a voltage-controlled current source.

For a fixed positive value of V_{GS} a single (I_D, V_{DS}) curve characterizes the device and has the form illustrated in Fig. 14.5. For small values of V_{DS} the n channel acts like a resistor with I_D increasing linearly with V_{DS}. At higher voltages, the amount of current which can be drawn through the channel saturates and the *I–V* curve flattens out. This is the so-called "active" region. At large values of V_{DS} the reverse-biased *pn* junction between the drain material and the substrate breaks down, as discussed in Section 13.1 for the diode. In this regime I_D increases drastically with increasing V_{DS}.

FIGURE 14.5.
Plot of I_{DS} versus V_{DS} for fixed V_{GS}.

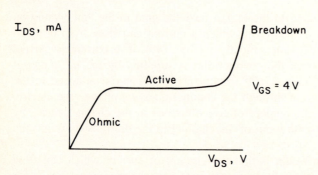

The gate terminal is coated with an oxide surface which is not a very good electrical conductor. Anyone who has tried to start an automobile with heavily oxidized battery terminals is aware of this fact. The coating reduces the current I_G into the gate terminal very nearly to zero. Thus, from Eq. 14.2, $I_D = I_S$ and all the current flowing into the drain point must flow out through the source. The notation may sound a little strange, but remember, it is the electrons which carry the current. Thus, we can say, "all the *electrons* which enter the source S flow out through the drain D," and the terms make more sense. We call this current I_{DS} or sometimes just I_D since $I_D = I_{DS}$. Henceforth, we shall use the approximation that $I_G = 0$.

With $I_G = 0$, the device may be described by just three mathematical variables. The current I_{DS} may then be considered to be a function of the two voltages:

$$I_{DS} = f(V_{GS}, V_{DS})$$

A function of two variables describes a surface in three-dimensional space where I_{DS} is the "elevation" of the surface above a point in the V_{GS}–V_{DS} plane. In other words, if we let V_{GS} and V_{DS} correspond to x and y in a Cartesian coordinate system, then I_{DS} corresponds to the "elevation," or z distance, above or below that point. Since it is three-dimensional, such a surface is hard to draw on two-dimensional paper. We describe it instead by using families of curves, all of which are similar to the curve in Fig. 14.5.

A set of these curves are presented in Fig. 14.6. The basic plot is that of I_{DS} versus V_{DS}, but note that each curve is labeled by a *different* value of V_{GS}. This is the way in which a "third dimension" is included in a two-dimensional graph. Each curve in Fig. 14.6 represents the intersection of a vertical plane through the V_{GS} point with which the curve is labeled and the surface described above [i.e., the function $I_{DS} = f(V_{GS}, V_{DS})$ evaluated with V_{GS} fixed]. In the active region increasing values of V_{GS} result in larger values of I_{DS} for the same value of V_{DS}. This occurs since the width of the channel is larger for larger V_{GS} which is due in turn to the larger electric field the voltage generates. Many devices of this same general type are manufactured and each will have a slightly different set of curves which determine its detailed operation. In Fig. 14.6 the breakdown region is not shown.

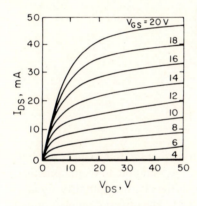

FIGURE 14.6.
Representative family of curves for an *n* channel enhancement MOSFET.

A passive device such as a transistor or diode must be attached to external circuitry to function usefully. To illustrate the operation of a MOSFET transistor as well as the analysis of a circuit using such a device, consider the circuit diagram of Fig. 14.7*a*. As with the diode, we need to find the operating point Q in the

FIGURE 14.7*a*.
Circuit which includes a MOSFET.

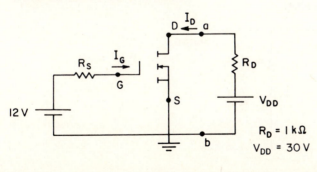

(a)

parameter space (V_{GS}, I_{DS}, V_{DS}) illustrated by the family of curves in Fig. 14.6. Since no current flows into the gate $(I_G = 0)$ there is no voltage drop across R_S and $V_{GS} = 12$ V. Thus, Q must fall on the curve labeled with this value. The exact position can now be found in a manner analogous to the analysis performed in diode circuits. The output circuit, or load, consists of a battery and resistor in series. No matter what is attached to the left, at points a and b, the $I–V$ pair (I_{DS}, V_{DS}) must be located on a straight line which is determined by the linear circuit to the right of the point $a–b$. As before, this line is called the load line. Since $V_{oc} = 30$ V and $I_{sc} = 30$ mA, the load line is the line connecting the two dots plotted in Fig. 14.7b. The point where it crosses the $V_{GS} = 12$ V curve then yields the two other unknown parameters of the operating point Q, $(I_{DS}, V_{DS}) = (14$ mA, 16 V$)$.

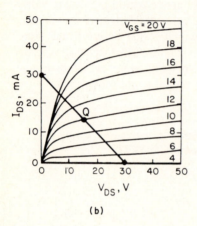

(b)

FIGURE 14.7b.
Finding the operating point for the MOSFET circuit.

The purpose of the batteries and the resistors surrounding the transistor is to place the point Q in a region of parameter space (V_{GS}, I_{DS}, V_{DS}) suitable for some practical application. The position in this case is appropriate for the circuit to amplify signals. Unlike the diode, which is purely passive, the transistor can convert dc power drawn from the attached batteries (often called power supplies) into time-varying power associated with some signal. As an example, suppose a signal, say, $v_i(t) = 1.0 \cos(\omega t)$ V, is included in series with the 12-V battery as shown in Fig. 14.7c. What would the output voltage of this elemental amplifier circuit be? Since the parameters

FIGURE 14.7c.
Elemental MOSFET signal amplifier.

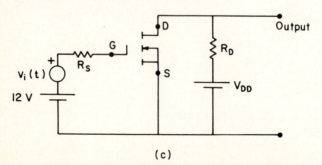

(c)

at the output of the device are unchanged, the load line remains the same. The operating point then will move along the load line to V_{GS} curves labeled 13 V, 12 V, 11 V, 12 V, and so on, as the voltage V_{GS} varies due to the signal. As the system oscillates about the point Q it also moves in the (I_{DS}, V_{DS}) plane, as shown by the arrows in Fig. 14.7d. Reading from the graph, we note that as V_{GS} varies from 11 V to 13 V, I_{DS} varies from 12 mA to 16 mA, and V_{DS} varies from 18 V to 14 V. We can write the output as

$$v_{DS}(t) = V_{DS} + v_o(t)$$

where $V_{DS} = 16$ V and $v_o(t)$ is the time-varying part of the output voltage.

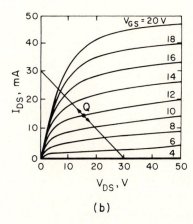

(b)

FIGURE 14.7d.
Plot showing the motion of the operating point along the load line as V_{GS} varies by ± 1.0 V about the mean value of 12 V.

If we only consider the oscillating part of the voltage, what we term the output *signal*, a variation of $1.0 \cos(\omega t)$ V at the input yields an output signal $v_o(t)$ that is approximately given by $v_o(t) = -2.0 \cos(\omega t)$ V. We say approximately, since the device will only yield a proportionality betwen v_o and v_i if it behaves in a linear manner. For a large signal like this, the actual output will most likely be distorted and therefore will not be a pure cosine wave. The sign of v_o is negative since when V_{GS} goes up, V_{DS} goes down. We define the voltage gain A_v of a transistor amplifier as the ratio of the output time-varying signal to the input time-varying signal. For this circuit, then,

$$A_v = \frac{v_o(t)}{v_i(t)} \approx -2.0$$

This circuit is termed an "inverting amplifier" since the sign of the signal has changed. More realistic amplifier circuits are discussed later in the text.

TEXT EXAMPLE 14.1 _____

The MOSFET used in the circuit illustrated here,

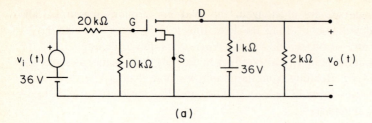

(a)

has the characteristic family of curves shown below.

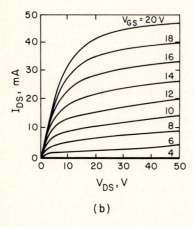

(b)

Find the operating point parameters V_{GS}, I_{DS}, and V_{DS}.

Solution: We are interested in the dc voltages and currents and hence we may suppress the small signal $v_i(t)$, replacing it by a short circuit. Since no current flows into the gate, the voltage V_{GS} is given by the voltage divider principle:

$$V_{GS} = \left(\frac{10}{20 + 10}\right) 36 = 12 \text{ V}$$

The appropriate curve in the family is then $V_{GS} = 12$ V. To find the rest of the operating point parameters we need to construct the load line. The Thevenin equivalent circuit for the load portion is given at the right below since the open-circuit voltage between D and S is $V_{oc} = [2/(1 + 2)]36 = 24$ V, whereas the short-circuit current is $I_{sc} = 36/1k = 36$ mA.

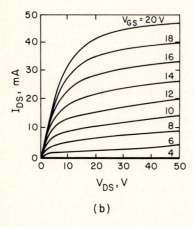

(c)

The load line here is found by connecting the two points $(I_{DS}, V_{DS}) = (36 \text{ mA}, 0)$ and $(I_{DS}, V_{DS}) = (0, 24 \text{ V})$:

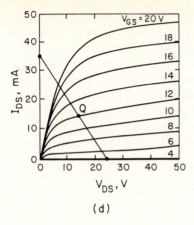

(d)

Intersection of this line with the $V_{GS} = 12$ V curve yields the operating point parameters $(I_{DS}, V_{DS}) = (14$ mA, 14 V), which completes the set. Since these parameters were determined graphically, they may be somewhat in error. To check, we note that KCL requires

$$\frac{36 - V_{DS}}{1k\Omega} = I_{DS} + \frac{V_{DS}}{2k\Omega}$$

Substituting into this expression using the values above yields 22 mA on the left-hand side and 21 mA on the right. These values agree to within 5%, which is as good as could be expected using a graphical method. Note that the same 36-V power supply voltage was used on the gate and drain sides of the transistor rather than two Áifferent voltage values. As discussed further in Section 14.5, this is of great economic advantage and is the usual case.

14.2 A LINEAR MODEL FOR THE MOSFET TRANSISTOR

The similarity of our MOSFET linear analysis to the diode analysis in Chapter 13 suggests that a linear model may also be constructed for a MOSFET. Such a model, then, may replace the actual device in applications using small time-varying currents and voltages. Since there are now three variables, two derivatives are needed at Q rather than the single derivative $(di/dv)_Q$ needed for the diode. Mathematically we again consider I_{DS} to be a function of V_{GS} and V_{DS}:

$$I_{DS} = f(V_{GS}, V_{DS})$$

which defines a surface of height I_{DS} over the (V_{GS}, V_{DS}) plane. For small variations about Q we use an analogy to the Taylor series expansion for a function of two variables:

$$I_{DS}(V_{GS} + \Delta V_{GS}, V_{DS} + \Delta V_{DS}) = I_{DS}(Q) + \left.\frac{\partial i_{DS}}{\partial v_{GS}}\right|_Q \Delta V_{GS} + \left.\frac{\partial i_{DS}}{\partial v_{DS}}\right|_Q \Delta V_{DS} + \cdots$$

In this expression $(\partial i_{DS}/\partial v_{GS})_Q$ is evaluated with V_{DS} held constant. Likewise, the quantity $(\partial i_{DS}/\partial v_{DS})_Q$ is evaluated with V_{GS} held constant. This type of derivative is called a partial derivative since it represents the slope "in one direction." For example, on the surface of the earth we might say that the height h of the ground "slopes upward" going in the east direction. This would represent a positive partial derivative of h in the east direction (e) holding the north position (n) constant, that is, $(\partial h/\partial e)_{n=\text{constant}}$ is positive. At that same point the height might increase, decrease, or remain the same in the northward direction. This slope would be characterized by the partial derivative $(\partial h/\partial n)_{e=\text{constant}}$. Both derivatives must be specified to describe how the elevation of the ground varies at a given point on the earth.

The two linear model parameters g_m and r_d which we use for MOSFET analysis are related to these partial derivatives by definition as follows:

$$g_m = \left.\left(\frac{\partial i_{DS}}{\partial v_{GS}}\right)\right|_Q \quad \text{mhos}$$

evaluated with V_{DS} held constant, and

$$r_d = \left.\left(\frac{\partial i_{DS}}{\partial v_{DS}}\right)^{-1}\right|_Q \quad \text{ohms}$$

evaluated with V_{GS} held constant. Then, for small deviations from the operating point Q we may drop the higher-order terms in the expansion, which yields

$$I_{DS} = I_{DS}(Q) + g_m\Delta V_{GS} + \left(\frac{1}{r_d}\right)\Delta V_{DS}$$

If a small time-varying signal is introduced at the gate (e.g., $v_i(t)$ in Text Example 14.1), then, since $I_{DS}(Q)$ is a constant, the time-varying part of I_{DS} is given by

$$i_d(t) = g_m v_{gs}(t) + \left(\frac{1}{r_d}\right)v_{ds}(t) \tag{14.3}$$

As usual, we use the convention that lowercase letters imply time variations in the quantity (i.e., v_{gs} and v_{ds}). Since g_m and $1/r_d$ are just numbers, this is a linear relationship relating small changes in i_d to the variations in V_{GS} and V_{DS}. For brevity we also have adopted the notation i_d rather than i_{ds}.

Graphically, g_m and r_d can be estimated as shown in Fig. 14.8. Returning to the plot in Fig. 14.6, the "transconductance" g_m is found using the vertical arrow construction about the operating point Q. For a 4 V change in V_{GS}, the current I_{DS} changes by 10 mA which yields $g_m = 2.5 \times 10^{-3}$ mhos. Note that the line through Q is vertical, corresponding to a constant value for V_{DS} as required by the definition of a partial derivative. To evaluate r_d, a triangle is constructed at Q tangent to the $V_{GS} = 12$ V curve through Q (Fig. 14.8b). The construction yields $r_d = 5.33$ kΩ. In this technique we are approximating the partial derivative with the appropriate tangents to the surface defined by the function $I_{DS} = f(V_{GS}, V_{DS})$.

Given these parameters, a linear small-signal model may now be constructed. The purpose of this model is to determine the small variations about Q due to a signal $v_i(t)$, not to determine the external power supply voltages and resistors

FIGURE 14.8a.
Graphical construction which yields an estimate of the derivative of I_{DS} with respect to V_{GS} at a constant value of V_{GS}.

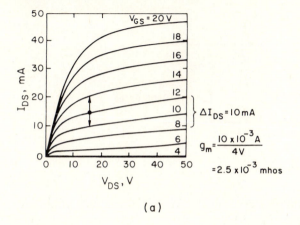

(a)

FIGURE 14.8b.
Graphical construction which yields an estimate of the reciprocal of the derivative of I_{DS} with respect to V_{DS} at a constant value of V_{GS}.

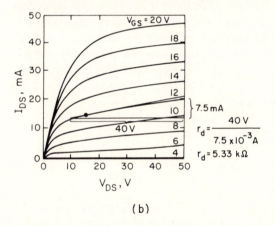

(b)

which yield the Q point. Due to the linearity of the model we may apply the superposition principle and determine the system response to $v_i(t)$ alone. That is, if we are interested only in small variations about Q in the circuit of Fig. 14.7c, we suppress the dc sources and study the resultant circuit for the linear response to $v_i(t)$ as shown schematically in Fig. 14.9a. Our goal is to create a linear model for the device which we insert within the dotted circle. At the input no current is drawn. This means that the input resistance r_i of the two-port linearized circuit model which described the transistor is infinite. We represent this by an open circuit. From Eq. 14.3, the current into the point d, i_d, is made up of the sum of two parts. The contribution to i_d from the term v_{ds}/r_d corresponds to the current that would be drawn through a resistor of value r_d between the point d and the point s. The contribution to i_d from the term $g_m v_{gs}$ corresponds to a voltage-controlled current source. The linear *circuit* model for the MOSFET is therefore as shown in Fig.

FIGURE 14.9a.
Circuit diagram for determining the response of a MOSFET to the small-signal $v_i(t)$ with the steady (dc) voltages suppressed.

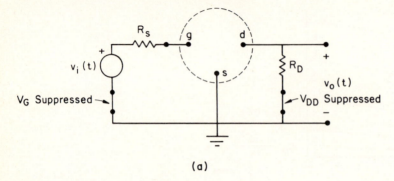

(a)

14.9b. Using this small-signal linear model for the MOSFET we may redraw the circuit of Fig. 14.7a in the form shown in Fig. 14.9c.

The model inside the dashed box in Fig. 14.9c is identical to model (d) of Fig. 4.12 under the condition that $R_i = \infty$. The reader may find it useful to review the use of dependent sources as discussed in detail in that chapter. Here, we make a brief review of the topic emphasizing those aspects important to small-signal transistor amplifier applications. In a circuit diagram such as in Fig. 14.9c the input and output portions of the system are only connected along the lower edge which is also at ground potential. In the real device there are other connections, of course, through various semiconductor materials. In the linear model the connection is made mathematically via the dependent source. That is, the current being driven by that source in the output portion of the device is equal to $g_m v_{gs}$, where v_{gs} is the voltage that exists in the input portion of the circuit.

If we are interested in the output voltage generated by the current source, we may use Ohm's law. For a single phasor component,

$$\hat{V}_o = -\hat{I}_d \mathbf{Z}_o$$

where in this case $\mathbf{Z}_o = [r_d \,//\, R_D]$ and $\hat{I}_d = g_m \hat{V}_{gs}$. The minus sign arises from the definition of the polarity of v_o in the figure and the direction of the controlled source.

FIGURE 14.9b.
The linear model we insert inside the dotted circle in place of the MOSFET.

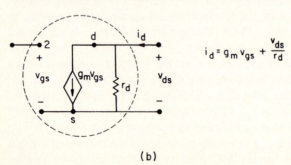

$$i_d = g_m v_{gs} + \frac{v_{ds}}{r_d}$$

(b)

FIGURE 14.9c.
The small-signal linearized circuit of the elemental MOSFET amplifier.

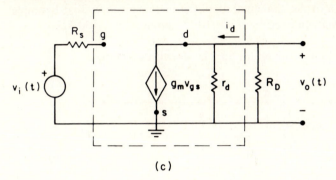

(c)

Since there are no capacitors or inductors in the circuit diagram at this juncture, a phasor analysis is not necessary, and we have

$$v_o(t) = -g_m v_{gs}(t)[r_d \,/\!/\, R_D]$$

But since the gate is an open circuit and the point s is grounded,

$$v_{gs}(t) = v_i(t)$$

and

$$v_o(t) = -g_m v_i(t)[r_d \,/\!/\, R_D]$$

The output is mathematically related to the input signal in this last equation. Finally, if we wish to find the voltage gain $A_v = v_o/v_i$, we have

$$A_v = -g_m[r_d \,/\!/\, R_D] = -\alpha_0 g_m R_D$$

where $\alpha_0 = r_d / (r_d + R_D)$. Note that since g_m has units of mhos and $[r_d \,/\!/\, R_D]$ has units of ohms, A_v is a dimensionless number, as must hold from its definition.

For the particular circuit studied here this expression may be evaluated using the values for g_m and r_d found graphically earlier in this section. Substituting these values and $R_D = 1 \text{ k}\Omega$, the output voltage is

$$v_o(t) = -g_m v_i(t)\left(\frac{R_D r_d}{R_D + r_d}\right) = -2.1 v_i(t)$$

and the gain of the circuit is

$$A_v = \frac{v_o}{v_i} = -2.1$$

This result is nearly the same as the value found above in Section 14.1 using a nonlinear graphical analysis of this same circuit.

TEXT EXAMPLE 14.2

Find the small-signal gain of the elemental amplifier circuit given in Text Example 14.1 if the transistor is characterized by $g_m = 10^{-2}$ mho and $r_d = 100$ kΩ.

Solution: The linearized circuit diagram may be drawn as follows:

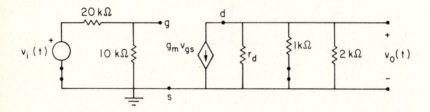

where suppressed dc power supplies are indicated by short circuits. At the input,

$$v_{gs}(t) = \left(\frac{10}{10 + 20}\right)v_i(t) = \left(\frac{1}{3}\right)v_i(t)$$

At the output,

$$v_o(t) = -g_m v_{gs}(t)[r_d \,//\, 1 \text{ k}\Omega \,//\, 2 \text{ k}\Omega]$$

Substituting values for g_m and r_d and the relationship between $v_{gs}(t)$ and $v_i(t)$ yields $A_v = -2.2$.

In Chapter 4 we commented that the resistance value R_i in a three-parameter model was not in general independent of the load resistor. However, in the case of the ideal MOSFET, since $R_i = \infty$, changing the value of the load will have no effect on R_i. In this sense, then, the linear model for the ideal MOSFET is exactly characterized by a three-parameter model, that is, model c of Fig. 4.12, with the three parameters (∞, g_m, r_d).

Before launching into a discussion of more realistic amplifier circuits, we introduce our second generic transistor, an *npn* bipolar junction device.

14.3 BIPOLAR JUNCTION TRANSISTORS

The second class of transistors with which we will deal is termed a bipolar junction transistor (BJT). All of our examples here will deal with *npn* transistors in which a thin layer of *p*-type semiconductor material separates two *n*-type materials. Very similar devices are constructed using the analogous *pnp* construction. As with the MOSFET, there are three electrical contacts made to a BJT transistor. These are termed the base B, the collector C, and the emitter E. Symbolically, we denote the *npn* transistor as shown in Fig. 14.10, where, as also done for the MOSFET, we have defined the three voltages and three currents which are needed to describe

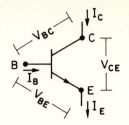

FIGURE 14.10.
Symbol for the *npn* transistor.

the device. The arrow on the emitter implies that conventional current flows in the direction shown during normal configurations of the transistor. The symbol for a *pnp* transistor is identical except the arrow is in the opposite direction. The direction of the arrow shows the forward direction of the "diode" corresponding to the semiconductor *pn* junction between the base and emitter.

Bipolar junction transistors (BJT) are made with the sandwich construction shown schematically in Fig. 14.11 for the *npn* device. The *n*-type material connected to the emitter is more heavily doped than the *n*-type semiconductor connected to the collector and, as noted above, the *p*-type region is very thin. At first glance, the BJT looks like two diodes constructed back to back. However, since the base is very thin, when the collector voltage is sufficiently positive, most of the electrons which enter the emitter actually pass right through into the collector (hence the name).

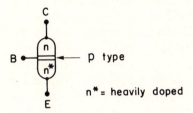

FIGURE 14.11.
Schematic diagram showing construc-tion of an *npn* transistor.

From Kirchhoff's current law,

$$I_E = I_B + I_C \tag{14.4}$$

and by KVL,

$$V_{CB} = V_{CE} - V_{BE} \tag{14.5}$$

As with the MOSFET we can always derive two of the parameters, say, I_E and V_{CB}, from the other four quantities in Eq. 14.4 and Eq. 14.5. However, for the BJT device we cannot set $I_B = 0$ as we did I_G in the MOSFET. This means that the family of curves needed to analyze the BJT cannot be constructed on a single graph, and two plots are required. In the next few paragraphs the construction and operation of the *npn* transistor are discussed in a manner analogous to the MOSFET analysis presentation above.

Many applications use the so-called common emitter configuration for the *npn* device. This means that the voltage of the emitter is used as the reference. In this text we will most closely investigate the device when configured as shown in Fig. 14.12. The base–emitter *pn* junction then behaves very much like a diode. When V_{BE} is positive, that diode is forward-biased and the curves of I_B versus V_{BE} appear

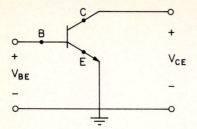

FIGURE 14.12.
**Common emitter configuration of an *npn*
transistor.**

as shown in Fig. 14.13*a*. The family of curves plotted in this figure displays very little variation with the voltage V_{CE}. The curves are very close together and are, in practice, almost indistinguishable from each other. For our applications we shall make the very good approximation represented in Fig. 14.13*b* in which all the curves for different V_{CE} values are collapsed into the single representative curve shown. As indicated, this approach is valid provided $V_{CE} \geq 1$ V which, in effect, just means that V_{CE} is greater than a typical diode threshold voltage.

When V_{CE} is large enough so that $V_{CB} = V_{CE} - V_{BE}$ is a positive number, the *pn* junction between the base and the collector is reverse-biased. This is the condition used in amplifier circuits, and we study that case here. To simplify the analysis, let us first assume that the potential $V_{CB} = 0$ so that $V_C = V_B$ and both are greater than the threshold voltage. The emitter current is in the direction of the arrow in Fig. 14.12 since the base–emitter junction is forward-biased. In the heavily doped *n*-type emitter material, this current is carried by electrons which are injected upward into the *p*-type base. Since this material is only lightly doped, it has a low conductivity and furthermore is very thin. The consequence is that most of the injected electrons travel all the way across the base and are "collected" by the electrode attached at point C, which is at a positive potential ($V_C = V_B > 0$). The amount of current which takes this path is greater than that which takes the path through the base electrode, and we have the important results that

$$I_C \gg I_B$$

Under this condition,

$$I_E = I_B + I_C \approx I_C$$

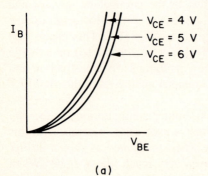

FIGURE 14.13*a*.
Family of curves for the base–emitter diode curve of I_B versus V_{BE}. Each curve is labeled by a different value for V_{CE}.

(a)

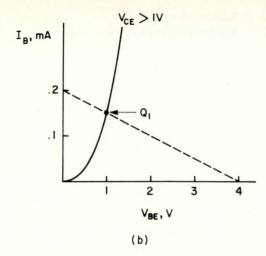

FIGURE 14.13b.
Typical base–emitter current–voltage profile for a BJT transistor. Dashed line is the load line for the analysis below.

As V_{CE} is increased holding V_{BE} fixed, more electrons will be attracted and the collector current I_C will increase. Eventually, a "saturation" level is reached corresponding to the reversed bias current of the *pn* junction between the base and collector. A representative curve of I_C versus V_{CE} for a fixed value of the base current (0.15 mA) is drawn in Figs. 14.14*a*. As with all diodes, the base–emitter voltage difference never gets much larger than the threshold voltage for the base–emitter diode. For the silicon device studied here, the V_{BE} threshold voltage is about 0.7 V. When V_{CE} is also equal to 0.7 V, we can determine the collector current from the expanded plot of I_B versus V_{CE} in Fig. 14.14*b*. The plot indicates that $I_C = 8$ mA when $V_{CE} = 0.7$ V $= V_{BE}$. Since $I_B = 0.15$ mA, the collector current is 53 times the base current. As V_{CE} is increased further keeping the base current constant, the collector current continues to rise but at a slower rate of increase. We say the collector current saturates at about $V_{CE} = 5$ V since increasing V_{CE} above that value does not increase the current very much (see Fig. 14.14*b*). The region where the curve is nearly flat is called the "active" region and is the region of operation for transistors when they are used in amplifier circuits.

For constant V_{CE}, increasing the base current increases the collector current since most of the electrons attracted from the emitter pass right through the base to the collector. The result is that in the I_C–V_{CE} plane there will be a family of curves

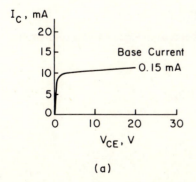

FIGURE 14.14a.
Plot of I_C versus V_{CE} for fixed I_B in an *npn* transistor.

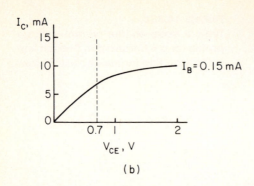

FIGURE 14.14b.
Expanded plot of I_C versus V_{CE}.

(b)

each labeled by a different value of the base current I_B (see Fig. 14.15). If the base current is zero, only a small collector current will flow, a current which corresponds to the reverse bias current across the collector–base junction. The two sets of curves given in Fig. 14.13b and Fig. 14.15 constitute the basis for analysis of the *npn* bipolar junction transistor.

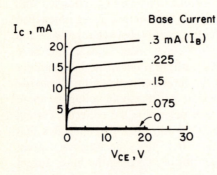

FIGURE 14.15.
Typical collector–emitter current–voltage profiles for different values of the base current. The collector current is very small but not zero when $I_B = 0$.

As a specific example of the method for determining the operating point of a BJT, consider the elemental amplifier circuit shown in Fig. 14.16, where the "signal" $v_i(t)$ is small compared to 4 V. To find the operating point we suppress the signal by letting $v_i(t) = 0$. We then analyze the base–emitter diode curve to find the base current. This can be done graphically as discussed in Chapter 13 using the Thevenin equivalent at the input to the circuit. The load line is the dashed line in Fig. 14.13b, and its intersection (labeled Q_1) with the diode curve yields $I_B = 0.15$ mA.

Using the curve $I_B = 0.15$ mA we can now find the operating point for the output portion of the circuit. Using the Thevenin equivalent for the output circuit, the load

FIGURE 14.16.
A BJT amplifier circuit using the transistor whose I–V curves in Figs. 14.13b and 14.15.

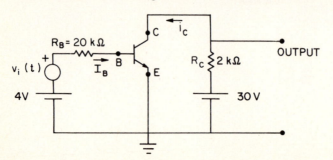

line can be constructed yielding Q_2, as shown in Fig. 14.17. The (I, V) coordinates of Q_2 are the dc values of I_C and V_{CE}, which in this example are 10.5 mA and 9.0 V, respectively.

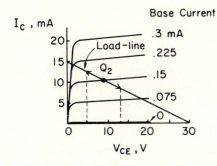

FIGURE 14.17.
Output load line, operating point Q₂, and the variation along that load line for an input voltage signal $v_i(t)$.

We now consider the effect of a time-varying voltage $v_i(t) = 0.7 \cos(\omega t)$ V in series with the power supply in Fig. 14.16. The parallel line construction in Fig. 14.18 shows how to determine the change in the base current using the intersection of the diode curve and the load lines corresponding to the extreme values of $v_i(t)$. Reading from the I_B axis, these extreme positions on the load line correspond to a variation in $i_b(t)$ of ± 0.03 mA. Using the $I_C–V_{CE}$ graph in Fig. 14.17, this variation in I_B, when projected along the output load line, yields the time variation of the collector current, which oscillates between 12.5 mA and 8.5 mA. The time-varying part of the collector current $i_c(t)$ exhibits excursions of ± 2 mA about the dc value of 10.5 mA. Note that the time-varying current to the transistor has been amplified because $i_c(t) \gg i_b(t)$. The ratio of $i_c(t)$ to $i_b(t)$ is termed the ac current gain which for this circuit is given by:

$$A_I = \frac{i_c(t)}{i_b(t)} = \frac{2 \cos(\omega t)}{0.03 \cos(\omega t)} = 67$$

At the operating point the dc ratio of I_C (10.5 mA) to I_B (0.15 mA) is 70. This

FIGURE 14.18.
Variation of the base–emitter parameters in the *I–V* plane when a ± 0.7-V signal is applied.

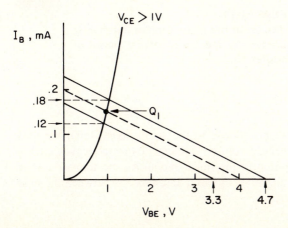

ratio is termed the dc current gain β_o and is slightly larger than the ac gain A_I. The output voltage can also be read from the graph which shows values that vary betweeen the extremes 4.5 V and 13.5 V. Note that when V_{BE} is increased, V_{CE} decreases. The circuit thus has an ac voltage gain given by

$$A_v = \frac{v_{ce}(t)}{v_i(t)} \approx \frac{-4.5 \, \cos(\omega t)}{0.7 \, \cos(\omega t)} = -6.4$$

We use an approximation sign since there is no guarantee that the time variation of the collector–emitter voltage is an exact cosine function since the device is nonlinear. The large variations we have used here are required to perform a graphical analysis. Usually such large-signal applications lead to distortion of the output signal which may be very undesirable.

More realistic amplifier circuits employ capacitors to couple signals into and out of the device. The effect of capacitors on the frequency response of such circuits is studied in detail in Chapter 15. Large amplification factors can be realized if individual units are "cascaded" so that they act in series. Furthermore, with modern technology entire circuits including resistors, capacitors, and transistors can be "integrated" into a single combination of metal, silicon, and resistive materials termed an integrated circuit. The operational amplifier used in previous chapters is an example of a device which uses a large number of transistors in its design.

14.4 LINEAR MODEL FOR THE BJT AMPLIFIER

As with the MOSFET, we can derive linear relations for the time-varying voltages and currents around the operating point. At the input we shall for the present continue to make the very good approximation that the base current depends only upon the base–emitter voltage. This corresponds to a single curve such as that shown in Fig. 14.13b rather than the *set* of curves shown in Fig. 14.13a. The base–emitter junction then acts exactly like a diode and we can use the linear diode model derived in Chapter 13. In Fig. 14.19 this linear model is inserted between the base (b) and emitter (e) in the input portion of the circuit in Fig. 14.16. In this model $r_i^{-1} = (dI_{BE}/dV_{BE})_{Q1}$ and V_T is the intercept on the V_{BE} axis of the line tangent to the curve at Q_1. Since we are ignoring the "family of base–emitter curves," the derivative in this expression is an ordinary derivative rather than a partial derivative.

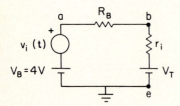

FIGURE 14.19.
At the input the base–emitter junction is replaced by the linear diode model of Chapter 13.

When considering the base–emitter response to the small signal $v_i(t)$, we may suppress the battery V_B and the threshold voltage "battery" V_T which leaves $v_i(t)$ in series with R_B and r_i. The contribution to the linear model for the transistor from the base–emitter diode is then just the series dynamic resistance r_i between the base and emitter, as seen in Fig. 14.20a.

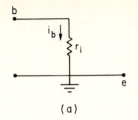

b

i_b

r_i

e

(a)

FIGURE 14.20a.
Contribution of the base–emitter diode
to the small-signal linear model of the
BJT transmitter.

Turning to the collector–emitter portion of the device, the collector current depends on both the base current and the collector–emitter voltage. At the operating point Q_2, for small time variations, we can use a series expansion and derive

$$i_c(t) = \left(\frac{\partial i_C}{\partial i_B}\right)_{Q_2} i_b(t) + \left(\frac{\partial i_C}{\partial v_{CE}}\right)_{Q_2} v_{ce}(t)$$

which we may write as the linear relationship

$$i_c(t) = \beta i_b(t) + \left(\frac{1}{r_o}\right)v_{ce}(t) \tag{14.6}$$

Using the identical reasoning which we applied for the MOSFET case, this linear *mathematical* model for small time-varying voltages and currents may be represented by the *circuit* of Fig. 14.20b. In this model the device acts as a current-controlled current source. Three parameters are used to specify its operation, r_i, β, and r_o. The fact that r_i takes on a finite value for the BJT device makes the analysis slightly more complicated than for the MOSFET for which $r_i = \infty$.

FIGURE 14.20b.
Linear circuit model for the *npn* transistor for small variations around the operating
points Q_1 and Q_2.

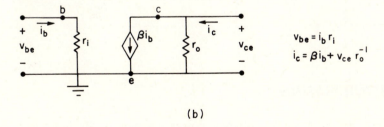

b

c

$+$ i_b

v_{be}

$-$

r_i

βi_b

r_o

i_c

$+$

v_{ce}

$-$

e

$v_{be} = i_b r_i$

$i_c = \beta i_b + v_{ce} r_o^{-1}$

(b)

The values of β and r_o at the operating point may be found using graphical methods (as for g_m and r_d with the MOSFET), or they might be "given" parameters listed in a specification sheet for the transistor. Remember, β is found by holding V_{CE} constant whereas r_o is found by holding I_B constant. In Example 14.2 at the end of the chapter, we work out values for r_i, β, and r_o in some detail by the graphical method.

The elemental amplifier circuit of Fig. 14.16 may now be analyzed using the model of Fig. 14.20b. Suppose that at the operating point we find, or are given, $r_i = 2.0\ \text{k}\Omega$, $\beta = 67$, and $r_o = 40\ \text{k}\Omega$. Then, for the small time-varying voltages and currents (those due to v_i), we may use superposition and suppress the dc sources (including V_T) by replacing them with short circuits. This yields the circuit model of Fig. 14.21, where once again the dashed box encloses the three-parameter linear model for the npn transistor. To find $A_V = v_o / v_i$, we note that by Ohm's law

$$i_b = \frac{v_i}{R_B + r_i} = \frac{v_i}{22 \times 10^3} = (4.5 \times 10^{-5})v_i$$

At the output the voltage is found by calculating the equivalent resistance of r_o and R_C in parallel and multiplying by the current βi_b:

$$v_o = -\beta i_b (r_o \,/\!/\, R_C) = -(1.27 \times 10^5)i_b$$

Combining these two expressions and evaluating $A_v = v_o / v_i$ yields $A_v = -5.7$. This result is 11% less than the graphical large-signal result found above. Recollect however that the latter result was estimated for a signal which was large enough that the transistor was in a nonlinear operating mode. For small signals the model result for the gain is more accurate than the graphical analysis if r_i, β, and r_o are known accurately.

We reiterate that although signals may be amplified, transistors do not generate power. They must be attached to the voltage sources that determine their operating point, and any increase in signal power must be drawn from these sources.

FIGURE 14.21.
Linear model for the amplifier of Fig. 14.16. Double dots represent suppressed dc sources.

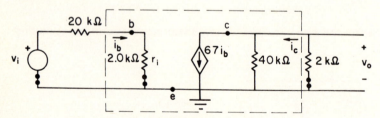

14.5 TRANSISTOR BIASING

For ease of analysis we have shown different power supplies in the input and output portions of both the MOSFET and BJT amplifiers. In practice this would be expensive and unnecessary. For example, in the MOSFET circuit of Fig. 14.7a we could just as easily use one power supply and a voltage divider to determine both the 12 V at the gate and the 30 V value of V_{DD} as shown in Fig. 14.22. Since no current is drawn by the device at the gate point G, no current flows through R_s. Then V_{GS} equals the voltage divider result $V_{GS} = [6/(6 + 9)]30\ \text{V} = 12\ \text{V}$. Schematically, the supply voltage is often just noted as a terminal labeled with

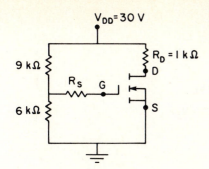

FIGURE 14.22.
MOSFET circuit to yield V_{GS} = 12 V using a 30-V battery.

some voltage V_{DD}. In a typical application this voltage would be routed or "bussed" to a number of transistors in a more complex circuit. This is the origin of the term "power bus" which is used in circuit design. Note that since $I_G = 0$ the value of R_s does not matter because no current is drawn through it. The negative side of the 30-V supply must of course be connected to ground potential, and it is always understood that this is the case when the notation of Fig. 14.22 is used.

For the BJT transistor, analysis is nearly identical except that the base current is not small enough to be ignored (as we did I_G in the MOSFET). As an example consider the more realistic BJT amplifier circuit of Fig. 14.23a. This circuit shows the use not only of one power supply (V_{CC}) but of two capacitors. These two capacitors are termed coupling capacitors since they couple the signal $v_i(t)$ into the transistor and couple the output of the transistor to the load resistor R_L. We discuss the detailed effect of these capacitors on the frequency response of the amplifier in Chapter 15. Of more interest here is the fact that these capacitors also *decouple* both the signal and the load from the dc operation of the transistor. This means that only the resistors R_1, R_2, and R_C will affect the dc operating point of the amplifier. This decoupling occurs since, at dc, capacitors like open circuits.

FIGURE 14.23a.
Circuit diagram for a BJT amplifier using one power supply.

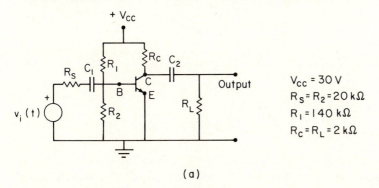

V_{cc} = 30 V
R_S = R_2 = 20 kΩ
R_1 = 140 kΩ
R_C = R_L = 2 kΩ

(a)

To find the dc operating point due to the power supply V_{CC}, remembering that the negative terminal of this supply is grounded, we redraw the dc circuit as in Fig. 14.23b. Using Thevenin's theorem, the circuit to the left of B and to the right of C may each now be reduced to a single source in series with a single resistor as shown in Fig. 14.23c. We may then proceed to find the operating point using the load line method discussed in Section 14.3.

FIGURE 14.23b.
Circuit diagram for $v_i = 0$.

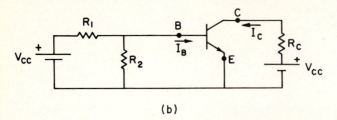

(b)

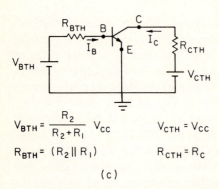

FIGURE 14.23c.
The "Thevenized" base–emitter and collector–emitter portions of the BJT amplifier circuit.

$$V_{BTH} = \frac{R_2}{R_2 + R_1} \, V_{CC} \qquad V_{CTH} = V_{CC}$$

$$R_{BTH} = (R_2 \| R_1) \qquad R_{CTH} = R_C$$

(c)

For a complete analysis of the amplifier, we must know how to treat the capacitors at all frequencies. We spend some time on this problem in Chapter 15. For our purposes here, we only need to understand how the coupling capacitors behave for the most important frequencies, namely, the frequencies we want to amplify. We argue that they should be replaced by *short circuits* at those frequencies. To see this, consider the base–emitter portion of Fig. 14.23a with the dc voltage suppressed and some phasor $\hat{V}_i$ at frequency ω applied as shown in Fig. 14.23d. The voltage which appears at the base is given by the voltage divider rule:

$$\hat{V}_B = \left[\frac{R_E}{R_E + R_s - (j/\omega C_1)} \right] \hat{V}_i$$

where $R_E = R_1 \, \| \, R_2$. Now if the phasor frequency is "high enough," the $1/\omega C_1$ term will be very small and

$$\hat{V}_B = \left[\frac{R_E}{R_E + R_s} \right] \hat{V}_i$$

This is the same result which would hold if C were replaced by a short circuit and our proof is complete. We study how high ω must be to satisfy this requirement in Chapter 15.

The small-signal response may now be determined at "high" frequencies by using a linear model of the BJT transistor. At the operating point we first need to know the values of r_i, β, and r_o. Assume that they are either found as demonstrated in the examples at the end of the chapter or are given to be $r_i = 2.5 \text{ k}\Omega$, $\beta = 67$, and $r_o = 7.5 \text{ k}\Omega$. To study the small-signal response the power supply is suppressed and the linear model for the resulting circuit is as shown in Fig. 14.23e.

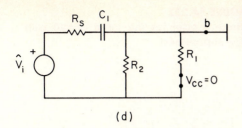

FIGURE 14.23d.
Base–emitter circuit with V_{CC} suppressed and with some phasor $\hat{V}_i$ applied.

(d)

The arrows show where the coupling capacitors have been replaced by the short circuits. The expression for $i_b(t)$ is given by

$$i_b(t) = \frac{v_i(t)}{R_s + (r_i//R_1//R_2)}\left(\frac{R_1//R_2}{r_i + R_1 + R_2}\right)$$

The output voltage may now be found. Since r_o, R_C, and R_L are in parallel,

$$\frac{1}{R_p} = \frac{1}{r_o} + \frac{1}{R_C} + \frac{1}{R_L}$$

and

$$v_o(t) = -\beta i_b(t)R_p$$

The final amplifier voltage gain is

$$A_v = \frac{v_o(t)}{v_i(t)} = \frac{-\beta R_p}{R_s + (r_i//R_1//R_2)}\left(\frac{R_1//R_2}{r_i + (R_1//R_2)}\right)$$

It is helpful to do the specific example from Fig. 14.23a, one step at a time. The parallel combination of the resistors R_1 and R_2 yields 17.5 kΩ so we have at the input the circuit shown in Fig. 14.23f. This parallel combination of 17.5 kΩ and 2.5 kΩ is approximately 2.2 kΩ. Then, the total current $i(t)$ is

$$i(t) = \frac{v_i(t)}{22.2 \text{ k}\Omega}$$

FIGURE 14.23e.
Linear model for small-signal response of the BJT amplifier with the power supply (V_{CC}) suppressed by shorting it to ground.

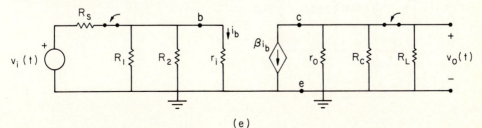

(e)

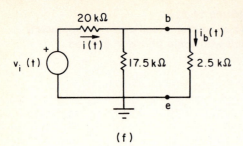

(f)

FIGURE 14.23f.
Linear model for the input circuit with
V_{CC} **suppressed.**

Using the current divider law,

$$i_b(t) = i(t)\frac{17.5 \text{ k}\Omega}{20 \text{ k}\Omega} = \frac{(0.875)v_i(t)}{22.2 \text{ k}\Omega}$$

At the output the linear model is as shown in Fig. 14.23g. The parallel combination 2 kΩ // 7.5 kΩ equals 1.58 kΩ, so the load current is

$$i_L(t) = \left(\frac{-1.58}{3.58}\right)\beta i_b(t) = -(0.44)\beta i_b(t)$$

Finally, since $v_o(t) = i_L(t)R_L$, we can substitute for β and $i_b(t)$ into the expression above. Then, dividing by $v_i(t)$ yields

$$A_v = \frac{v_o}{v_i} = -2.32$$

To summarize, the reduction of the number of power supplies necessary to operate the amplifier has not been accomplished without a price. Both the analysis and the design are more complicated because the resistors associated with the power supplies must be taken into account in calculating the gain.

Finally, we make note of an approximation which is often used to quickly determine the dc characteristics of the base–emitter diode. Usually, $r_i << R_B$, where R_B is the equivalent resistance in series with the base, and V_T is in the range

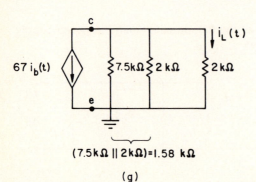

(g)

FIGURE 14.23g.
Linear model for the output circuit.

0.55 to 0.8 V for a silicon device. Taking $V_T \simeq 0.7$ V, which is in the middle of this range and ignoring r_i as compared with R_B, we can estimate I_B as

$$I_B \simeq \frac{V_B - 0.7}{R_B}$$

For example, applying this approximation to the circuit in Fig. 14.16 yields

$$I_B \simeq \frac{4 - 0.7}{20 \text{ k}\Omega} = 0.165 \text{ mA}$$

This is quite close to the 0.15 mA value found by the load line approach earlier but requires no curve plotting. If r_i is known or can be estimated, then a more precise I_B may be found analytically. For example, from Fig. 14.18 the student can verify that $r_i \simeq 2.0$ kΩ. Then, the base current is given by

$$I_B = \frac{4 - 0.7}{22 \text{ k}\Omega} = 0.15 \text{ mA}$$

which is a more accurate value for I_B. Since a good amplifier design should not depend critically upon the exact location of the dc operating point, such an approximation is usually justified. In fact, the designer usually picks a desired base current I_B and then determines R_B from the approximation

$$R_B = \frac{V_B - 0.7}{I_B}$$

A second approximation is also often used in the output portion of the circuit. For most transistors the dc values of I_C and I_B have a ratio which is roughly equal to the linear model value for β, which in turn is usually supplied by the manufacturer. If we define the ratio

$$\beta_o = \frac{I_C}{I_B}$$

then combining these two approximations,

$$I_C \simeq \frac{\beta_o(V_B - 0.7)}{R_B}$$

If we then approximate β_o by β, I_C can be determined as can V_{CE} from $V_{CE} = V_{CC} - I_C R_C$.

TEXT EXAMPLE 14.3 _____

Use the approximation approach outlined above to estimate the operating points Q_1 and Q_2 in the circuit given in Fig. 14.23a if the transistor used is as described by the curves in Figs. 14.13b and 14.15. Let $\beta_o = 67$.

Solution: Since $R_2 = 20$ kΩ and $R_1 = 140$ kΩ, the Thevenin equivalent voltage (see Fig. 14.23c) is $V_{BTH} = 30$ V/8 $= 3.75$ V. Then, since $R_{BTH} = R_1 // R_2 = 17.5$ kΩ,

$$I_B = \frac{3.75 - 0.7}{17.5 \text{ k}\Omega} = 0.17 \text{ mA}$$

Since $\beta_o = 67$, $I_C = 11.7$ mA and

$$V_{CE} = 30 - I_C R_C = 6.65 \text{ V}$$

This approximation is checked against the load line approach in the problem set.

14.6 BIAS STABILIZATION CIRCUIT: MOSFET

All transistor devices used in amplifiers have parameters that vary with temperature and that also vary from one manufacturing batch to another. To compensate for these variations the dc bias circuit often incorporates the self-compensating element R_K in the manner shown in Fig. 14.24a. The symbol R_K is used instead of R_s to avoid confusion with a signal source resistance. Since the gate is an open circuit, the voltage of the gate with respect to ground potential is given by the expression

$$V_G = \left(\frac{R_1}{R_1 + R_2}\right) V_{DD}$$

However, the gate to source voltage V_{GS} is now not simply V_G because of the source voltage $V_S = I_{DS} R_K$. The circuit equations are

$$V_{DD} = I_{DS}(R_D + R_K) + V_{DS} \tag{14.7a}$$

$$V_{GS} = V_G - I_{DS} R_K \tag{14.7b}$$

To find the dc operating point, both Eq. 14.7a and Eq. 14.7b must be satisfied. Even if we know the circuit parameters $(V_{DD}, R_1, R_2, R_D,$ and $R_K)$, the solution still requires an iterative approach since there are only two equations but three unknowns. The third "equation" corresponds to the set of curves for the device. First, I_{DS} is guessed and V_{GS} determined from Eq. 14.7b. Armed with V_{GS} and I_{DS}, V_{DS} is specified by the graph. If this value satisfies Eq. 14.7a, the solution is found. If not, the process is repeated.

More commonly, we start with some *desired* operating point and choose values of the circuit elements to achieve that result. This approach is actually much easier. For example, for the MOSFET of Fig. 14.24a, suppose we choose an operating point: $V_{GS} = 8$ V, $I_{DS} = 4$ mA, and $V_{DS} = 5.5$ V. If the dc source V_{DD} is 24 V, we must lower V_G somewhat to yield sensible V_{GS} values. If $R_1 = R_2 = 100$ kΩ, then a good start is given by

$$V_G = \frac{(24)(100)}{200} = 12 \text{ V}$$

FIGURE 14.24.
(a) A MOSFET bias circuit including a resistor R_K between source and ground. (b) Characteristics of a typical MOSFET.

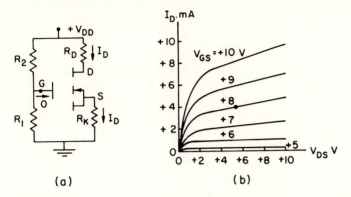

(a)   (b)

Since $V_{GS} = 8 = V_G - I_D R_K$, we deduce that $I_D R_K = 4$ V, and R_K must therefore be set equal to 1 kΩ. To find R_D we use Eq. 14.7a:

$$I_{DS}(R_D + R_K) = V_{DD} - V_{DS} = 24 - 5.5 = 18.5 \text{ V}$$

Since $I_{DS} = 4$ mA, it follows that $R_D + R_k = 4.6$ kΩ and that R_D must be set equal to 3.6 kΩ. Resistor values can be easily purchased to within 1%, which is accurate enough for most circuit applications.

The introduction of the "self-bias resistor" R_K will also affect the small-signal response of the amplifier. If the signal is coupled into the transistor through a capacitor, the linear model becomes the circuit shown in Fig. 14.24c. At "high enough" frequencies, the coupling capacitor will act like a short circuit, $v_g(t)$ will be equal to $v_i(t)$, and we may proceed as follows. To find the voltage gain v_o / v_g, we have

$$v_{gs} = v_g - i_d R_K \qquad (14.8a)$$

Using the current divider principle,

$$i_d = \frac{r_d(g_m v_{gs})}{r_d + R_K + R_D} = \alpha g_m v_{gs} \qquad (14.8b)$$

FIGURE 14.24(c).
Linear circuit model of Fig. 14.23a with small input signal.

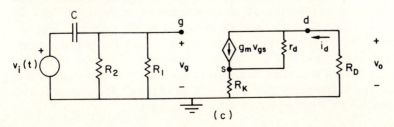

(c)

since the series combination $R_K + R_D$ is in parallel with r_d. For another way to derive this result see Text Example 14.4. In Eq. 14.8b we have defined the parameter α by the expression

$$\alpha = \frac{r_d}{r_d + R_K + R_D}$$

Combining Eqs. 14.8a and 14.8b yields

$$i_d = \alpha g_m(v_g - i_d R_K) \tag{14.8c}$$

and, solving for i_d,

$$i_d = \frac{\alpha g_m v_g}{1 + \alpha g_m R_K}$$

Finally, since $v_o = -i_d R_D$ and $v_g = v_i$ at "high" frequencies,

$$A_v = \frac{v_o}{v_i} = \frac{-\alpha g_m R_D}{1 + \alpha g_m R_k} \tag{14.8d}$$

From Eq. 14.8d we see that the effect of R_K is to decrease the magnitude of i_d and thus of the voltage gain. For $R_K = 0$, the gain A_v can be written as $A_v = v_o/v_i = -\alpha_o g_m R_D$, where $\alpha_o = r_d/(r_d + R_D)$ which is the result found in Section 14.2. For R_K not equal to zero, the value of α is decreased slightly from α_o, but the main effect is the introduction into the denominator of the factor $1 + \alpha g_m R_K$, which lowers the gain more drastically. Although the gain is decreased, it is also made less sensitive to changes in g_m, since g_m now appears in both the numerator and denominator. This is one of the most important reasons to use R_K. In the limit, if we let $\alpha g_m R_K \to \infty$, the gain will be independent of g_m and $A_v = -R_D/R_K$. This is important because a good amplifier design should not depend upon the exact value of g_m (or β in a BJT device) and because for typical commercial products such a parameter is variable from one device to the next. Furthermore, g_m and β depend upon temperature. Since it is very undesirable for an amplifier to change its gain with temperature, circuits such as the ones studied here reduce this tendency.

The reader may note a certain similarity in this analysis and the Op-Amp study done in Chapter 4. In Eq. 14.8c, for example, i_d appears on both sides of the equation. This algebraic form also occurred in the Op-Amp analysis. Similarly, in the latter case we took a limit as $A \to \infty$ and found that the gain was determined solely by the external resistors. For example, the gain of the inverting amplifier was $A_V = -R_2/R_1$. Here, we noted that if $\alpha g_m R_K \to \infty$, the gain is just $-R_D/R_K$, a form independent of the transistor properties. Both systems employ negative feedback. In the present case the output current i_d appears in Eq. 14.8a for v_{gs} with a negative sign. This means that the input circuit voltage depends upon the output current. Since i_d in turn depends upon v_{gs}, a closed loop is formed. Although a detailed study of negative feedback is beyond the scope of this text, it is important to note examples of this type for reference in more advanced courses. The student is also referred to Section 4.6 for a discussion of negative feedback.

The intrinsic value of the decrease in sensitivity of the amplifier gain to changes in the transistor characteristics also applies to the dc operating point. A MOSFET temperature increase tends to an increase in the current I_D. However, in a self-biased circuit an increase in I_D will cause an increase in the self-bias voltage $I_D R_K$. This leads to a *decrease* in the voltage V_{GS}. This decrease in V_{GS} will tend to decrease I_D, counterbalancing the effect of the temperature increase. This benefit is the major reason for including the resistor R_K.

We conclude this discussion by computing the voltage gain for the particular operating point and circuit elements used in the example above. At the operating point in part (a) of Fig. 14.24, $I_D = 4$ mA, $V_{GS} = 8$ V, and $V_{DS} = 5.5$ V. The small-signal parameters may be shown by the student to be $g_m = 2 \times 10^{-3}$ mhos and $r_d = 6$ kΩ. We also found above that $R_K = 1$ kΩ. Then,

$$\alpha = \frac{6}{6 + 1 + 3.6} = 0.57$$

and

$$A_v = \frac{v_o}{v_g} = -\left[\frac{(0.57)(2 \times 10^{-3})(3.6 \times 10^3)}{1 + (0.57)(2 \times 10^{-3})(1 \times 10^3)}\right] = -1.9 \qquad (14.8e)$$

If $R_K = 0$, the gain v_o/v_g would be about -4.1, so in this example the effect of R_K is to reduce the ac gain by a factor of about 2.1. In Chapter 15 we show how some of the benefits of R_K may be realized without affecting the signal gain.

TEXT EXAMPLE 14.4

Find the small-signal amplifier gain $A_v = v_o/v_i$ for the circuit below if the device is characterized by the linear model parameters $g_m = 5 \times 10^{-2}$ mho and $r_d = 50$ kΩ.

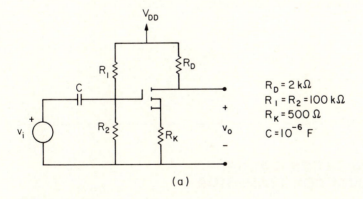

$R_D = 2$ kΩ
$R_1 = R_2 = 100$ kΩ
$R_K = 500$ Ω
$C = 10^{-6}$ F

(a)

Solution: First we note that the circuit diagram for the linear model is identical to part (c) of Fig. 14.24. At high enough frequencies that the coupling capacitor may be replaced by a short circuit, $v_i(t)$ directly appears at the gate and

$$v_g(t) = v_i(t)$$

Since $v_{gs}(t) = v_g(t) - v_s(t)$ and $v_s = i_d R_K$,

$$v_{gs} = v_g - i_d R_K$$

Redrawing the output circuit somewhat we have

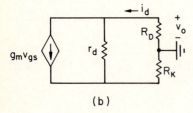

(b)

From the current divider rule,

$$i_d = \left(\frac{r_d}{r_d + R_K + R_D} \right) g_m v_{gs} = \alpha g_m v_{gs}$$

In this example $\alpha = 0.97$. Continuing we can solve for i_d in terms of v_i by substituting for v_{gs}:

$$i_d = \alpha g_m (v_i - i_d R_K)$$

Gathering terms,

$$i_d = \frac{\alpha g_m v_i}{1 + \alpha g_m R_K}$$

which is the same result found in the text. Since $v_o = -i_d R_D$ we have

$$A_V = \frac{-\alpha g_m R_D}{1 + \alpha g_m R_K}$$

Substitution yields $A_v = -3.84$.

14.7 BIAS STABILIZATION CIRCUIT: BIPOLAR JUNCTION TRANSISTOR

For the common emitter amplifier circuits we have discussed so far, we have used bias circuits such as the one shown in Fig. 14.23a. Since the coupling capacitors are open circuits at dc, the dc circuit can be redrawn as in Fig. 14.25. The equations in the figure show approximations that can be used to determine the operating point for a silicon device. Once again, the problem with such a "fixed bias" circuit is the

FIGURE 14.25.
"Fixed bias" circuit for junction transistor, with approximate equations for operating point. These equations assume that $r_i = 0$ and use the parameter β_0, the ratio of I_C to I_B at dc.

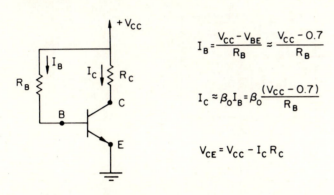

$$I_B = \frac{V_{CC} - V_{BE}}{R_B} \simeq \frac{V_{CC} - 0.7}{R_B}$$

$$I_C \approx \beta_0 I_B = \beta_0 \frac{(V_{CC} - 0.7)}{R_B}$$

$$V_{CE} = V_{CC} - I_C R_C$$

variability of the transistor parameter β_o with the manufacturing process and with temperature. The same base current I_B may result in a larger or smaller collector current than desired. The bias stabilization circuit in part (a) of Fig. 14.26 is designed to keep the dc collector current nearly constant even as β_o changes with temperature.

In part (b) of Fig. 14.26 we have simplified the circuit by using Thevenin's theorem in the base-emitter portion of the circuit. The circuit equations then become

$$V_B = I_B R_B + V_{BE} + (I_B + I_C)R_E \tag{14.9a}$$

$$V_{CC} = I_C R_C + V_{CE} + (I_B + I_C)R_E \tag{14.9b}$$

If, as noted in the figure, we let $I_C = \beta_o I_B$,

$$V_B \simeq I_B R_B + V_{BE} + I_B(\beta_o + 1)R_E$$

Solving for I_B and I_C,

$$I_B = \frac{V_B - V_{BE}}{R_B + (\beta_o + 1)R_E} \tag{14.9c}$$

$$I_C = \beta_o I_B = \frac{\beta_o(V_B - V_{BE})}{R_B + (\beta_o + 1)R_E} \tag{14.9d}$$

We see from Eqs. 14.9c and 14.9d that as β_o increases, I_B will decrease and I_C will tend to stay constant. In the limit, if $(\beta_o + 1)R_E >> R_B$, I_C will be almost independent of β_o. Note also we can usually set $V_{BE} \simeq 0.7$ V in these expressions.

The effect of introducing the "emitter resistance" R_E into the circuit is thus to keep the dc collector current constant if β_o changes. The emitter resistance will have the same effect on the signal and, as found above for the MOSFET, will result in a reduction of the gain of the amplifier.

FIGURE 14.26.

(a) Bias stabilization circuit for junction transistors. (b) The circuit is redrawn using Thevenin's theorem in the base–emitter portion of the circuit.

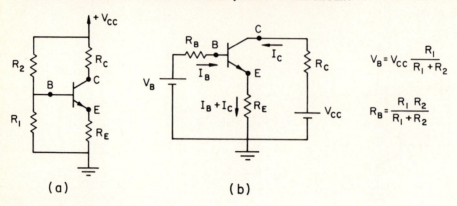

(a) (b)

TEXT EXAMPLE 14.5 _____

Suppose the same transistor whose curves are given in Figs. 4.13b and 14.17 are used in the circuit of Fig. 14.26. We desire the same operating points Q_1 and Q_2. Find suitable values of R_1, R_2, and R_C if $V_{CC} = 12$ V. You may use the base–emitter approximation described above (e.g., $V_{BE} = 0.7$ V) and the β_o approach.

Solution: From Figs. 4.13b and 14.17, $I_B = .15$ mA and $I_c = 10$ mA so $\beta_o = 67$. Using Eq. 14.9b and $V_{CE} = 9$ V,

$$12 = \beta_o I_B R_C + 9 + (\beta_o + 1)I_B R_E$$

As a starting point let $R_C = R_E$. Then

$$R_C = \frac{3}{(67 + 68)I_B} = 148 \ \Omega$$

From Eq. 14.9a, noting $R_B = R_1//R_2$ and $V_B = 12R_1/(R_1 + R_2)$,

$$\frac{R_1}{R_1 + R_2}(12) = \frac{I_B R_1 R_2}{R_1 + R_2} + 0.7 + (\beta_o + 1)I_B R_E$$

The last term equals 1.51 V,

$$12 = R_2 I_B + \left(\frac{R_1 + R_2}{R_1}\right)(2.21)$$

Suppose $R_1 = R_2$, then

$$R_2 = \frac{12 - 4.42}{0.00015} = 50.5 \text{ k}\Omega$$

Rounding off we could use the solution,

$$R_E = R_C = 150 \; \Omega$$

$$R_1 = R_2 = 50 \; k\Omega$$

Of course many other solutions could be found and some fine tuning might be necessary, but this set is a reasonable place to begin the design.

14.8   THE *h*-PARAMETER MODEL*

The linear model for a BJT transistor uses the three parameters r_i, β, and r_o, while the three parameters characterizing a MOSFET are ∞, g_m, and r_d. The former uses a controlled current source and the latter a voltage-controlled current source. As noted earlier, the fact that $r_i = \infty$ effectively decouples the load from the input circuit for a MOSFET device. However, this decoupling is less realistic for a BJT device. In effect, we have approximated the BJT circuit by assuming that a single diodelike curve characterizes the input (Fig. 14.13*b*) rather than the actual family of curves in Fig. 14.13*a*. Before considering the BJT in the general case, a more detailed discussion of two-port parameters is in order.

A two-port system that has no independent sources such as those described here for transistors requires four parameters for a complete characterization which is independent of the outside world. This can be understood as follows. A two-port, as illustrated in Fig. 14.27, involves the relationship between two currents and two voltages. If the box contains only linear passive and active elements and *no independent sources*, we could write the relation of the voltages to the currents using superposition in the form

$$v_1 = r_{11}i_1 + r_{12}i_2$$

$$v_2 = r_{21}i_1 + r_{22}i_2$$

The values of the four coefficients all have dimensions of resistance, are determined by the elements in the box, and do not depend on the circuits connected outside the box. It is important to note for our case that the voltages v_1 and v_2 and the currents i_1 and i_2 refer only to the small variations in voltage and current due to an imposed small signal and not to the dc voltages and currents which determine the operating point. The superposition principle allows us to do this and depends upon the linear (small-signal) applications we are studying.

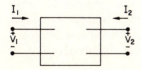

FIGURE 14.27.
Current and voltage conventions for a two-port.

---

* Sections marked with an asterisk are not required for the logical flow of the text.

The equations given above can also be solved for v_1 and i_2 in terms of i_1 and v_2, and a more common way to express the resulting set is through the so-called "h," or "hybrid," parameters:

$$v_1 = h_i i_1 + h_r v_2 \tag{14.10}$$

$$i_2 = h_f i_1 + h_o v_2 \tag{14.11}$$

The parameters are called "hybrid" because they have different units. For example, from the definitions, h_r and h_f must be dimensionless, while h_i represents a resistance in ohms and h_o represents a reciprocal resistance or conductance in mhos. These equations may be interpreted in terms of circuit elements with reference to Fig. 14.28. That this model is valid can be seen by analyzing the circuit at the two ports. At the input port, KVL yields

$$\frac{v_1 - h_r v_2}{h_i} = i_1$$

FIGURE 14.28.
General model of a two-port using hybrid parameters.

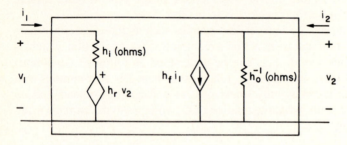

which can be written in the form of Eq. 14.10. At the output port, KCL yields

$$i_2 = h_f i_1 + h_o v_2$$

which is Eq. 14.11. The model given in Fig. 14.28 can be used to represent any two-port system containing linear passive elements.

Now there is a distinct similarity between this four-parameter model and the three-parameter model we have used for BJT transistors. The only difference is that in place of the controlled voltage source $h_r v_2$, we had a fixed battery V_T representing the base–emitter diode potential drop. Then, when superposition is applied, V_T is replaced with a short circuit and the only remaining parameter in the linear small signal model is r_i.

The parameter h_r, which we have ignored so far in the text, models the weak interaction that exists between the output and the input since it couples the output voltage v_2 to the input current i_1. The parameter h_r may be understood with reference to Fig. 14.29. The fourth parameter h_r corresponds to the partial derivative

$$h_r = \frac{\partial V_{BE}}{\partial V_{CE}} \qquad \text{with } I_B = \text{constant}$$

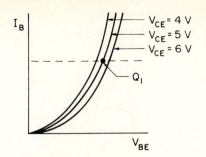

FIGURE 14.29.
Family of curves for I_B versus V_{BE} when V_{CE} is varied.

In the graph this corresponds to variations in the collector–emitter voltage between the family of curves for a constant value of I_B, that is, along the horizontal dashed line (I_B = constant).

While it is true that the h-parameter model in Fig. 14.28 is a more accurate representation of the BJT transistor, in most applications the three-parameter model is quite adequate. For reference, the h parameters have the following rough equivalence to the BJT model used in this text:

$$h_i \longleftrightarrow r_i$$

$$h_r \longleftrightarrow \text{set equal to zero}$$

$$h_f \longleftrightarrow \beta$$

$$(h_o)^{-1} \longleftrightarrow r_o$$

We use the term rough equivalence since h_i is only approximately equal to the r_i in a three-parameter model.

14.9   SUMMARY

In this chapter we introduced two types of transistors and generated linear models which described their response to small signals. Some simple amplifier circuits were presented and analyzed using these linear models. These circuits have an input signal $v_i(t)$ which is modified by the amplifier to yield the output signal $v_o(t)$. Although the amplifier circuits used in this chapter are not very practical, some realistic transistor biasing schemes were given and will be incorporated in the circuits of Chapter 15.

PRACTICE PROBLEMS
AND
ILLUSTRATIVE EXAMPLES

Some representative device characteristics which are used in the following problems are given below.

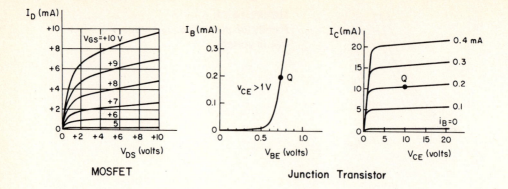

MOSFET   Junction Transistor

Problem 14.1

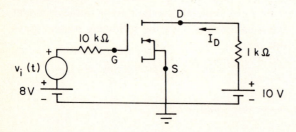

The MOSFET in the circuit has the characteristics given at the left above. Find I_D and V_{DS} for $v_i(t) = 0$.

$$\text{Ans.: } I_D = 4 \text{ mA}$$
$$V_{DS} = 6 \text{ V}$$

Problem 14.2

In the circuit of Problem 14.1, suppose $v_i(t)$ is a signal given by $1.0 \cos(\omega t)$ V. Using the load line and the MOSFET characteristics, find the time-varying part of the output voltage $v_{DS}(t) = v_o(t)$ by graphical analysis. Estimate the gain of the circuit $A_v = v_{ds}/v_i$.

$$\text{Ans.: } v_o(t) \text{ is approx. } - 1.6 \cos(\omega t) \text{ V}$$
$$(\text{Note inverse sign of } v_o(t) \text{ with}$$
$$\text{respect to } v_i(t).) \text{ So, } A_v \simeq -1.6$$

Problem 14.3

The 1-kΩ resistor in the circuit of Problem 14.1 is replaced by a resistor of value R_1. Find the value of R_1 if I_D is to be made equal to 3.75 mA and $V_{DS} = 4$ V. If R_1 is not changed, that is, it remains 1 kΩ, to what value must the 10-V battery be changed to yield the same operating point (i.e., $I_D = 3.75$ mA, $V_{DS} = 4$ V)?

Problem 14.4

Find the new values of I_D and V_{DS} in the circuit of Problem 14.1 if a 30-kΩ resistor is inserted between the point G and ground potential. If $v_i(t) = 1.33 \cos(\omega t)$ V, find the two maximum excursions of v_{ds} graphically. Explain why this result implies that the amplifier is more nonlinear in this portion of (I_D, V_{DS}) space than it is in the "center" of the graph (e.g., in the region associated with Problem 14.2).

EXAMPLE 14.1

Solve Problem 14.2 replacing the MOSFET with a small-signal model.

Solution: At the operating point we can represent the MOSFET for small signals by the model

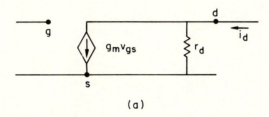

(a)

From the characteristic curves of the MOSFET, at the operating point $(I_D, V_{DS}) =$ (4 mA, 6 V) we can estimate g_m and r_d.

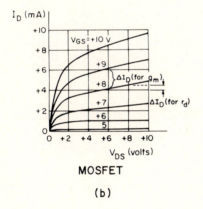

MOSFET

(b)

$$g_m = \frac{\partial i_D}{\partial v_{GS}} \approx \frac{\Delta I_D}{\Delta V_{GS}} \qquad \text{(for } V_{DS} = 6 \text{ V)}$$

$$= \frac{2 \times 10^{-3} \text{ mA}}{1 \text{ V}} = 2 \times 10^{-3} \text{ mhos}$$

Likewise,

$$\frac{1}{r_d} = \frac{\partial i_D}{\partial v_{DS}} \approx \frac{\Delta I_D}{\Delta V_{DS}} \qquad \text{(for } V_{GS} = 8 \text{ V)}$$

$$= \frac{0.5 \times 10^{-3} \text{ mA}}{2 \text{ V}} = 0.25 \times 10^{-3} \text{ mhos}$$

and

$$r_d = 4000 \ \Omega$$

The small-signal model for the circuit is, then, after suppressing the dc sources,

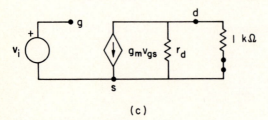

(c)

Since $v_{gs} = v_i$ and $r_d \mathbin{/\!/} 1 \text{ k}\Omega = 800 \ \Omega$, we have, from Ohm's law,

$$v_{ds} = -g_m v_i (800) = -1.6 v_i$$

which is the same result obtained using the load line method in Problem 14.2.

Problem 14.5

For a MOSFET with the characteristics given in Example 14.1, find the values in the small-signal linear model at an operating point $I_D = 4.5$ mA and $V_D = 2$ V.

Problem 14.6

Determine the linear model parameters g_m and r_d for the operating point in Problem 14.4. Use this result to find the theoretical small-signal gain of the circuit. Compare this result to the nonlinear voltage excursions found graphically in Problem 14.4.

Problem 14.7

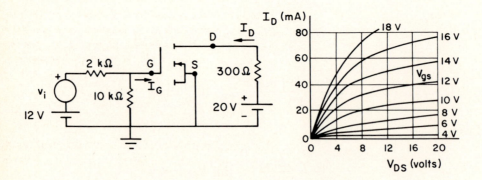

The MOSFET in the circuit above has the characteristics shown and, as usual, $I_G = 0$.

(a) For $v_i = 0$, find the dc operating
point of the MOSFET, I_D and V_{DS}.   Ans.: $I_D \approx 25$ mA; $V_{DS} \approx 12.5$V

(b) Find the approximate values of g_m
and r_d at the operating point.   Ans.: $g_m \approx 6 \times 10^{-3}$ mhos
$$r_d \approx 1400 \ \Omega$$

(c) For a small time-varying voltage v_i,
find the ratio of v_d / v_i where v_d is the
time-varying component of V_{DS}.   Ans.: $v_d / v_i \approx -1.23$

Problem 14.8

Find the new gain of the circuit in Problem 14.7 if the 10-kΩ resistor is removed
and if the linear model parameters quoted above are not affected by the shift in the
operating point.

Problem 14.9

Consider the elemental MOSFET amplifier in Fig. 14.7c. If V_{DD} remains at 30 V,
what resistor R_D is required to yield $V_{DS} = 20$ V? (No other elements in the circuit
change.) Assume that the parameters $g_m (2.5 \times 10^{-3}$ mhos) and r_d (5.33 kΩ) do not
change significantly from those values found in Section 14.2 for the same MOSFET
at a different but nearby operating point. Find the new circuit gain.

Problem 14.10

Find the gain in Text Example 14.2 if the MOSFET linear parameters g_m and r_d
are equal to those values given in Problem 14.9 above.

Problem 14.11

In Text Example 14.2, suppose the output of the circuit is loaded by a 5-kΩ resistor
placed between the output and ground. Find the new system gain if the linear model
parameters remain the same as in the text example.

Problem 14.12

A MOSFET transistor has the characteristics shown in the figure below and is used
in the circuit below where $R_s = 10$ kΩ.

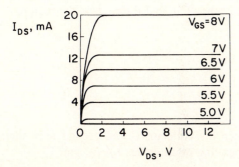

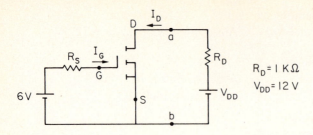

$R_D = 1 \ K\Omega$

$V_{DD} = 12 \ V$

(a) Find the operating point.

Ans.: $I_{DS} = 7 \ \text{mA}$
$V_{DS} = 5.0 \ V$

(b) Find the parameters g_m and r_d at the operating point.

Ans.: $g_m = 6 \times 10^{-3} \ \text{mhos}$
$r_d \simeq \infty$

(c) For a small voltage v_i in series with the 6-V battery, use the linear model to find the voltage gain $A_v = v_o/v_i$ where v_o is measured across the terminal a–b.

Ans.: $A_v = -6.0$

Problem 14.13

Suppose in the circuit of Problem 14.12 above that a 2-kΩ resistor is inserted between points a and b. Find the new operating point values I_{DS} and V_{DS}. Assuming that g_m and r_d remain unchanged, from the values given above find the new gain of the circuit.

Problem 14.14

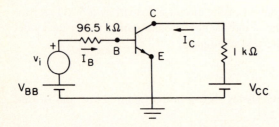

The junction transistor in the circuit has the characteristics shown at the beginning of the problem set. For the dc operating point Q shown in that figure, find the required values of V_{BB} and V_{CC}.

Ans.: $V_{BB} = 20 \ V$
$V_{CC} = 20 \ V$

Problem 14.15

Another 1-kΩ resistor is connected between the collector C and ground E in the circuit of Problem 14.14. Find the new operating point values I_B, I_C, and V_{CE}.

Problem 14.16

Suppose v_i in Problem 14.14 is such that $i_b(t) = 0.1 \cos(\omega t)$ mA. Use this base current and the load line to graphically find the time-varying part of the collector current $i_c(t)$ and of the collector-emitter voltage $v_{ce}(t)$.

$$\text{Ans.: } i_c(t) \simeq 5 \cos(\omega t) \text{ mA}$$
$$v_{ce}(t) \simeq -5 \cos(\omega t) \text{ V}$$

EXAMPLE 14.2

For the circuit of Problem 14.14 let $v_i = 0.1 \cos(100t)$. Find $i_b(t)$, $i_c(t)$, and $v_{ce}(t)$ using a linear model.

Solution: Since i_b, i_c, and v_c will vary only slightly from their values at Q, we may write the total current and voltage as

$$I_B(t) = I_B + i_b(t) = [0.2 \times 10^{-3} + i_b(t)] \text{ A}$$

$$I_C(t) = I_C + i_c(t) = [10 \times 10^{-3} + i_c(t)] \text{ A}$$

$$V_{CE}(t) = V_{CE} + v_{ce}(t) = [10 + v_c(t)] \text{ V}$$

To find $i_b(t)$, we use the linear model element for the base–emitter diode, which is just a resistance r_i, where r_i is the reciprocal of the slope of the characteristic at Q.

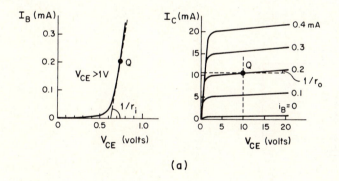

(a)

From the figure, $r_i \approx 500 \ \Omega$. Then,

$$i_b = \frac{v_i(t)}{96.5 \times 10^3 + 500} = 1.03 \cos(100t) \ \mu\text{A}$$

For the collector–emitter portion of the circuit, we may use the small-signal model below:

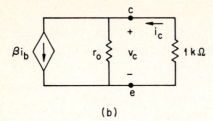

(b)

We estimate β using a vertical line through the operating point, that is, $\beta \approx \Delta i_C / \Delta i_B$ for $V_{CE} = 10$ V. From the figure, $\beta \approx (15 - 5) \times 10^{-3}/(0.3 - 0.1) \times 10^{-3} = 50$. The parameter $r_o \approx (\Delta i_C / \Delta v_C)^{-1}$ for $I_B = 0.2$ mA. From the graph, $r_o \approx 20/1 \times 10^{-3} = 20$ kΩ. Using the current divider law,

$$i_c(t) = \left(\frac{r_o}{r_o + 10^3} \right) \beta i_b(t) = 48 \cos(100t) \ \mu\text{A}$$

and

$$v_{ce}(t) = -i_c(t)(10^3) = -48 \cos(100t) \ \text{mV}$$

Problem 14.17

Find the circuit gain in the configuration of Problem 14.15 (i.e., including the extra 1-kΩ resistor) if the parameters r_i, β, and r_o do not change from those deduced in Example 14.2.

Problem 14.18

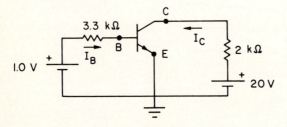

The junction transistor in the circuit above has the characteristics shown in the figure for Example 14.2.

(a) Find the operating point. Ans.: $I_B = 0.1$ mA, $V_{BE} = 0.67$ V
$I_C = 5.5$ mA, $V_{CE} = 9$ V

(b) Find the linear model parameters r_i,
β, and r_o. Ans.: (Almost the same as in Example 14.2.)

$r_i \approx (500 \ \Omega; \ \beta \approx 50; \ r_o \approx 20$ kΩ

(c) Find the voltage gain A_v if a small
signal $v_i(t)$ is inserted in series with
the 1.0-V battery.   Ans.: $A_v = -23.9$

Problem 14.19

In the circuit diagram below, let $V_{DD} = 10$ V. Find a set of values for R_1, R_2, and R_D which yield the same operating point as found in Problem 14.1.

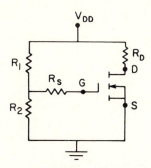

(In principle, correct values for R_1 and R_2 may be multiplied by any constant. In practice, R_1 and R_2 should not be too small or excessive power will be drawn from the supply. They should not be too large or the assumption that the gate is an open circuit will break down. Therefore, use resistors in the tens of kΩ range.)

Problem 14.20

A single 20-V power supply value is to be used in the circuit of Problem 14.7. That is, the 12-V supply is replaced by 20 V in the circuit. Choose a new value of the 2-kΩ resistor that yields the same operating point. Find the new linear gain of the amplifier.

Problem 14.21

Design a circuit using the same transistor and operating point as used in Problems 14.7 and 14.20 but which uses only a single 30-V power supply. Find the small-signal gain A_v of the new circuit and compare it to the answer to part c of Problem 14.7.

Problem 14.22

Design a system using only one battery which has the same operating point as the circuit in Problem 14.18.

Problem 14.23

Design a BJT biasing circuit for an *npn* transistor in the common emitter configuration using a single 10-V battery which has the operating points shown by the points labeled Q_1 below.

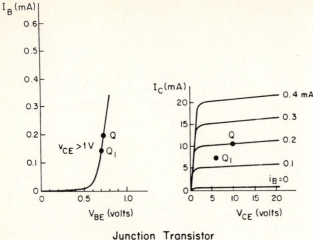

Junction Transistor

Problem 14.24

Estimate values for r_i, β, and r_o at the operating point Q in Problem 14.23.

EXAMPLE 14.3

An *npn* transistor characterized by the curves in Problem 14.23 is used in the circuit below.

(a)

Find the operating point of the device in this configuration. If the transistor is characterized by the linear model parameters $(r_i, \beta, r_o) = (500\ \Omega, 50, 20\ \text{k}\Omega)$ find the gain if the circuit $A_v = v_o(t)/v_i(t)$ at frequencies high enough that the two capacitors may be replaced by short circuits.

Solution: At dc the capacitors act like open circuits. At the base–emitter junction, then, the 10-V source and the two resistors may be replaced by their Thevenin equivalent circuit which consists of a 1-V battery in series with a 3.6-kΩ resistor.

Using these values a load line may be constructed. This intersects the base–emitter curve in Problem 14.23 at the approximate values $I_B = 0.1$ mA and $V_{BE} = 0.68$ V. At dc the collector–emitter side has the Thevenin equivalent circuit 10 V in series with 500 Ω. The corresponding loadline intersects the $I_B = 0.1$ mA curve at the approximate values $I_C = 5$ mA and $V_{CE} = 7$ V.

The circuit diagram associated with the linear model has the form shown below for frequencies high enough that the capacitors may be replaced by short circuits (highlighted by the arrows).

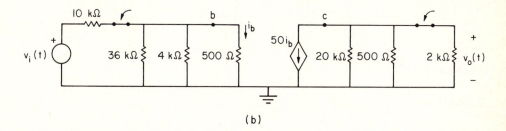

(b)

Analysis for the base current yields the equation

$$i_b = (8.4 \times 10^{-5})v_i(t)$$

This is amplified by 50 in the output side of the circuit. Using Ohm's law

$$v_o(t) = -(50)(8.4 \times 10^{-5})(2 \text{ k}\Omega \text{ // } 500 \text{ }\Omega \text{ // } 20 \text{ k}\Omega)v_i(t)$$

and

$$A_y = -1.65$$

Problem 14.25

The amplifier studied in Example 14.3 is loaded down by a 20-kΩ resistor which is in parallel with the 2-kΩ resistor at the output. Find the gain of the system in this configuration. If a requirement exists to maintain the gain at -1.65, determine a change in the value of the 4-kΩ resistors which would compensate for the decrease in gain found above. (Do not change the 36-kΩ or the 10-kΩ source resistor.) You may assume that r_i, β, and r_o do not change. *Hint:* Ignore the 36 kΩ resistor in your initial calculation and adjust the 4 kΩ to make v_b *larger*.

Problem 14.26

Draw the small-signal linear model for the circuit below if at the operating point the dynamic resistance r_i of the base–emitter diode is 1 kΩ, the current gain $\beta = 100$, and the dynamic resistance of the collector–emitter curve r_o is 10 kΩ. Find the voltage gain $A_v = v_o/v_i$ at frequencies high enough that C_1 and C_2 may be considered short circuits.

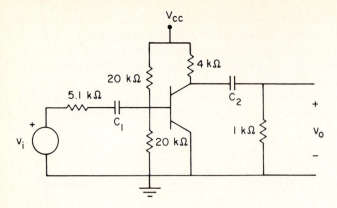

Problem 14.27

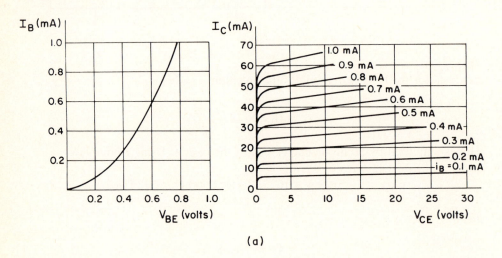

(a)

The transistor in the amplifier circuit below has the characteristics shown above.

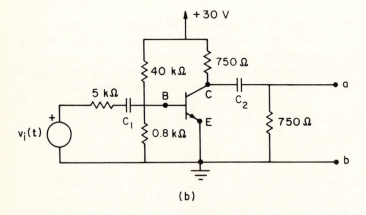

(b)

(a) Find the operating point of the transistor.

(b) At the operating point, find the parameters in the small-signal model of the transistor.

(c) Draw the small-signal circuit of the amplifier.

(d) Find the voltage gain $A_v = v_o/v_i$ if the frequency is high enough that the capacitors may be replaced by short circuits.

Problem 14.28

The circuit shown on the left below uses the MOSFET which is characterized by the curves shown on the right.

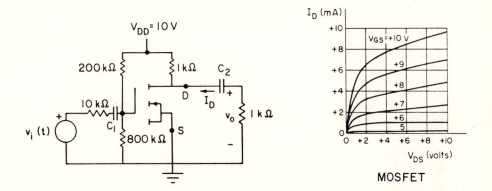

Find the operating point Q and draw the linear small-signal circuit diagram for the circuit. Evaluate the gain of the circuit $A_v = v_o/v_i$ if the capacitors are replaced by short circuits. (To do this you must evaluate g_m and r_d at Q.)

Problem 14.29

Suppose the 800-kΩ resistor in Problem 14.28 breaks, leaving an open circuit. Find the new operating point Q_1 and the gain of the new circuit. You may assume that g_m and r_d remain the same.

Problem 14.30

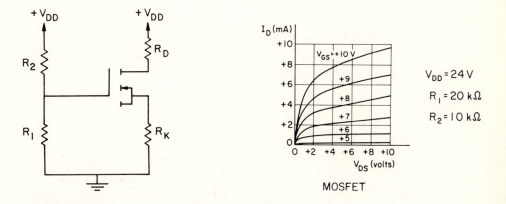

The n channel enchancement MOSFET in the circuit above has the characteristics shown. The desired operating point is at $V_{GS} = 8$ V, $I_D = 4$ mA. Find the required values of R_K and R_D if $V_{DD} = 24$ V, $R_1 = 20$ kΩ, and $R_2 = 10$ kΩ.

$$\text{Ans.: } R_K = 2 \text{ k}\Omega$$
$$R_D = 2.75 \text{ k}\Omega$$

Problem 14.31

Repeat Problem 14.30 without changing V_{DD}, R_1, and R_2 but making the requirement that $V_{DS} = 8$ V and $V_{GS} = 8$ V (i.e., find new values of R_K and R_D which yield this new operating point).

Problem 14.32

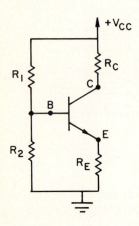
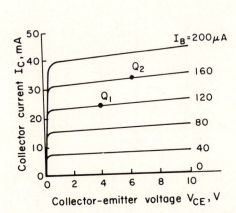

In the circuit shown the desired quiescent operating point Q_1 is shown on the collector–emitter characteristic. From the graph at Q_1, $I_C = 24$ mA and $I_B = 120$ μA. For $V_{CC} = 40$ V and $R_1 = R_2 = 120$ kΩ find the values of R_E and R_C. Assume that $V_{BE} = 0.7$ V.

$$\text{Ans.: } R_E = 502 \ \Omega \approx 0.5 \text{ k}\Omega$$
$$R_C = 996 \ \Omega \approx 1 \text{ k}\Omega$$

Problem 14.33

V_{CC}, R_1, and R_2 remain the same as given above in Problem 14.32 and the transistor remains the same. Find values of R_E and R_C which yield the operating point Q_2 shown above. You may assume $V_{BE} = 0.7$ V.

Problem 14.34

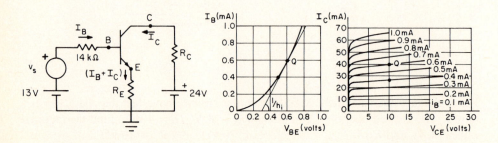

The characteristics of the transistor in the amplifier circuit above are as shown at the right. At the desired dc operating point Q ($v_s = 0$), $I_B = 0.6$ mA, $V_{BE} = 0.6$ V, $I_C = 40$ mA, and $V_{CE} = 10$ V as shown. Find the required values of R_E and R_C.

$$\text{Ans.:} \quad R_E \approx 100 \ \Omega$$
$$R_C \approx 250 \ \Omega$$

Problem 14.35

Find the values of R_E and R_C which yield the operating points indicated by the unlabeled dots in the two figures in Problem 14.34 if everything else is the same.

Problem 14.36

Draw the small-signal linear model of the amplifier circuit of Problem 14.34 from which the response to v_s could be found. Find the value of every element in the circuit.

Ans.:

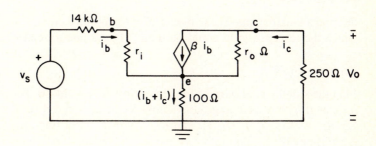

$$r_i \approx 500 \ \Omega, \ \beta \approx 70, \ r_o \approx 2500 \ \Omega$$

Problem 14.37

Find the gain v_o / v_s in Problem 14.36 if $r_o = \infty$ rather than 2500 Ω, where v_o is the voltage across the 250-Ω resistor. Find the gain if the 100-Ω resistor were short-circuited.

$$\text{Ans.:} \quad A_v = -0.81 \ (\text{with } R_E)$$
$$A_v = -1.21 \ (\text{without } R_E)$$

Problem 14.38

A signal $v_i(t)$ from an ideal source is coupled to point B in Problem 14.32 through a capacitor. Let the base emitter linear model resistance r_i equal 600 Ω, $\beta = 200$, and $r_o = \infty$. Find the voltage gain $A_v = v_o / v_i$ if v_o is measured at point C with respect to ground. Assume that the signal frequency is high enough that the coupling capacitor may be replaced by a short circuit.

Problem 14.39

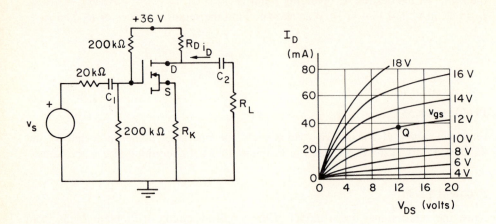

The characteristics of the MOSFET in the amplifier circuit above are as shown on the right. At the desired quiescent operating point, $V_{GS} = 12$ V, $V_{DS} = 12$ V, and $I_D = 36$ mA. Find the required values of R_K and R_D.

Problem 14.40

Draw the small-signal linear model of the amplifier circuit of Problem 14.39 from which the response to v_s could be found. Determine g_m and r_d at Q.

Problem 14.41

Find the small-signal gain of the circuit in Problem 14.40, $A_v = v_o/v_s$, where v_o is measured with respect to ground across R_L, if $g_m = 6.5 \times 10^{-3}$ mho and $r_d = 1$ kΩ. Let $R_L = 5$ kΩ, $R_K = 167$ Ω, and $R_D = 500$ Ω. Also, find the gain if the resistor R_K is short-circuited. (You may assume that the linear model parameters g_m and r_d do not change in this last step and that in both cases the signal frequency is sufficiently high that the capacitors act like short circuits.)

$$\text{Ans.: } A_V = -0.44 \text{ (with } R_K)$$
$$A_V = -0.93 \text{ (without } R_K)$$

Problem 14.42

A signal $v_i(t)$ is capacity-coupled from an ideal source to the gate of the MOSFET circuit in Problem 14.30. Find the gain $A_v = v_o/v_i$ if v_o is measured between point D and ground. Let $g_m = 2 \times 10^{-3}$ mho and $r_d = 5$ kΩ.

Problem 14.43

In the inverting transistor amplifier circuit below we may choose from a variety of transistors, each of which has the three-parameter model $(r_i, \beta, r_o) = (1 \text{ k}\Omega, \beta, 50 \text{ k}\Omega)$. We wish to have a voltage gain A_v at high frequencies such that $|A_v| > 4$. What is the minimum acceptable value for β? By high frequencies we mean frequencies such that the coupling capacitors may be considered to be short circuits. Here, as usual $A_v = v_o/v_i$, where v_o is the voltage across the resistor R_L.

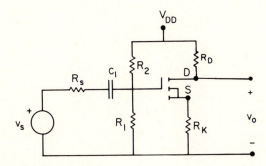

$$V_{CC} = 30\,V$$
$$R_S = R_2 = 20\,k\Omega$$
$$R_1 = 140\,k\Omega$$
$$R_C = R_L = 2\,k\Omega$$

Problem 14.44

The n channel enhancement MOSFET circuit shown below has the three-parameter model $(r_i, g_m, r_d) = (\infty, g_m, 100\,k\Omega)$. If $R_S = 1k\Omega$, $R_1 = R_2 = 100\,k\Omega$, $R_K = 500\Omega$ and $R_D = 1\,k\Omega$, find the minimum value of g_m such that the absolute value of the small-signal gain $A_v = v_o/v_s$ is greater than 1.7. What is the gain in the limit as $g_m \to \infty$? What gain will the same circuit have if R_K is set equal to zero and g_m equals the minimum value calculated above?

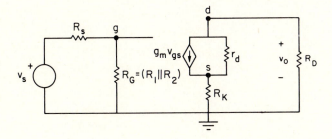

Problem 14.45

Make a table showing the absolute value of the gain $A_v = v_o/v_s$ for the circuit shown below as a function of R_K for values between $R_K = 0\,\Omega$ and $R_K = R_D$. Let $R_D = 2.5\,k\Omega = R_S$, $R_1 = R_2 = 100\,k\Omega$ and $(g_m, r_d) = (10^{-2}\,mho, 10\,k\Omega)$. Include at least four points in the table and make a sketch of $|A_v|$ versus R_K.

Chapter 15

An Introduction to Realistic Transistor Amplifiers

In this chapter we build upon the elements of transistor operation presented in Chapter 14. Several amplifier circuits are presented and analyzed. In particular, the use of coupling capacitors and the effects of stray capacity are included in the frequency response analysis presented. Self-biasing and multistage amplifiers are discussed as is the emitter follower circuit and its analysis. The presentation is not exhaustive but is meant as an introduction to techniques used in constructing small-signal amplifiers.

15.1 FREQUENCY RESPONSE OF TRANSISTOR AMPLIFIERS: COUPLING CAPACITORS

Capacitors allow us to "inject," or "couple," a signal into an amplifier and out to a load with minimal interaction between the source or load and the dc power supplies necessary to run the device. This coupling is accomplished for a MOSFET circuit in the manner shown in Fig. 15.1. As discussed in Chapter 14, to determine the dc operating point, we may use the fact that at dc the capacitors C_1 and C_2 are open circuits. They then "decouple" the resistors R_s and R_L from the transistor at dc. The

FIGURE 15.1.
(*a*) An amplifier stage using coupling capacitors C_1 and C_2 to isolate v_s and R_L from the dc bias. (*b*) The dc circuit for (*a*).

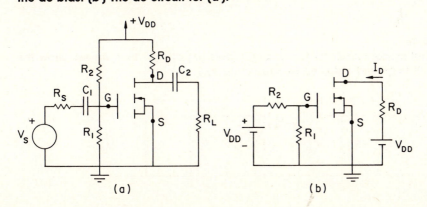

(a)   (b)

resulting dc circuit is presented in part (b) of Fig. 15.1, and the source and load resistance have no effect in determining the operating point. To study the ac response to small signals, we may represent the transistor by its linear model. The resulting circuit model is linear and to study small signals we suppress the voltage V_{DD}. The resulting small-signal circuit becomes that shown in Fig. 15.1c. (Remember, suppressing V_{DD} creates a short circuit to ground so that both R_2 and R_D go to ground in the ac circuit.)

In Chapter 10 we developed a formal analysis procedure involving the poles and zeros of the transfer function for a given circuit. In this circuit, the transfer function of most interest is

$$H(s) = A_v = \frac{V_L}{V_s} \tag{15.1}$$

This function can be constructed in two steps by determining

$$H_1(s) = \frac{V_{gs}}{V_s}$$

and

$$H_2(s) = \frac{V_L}{V_{gs}}$$

Then $H(s)$ from Eq. 15.1 can be constructed from the product

$$H = H_1 H_2$$

If we let $R_G = R_1 R_2 /(R_1 + R_2)$ then from the voltage divider principle,

$$H_1(s) = \frac{V_{gs}}{V_s} = \frac{R_G}{R_s + R_G + (1/sC_1)} = \frac{sC_1 R_G}{sC_1(R_s + R_G) + 1}$$

Writing this in standard polynomial form,

$$H_1(s) = \frac{sR_G}{(R_s + R_G)(s + \omega_{g1})} \tag{15.2a}$$

FIGURE 15.1c.
Small-signal model circuit for the circuit in part (a) of Fig. 15.1. Arrows show the short circuit to ground obtained by supressing V_{DD}.

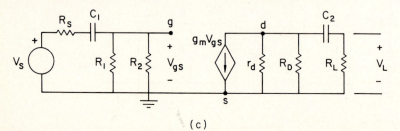

(c)

where

$$\omega_{g1} = \frac{1}{(R_s + R_G)C_1} \tag{15.2b}$$

Also, defining $R_A = \dfrac{r_d R_D}{r_d + R_D} = r_d // R_D$,

$$H_2(s) = \frac{V_L}{V_{gs}} = \frac{-g_m R_A R_L}{R_A + R_L + (1/sC_2)} = \frac{-sC_2 g_m R_A R_L}{sC_2(R_A + R_L) + 1}$$

which in polynomial form is given by

$$H_2(s) = \frac{V_L}{V_{gs}} = \frac{-g_m R_p s}{s + \omega_{d1}} \tag{15.3a}$$

where

$$\omega_{d1} = \frac{1}{(R_A + R_L)C_2} \tag{15.3b}$$

and

$$R_p = \frac{R_A R_L}{R_A + R_L}$$

Finally, the transfer function is the product of Eqs. 15.2a and 15.3a:

$$H(s) = \frac{-Ks^2}{(s + \omega_{g1})(s + \omega_{d1})} \tag{15.4}$$

where we show the minus sign explicitly and define K by

$$K = \frac{R_G}{R_s + R_G} g_m R_p$$

The full transfer function has two zeros at $s = 0$ and two poles. In practice, R_1 and R_2 are normally large compared to R_A and R_L so that $\omega_{g1} << \omega_{d1}$. For that case the asymptotic frequency response (Bode plot) is shown in Fig. 15.1d. If $\omega_{g1} > \omega_{d1}$, the curve would be similar but the frequency labels for the poles would be reversed. Each of the portions of the circuit involving a coupling capacitor C_c contributes a zero at $s = 0$ and a pole of the form $s = -1/C_c R_{EQ}$ to the overall transfer function. At frequencies above the maximum pole (in this case ω_{d1}) the magnitude of the voltage gain is constant.

We now show analytically why at these high frequencies the coupling capacitors can be replaced by short circuits in the analysis. This is the analysis method used in Chapter 14 where we first introduced coupling capacitors. That this is true is proved as follows. For oscillatory signals the impedance of a capacitor is $-j/\omega C$. When

FIGURE 15.1d.

Asymptotic frequency response for the small-signal model of the circuit in part (a) of Fig. 15.1. $H(s)$ has a double-zero at $s = 0$ and poles at $s = -\omega_{d1}$ and at $s = -\omega_{g1}$.

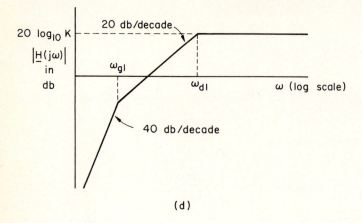

(d)

such an impedance is in series with one or more resistors as is the case here (C_1 is in series with R_s and $R_G = R_1 \mathbin{//} R_2$), then the total impedance Z_T is of the form

$$\mathbf{Z}_T = R_{EQ} - \frac{j}{\omega C}$$

This may also be written

$$\mathbf{Z}_T = R_{EQ}\left[1 - \frac{j}{\omega R_{EQ}C}\right]$$

or

$$\mathbf{Z}_T = R(1 - j\epsilon)$$

where $\epsilon = 1/\omega R_{EQ}C$. For $\omega \gg (R_{EQ}C)^{-1}$, ϵ is a very small number. This means that $\mathbf{Z}_T$ has a very small imaginary part. Furthermore, the magnitude of $\mathbf{Z}_T$ is given by $|\mathbf{Z}_T| = R_{EQ}(1 + \epsilon^2)^{1/2}$. Using the well-known approximation

$$(1 + x)^n \simeq 1 + nx$$

which is valid when $x \ll 1$, we have

$$|\mathbf{Z}_T| = R_{EQ}(1 + \epsilon^2)^{1/2} \simeq R_{EQ}\left(1 + \frac{\epsilon^2}{2}\right) \simeq R_{EQ}$$

The magnitude of $\mathbf{Z}_T$ is very nearly equal to R_{EQ} since a number ϵ^2 is even smaller than ϵ when ϵ is less than one. Because of this we shall often replace capacitors in series with a resistance R with short circuits when $\omega \gg (RC)^{-1}$. In the present case for frequencies $\omega \gg \omega_{d1}$, C_1 and C_2 can both be replaced by short circuits (since $\omega_{d1} > \omega_{g1}$). The resulting circuit is illustrated in Fig. 15.1e, where arrows show

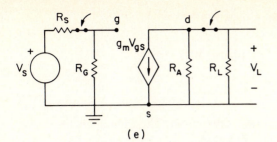

FIGURE 15.1e.
Small-signal model when $\omega >> \omega_{d1}$.

(e)

the coupling capacitors replaced by short circuits. For this circuit only resistors and sources exist and

$$\frac{V_L}{V_s} = \frac{-g_m R_G (R_A R_L)}{(R_A + R_L)(R_G + R_s)} = \frac{-g_m R_p R_G}{R_G + R_s} = -K \qquad (15.5)$$

Note that in the limit when s is much greater than ω_{g1} and ω_{d1}, Eq. 15.4 also reduces to this result since for large s

$$H(s) \simeq \frac{-Ks^2}{s^2} = -K.$$

TEXT EXAMPLE 15.1

Suppose in the Figs. 15.1a–e that $C_1 = C_2 = 1.0 \ \mu F$, $R_1 = R_2 = 200$ kΩ, $R_s = 1$ kΩ, $R_D = r_d = 10$ kΩ, $g_m = 5 \times 10^{-3}$ mho, and $R_L = 5$ kΩ. Find the poles and zeros of the circuit. What is the gain at high frequencies? Find the -3db point.

Solution: We already know that there are two zeros at $s = 0$. The parallel resistance $R_G = 100$ k$\Omega >> R_s$ so $R_G + R_s \simeq R_G$ and the pole associated with the input is given by Eq. 15.2b

$$\omega_{g1} \simeq \frac{1}{R_G C_1} = \frac{10^6}{10^5} = 10 \text{ rad/sec}$$

For the pole associated with the output, $R_A = 5$ k$\Omega = R_L$ and from Eq. 15.3b

$$\omega_{d1} = \frac{10^6}{10^4} = 100 \text{ rad/sec}$$

The high-frequency gain is found as follows. After setting all the coupling capacitors to short circuits resistive circuit analysis methods may be used to show

$$V_g = \frac{100}{101} V_s$$

$$V_L = -g_m V_{gs} \frac{R_L}{2}$$

and finally

$$A_v = \frac{V_L}{V_s} = -(5 \times 10^{-3})(2500)\left|\frac{100}{101}\right|$$

$$A_v = -12.4$$

Substitution into Eq. 15.5 for $-K$ yields this same result. This gain holds for $\omega \gg$ 100 rad/sec. The -3db point occurs at 100 rad/sec.

Similar considerations apply to amplifier circuits employing BJT transistors. An example is worked out in the next section and others in the problem set at the end of the chapter.

15.2 HIGH-FREQUENCY RESPONSE OF TRANSISTOR AMPLIFIERS: STRAY AND PARALLEL CAPACITORS

We turn now to the high-frequency characteristics of transistor amplifier circuits. Amplifiers of the type shown in part (a) of Fig. 15.1 are called "R–C amplifiers," where the R refers to the desired resistive character of the circuit, such as illustrated in Fig. 15.1e, and the C refers to the coupling capacitors. As we have noted above, at high frequencies the coupling capacitors may be considered to be short circuits. However, when the frequency is high enough, the effect of other capacitances in the system cannot be ignored. Whenever a voltage exists between any conductors there will be a charge on the conductors and they will constitute a capacitance. Thus, across *any* pair of terminals in the system there is always some "stray" capacitance. This capacitance C_s is often very small so that over a broad range of frequencies its reactance $(1/\omega C_s)$ may be very large compared to any resistance with which it is in parallel. Under those conditions the current through C_s will be negligibly small. However, as the signal frequency increases the capacitive reactance associated with stray capacity decreases and at some frequency it must be taken into account. These comments may be shown quantitatively as follows. In Fig. 15.2, C_s represents a stray capacitance in parallel with some resistor R. The impedance is worked out on the right-hand side of the figure as a function of s and ω.

FIGURE 15.2.
The impedance of R and C_s in parallel.

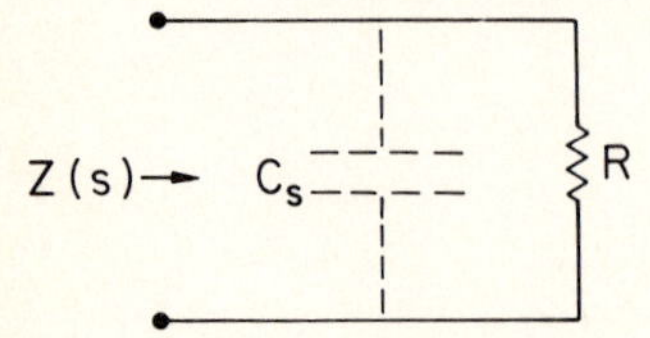

$$Z(s) = \frac{(R)(1/sC_s)}{R + 1/sC_s} = \frac{R}{sC_sR + 1} = \frac{1}{C_s}\left(\frac{1}{s + 1/RC_s}\right)$$

$$= \frac{1}{C_s}\left(\frac{1}{s + \omega_2}\right)$$

$$\underline{Z}(j\omega) = \frac{R}{1 + j\omega RC_s} = \frac{R}{1 + j\omega/\omega_2} \qquad \omega_2 = 1/RC_s$$

Since $|\mathbf{A/B}| = |\mathbf{A}| \, / \, |\mathbf{B}|$,

$$|\mathbf{Z}_T(j\omega)| = \frac{R}{(1 + \omega^2/\omega_2^2)^{1/2}}$$

For frequencies low enough that $\omega << \omega_2$, $|\mathbf{Z}_T| \simeq R$ and the capacitor acts as an open circuit. For $\omega >> \omega_2$,

$$|\mathbf{Z}_T| \simeq \frac{R\omega_2}{\omega} = \frac{1}{\omega C_s}$$

and as ω goes to infinity $|\mathbf{Z}_T| \rightarrow 0$. At high frequencies the stray capacity acts as a short circuit. Consider the specific values $R = 10$ kΩ and $C_s = 10$ pF. Then, $\omega_2 = 10^7$ rad/sec. In this case, if the signal has a radian frequency ω which is less than, say, 10^6 rad/sec, the stray capacitance may be neglected. However, at a signal radian frequency greater than about 10^8 rad/sec, the impedance is almost solely equal to the capacitive reactance.

Stray capacity usually limits the ability of a given circuit to operate at arbitrarily high frequencies. In addition, capacitors are often deliberately inserted across the load resistor to limit the frequency range in which the amplifier operates. In the notation below, we shall use the symbol C_s to stand both for stray capacitors and for parallel capacitors which are purposely introduced to limit the frequency response. A stereo system, for example, need not be sensitive at frequencies above the response of the human ear. We discussed other applications where frequency response limiting is necessary in Chapter 10 with regard to the interface between analog and digital electronics in applications of operational amplifier circuits.

The small-signal circuit of Fig. 15.1c for the amplifer of part (a) in Fig. 15.1 must be modified at high frequencies to take into account either the stray capacitances or those capacitors deliberately introduced to limit the frequency response. The resulting high-frequency circuit is shown in Fig. 15.3a. In the small-signal model circuit of this figure we have continued to assume, as shown by the arrows, that the coupling capacitors are short circuits. This circuit model may therefore be used only at frequencies well above ω_{g1} and ω_{d1}. The capacitance C_{s1} includes all the stray capacitance across the resistors R_1 and R_2, the capacitance between the gate and source of the MOSFET, *and* the capacitance between the gate and drain. (Note that

FIGURE 15.3a.
The (high-frequency) small-signal circuit of the amplifier of part (a) of Fig. 15.1 is modified to include the parallel capacitors C_{s1}, C'_{s2}, and C''_{s2}. The coupling capacitors are replaced by short circuits.

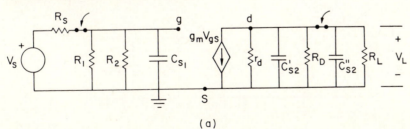

(a)

since the capacitance between gate and drain is not directly in parallel with R_1 and R_2, it is some equivalent or effective input capacitance that is represented by C_{s1}.) The capacitance C'_{s2} includes all the stray capacitance across R_D and the capacitance between drain and source of the MOSFET. The capacitance C''_{s2} includes any stray capacity across the resistor R_L and any capacity deliberately introduced across the load. If the voltage across R_L is used as the input to another amplifier, C''_{s2} would also include the equivalent input capacitance of that amplifer.

To find the high-frequency transfer function V_L/V_s for the circuit of Fig. 15.3a we may simplify the input circuit by the use of Thevenin's theorem as illustrated in Fig. 15.3b. In the output circuit all the resistors on the right-hand side may be lumped into the single equivalent resistor R_p and the two parallel capacitors may be combined into $C_{s2} = C'_{s2} + C''_{s2}$. From Fig. 15.3$b$,

$$\frac{V_{gs}}{V_T} = \frac{1/(sC_{s1})}{R_T + 1/(sC_{s1})} = \frac{1}{sC_{s1}R_T + 1} = \frac{\omega_{g2}}{s + \omega_{g2}}$$

where $\omega_{g2} = 1/C_{s1}R_T$. Then,

$$H_1(s) = \frac{V_{gs}}{V_s} = \frac{R_G}{R_G + R_s} \frac{\omega_{g2}}{s + \omega_{g2}} \tag{15.6a}$$

and the input portion of the circuit has a pole at $s = -\omega_{g2}$. For the output circuit,

$$H_2(s) = \frac{V_L}{V_{gs}} = \frac{-g_m R_p}{sR_p C_{s2} + 1} = \frac{-g_m}{C_{s2}(s + \omega_{d2})} \tag{15.6b}$$

where $\omega_{d2} = 1/R_p C_{s2}$. The output portion of the circuit has a pole at $s = -\omega_{d2}$. Finally, the overall transfer frequency at high frequencies is the product of Eqs. 15.6a and 15.6b:

$$H(s) = \frac{-K'\omega_{g2}}{(s + \omega_{g2})(s + \omega_{d2})}$$

FIGURE 15.3b.
The input and output of Fig. 15.3a is simplified by use of Thevenin's theorem, the definition of $R_P = r_d // R_D // R_L$, and the introduction of $C_{s2} = C'_{s2} // C''_{s2}$.

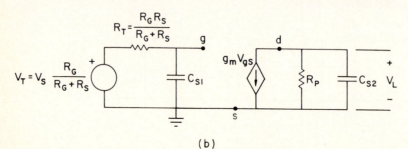

(b)

where

$$K' = \frac{g_m R_G}{C_{s2}(R_G + R_s)}$$

Notice that for $s \ll \omega_{ga}$ or ω_{d2}, H(s) $= -K$ as in Eq. 15.5.

The high-frequency asymptotic frequency response is shown in Fig. 15.3c, where it has been assumed that ω_{d2} is larger than ω_{g2} in constructing the plot. Three features of the asymptotic frequency response are to be noted:

1. The response applies only at high frequencies, since in the circuit of Fig. 15.3a we have assumed the coupling capacitors to be short circuits.

2. The magnitude of the transfer function for frequencies below ω_{g2} is exactly the same as the magnitude of the transfer function for "high" frequencies in the circuit of Fig. 15.1c which is shown in Fig. 15.1d. This holds since at frequencies well below ω_{g2} the stray capacitances act like open circuits and the full circuit model reduces to the circuit of Fig. 15.1e, where only resistances contribute.

3. The -3db point occurs at ω_{g2} in this case.

FIGURE 15.3c.
Asymptotic frequency response for the circuit of Fig. 15.3a.

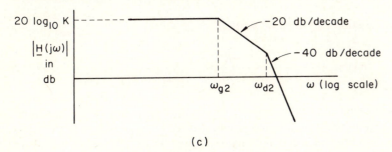

(c)

15.3 A COMPLETE FREQUENCY RESPONSE ANALYSIS OF R–C TRANSISTOR AMPLIFIERS

The full small-signal model of the R–C amplifier of part (a) in Fig. 15.1 must include *both* the coupling capacitors and the stray (or intentional) capacitors as shown in Fig. 15.4a. The overall transfer function H(s) $= V_L / V_s$ of the model above may be written as before, H(s) $= H_1(s)H_2(s)$, where $H_1(s) = V_{gs}/V_s$ and $H_2(s) = V_L/V_{gs}$. Although very tedious to derive, the two functions $H_1(s)$ and $H_2(s)$ are as shown below:

$$H_1(s) = \frac{sR_G[(1/C_{s1}R_G) + (1/C_{s1}R_s)]}{(R_s + R_G)[s^2 + s[(1/C_1R_s) + (1/C_{s1}R_G) + (1/C_{s1}R_s)] + (1/C_{s1}C_1R_sR_g)]}$$

(15.7a)

FIGURE 15.4a.

Small-signal model of the circuit of part (a) in Fig. 15.1 including the coupling and stray capacitances.

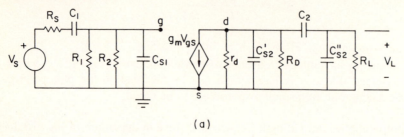

(a)

where $R_G = R_1 // R_2$ and

$$H_2(s) = \frac{-sg_m K''/C_{s2}''}{s^2 + sK''[(1/R_pC_s) + (1/R_LC_{s2}'') + (1/R_pC_{s2}'') + (C_{s2}''/R_LC_2C_{s2}'')] + (K''/R_pR_LC_2C_{s2}'')}$$

(15.7b)

where

$$K'' = 1 + \frac{C_{s2}'}{C_2} + \frac{C_{s2}'}{C_{s2}''}$$

The full transfer function may then be written

$$H(s) = \frac{-K'''s^2}{(s + \omega_{g1}')(s + \omega_{d1}')(s + \omega_{g2}')(s + \omega_{d2}')}$$

(15.8)

where K''' is a constant and the set $(-\omega_{g1}', -\omega_{d1}')$ are the roots of the quadratic equation in the denominator of Eq. 15.7a and the set $(-\omega_{g2}', -\omega_{d2}')$ are the roots of the quadratic equation in the denominator of Eq. 15.7b. Although not at all obvious it is nonetheless true that if the various roots are well separated in frequency value, a broadband system, the primed break points will be nearly identical to those found using the approximations for the various capacitors discussed above. Using the analysis methods learned for Bode plots, the asymptotic form for $|\mathbf{H}(\omega)|$ is as shown in Fig. 15.4b where we have dropped the primes. Again, to be definite we have chosen $\omega_{d2} > \omega_{g2}$, and $\omega_{d1} > \omega_{g1}$.

To summarize, since the derivation of $H_1(s)$ and $H_2(s)$ is somewhat painful, it is much easier to separate the amplifier response conceptually into the three frequency ranges shown in Fig. 15.4b. In these ranges some useful simplifications occur. In the low-frequency range all the stray (or parallel) capacitors may be replaced by open circuits. In the high-frequency range all the coupling capacitors may be replaced by short circuits. In the midfrequency range both of these simplifications apply. To illustrate, consider the transfer function $H_2(s)$ corresponding to the output portion of the circuit under discussion. The circuits appropriate to each of the three frequency ranges are shown in Fig. 15.4c. The functional form for $H_2(s)$ in these three cases is much easier to derive than the full algebraic result given in Eq. 15.7b.

FIGURE 15.4b.
The asymptotic frequency response of the circuit of Fig. 15.4a when $\omega_{d2} \gg \omega_{d1}$, $\omega_{g2} \gg \omega_{g1}$, $\omega_{g1} < \omega_{d1}$, and $\omega_{g2} < \omega_{d2}$.

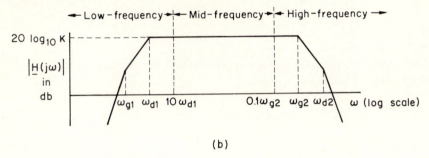

(b)

At *low frequencies,*

$$V_L = -\left(\frac{R_L}{Z'}\right)(g_m V_{gs})(r_d \; // R_D \; // Z')$$

where $Z' = R_L + (1/sC_2)$. In Problem 15.21 the student is asked to show that

$$H_2(s) = \frac{V_L}{V_{gs}} = \frac{-g_m R_p s}{s + \omega_{d1}} \qquad (15.9a)$$

where

$$\omega_{d1} = \frac{1}{(R_A + R_L)C_2}$$

$$R_A = \frac{r_d R_D}{r_d + R_D}$$

$$R_p = \frac{R_A R_L}{R_A + R_L}$$

FIGURE 15.4c.
Ouput circuit models for the low-, middle-, and high-frequency ranges. At low frequencies C_{s2} is an open circuit; at high frequencies C_2 is a short circuit; and at midfrequencies both conditions hold.

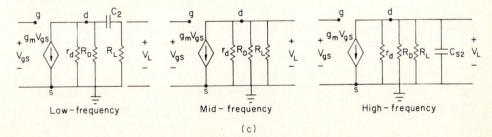

(c)

In the *midfrequency range*,

$$H_2(s) = \frac{V_L}{V_{gs}} = -g_m R_p \tag{15.9b}$$

Note that $R_p = (r_d // R_D // R_L)$.

At *high frequencies*, the load voltage appears across the parallel combination of three resistors and the capacitor C_{s2} and

$$V_L = -g_m V_{gs}[r_d // R_D // R_L // (1/sC_{s2})]$$

and (see Problem 15.21) we have

$$H_2(s) = \frac{V_L}{V_{gs}} = \frac{-g_m}{C_{s2}(s + \omega_{d2})} \tag{15.9c}$$

where

$$\omega_{d2} = \frac{1}{R_p C_{s2}}$$

The assumption that $\omega_{d2} >> \omega_{d1}$ can be seen to mean that

$$\frac{1}{R_p C_{s2}} >> \frac{1}{(R_A + R_L)C_2}$$

or

$$R_p C_{s2} << (R_A + R_L)C_2$$

Since R_p is the parallel combination of R_A and R_L it must be less than $R_A + R_L$. Therefore, the assumption is satisfied if $C_{s2} << C_2$; that is, if the stray or load capacitance is less than the inserted coupling capacitance, the condition $\omega_{d2} >> \omega_{d1}$ will hold. This condition is very easy to satisfy. The frequency response of $H_2(s)$ has the following characteristics:

1. The absolute value of the gain at midband frequencies is $g_m R_p$.
2. The low-frequency −3db point occurs at ω_{d1}.
3. The high-frequency −3db point occurs at ω_{d2}.

Similar sets of approximations may be used to study the input portion of the circuit. The middle frequency range is often termed the midband.

In the discussion thus far we have concentrated entirely upon MOSFET amplifiers. In the problems at the end of the chapter bipolar junction transistors are also used. The major modification to the analysis is that the linear model must correspond to the BJT device, which is repeated for reference in Fig. 15.5, and the

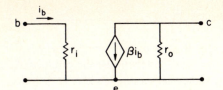

FIGURE 15.5.
Linear model for a BJT transistor.

fact that the base current cannot be ignored as we do with the gate current for a MOSFET. An *R–C* amplifier circuit using a BJT might look as shown in Fig. 15.6*a*.

FIGURE 15.6a.
A BJT amplifier circuit with coupling capacitors.

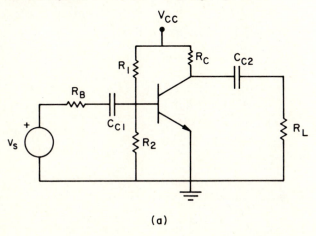

(a)

The small-signal linear model for this circuit including the various stray capacitors is given in Fig. 15.6*b*. The analysis method is analogous to the MOSFET case and is worked out in Text Example 15.2. A circuit such as the one illustrated in Fig. 15.6*a* is in the common emitter mode since the input and output signals are measured with respect to the emitter which is at ground potential.

FIGURE 15.6b.
Small-signal linear model for the circuit in Fig. 15.6a.

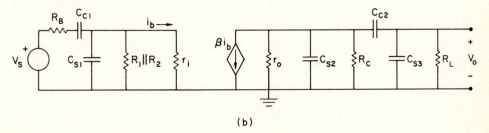

(b)

TEXT EXAMPLE 15.2

For the circuit in Fig. 15.6b derive an expression for the transfer functions $H_1 = i_b/V_s$ and $H_2 = V_o/i_b$. The total transfer function V_o/V_s is then just the product of these two. Find the midband gain if all the external resistors are 4.9 kΩ, $\beta = 100$, $r_i = 100\ \Omega$, and $r_o = 100$ kΩ.

Solution: Straightforward circuit analysis yields

$$i_b = \left(\frac{V_s}{Z_1}\right)\left(\frac{Z_2}{Z_2 + r_i}\right)$$

where Z_1 is the total impedance seen by the source and the second expression is derived from the current divider principle. In this equation

$$Z_1 = R_B + \frac{1}{sC_{c1}} + [(1/sC_{s1})//(R_1//R_2)//r_i]$$

and

$$Z_2 = \frac{(1/sC_{s1})(R_1//R_2)}{(1/sC_{s1}) + (R_1//R_2)}$$

Dividing both sides by V_s yields

$$H_1 = \frac{Z_2}{Z_1(Z_2 + r_i)}$$

For the output portion of the circuit, let

$$Z_3 = r_o//(1/sC_{s2})//R_C$$

and

$$Z_4 = \frac{1}{sC_{c2}} + [(1/C_{s3})//R_L]$$

then using the current divider and Ohm's laws

$$H_2 = \frac{V_o}{i_b} = -\beta\left(\frac{Z_3}{Z_3 + Z_4}\right)[(1/sC_{s3})//R_L]$$

To find the midband gain we may use some shortcuts. First of all, the C_C are short circuits and all the C_s are open circuits. Also, since $R_1//R_2$ is much larger than r_i,

$$i_b \simeq \frac{V_s}{R_B + r_i}$$

and

$$\frac{i_b}{V_s} \simeq \frac{1}{5000}$$

At the output, r_o is much greater than either R_C or R_L, so

$$V_o = -\beta i_b \left(\frac{R_L}{2} \right)$$

since $R_C = R_L$. Substituting for i_b from above,

$$\frac{V_o}{V_s} = \frac{-(100)(2450)}{5000}$$

and

$$A_v = -49$$

15.4 BIAS STABILIZATION REVISITED

At the end of Chapter 14 we pointed out that adding a resistor from the source of a MOSFET to ground or from the emitter of a BJT to ground has a number of benefits. For example, such a resistor reduces the sensitivity of the amplifier circuit to variations of the parameters β or g_m from one device to another. It also minimizes the effect of the temperature dependence of a given transistor. The one drawback of the bias resistor is that the overall gain of the amplifier is reduced from the value of the same circuit constructed with a short circuit to ground. However, many applications of amplifier circuits deal with small oscillating signals. In such a case the dc or low-frequency gain is of little importance and only the midband properties are crucial. Since at high frequencies the impedance of a capacitor is zero, if we place a capacitor in parallel with the bias resistor we retain the advantages of the dc bias stabilization but "short out" the resistor in the midband range. This yields a high gain where we may want it and yet retains the stabilizing features of the bias resistor.

To show this mathematically we use an analysis similar to that in Section 14.6. We concentrate for now on the effect of the capacitor C_K on the transfer function $H = V_o / V_i$ for the circuit in Fig. 15.7. The addition of C_K in the circuit does not change the dc bias or the operating point since the capacitance is an open circuit for dc. In the linear model for the amplifier shown in Fig. 15.7b however, the equations for the output portion of the transfer function (see Eqs. 14.8a–d) are modified by the substitution of Z_K [which is equal to $R_K // (1/sC_K)$] for R_K. Then,

$$H_2(s) = \frac{V_o}{V_g} = -\frac{\alpha g_m R_D}{1 + \alpha g_m Z_K} \qquad (15.10)$$

FIGURE 15.7.
A capacitance C_K is put in parallel with R_K to make the impedance small when the frequency is above some minimum value.

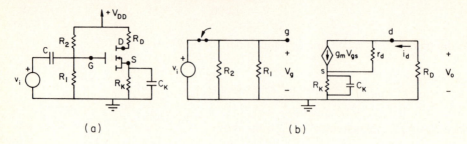

(a)   (b)

where

$$\alpha = \frac{r_d}{r_d + Z_K + R_D}$$

and

$$Z_K = \frac{R_K}{1 + sR_KC_K}$$

At dc, $s = 0$ and $Z_K = R_K$ as before. If we define $\omega_{min} = 1/R_KC_K$, then at frequencies above ω_{min}, $sC_K >> 1/R_K$, and $Z_K = 1/sC_K$. This in turn implies that for frequencies greater than ω_{min}, Z_K goes to zero. We say that at high frequencies the capacitor shorts out the resistor R_K and the amplifier gain increases. To make this result more quantitative, we substitute the high-frequency limit of Z_K into Eq. 15.10:

$$H(s) = -\frac{\alpha g_m R_D}{1 + (\alpha g_m/sC_K)}$$

where α now has the form

$$\alpha = \frac{r_d}{r_d + R_d + (1/sC_K)}$$

If we design the circuit so that in addition

$$\omega_{min}C_K >> (r_d + R_D)^{-1}$$

then a considerable simplification occurs since then

$$\alpha \simeq \frac{r_d}{r_d + R_D}$$

We must check this later, but assuming that it holds we have

$$H_2(s) = -\frac{g_m r_d R_D}{(r_d + R_D)[1 + (\alpha g_m /sC_K)]} \tag{15.11}$$

To achieve the maximum possible gain, we wish to make the absolute value of the $\alpha g_m /\omega_{min} C_K$ term much less than unity. For design purposes we may use

$$\frac{\alpha g_m}{\omega_{min} C_K} \leq 0.1$$

and deduce a required value for C_K, for example,

$$C_K \geq \frac{10\alpha g_m}{\omega_{min}} \tag{15.12}$$

For the example above, using the same values as in Section 14.6, $\alpha \simeq r_d/(r_d + R_D) = 6/(6 + 3.6) = 0.625$. Let $\omega_{min} = 100$ rad/sec ($f_{min} \simeq 16$ Hz). Then, the requirement becomes

$$C_K \geq \frac{(10)(0.625)(2 \times 10^{-3})}{100} = 125\mu\text{F}$$

Using $C_K = 125\ \mu\text{F}$, $1/\omega_{min} C_K = 80\ \Omega$, which confirms the assumptions that $1/\omega_{min} C_K << R_K = 1\ \text{k}\Omega$ and $1/\omega_{min} C_K << (r_d + R_D) = 9.6\ \text{k}\Omega$. For this example, then, if $\omega > \omega_{min} = 100$ rad/sec,

$$H_2 = \frac{(0.625)(2)(3.6)}{1} = -4.5$$

while for the very low frequencies $\omega < \omega_{min}$,

$$H_2(\omega) = -1.9$$

These two results show that, including the capacitor C_K increases the gain in the midband frequency range. The algebraic solution for the full transfer function which is valid at all frequencies is very cumbersome and beyond our needs in this text.

As noted above, addition of the capacitor C_K allows the benefits of the bias resistor R_K for the dc operation of the amplifier but gives the full ac gain in the center of the amplifier frequency response. Similar results hold for BJT amplifiers, examples of which are treated in the problems at the end of this chapter. This is a rare example of a case where the designer can "have his cake and eat it too." The high-gain characteristics of a grounded emitter are attained for the ac response while the stabilization characteristics of self-biasing are also retained.

A slightly more realistic circuit diagram for a MOSFET amplifier including a load resistor as well as a self-biasing network is shown in Fig. 15.8. Here, the coupling capacitors C_1 and C_2 transfer the signal into and out of the transistor so that the resistors R_s and R_L do not affect the dc operating point. Likewise, the bias resistor R_K stabilizes the dc operating characteristics but is shorted out or "bypassed"

FIGURE 15.8.
A realistic MOSFET amplifier including a load R_L.

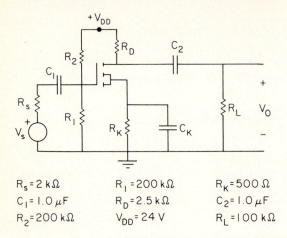

$R_s = 2\,k\Omega$	$R_1 = 200\,k\Omega$	$R_K = 500\,\Omega$
$C_1 = 1.0\,\mu F$	$R_D = 2.5\,k\Omega$	$C_2 = 1.0\,\mu F$
$R_2 = 200\,k\Omega$	$V_{DD} = 24\,V$	$R_L = 100\,k\Omega$

by C_K in the midband. Define ω_1 and ω_2 as the poles for the high-pass response of the coupling capacitors. Then the full (midband) gain of the amplifier is attained for ω greater than the largest of the set ($\omega_1, \omega_2, \omega_{min}$).

One final comment is in order since there is a potentially confusing aspect of transistor biasing. The capacitor C_K is in parallel with resistor R_K. In our discussions of stray or other parallel capacitors we commented that they could be replaced by open circuits in the midband. This is *not* the case for C_K since it acts as a short circuit in the midband. In fact, the whole idea is to "short out" R_K at frequencies to be amplified but to let it function as on open circuit at dc.

TEXT EXAMPLE 15.3 _____

Find the midband gain of the circuit in Fig. 15.8 if the linear model for the MOSFET is characterized by $g_m = 10^{-2}$ mho and $r_d = 100\,k\Omega$.

Solution: In the midband the coupling capacitors C_1 and C_2 may be replaced by short circuits and any stray capacity is ignored (stray capacitors replaced by open circuits). The capacitor C_K is also replaced by a short circuit since its purpose is to bypass the resistor R_K. We assume here without proof that R_K and C_K are chosen appropriately to accomplish the resulting gain increase in the frequency range desired for this particular amplifier. Our simplified approach also assumes that the high-frequency cutoff due to stray capacity is at high enough frequencies that a true midband exists. In the midband then we have the circuit shown below,

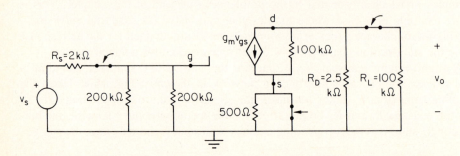

where arrows show the shorted capacitors C_1, C_2, and C_K. Using the voltage divider rule and $R_1 // R_2 = 100 \text{ k}\Omega$ yields

$$v_g = \frac{100}{102} v_s = 0.98 v_s$$

Since the capacitor C_K shorts out R_K, $v_g = v_{gs}$. The current $g_m v_{gs}$ creates a voltage across the load resistor which is in parallel with R_D and r_d. The total resistance is

$$R_T = 100 \text{ k}\Omega // 100 \text{ k}\Omega // 2.5 \text{ k}\Omega = 2.4 \text{ k}\Omega$$

The total gain is

$$A_v = \frac{v_o}{v_s} = -(0.98)(g_m)(R_T)$$

and

$$A_V = -23.5$$

15.5 THE EMITTER FOLLOWER

Often in electronics design it proves useful to isolate one circuit from another. This is particularly true when the output impedance of one portion of a network is large compared to the input resistance of the next part of the circuit and we say the former cannot "drive" the latter. In such a case we may need to insert a device which reproduces the voltage accurately but which can supply the current needed for the remainder of the circuit. This is called a voltage follower, or a buffer amplifier. We have already discussed how Op-Amps may be used for this purpose in Chapter 4. However, these are treated as ideal devices in this text, and it is important to see how transistors may be used to accomplish a similar function.

The BJT circuit illustrated in Fig. 15.9a is termed an emitter follower and may be used for this purpose. The small-signal linear model for this circuit in the midband frequency range is given in Fig. 15.9b where C_{C1} and C_{C2} have been replaced by

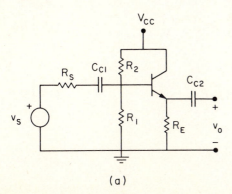

FIGURE 15.9a.
Circuit diagram for an emitter follower.

(a)

FIGURE 15.9b.
Small-signal model for the emitter follower.

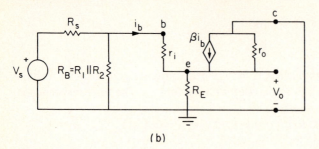

(b)

short circuits. In this configuration the collector is at ground potential insofar as the small signal model is concerned, so we may reconfigure the circuit diagram as in Fig. 15.9c.

FIGURE 15.9c.
An equivalent way to draw the circuit diagram of Fig. (15.9b).

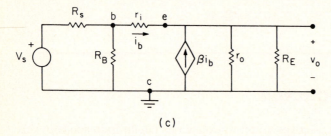

(c)

To analyze the circuit we first use KCL at node b,

$$\frac{v_s - v_b}{R_s} = \frac{v_b}{R_B} + i_b$$

Rearranging

$$\frac{v_b}{R_s} + \frac{v_b}{R_B} = \frac{v_s}{R_s} - i_b$$

Defining $R' = R_s /\!/ R_B$,

$$v_b = \frac{R'}{R_s} v_s - R' i_b \tag{15.13a}$$

At the output, if we let $R_L = R_E /\!/ r_o$,

$$v_0 = (\beta + 1) i_b R_L \tag{15.13b}$$

Finally, we use

$$i_b = \frac{v_b - v_o}{r_i} \tag{15.13c}$$

Combining these three equations,

$$\frac{v_o}{v_s} = \frac{(\beta + 1)R_L R'}{(r_i + R')R_s + (\beta + 1)R_s R_L} \tag{15.13d}$$

Now if we let $R_B \gg R_s$, $r_o \gg R_E$, and $r_i \ll R_s$, then

$$\frac{v_o}{v_s} = \frac{(\beta + 1)R_E R_s}{R_s^2 + (\beta + 1)R_s R_E} \tag{15.13e}$$

Finally, if $(\beta + 1)R_E \gg R_s$ and $\beta \gg 1$, then

$$\frac{v_o}{v_s} \approx 1 \tag{15.13f}$$

This result shows that if β is large and the resistors are chosen properly, a "voltage follower" can be built for which $v_o \approx v_s$. The circuit is also called an emitter follower since the emitter voltage "follows" the source voltage.

The output resistance may be determined by considering the Thevenin equivalent resistance given by

$$R_o = \frac{v_{oc}}{i_{sc}}$$

A short circuit at the output draws the full current $i_{sc} = (\beta + 1)i_b$. For $R_B \gg r_i$, $i_b = v_s/(R_s + r_i)$. Then, using the fact that $v_{oc} = v_o = v_s$, we have

$$R_o = \frac{v_s}{(\beta + 1)i_b} = \frac{v_s(R_s + r_i)}{(\beta + 1)v_s}$$

and finally,

$$R_o \simeq \frac{R_s + r_i}{\beta + 1} \tag{15.14}$$

This result shows that the resistance seen by the outside world is considerably less than the source resistance R_s if β is large. The circuit therefore presents virtually the same voltage at the output as presented by the source ($v_o \approx v_s$) but has a greater capability to drive current into a load.

Note that the circuit diagram in Fig. 15.9c shows the *collector* as the reference voltage. This is termed the common collector configuration since the signal voltage and output voltage are both referenced to the collector in the ac circuit. The third possibility, a common base configuration, is also used in some applications, for example, in some digital logic gates.

Finally, we note that MOSFET source followers are even more effective devices since they have a much higher input resistance than the BJT device discussed here. Because of this property they are very effective as the first transistor stage of an operational amplifier. This feature is very important since one of the Op-Amp laws, $I^+ = I^- = 0$, requires that very little current flows into the positive or negative inputs to the device. A MOSFET circuit that can be used as a source follower is discussed in the problem set.

A number of computer programs exist which deal with real transistor characteristics, stray capacity, temperature effects and so forth. Most modern circuit designers use these as tools to simulate circuits, particularly very complicated and expensive integrated circuits, before they are built. Our purpose here is to show some of the principles and subtleties in such an analysis.

PRACTICE PROBLEMS
AND
ILLUSTRATIVE EXAMPLES

EXAMPLE 15.1

Consider the following transistor amplifier circuit

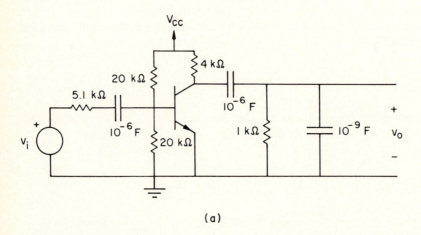

(a)

where the linear model of the transistor is characterized by $r_i = 1 \text{ k}\Omega$, $\beta = 100$, and $r_o = 10 \text{ k}\Omega$. You may assume that the stray capacitors are so small that they may be ignored.

(a) Find the midband gain of the amplifier.

Solution: First we draw the linear model for the small-signal response (suppressing V_{CC} by shorting it to ground):

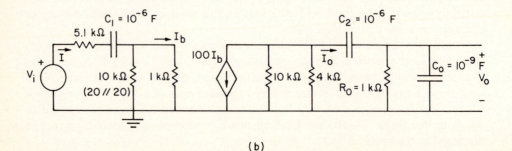

(b)

The two 1-μF capacitors C_1 and C_2 are coupling capacitors, coupling the signal into the transistor and out to the load. In the midband they act as short circuits. The 10^{-9} −F capacitor has been used to limit the frequency response of the amplifier for reasons discussed in the text. When ω is very high, the 10^{-9} −F capacitor will act like a short circuit and severely attenuate the output signal. At midband however, ω is small enough so that the latter capacitor acts like an open circuit. The midband circuit is thus

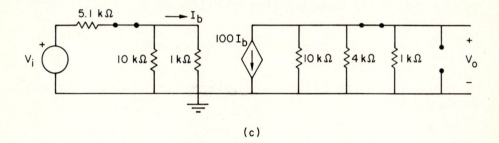

(c)

Note that this is identical to Example 4.5 except that the coupling capacitors are specifically shown as short circuits and the parallel capacitor as an open circuit. The solution in the midband frequency range (frequencies above the low-frequency cutoff due to the coupling capacitors and below the high-frequency cutoff due to the parallel capacitor across the load) is identical to Example 4.5:

$$A_V = \frac{v_o}{v_i} = -11.2$$

(b) Find the low-frequency "−3 db" point of the amplifier.

Solution: From the linear model for the input circuit we see that the total current I is given by

$$I = \frac{V_i}{R_T + (1/sC_1)} = \left(\frac{V_i}{R_T}\right)\left(\frac{s}{s + \omega_1}\right)$$

where $R_T = 5.1 \text{ k}\Omega + (10 \text{ k}\Omega //1\text{k}\Omega) = 6 \text{ k}\Omega$ and $\omega_1 = 1/R_T C_1$. Using the current divider law, $I_b = (10/11)I$ and

$$I_b = \left(\frac{10V_i}{11R_T}\right)\left(\frac{s}{s + \omega_1}\right)$$

This portion of the circuit has the transfer function

$$H_1(s) = \frac{I_b}{V_i} = \left(\frac{10}{11R_T}\right)\left(\frac{s}{s + \omega_1}\right)$$

which has a single zero at $s = 0$ and a pole at $\omega_1 = 1/R_T C_1 = 1/(6 \times 10^{-3}) = 160$ rad/sec. The frequency $\omega = 160$ rad/sec is only one candidate for the low-frequency cutoff, however, since there are two coupling capacitors. Considering the output portion of the circuit the transfer function of interest is

$$H_2 = \frac{V_o}{I_b}$$

since then $H = V_o/V_i = H_1 H_2$. In the midband the 10^{-9} $-$F capacitor may be considered to be an open circuit. Combining in parallel the 10-kΩ and the 4-kΩ resistor yields 2.85 kΩ. Using the current divider principle,

$$I_o = \frac{-2.85 \text{ k}\Omega}{2.85 \text{ k}\Omega + 1 \text{ k}\Omega + (1/sC_2)} 100 I_b$$

The voltage V_o across the 1-kΩ resistor is given by Ohm's law $V_o = I_o (1 \text{ k}\Omega)$. Combining terms,

$$I_o = -\left(\frac{2.85}{3.85}\right)\frac{100 I_b s}{s + \omega_2}$$

where $\omega_2 = 1/(3.85 \text{ k}\Omega)(10^{-6} \text{ F}) = 259$ rad/sec. The expression for I_o has a zero at $s = 0$ and a pole at $\omega_2 = 259$ rad/sec. Ignoring the load and stray capacity then, the amplifier response obtained by multiplying H_1 and H_2 is of the form

$$H = \frac{-Ks^2}{(s + \omega_1)(s + \omega_2)}$$

As we shall see, the 10^{-9} $-$F capacitor will add another pole to the response function creating a bandpass amplifier. Finally, note that since ω_2 is higher than ω_1, it determines the low-frequency -3-db point which occurs at 259 rad/sec.

(c) Find the high-frequency -3-db point of the amplifier.

Solution: The high-frequency cutoff can be found at the output by setting the 10^{-6} $-$F capacitor equal to a short circuit since the desired frequency will be well above the midband. The three resistors (10 kΩ, 4 kΩ, and 1 kΩ) are all in parallel with the 10^{-9} $-$F capacitor. These three resistors are equivalent to a 0.74-kΩ resistor. The output voltage is

$$V_o = -100 I_b \left[\frac{(0.74 \text{ k}\Omega)(1/sC_o)}{0.74 \text{ k}\Omega + (1/sC_o)}\right]$$

and the output transfer function $H_o = V_o/I_b$ may be written

$$H_o = -\left(\frac{100}{C_o}\right)\left(\frac{1}{s + \omega_3}\right)$$

where $\omega_3 = 1/(0.74 \text{ k}\Omega)(C_o) = 10^9/0.74 \times 10^3 = 1.35 \times 10^6$ rad/sec. This expression has a single pole at ω_3 which determines the high-frequency response of the amplifier.

(d) Sketch the voltage gain as a function of frequency in the Bode plot form.

Solution: The overall transfer function has two zeros at $s = 0$ and poles at ω_1, ω_2, and ω_3. The asymptotic Bode plot is shown in solid line. The approximate actual response is shown by the dashed line:

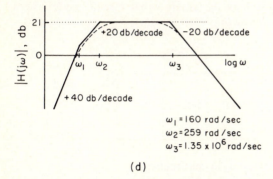

$$\omega_1 = 160 \text{ rad/sec}$$
$$\omega_2 = 259 \text{ rad/sec}$$
$$\omega_3 = 1.35 \times 10^6 \text{ rad/sec}$$

(d)

One final comment is needed. We have derived the response by breaking it into regimes where the algebra is tractable. If we had set out to derive the full transfer function the regulating expression would be quite complicated (see, for example, Eqs. (15.7a) and (15.7b)). In the three appropriate frequency ranges, however, the expressions here would be recovered.

Problem 15.1

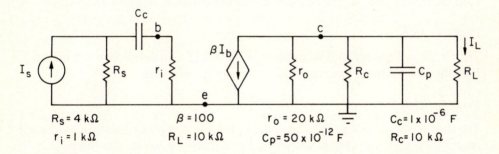

$$R_s = 4 \text{ k}\Omega \qquad \beta = 100 \qquad r_0 = 20 \text{ k}\Omega \qquad C_c = 1 \times 10^{-6} \text{ F}$$
$$r_i = 1 \text{ k}\Omega \qquad R_L = 10 \text{ k}\Omega \qquad C_p = 50 \times 10^{-12} \text{ F} \qquad R_c = 10 \text{ k}\Omega$$

The circuit shown above may be used as the small-signal circuit model for a BJT transistor current amplifier.

(a) Show that the midband circuit is identical in form to Problem 4.28 if we define a new $R'_L = R_L \,/\!/\, R_C$.

(b) Find the transfer function $A_I(s) = I_L/I_s$.

$$\text{Ans.: } A_I(s) = -\frac{(16 \times 10^7)s}{(s + 5 \times 10^6)(s + 200)}$$

(c) Find the midband gain and the upper and lower frequencies where the gain is reduced by 3 db.

$$\text{Ans.: } A_I = -32$$
$$\omega_1 = 200 \text{ rad/sec (32 Hz)}$$
$$\omega_2 = 5 \times 10^6 \text{ rad/sec (795 kHz)}$$

Problem 15.2

Suppose an additional stray capacity between the collector and ground in Problem 15.1 is 30 pF. Find the new -3-db point at high frequencies.

Problem 15.3

The circuit shown below may be used as the small-signal circuit model for a transistor current amplifier. Find the following:

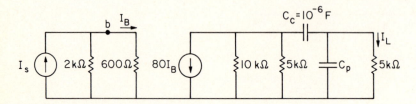

(a) The small-signal current gain $A_I = I_L / I_s$ at midfrequencies.

(b) The value of the capacitance C_p so that the absolute value of the current gain is equal to $\sqrt{2}/2$ times the midband gain at a frequency 10^6 rad/sec (that is, the high frequency -3-db point should be at 10^6 rad/sec).

Ans.: (a) -24.6, (b) 500 pf

Problem 15.4

Find the -3-db point at low frequencies for the circuit in Problem 15.3.

Problem 15.5

The circuit in Problem 15.3 is attached to a 100-kΩ load rather than a 5-kΩ load. Find the new midband gain and high-frequency -3-db point.

Problem 15.6

The circuit shown below may be used as the low-frequency small-signal circuit model for a transistor current amplifier. For a transistor having the values $\beta = 90$ and $r_o = 10$ kΩ at the operating point, find the following:

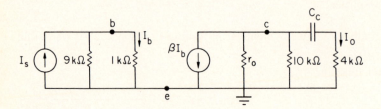

(a) The small-signal current gain $A_I = I_o / I_s$ at midfrequencies.

(b) The value of C_c required so that the absolute value of the current gain is equal to $\sqrt{2}/2$ times the midband gain at an angular frequency of 200 rad/sec. (That is, the low-frequency -3-db point should be at 200 rad/sec.)

Problem 15.7

Suppose a stray capacity of 10^{-10} F exists between the collector and ground in the circuit of Problem 15.6. Find the high-frequency -3-db point of the amplifier.

Problem 15.8

Two coupling capacitors were used in Example 14.3 of the Chapter 14 problem set. Derive the transfer functions $H_1 = I_b/V_i$ and $H_2 = V_o/I_b$ for the circuit. Determine the -3-db point at low frequency for $H = H_1H_2$. Which capacitor determines this frequency? Verify that for high frequencies your expression yields the same gain derived in that example, $V_o/V_i = -1.65$.

Problem 15.9

We wish to limit the high-frequency response of the amplifier circuit in Problem 15.8 such that the -3-db point to high frequencies is at 20 kHz (125,600 rad/sec). Find the appropriate capacitor C_L to place across the 2-kΩ load resistor.

Problem 15.10

How will the high-frequency response change in Problem 15.9 if the additional 20-kΩ load resistor discussed in Problem 14.25 is again inserted across the 2-kΩ load. Assume C_L remains the same as found in Problem 15.9.

Problem 15.11

Derive an expression for the amplifier gain $H = A_v = V_{ab}/V_i$ in Problem 14.27. Find the mid-frequency gain if $(r_i, \beta, r_o) = (500\ \Omega, 100, 10\ \text{k}\Omega)$. Sketch the asymptotic Bode Plot for $|\mathbf{H}(j\omega)|$ using symbols ω_1 and ω_2 for the low-frequency break points. Determine ω_1 and ω_2 specifically for $C_1 = C_2 = 1.0 \times 10^{-6}$F.

Problem 15.12

Suppose the amplifier in Problem 15.11 is attached to a digital computer which samples the signal at a rate of 20,000 samples per second. Choose a load capacitor such that the transfer function A_v has a -3-db point at the Nyquist frequency for this sample rate (i.e., at $20,000/2 = 10,000$ Hz).

Problem 15.13

The circuit shown below may be used as the low-frequency small-signal circuit model for a certain transistor amplifier. Find the low-frequency 3-db point.

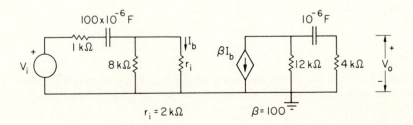

Problem 15.14

Sketch the asymptotic Bode plot for $|\mathbf{H}(j\omega)|$ for the circuit in Problem 15.13 if an additional 10^{-9} $-$F capacitor is placed in parallel across the 4-kΩ resistor.

Problem 15.15

The circuit shown below was studied in Problem 14.28.

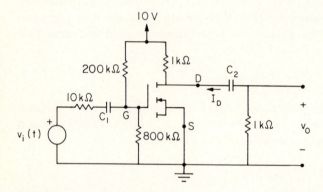

Find the transfer function and the value of its poles and zeros if $C_1 = 1$ μF and $C_2 = 0.1$ μF. How would the transfer function change if a 0.001-μF capacitor is placed in parallel with the 1-kΩ resistor? Find the midband gain and the high-frequency -3-db point.

Problem 15.16

Draw the small-signal linear model for the circuit in Problem 14.26 if at the operating point the dynamic resistance r_i of the base emitter diode is 1 kΩ, if the current gain $\beta = 100$, and if the dynamic resistance of the collector–emitter curve, r_o, is 10 kΩ. Find the voltage gain $A_v = v_o/v_i$ in the midband as well as its poles and zeros if $C_1 = C_2 = 1$ μF and a capacitor $C_3 = 100$ pF is placed across the 1-kΩ load resistor as shown here.

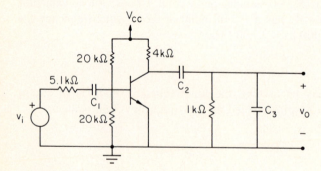

Problem 15.17

Find the midband gain and the poles and zeros of the frequency response for the transistor amplifier shown below if $C_1 = 1$ μF, $C_2 = 0.2$ μF, and a stray 100-pF capacitor exists in parallel with the 0.8-kΩ resistor. The transistor linear model is characterized by the parameters $(r_i, \beta, r_o) = (0.8 \text{ k}\Omega, 80, 100 \text{ k}\Omega)$. The stray capacitor is not shown.

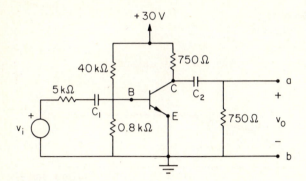

Problem 15.18

In the circuit for Text Example 15.1 it is desired to limit the high-frequency response with a pole at 10^5 rad/sec. Choose an appropriate load capacitor in parallel with R_L to yield this pole.

Problem 15.19

A two-stage MOSFET amplifier may be analyzed by straightforward application of the analysis techniques developed in Chapters 14 and 15. Find and sketch a log–log plot of the full transfer function for the circuit below by following the procedure outlined.

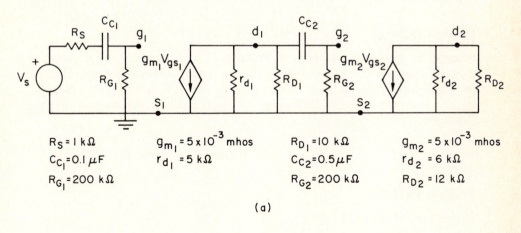

$R_S = 1 \text{ k}\Omega$ $\qquad$ $g_{m_1} = 5 \times 10^{-3}$ mhos $\qquad$ $R_{D_1} = 10 \text{ k}\Omega$ $\qquad$ $g_{m_2} = 5 \times 10^{-3}$ mhos

$C_{C_1} = 0.1 \mu\text{F}$ $\qquad$ $r_{d_1} = 5 \text{ k}\Omega$ $\qquad$ $C_{C_2} = 0.5 \mu\text{F}$ $\qquad$ $r_{d_2} = 6 \text{ k}\Omega$

$R_{G_1} = 200 \text{ k}\Omega$ $\qquad\qquad\qquad$ $R_{G_2} = 200 \text{ k}\Omega$ $\qquad$ $R_{D_2} = 12 \text{ k}\Omega$

(a)

(a) Find the transfer function V_{g1}/V_s. $\qquad$ Ans.: $V_{g1}/V_s =$
$$\frac{200s}{201(s + 49.8)} \approx s/(s + 50)$$

(b) Find the transfer function V_{g2}/V_{g1}. $\qquad$ Ans.: $V_{g2}/V_{g1} = -16s/(s + 9.8)$
$$\approx -16s/(s + 10)$$

(c) Find the transfer function V_{d2}/V_{g2}.   Ans.: $V_{d2}/V_s = -20$

(d) Find the full transfer function V_{d2}/V_s. Ans.: $V_{d2}/V_s = 320s^2/[(s+50)(s+10)]$

(e) Sketch the magnitude (in db) of the frequency response versus frequency (on a log scale). Find the low-frequency 3-db point.   Ans.: $\omega \approx 50$ rad/sec

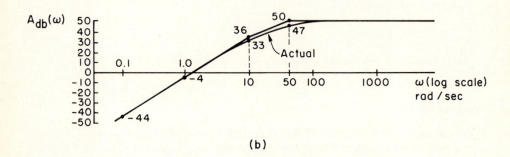

(b)

(f) Sketch on the same plot as part e the actual response (see sketch above).

Roughly speaking, the gain of the first stage is -16 and the gain of the second stage is -20, yielding a total gain of $+320$. This is shown by the answer for part d as $|\omega|$ gets large compared to 50.

Problem 15.20

Suppose the two-stage amplifier in Problem 15.19 was supplying a signal to a digital computer which sampled the signal at a rate of 2000 samples/sec. Determine the capacitor C_p which must be placed in parallel to resistors R_{D2} and r_{d2} to yield a one-pole low-pass filter with a -3-db point at the Nyquist frequency (1000 Hz).

Ans.: $C_p = 3.9 \times 10^{-8}$ F

Problem 15.21

Derive Eqs. 15.9a and 15.9c

Ans.: Given.

Problem 15.22

In the circuit shown, the MOSFET is such that $g_m = 2 \times 10^{-3}$ mhos and $r_d = 6$ kΩ. Signals above 1000 Hz are to be amplified.

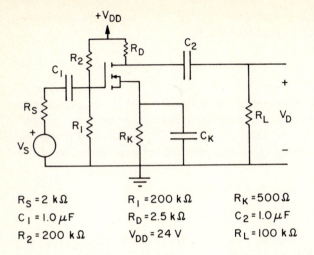

$$R_S = 2 \text{ k}\Omega \qquad R_1 = 200 \text{ k}\Omega \qquad R_K = 500 \,\Omega$$
$$C_1 = 1.0 \,\mu F \qquad R_D = 2.5 \text{ k}\Omega \qquad C_2 = 1.0 \,\mu F$$
$$R_2 = 200 \text{ k}\Omega \qquad V_{DD} = 24 \text{ V} \qquad R_L = 100 \text{ k}\Omega$$

(a) Find the midband gain when C_K is included in the circuit and $\omega \gg \omega_1, \omega_2$, and ω_{min}. The frequencies ω_1 and ω_2 are the poles due to the coupling capacitors and ω_{min} is due to C_K.

Ans.: $V_o / V_s = -3.4$

(b) Find the midband gain ($\omega > \omega_1, \omega_2$) of this amplifier if C_K is removed from the circuit.

Ans.: $V_o / V_s = -2.0$

(c) If C_K is in the circuit and we desire to achieve the midband gain of part a for $\omega > 100$ rad/sec (16 Hz), use the criterion in the text (Eq. 15.12) to determine the minimum required value of C_K.

Ans.: $C_K > 140 \,\mu F$

(d) Check ω_1 and ω_2 to verify that 100 rad/sec is in the midband region.

Ans.: Yes, since ω_1 and $\omega_2 \approx 10$ rad/sec

Problem 15.23

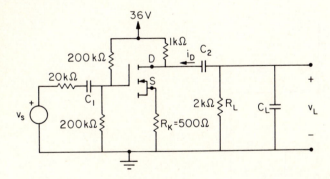

For the circuit above, find the midband gain and the poles and zeroes of the transfer function $A_v = v_L / v_s$ if the transistor is characterized by the parameters $g_m = 8 \times 10^{-3}$

mho and $r_d = \infty$, and if $C_1 = C_2 = 1\ \mu F$ and $C_L = 10^{-9}$ F. Find the midband gain if a large capacitor C_K is placed in parallel with R_K.

Problem 15.24

The circuit shown below is the approximate small-signal model for an amplifier in the frequency range where stray capacitors can be considered to be open circuits. It has also been assumed that r_o is so large compared to R_E and R_L that it may be omitted from the model. No coupling capacitors are used to isolate the effect of the capacitor C_E.

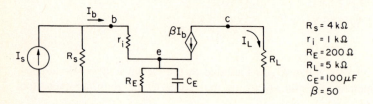

$R_s = 4\ k\Omega$
$r_i = 1\ k\Omega$
$R_E = 200\ \Omega$
$R_L = 5\ k\Omega$
$C_E = 100\ \mu F$
$\beta = 50$

(a) Find $H(s) = I_L/I_s$.   Ans.: $H(s) = -40[(s + 50)/(s + 150)]$

(b) Find the magnitude of the frequency
response of I_L/I_s at $\omega = 0$ and $\omega \gg$
150 rad/sec   Ans.: $|\mathbf{H}(0)| \approx 13.3$
$|\mathbf{H}(\omega > 150)| \approx 40$

Problem 15.25

For the circuit below let $C_{C1} = 10^{-6}$ F $= C_{C2}$, $C_E = 10^{-5}$ F, $R_1 = R_2 = 4\ k\Omega$, $R_L = 10\ k\Omega$, $R_E = R_C = R_S = 1\ k\Omega$, $\beta = 100$, $r_i = 2\ k\Omega$, and $r_o = \infty$.

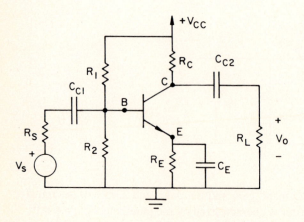

(a) Find the voltage gain $A_v = V_o / V_s$ at
high frequencies.　　　　　　　　　Ans.: $A_v = -22.8$

(b) Find the value of the pole correspond-
ing to the capacitor C_{C1}.　　　　　Ans.: $\omega_1 = 500$ rad/sec

(c) Find the value of the pole correspond-
ing to capacitor C_{C2}.　　　　　　Ans.: $\omega_2 = 91$ rad/sec

(d) Find the frequency at which $|\mathbf{Z}_E| = 0.707 R_E$.　　　　　　Ans.: $\omega = (R_E C_E)^{-1} = 10^2$ rad/sec

Problem 15.26

In the amplifier circuit of Problem 15.25, the linear model parameters are
$(r_i, \beta, r_o) = (1\text{ k}\Omega, 200, \infty), R_1 = R_2 = 120\text{ k}\Omega, R_s = R_C = 1\text{ k}\Omega, R_L = 10k\Omega$, and
$R_E = 500\ \Omega$. Find the high-frequency gain with and without C_E. Suppose C_E is
included and is large enough that ω_{min} is well below the poles due to C_{C1} and C_{C2}.
If $C_{C1} = 10^{-6}$ F and $C_{C2} = 10^{-7}$ F, find the low-frequency -3-db point. (You may
assume C_E shorts out R_K in the latter analysis.)

Problem 15.27

Choose a load capacitor C_L parallel to R_L in Problem 15.26 such that the -3-db
point at high frequencies occurs at 10^6 rad/sec.

Problem 15.28

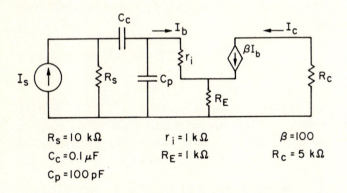

$R_s = 10\text{ k}\Omega$　　　　$r_i = 1\text{ k}\Omega$　　　　$\beta = 100$
$C_c = 0.1\ \mu F$　　　　$R_E = 1\text{ k}\Omega$　　　　$R_c = 5\text{ k}\Omega$
$C_p = 100\text{ pF}$

For the transistor used in the circuit whose linear model is shown above, $r_o = \infty$
and has not been included in the circuit diagram.

(a) Find the magnitude of the midband
gain I_c / I_s.　　　　　　　　　Ans.: $I_c / I_s = 8.9$

(b) Find the frequency of the 3-db point

for I_c/I_s in the high-frequency region of the
curve.   Ans.: $\omega_2 = 1.1 \times 10^6$ rad/sec

(c) Find the frequency of the 3-db point
for I_c/I_s in the low-frequency region of the
curve.   Ans.: $\omega_1 = 89$ rad/sec

Problem 15.29

Find the midband gain of the circuit in Problem 15.28 if a large capacitor is placed across R_E.

Problem 15.30

(a) The circuit shown below may be used as the linear model of a certain transistor amplifier. Find the value of the transfer function I_o/V_s at midband frequencies.

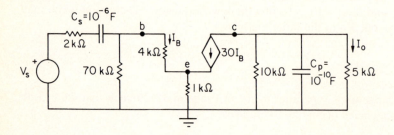

Ans.: $I_o/V_s = -0.53$ mA/V

(b) Find the high frequency at which the transfer function i_o/v_s is at its -3-db point relative to the midband value.

Ans.: $\omega = 3 \times 10^6$ rad/sec

Problem 15.31

The circuit diagrammed in Fig. 15.9a is characterized by $R_s = 1$ kΩ, $R_1 = R_2 = 100$ kΩ, and $R_E = 500$ Ω while the transistor is characterized by $(r_i, \beta, r_o) = (1$ k$\Omega, 100, \infty)$. The Thevenin equivalent circuit at the output has a resistance given by R_o (Eq. 15.14).

(a) Find the voltage across a 50 Ω resistor placed as a load at the output in terms of the source V_s.

(b) Compare this voltage to the voltage across the same 50 Ω resistor placed as a load across the original source (including the source resistance R_s).

Problem 15.32

Draw the linear small-signal model for the two-stage BJT amplifier shown below if the transistors are both characterized by the three parameters (r_i, β, r_o).

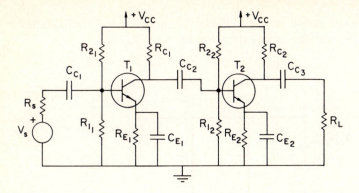

Problem 15.33

Find the total midband gain in the circuit of Problem 15.32 if both transistors have $(r_i, \beta, r_o) = (500\ \Omega, 100, \infty)$, and the two circuits are identical with $R_1 = R_2 = 40\ k\Omega$, $R_C = 1\ k\Omega$, $R_S = R_L = 1\ k\Omega$ and $R_E = 500\ \Omega$. You may assume that the C_E are large.

Problem 15.34

Construct a small-circuit linear model for the MOSFET source follower circuit shown below in the midband frequency range.

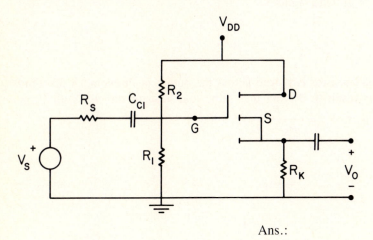

Ans.:

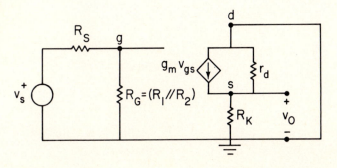

Problem 15.35

Re-draw the circuit model above in the so-called "common-drain" configuration analogous to Fig. 15.19c and derive an expression for $A_v = V_o/V_s$.

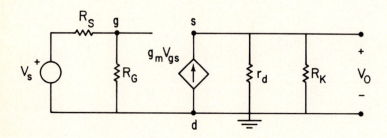

$$\text{Ans.:} \quad A_v = \frac{R_G}{R_G + R_S}\left(\frac{g_m(r_d\ //R_K)}{1 + g_m(r_d\ //R_K)}\right)$$

Problem 15.36

In the circuit above let $R_S = 2.5\text{ k}\Omega$, $R_1 = R_2 = 500\text{ k}\Omega$, $g_m = 5 \times 10^{-3}$mhos, $R_K = 10\text{ k}\Omega$ and $r_d = 100\text{ k}\Omega$. Find A_v.

$$\text{Ans.:} \quad A_v = 0.97$$

Problem 15.37

Suppose a 500 Ω load resistor is placed across the output in Problem 15.35. Find V_o/V_s. Compare this result to driving that same load resistor directly with V_s in series with R_S, that is, to $V_o = (\dfrac{R_L}{R_S + R_L})V_s$.

Index